MOLECULAR RECOGNITION AND POLYMERS

MOLECULAR RECOGNITION AND POLYMERS

Control of Polymer Structure and Self-Assembly

Edited by

VINCENT M. ROTELLO

S. THAYUMANAVAN

 **WILEY**

A JOHN WILEY & SONS, INC., PUBLICATION

Published by John Wiley & Sons, Inc., Hoboken, New Jersey.
Published simultaneously in Canada

For general information on our other products and services or for technical support, please contact our Customer Care Department within the United States at (800) 762-2974, outside the United States at (317) 572-3993 or fax (317) 572-4002.

Wiley also publishes it books in a variety of electronic formats. Some content that appears in print may not be available in electronic formats. For more information about Wiley products, visit our web site at www.wiley.com.

Library of Congress Cataloging-in-Publication Data:

Molecular recognition and polymers: control of polymer structure and
self-assembly / [edited by] Vincent Rotello, Sankaran Thayumanavan.
    p. cm.
  Includes index.
  ISBN 978-0-470-27738-6 (cloth)
1. Biomimetic polymers. 2. Molecular recognition. 3. Supramolecular chemistry.
I. Rotello, Vincent M. II. Thayumanavan, Sankaran.
  QD382.B47M65 2008
  547′.1226–dc22

                              2008007590

Printed in the United States of America

10 9 8 7 6 5 4 3 2 1

This book is dedicated to the memory of Dmitry Rudkevich, and to his wife, Sasha, and sons, Dmitry Jr. and Eric.

CONTENTS

9. SEQUENCE-SPECIFIC HYDROGEN BONDED UNITS FOR DIRECTED ASSOCIATION, ASSEMBLY, AND LIGATION   207

Bing Gong

PART III   BIOMOLECULAR RECOGNITION USING POLYMERS

12. COLORIMETRIC SENSING AND BIOSENSING USING FUNCTIONALIZED CONJUGATED POLYMERS

Amit Basu

13. GLYCODENDRIMERS AND OTHER MACROMOLECULES BEARING MULTIPLE CARBOHYDRATES

Mary J. Cloninger

PREFACE

Inter- and intramolecular networks of non-covalent interactions are responsible for a wide array of phenomena in fields of biology and chemistry. Biological systems use specific patterns of complementary functionality to provide exquisite control over biopolymer recognition processes such as protein–protein and protein–polynucleic acid binding. In Nature, these specific supramolecular interactions play many key roles, including stabilization of structure, information storage and transfer, catalysis and self-assembly. Likewise, controlled application of non-covalent interactions provides an effective tool for fabrication of man-made systems, allowing the creation of higher-order architecture required for devices and materials, as well as the dynamic properties required for efficient utilization of these attributes.

The use of specific interactions to control polymer structure and properties is a rapidly emerging field. We have assembled a group of authors at the forefront of this field that are studying both the fundamental science inherent in polymer self-assembly and applications of this strategy to functional systems. This book is designed for researchers in a wide range of areas, and features both fundamental aspects and applications of these fascinating systems.

The book is divided into three sections. The first section provides a general overview of the fundamentals of supramolecular polymers. In Chapter 1, Thibault and Rotello provide a brief introduction to these systems and in Chapter 2, Azagarsamy, Krishnamoorthy, and Thayumanavan describe the rapidly emerging area of amphiphilicity in polymer and dendrimers self-assembly. Interactions at interfaces are sometimes similar but often quite different than those in solution, a topic covered by Loveless, Kersey, and Craig in Chapter 3.

The second section of the book provides a wide variety of examples of the self-assembly of polymer systems. Aspects covered include hydrogen bond-mediated recognition and self-assembly using block copolymers and telechelic oligomers, as described in Chapter 4 by Mather and Long. Chapter 5 covers the highly versatile "plug and play" non-covalent sidechain modification of polymers, as described by Nair and Weck. Extension of this polymer-mediated assembly to nanoparticles is the focus of Chapter 6 by Chen, Ofir, and Rotello, while Chapter 7 by McKenzie and Rowan describes metallo-supramolecular systems. In Chapter 8, Mason, Steinbacher, and McQuade provide an overview of capsule formation using polymers and biopolymers. Chapter 9 by Gong features the efforts of synthetic chemists to replicate the specific hydrogen bonding patterns found in biology. Chapter 10 focuses on function, with Guan covering the use of supramolecular polymer systems to tailor mechanical properties. Shao and Parquette outline the use of hydrophobicity to control dendrimers structure and dynamics in Chapter 11.

The third section of the book covers the area of biomolecular recognition using polymer systems. The creation of colorimetric sensors using polymers is presented by Basu in Chapter 12. Chapter 13 by Cloninger focuses on glycopolymers and glycodendrimers, while in Chapter 14, Dong, Yuwono, and Hartgerink cover the creation of nanofibers via peptide self-assembly. Finally, in Chapter 15, Wu and Shimizu provide an overview of the field of molecularly-imprinted polymers, describing the formation of these systems and their applications.

Supramolecular chemistry is a beautiful field, featuring modularity, tenability, and versatility. We hope that this book fires your imagination for this emerging field.

VINCENT ROTELLO
SANKARAN "THAI" THAYUMANAVAN

Department of Chemistry
University of Massachusetts, Amherst

ACKNOWLEDGMENTS

The Editors would like to acknowledge Carol Greene, Denia Fraser, and Denise Schwartz, without whose help this book would not have been possible.

LIST OF CONTRIBUTORS

Editors

VINCENT M. ROTELLO, Department of Chemistry, Program in Molecular and Cell Biology, University of Massachusetts, Amherst, 710 North Pleasant Street, Amherst, MA 01003, USA

S. THAYUMANAVAN, Department of Chemistry, Program in Molecular and Cell Biology, University of Massachusetts, Amherst, 710 North Pleasant Street, Amherst, MA 01003, USA

Contributors

MALAR A. AZAGARSAMY, Department of Chemistry, University of Massachusetts, Amherst, 710 North Pleasant Street, Amherst, MA 01003, USA

AMIT BASU, Department of Chemistry, Brown University, Box H, 324 Brook St., Providence, RI 02912, USA

HUNG-TING CHEN, Department of Chemistry, University of Massachusetts, Amherst, 710 North Pleasant Street, Amherst, MA 01003, USA

MARY J. CLONINGER, Department of Chemistry and Biochemistry, Montana State University, 108 Gaines Hall, Bozeman, MT 59717, USA

STEPHEN L. CRAIG, Department of Chemistry, Duke University, Box 90354, Durham, NC 27708-0354, USA

HE DONG, Department of Chemistry, Rice University, Mail Stop 60, 6100 Main Street, Houston, TX 77005

BING GONG, Department of Chemistry, State University of New York, 811 Natural Sciences Complex, Buffalo, NY 14260, USA

ZHIBIN GUAN, Department of Chemistry, University of California, Irvine, 1102 Natural Sciences II, Irvine, CA 92697-2025, USA

JEFFREY D. HARTGERINK, Departments of Chemistry and Bioengineering, Rice University, MS 60, 6100 Main Street, Houston, TX 77005, USA

FARRELL R. KERSEY, Department of Chemistry, Duke University, Box 90354, Durham, NC 27708-0354, USA

K. KRISHNAMOORTHY, Department of Chemistry, University of Massachusetts, Amherst, 710 North Pleasant Street, Amherst, MA 01003, USA

TIMOTHY E. LONG, Department of Chemistry, Macromolecules and Interfaces Institute, Virginia Tech, 107 Davidson Hall, Blacksburg, VA 24061-0001, USA

DAVID M. LOVELESS, Department of Chemistry, Duke University, Box 90354, Durham, NC 27708-0354, USA

BRIAN P. MASON, Baker Laboratory, Department of Chemistry and Chemical Biology, Cornell University, Ithaca, NY 14853, USA

BRIAN D. MATHUR, Hewlett-Packard Co., 16399 W. Bernardo Dr., San Diego, CA 92127, USA

BLAYNE M. MCKENZIE, Department of Macromolecular Science and Engineering, Case Western Reserve University, 2100 Adelbert Road, Cleveland, OH 44 106-7202, USA

D. TYLER MCQUADE, Department of Chemistry and Biochemistry, Florida State University, Tallahassee, FL 32306, USA

KAMLESH P. NAIR, School of Chemistry and Biochemistry, Georgia Institute of Technology, Atlanta, GA 30332-0400, USA

YUVAL OFIR, Department of Chemistry, University of Massachusetts, Amherst, 710 North Pleasant Street, Amherst, MA 01003, USA

JON R. PARQUETTE, Department of Chemistry, Ohio State University, 100 West 18th Avenue, Columbus, OH 43210, USA

STUART J. ROWAN, Department of Macromolecular Science and Engineering, Case Western Reserve University, 2100 Adelbert Road, Cleveland, OH 44106-7202, USA

HUI SHAO, Department of Chemistry, Ohio State University, 100 West 18th Avenue, Columbus, OH 43210, USA

KEN D. SHIMIZU, Department of Chemistry and Biochemistry, University of South Carolina, Columbia, SC 29208, USA

JEREMY L. STEINBACHER, Baker Laboratory, Department of Chemistry and Chemical Biology, Cornell University, Ithaca, NY 14853

RAYMOND J. THIBAULT, Epoxies Research and Development, The Dow Chemical Company, 2301 N Brazosport Blvd, Freeport, TX 77541-3257, USA

MARCUS WECK, Molecular Design Institute and Department of Chemistry, 100 Washington Square East, New York University, New York, NY 10003-6688, USA

XIANGYANG WU, University of South Carolina, Department of Chemistry and Biochemistry, Columbia, SC 29208, USA

VIRANY M. YUWONO, Department of Chemistry, Rice University, Mail Stop 60, 6100 Main Street, Houston, TX 77005, USA

LIST OF FIGURES

Figure 6.3   UV visible spectra of assembled gold nanoparticle thin films by dendrimers of different generations ranging from G0 to G4. Reprinted with permission from Srivastava, Frankamp, et al. (2005). Copyright 2005 American Chemical Society.

Figure 6.4   (a) Self-assembly of magnetic FePt nanoparticles by DNAs and (b) a TEM micrograph of network-like FePt–DNA aggregates; scale bar = 100 nm. Reprinted with permission from Srivastava, Samanta, Arumugam, et al. (2007). Copyright 2007 RSC Publishing.

Figure 6.5   (a) The formation of ferritin-mediated self-assembly of FePt nanoparticles via electrostatic interactions, (b) magnetic dipole–dipole interaction of ferritins assembled with FePt nanoparticles, and (c) zero field cooling and field cooling results for the ferritin–FePt nanoparticle composite film and individual components. Reprinted with permission from Srivastava, Samanta, Jordan, et al. (2007). Copyright 2007 American Chemical Society.

Figure 6.6   (a) Schematic representation of PEI-mediated assembly of gold nanoparticles. Transmission electron micrographs of (b) hexagonal and (c) cubic packing arrangements of nanoparticles. Reprinted with permission from Schmid et al. (2000). Copyright 2000 Wiley InterScience.

Figure 6.7   Illustration of multipoint hydrogen bonding based self-assembly: (a) hydrogen bond formation between barbituric acid functionalized gold nanoparticles and Hamilton receptor functionalized block copolymers and (b) selective deposition of nanoparticles on a microphase-separated block copolymer film. Reprinted with permission from Binder et al. (2005). Copyright 2005 American Chemical Society.

Figure 6.8   TEM micrographs of (a) "dots" patterned samples formed through affinity of Terpy-functionalized gold nanoparticles on PS domain, (b) Fe-treated cross-linked samples, and (c) ethanol-treated samples after swelling in chloroform vapor. Reprinted with permission from Shenhar et al. (2005). Copyright 2005 Wiley InterScience.

Figure 6.9   A schematic representation of orthogonal process for nanoparticles self-assembly: (a) a patterned silicon wafer with Thy-PS and PVMP polymers fabricated through photolithography and (b) orthogonal surface functionalization through Thy-PS/DP-PS recognition and PVMP/acid–nanoparticle electrostatic interaction. Reprinted with permission from Xu et al. (2006). Copyright 2006 American Chemical Society.

Figure 6.10   A procedural schematic representation for deposition of two kinds of particle arrays on the patterned template. Reprinted with permission from Zheng et al. (2002). Copyright 2002 Wiley InterScience.

Figure 6.11   (a) Scheme of patterning nanoparticle thin films and (b) scanning electron micrograph of self-assembly patterns after lift-off process. Reprinted with permission from Hua et al. (2002). Copyright 2002 American Chemical Society.

Figure 9.3   (a) Duplex **3•4** consisting of two different strands with complementary H bonding sequences. (b) Duplex **5•5** consisting of two identical single strands with a self-complementary H bonding sequence. Interstrand NOE contacts revealed by NOESY are indicated by arrows.

Figure 9.4   Duplex **3•4** and two duplexes **3•6** and **3•7** containing mismatched-binding sites (red arrows). The repulsive mismatched binding sites in **3•6** (acceptor–acceptor) and in **3•7** (donor–donor) cause more than a 40-fold decrease in the stabilities of these duplexes compared to **3•4**.

Figure 9.5   GPC studies (CHCl$_3$, 1 mL/min) on mixtures of (a) **8•8 + 9•9 + 3′•4′**, (b) **3′•4′ + 11•9•9**, and (c) **8•8 + 9•9**. The retention times of the individual duplexes are identical to those shown here.

Figure 9.6   Oligoamide strands **10**, **10′**, **11**, and **12** containing two quadruply H bonding units linked in a head-to-head fashion. Control strand **12′** is essentially half of **12**.

Figure 9.7   If adopting an extended conformation, (a) strands **10** and **11** should form H bonded polymer chains, (b) strands **10′** and **11** should form an 8-H bonded duplex, and (c) strand **12** should form H bonded polymer chains.

Figure 9.8   Hydrogen bonded duplex consisting of folded strands: (a) heteroduplex **10•11** and (b) homoduplex **12•12**.

Figure 9.9   When attached to a duplex template, two otherwise flexible peptide chains are directed to form a stably folded β-sheet.

Figure 9.10   Design of supramolecular block copolymers based on the 6-H bonded heteroduplex **15•16**. Mixing three templated PS chains with three complementarily templated PEG chains leads to nine block copolymers.

Figure 9.11   GPC traces of **15c•16c**, **15c**, and **16c** eluted with DMF/toluene (10/90, v/v, left) and DMF (right) at 60 °C.

Figure 9.12   (a) AFM image of spin-cast **15d•16d** from benzene showing cylindrical nanodomains from the microphase separation of the supramolecular block copolymer and (b) AFM image of spin-cast **15b•16b** from benzene showing self-assembled fibers.

Figure 9.13   Templated olefin cross-metathesis of two groups of olefins tethered to the two duplex strands.

Figure 9.14   (a) Complementary strands **19** and **20** carrying trityl-protected thio groups can be cross-linked when subjected to reversible redox conditions. (b) MALDI spectra show that **19** and **20** were sequence specifically cross-linked into **19–20** in both methanol and water.

Figure 9.15   (a) Complementary strands **21** and **22** carrying trityl-protected thio groups were cross-linked in the presence of iodine. (b) MALDI spectra show that **21** and **22** were sequence specifically cross-linked into **21–22** in water containing

polymer. The solid line is the fitting with the WLC model at a 0.55-nm persistence length (L is the contour length during stretching). Adapted from Roland and Guan (2004). Copyright 2004 American Chemical Society.

a collagen-like domain and a carbohydrate-binding domain. (d) Galectins with two covalently linked carbohydrate-binding domains.

Figure 13.4   A schematic showing some of the frameworks that have been reported for the study of protein–carbohydrate interactions. Carbohydrates are represented as cyclohexane. (Top) Glycodendrimer, carbohydrate-functionalized nanoparticle, and star polymer. (Bottom) Linear glycopolymer, carbohydrate-functionalized protein, and carbohydrate-functionalized surface.

Figure 13.5   The structure of the G(2)-poly(amido amine) dendrimer.

Figure 13.6   The structure of the G(3)-poly(propylene imine) dendrimers.

Figure 13.7   (a) A poly(aryl ether) dendron and (b) a polyester dendrimer.

Figure 13.8   A polylysine dendrimer.

Figure 13.9   Phosphorus-containing dendrimer with 96 terminal chlorides; R is indicated by the bold type in the structure.

Figure 13.10   Synthesis of mannose-/glucose-functionalized dendrimers. For dendrimer loading, $x + y = 50\%$ loading was used because optimal activity for glycodendrimers with concanavalin A was previously determined to occur at 50% functionalization. Remaining amines from the poly(amido amine) (PAMAM) substrate were capped as alcohols.

Figure 13.11   Hemagglutination inhibition assay results for the interaction of mannose-/glucose-functionalized dendrimers with concanvalin A. (●) G(3), (■) G(4), (◆) G(5), and (▲) G(6).

Figure 13.12   A synthetic vaccine bearing multiple carbohydrate antigens; KLH, keyhole limpet hemocyanin.

Figure 13.13   Multifunctional anticancer poly(amido amine) (PAMAM) dendrimers. Clockwise from top: diol, taxol, fluorescein isothiocyanate, folic acid, and acetyl groups.

Figure 13.14   Fucose-/galactose-functionalized antibacterial dendrimers.

Figure 13.15   Synthesis of 2,2,6,6-tetramethylpiperidine N-oxide (TEMPO) functionalized dendrimers.

Figure 13.16   Stackplot of electron paramagnetic resonance (EPR) spectra of 2,2,6,6-tetramethylpiperidine N-oxide (TEMPO)-functionalized dendrimers with 5, 10, 25, 50, 75, 90, and 95% TEMPO.

Figure 13.17   Graph of line-broadening effects for 2,2,6,6-tetramethylpiperidine N-oxide (TEMPO) and R-4-isothiocyanato (R-NCS) functionalized dendrimers from Figure 13.15. NCS-TEMPO was added first half of the time and R-NCS was added first half of the time.

Figure 13.18   Comparison of calculated and experimental results: (●) experimental results from Figure 13.17 and (—) calculated results for a 16 Å volume shell.

SCHEMES

CHARTS

LIST OF TABLES

EDITOR BIOGRAPHIES

Vincent M. Rotello received his B.S. from Illinois Institute of Technology in 1985. He obtained his Ph.D. in 1990 from Yale University with Harry Wasserman in the area of natural products synthesis. From 1990–93, he was an NSF postdoctoral fellow with Julius Rebek Jr. at M.I.T. in the area of host-guest chemistry. Since 1993, Professor Rotello has been at the University of Massachusetts at Amherst as an Assistant Professor from 1993–1998, Associate Professor (1998–2001), Professor (2001–2005), and Charles A Goessmann Professor of Chemistry (2005–). He has been the recipient of the NSF CAREER, and Cottrell Scholar award, as well as the Camille Dreyfus Teacher-Scholar, and the Sloan Fellowships and is currently a Fellow of the Royal Society of Chemistry (U.K.) His research program spans the areas of devices, polymers, nanotechnology, and biological systems, with over 245 papers and one book (*Nanoparticles: Building Blocks for Nanotechnology*, Kluwer: New York, 2004) published to date.

S. Thayumanavan ("Thai") obtained his B.Sc. and M.Sc. degrees from The American College, Madurai, India. He obtained his Ph.D. from the University of Illinois at Urbana-Champaign with Professor Peter Beak in 1996. After a postdoctoral stint with Professor Seth Marder at the California Institute of Technology, he started his independent career at Tulane University in 1999. His group moved to the University of Massachusetts at Amherst in 2003. He has been the winner of NSF-CAREER award, Cottrell Scholar award, and 3M Non-tenured Faculty awards. His research interests involve design and syntheses of macromolecules (dendrimers and polymers) to achieve controllable nanoscale assemblies that are of interest in chemistry, materials, and biology.

PART I

FUNDAMENTALS OF SUPRAMOLECULAR POLYMERS

CHAPTER 1

A BRIEF INTRODUCTION TO SUPRAMOLECULAR CHEMISTRY IN A POLYMER CONTEXT

RAYMOND J. THIBAULT and VINCENT M. ROTELLO

1.1. INTRODUCTION AND BACKGROUND

Self-assembly of molecular and macromolecular systems is a versatile and modular tool for the creation of higher order structures (Lehn 1993). Nature employs self-assembly extensively using both phase segregation and "lock and key" specific interactions to generate the diverse range of highly ordered systems observed in living organisms. Applying biologically inspired self-assembly strategies to synthetic macromolecules provides access to a wide range of desirable structural and dynamic properties. Controlled noncovalent interactions are a particularly attractive strategy for controlling polymer aggregation, using the array of recognition elements developed by supramolecular chemists. The modularity and tunability of these recognition elements makes this approach versatile, because the assembling units can be synthetically tuned to enhance or minimize selectivity, directionality, and the affinity of the interaction. Likewise, the affinity of noncovalent interactions is thermally dependent, imparting reversibility to the assembly process and providing unique material properties such as defect correction and self-healing capabilities.

The toolkit of interactions available for supramolecular polymers can be divided into the six categories listed in Table 1.1 that includes hydrophobic interactions, which are dealt with in more detail in Chapter 2.

In this chapter, we present a brief overview of supramolecular polymers and polymerization and supramolecular interactions of polymer side chains. We provide examples of the control over the solution state polymer structure that can be achieved at the molecular level and then extended to micro- and macroscale assemblies.

Molecular Recognition and Polymers: Control of Polymer Structure and Self-Assembly.
Edited by V. Rotello and S. Thayumanavan
Copyright © 2008 John Wiley & Sons, Inc.

TABLE 1.1   Six Categories of Noncovalent Intermolecular Interactions

Interaction	Description and Bond Strengths	Selected Example
(London) dispersion forces	$>1\,\mathrm{kcal\,mol^{-1}}$; dynamic induced dipole–dipole interactions	
Stacking	$2-3\,\mathrm{kcal\,mol^{-1}}$ (face–face), $3-5\,\mathrm{kcal\,mol^{-1}}$ (edge–face); attractive forces between electron-rich interior with electron-poor exterior	
Hydrophobic	$1-10\,\mathrm{kcal\,mol^{-1}}$, association of nonpolar complements in aqueous or polar media	
Hydrogen bonding	$>1\,\mathrm{kcal\,mol^{-1}}$ (weak), $1-4\,\mathrm{kcal\,mol^{-1}}$ (moderate), $5-10\,\mathrm{kcal\,mol^{-1}}$ (strong); donor–acceptor interaction involving hydrogen atom as donor and base (electron pair) as proton acceptor	
Electrostatic	$1-10\,\mathrm{kcal\,mol^{-1}}$ (dipole–dipole), $10-30\,\mathrm{kcal\,mol^{-1}}$ (ion–dipole), $>45\,\mathrm{kcal\,mol^{-1}}$ (ion–ion); Coulombic attraction between opposite charges, highly dependent on media	
Dative bonding	$5-90\,\mathrm{kcal\,mol^{-1}}$; metal–ligand coordination, ligand donates electron pair(s) to center	

Association strengths are for systems in chloroform.

1.2. MAIN-CHAIN VERSUS SIDE-CHAIN SUPRAMOLECULAR POLYMERS

The concept of supramolecular polymers containing multiple hydrogen bonding units was introduced over a decade ago by Jean-Marie Lehn (Lehn 1993). In this study, three-point hydrogen bonding between bifunctional diamidopyridine and thymine derivatives results in the formation of supramolecular polymers featuring liquid crystalline ordering (Fig. 1.1). Lehn and coworkers later extended this strategy to include bifunctional molecules joined by chiral tartaric acid spacers (Gulikkrzywicki et al. 1993) and rigid anthracene-based linkers (Kotera et al. 1995). This approach is quite general, as can be seen in later chapters of this book.

The work done by E. W. Meijer using self-complementary ureidopyrimidinones builds upon Lehn's supramolecular polymers (Sijbesma et al. 1997). The quadruple hydrogen bonding system employed in these studies has two major differences from Lehn's polymers: 1) a high degree of association ($K_{\mathrm{dim}} > 10^6$ M^{-1}) and 2) self-complementarity that eliminates stoichiometric concerns. The high dimerization constant of ureidopyrimidinones makes this recognition element an excellent choice for supramolecular polymerizations, providing a high degree of polymerization in solution.

An alternative approach to supramolecular polymers is provided by covalently attaching recognition elements to the polymer backbone. These polymers can then be used as macromonomers for higher level polymer assembly or for "plug and

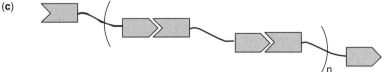

Figure 1.1 (a) Supramolecular polymers developed by Lehn using three-point hydrogen bonds between diamidopyridine and thymine residues and (b) analogous polymers by Meijer employing self-complementary, quadruple hydrogen bonds. (c) A schematic depiction of the extended chain of repeating bisfunctional monomers forming the backbone of supramolecular polymers.

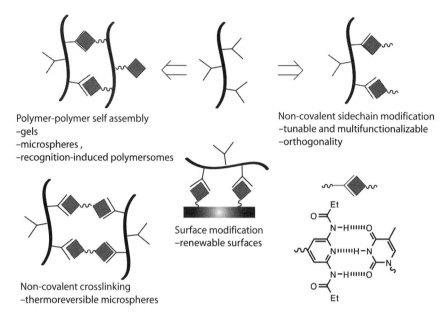

Polymer-polymer self assembly
–gels
–microspheres,
–recognition-induced polymersomes

Non-covalent sidechain modification
–tunable and multifunctionalizable
–orthogonality

Surface modification
–renewable surfaces

Non-covalent crosslinking
–thermoreversible microspheres

Figure 1.2   A schematic representation of the versatility of reversible, supramolecular side-chain modification and selected examples of interactions that can be employed.

play" noncovalent side-chain modification. Stadler (Stadler and Burgert 1986) initially investigated these systems for their elastomeric properties, serving to increase the miscibility between incompatible polybutadiene and polyisoprene blends using the dimerization of urazole moieties. These motifs have been investigated by a number of researchers, including Weck (Pollino et al. 2004) and Rotello (Deans et al. 1999; Ilhan et al. 2001), for a wide variety of applications (Fig. 1.2).

As you will see throughout this book, noncovalent interactions provide an elegant means to reversibly control polymer structures on the nano- and microscale. The lock and key nature, high directionality, and thermal response of these interactions make supramolecular polymer systems an attractive alternative for the fabrication of novel, functional materials. In addition, the wealth of available interactions allows the tuning of the form, function, and interaction strength of the assembling units, providing control in materials processing. Many investigators are discovering the versatility of supramolecular interactions for "bottom-up" methodology and "top-down" techniques in nanoscience and nanoscale engineering. Polymer scientists will likewise realize the expansive field available to them for the creation of novel plastics.

REFERENCES

Deans R, Ilhan F, Rotello VM. Recognition-mediated unfolding of a self-assembled polymeric globule. Macromolecules 1999;32:4956–4960.

Gulikkrzywicki T, Fouquey C, Lehn JM. Electron-microscopic study of supramolecular liquid-crystalline polymers formed by molecular-recognition-directed self-assembly from complementary chiral components. Proc Natl Acad Sci USA 1993;90:163–167.

Ilhan F, Gray M, Rotello VM. Reversible side chain modification through noncovalent interactions. "Plug and play" polymers. Macromolecules 2001;34:2597–2601.

Kotera M, Lehn JM, Vigneron JP. Design and synthesis of complementary components for the formation of self-assembled supramolecular rigid rods. Tetrahedron 1995;51:1953–1972.

Lehn JM. Supramolecular chemistry. Science 1993;260:1762–1763.

Lehn JM. Supramolecular chemistry—concepts and perspectives. Weinheim: VCH; 1995.

Pollino JM, Stubbs LP, Weck M. One-step multifunctionalization of random copolymers via self-assembly. J Am Chem Soc 2004;126:563–567.

Sijbesma RP, Beijer FH, Brunsveld L, Folmer BJB, Hirschberg JHKK, Lange RFM, Lowe JKL, Meijer EW. Reversible polymers formed from self-complementary monomers using quadruple hydrogen bonding. Science 1997;278:1601–1604.

Stadler R, Burgert J. Influence of hydrogen-bonding on the properties of elastomers and elastomeric blends. Makromol Chem Macromol Chem Phys 1986;187:1681–1690.

CHAPTER 2

MOLECULAR RECOGNITION USING AMPHIPHILIC MACROMOLECULES

MALAR A. AZAGARSAMY, K. KRISHNAMOORTHY, and
S. THAYUMANAVAN

2.1. INTRODUCTION

Molecular recognition in living systems, which is responsible for function and malfunction or dysfunction, involves macromolecular scaffolds as the recognition partners. The macromolecules that Nature uses for this purpose involve proteins or nucleic acids. Nature uses noncovalent interactions such as hydrogen bonding, electrostatic attraction, metal–ligand binding, hydrophobic, and $\pi-\pi$ interactions for recognition events. Most of these noncovalent interactions are quite nonspecific by themselves. However, Mother Nature has solved this issue by invoking multivalent interactions using a combination of the above interactions. Of more importance, when macromolecular scaffolds are used in Nature, a combination of the interaction features are precisely placed in three-dimensional space in order to be specific to the recognition partner for which these have evolved. Supramolecular chemists have been interested in mimicking the recognition events in nature (Cram 1988; Philp and Stoddart 1996; Wallimann et al. 1997; Zeng and Zimmerman 1997; Davis and Wareham 1999; Mueller-Dethlefs and Hobza 2000; Adams et al. 2001; Prins et al. 2001; Hof et al. 2002; Sun et al. 2002; Lehn 2005; Hannon 2007; Oshovsky et al. 2007). It is interesting, however, that most of the custom-designed systems for this purpose involve small molecule mimics, which rarely capture the essence of the complexities in biological systems. In contrast, when designing large macromolecules (polymers), synthetic methodologies are simply not available to match the efficiency of the cellular synthetic machinery that can program the assembly of a complex protein with a specific sequence. Moreover, we are only beginning to understand the factors that control the folding

Molecular Recognition and Polymers: Control of Polymer Structure and Self-Assembly.
Edited by V. Rotello and S. Thayumanavan
Copyright © 2008 John Wiley & Sons, Inc.

of artificial molecules to produce secondary structures, which are responsible for the spatial disposition of functional groups (Gellman 1998; Stigers et al. 1999; Cubberley and Iverson 2001; Hill et al. 2001; Huc 2004; Licini et al. 2005; Li and Yang 2006; Li et al. 2006; Goodman et al. 2007). Nonetheless, even when we focus on only a small part of the noncovalent interactions in macromolecules, fascinating sets of complex structures emerge. This chapter will focus on amphiphilicity as the driving force for the formation of interesting supramolecular assemblies from macromolecules in solution.

As the name suggests, amphiphilic is a property that presents the possibility of a molecule exhibiting a low-energy interaction with two different environments. The most commonly used amphiphilic molecules involve components that are hydrophilic (like an aqueous environment) and lipophilic (like a lipidlike environment, often referred to as hydrophobic). Note that these definitions are based on the interaction of the functionalities with the external environment. In addition, interesting supramolecular structures arise, depending on the mutual interaction energies between the hydrophilic and lipophilic functionalities. If these functionalities are mutually incompatible, then the driving force will be high for the lipophilic functionalities to aggregate and minimize their interaction with the hydrophilic functionalities and with the aqueous environment. In such a case, assemblies such as micelles or vesicles could be formed in aqueous solutions from these amphiphilic molecules. Morphologies exhibited by the self-assembly of amphiphilic molecules include spheres, rods, and vesicular assemblies (van Hest et al. 1995; Alexandridis et al. 2000; Discher and Eisenberg 2002; Lazzari and Lopez-Quintela 2003; Ruzette and Leibler 2005; Nakashima and Bahadur 2006). Occasionally lamellae, tubes, compound vesicles, hexagonally packed hollow hoops, compound micelles, and onionlike structures have also been obtained (Zhang et al. 1996; Zhang et al. 1997; Allen et al. 1999; Won et al. 1999; Choucair and Eisenberg 2003; Yan et al. 2004; Chen and Jiang 2005). It is also possible to obtain these types of assemblies using small molecule amphiphiles (Israelachvili 1991; Davidson and Regen 1997; Nagarajan 1997; Menger and Keiper 2000; Luk and Abbott 2002; Menger et al. 2005; Zana 2005; Lu et al. 2007). Although impressive progress has been made in understanding the structure–property relationship in small molecule amphiphiles, the poor mechanical stability and higher critical aggregate concentration (CAC) have restricted the scope of their applications. Amphiphilic polymers form interesting supramolecular assemblies at low concentrations and exhibit higher mechanical stability. These types of polymers have also been utilized in forming solid-state thin film nanoscale assemblies. This is not the focus of this chapter; there are several sources for obtaining details on these topics. We will focus here on solution-based structures and their utility in molecular recognition.

Amphiphilic polymers studied thus far for this purpose can be broadly classified into amphiphilic block copolymers and amphiphilic homopolymers. We will discuss both of these types of linear polymer architectures. Another interesting class of polymeric amphiphiles is based on branched architectures, known as dendrimers. The most interesting aspect of dendrimers is that their molecular weight and polydispersity can be precisely controlled; hence, these systems have the potential to be moved toward a truly biomimetic macromolecule. The purpose of this chapter is to provide a brief overview of supramolecular assemblies formed by amphiphilic macromolecules using a few selected examples.

2.2. AMPHIPHILIC BLOCK COPOLYMERS

A polymer is considered to be a copolymer when more than one type of repeat unit is present within the chain. There are a variety of copolymers, depending on the relative placement of the different types of repeat units. These are broadly classified as random, block, graft, and alternating copolymers (see Fig. 2.1 for structural details; Cheremisinoff 1997; Ravve 2000; Odian 2004). Among these structures, block copolymers have attracted particular attention, because of their versatility to form well-defined supramolecular assemblies. When a block copolymer contains two blocks (hydrophobic and hydrophilic), it is called an amphiphilic diblock copolymer. The immiscibility of the hydrophilic and lipophilic blocks in the polymers provides the ability to form a variety of assemblies, the structures and morphologies of which can be controlled by tuning the overall molecular weight and molar ratios of the different blocks (Alexandridis et al. 2000).

These polymers are capable of forming self-assembled structures such as micelles, vesicles, and bilayers (Figs. 2.2 and 2.3; van Hest et al. 1995; Allen et al. 1999; Alexandridis et al. 2000; Discher and Eisenberg 2002; Choucair and Eisenberg 2003; Lazzari and Lopez-Quintela 2003; Chen and Jiang 2005; Ruzette and Leibler 2005; Nakashima and Bahadur 2006). From a thermodynamic viewpoint, formation of these amphiphilic assemblies is often thought to be an entropy-driven process (Tanford 1980; De Maeyer et al. 1998). The entropy gain arises from 1) the hydrophobic effect: the ordered water around the hydrophobic segments prior to the formation of assembly are expelled into the bulk aqueous phase, resulting in a large gain in entropy of water molecules; and 2) the entropy gain for the hydrophobic segments: when the hydrophobic segments are at the core of a micelle, for

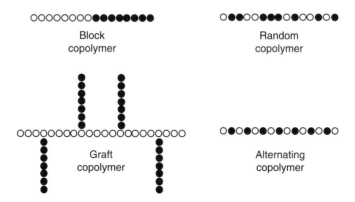

Figure 2.1   Representation of different types of copolymers.

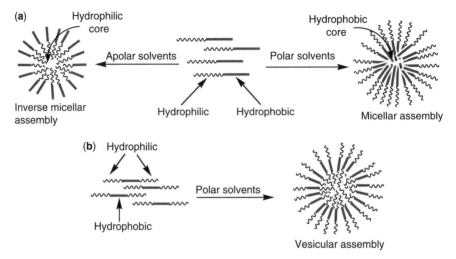

Figure 2.2 Representation of formation of (a) micellar and (b) inverse micellar vesicular assemblies from diblock and triblock copolymers, respectively.

example, the flexibility of the hydrophobic blocks is thought to be enhanced and thus there is an entropic gain.

The propensity to form supramolecular assemblies by amphiphilic molecules is estimated by their CAC, which is defined as the concentration below which only single chains of polymer exist. Above the CAC, both single chains and aggregates are observed. When studying a micelle, CAC is referred to as the critical micelle concentration (CMC). The CMC is much lower in amphiphilic block copolymers compared to small molecule amphiphiles. The most reliable method for determining CMC involves surface tension measurements, although other techniques such as absorption or fluorescence spectroscopy are also utilized (Alexandridis et al. 1994; Holland et al. 1995; Lopes and Loh 1998). Several factors can influence the CAC of an amphiphilic species, the most important component of which involves the nature of the block that forms the core of the micelle. The nature of the hydrophilic

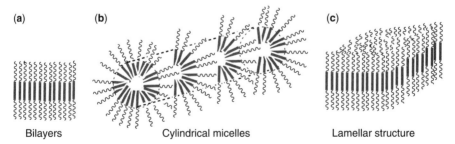

Figure 2.3 Representation of organized assemblies (a) bilayers, (b) cylindrical micelles, and (c) lamellar structure.

block, hydrophobic solubilizers, and ionic strength of the solution also influence the thermodynamic stability of the micelles at a certain concentration.

Here, we are most interested in the interaction of these organized assemblies with another molecule, that is, molecular recognition. We will broadly classify these interactions into nonspecific and specific ones. In addition to the chemical interaction between the molecules, there has been a significant amount of work on achieving amphiphilic block copolymer assemblies that respond to physical stimuli, for example, temperature. We will not discuss this in any detail, except when relevant to the focused topic of chemical recognition.

2.2.1. Nonspecific Interactions

The most common nonspecific interaction observed in amphiphilic polymer assemblies involves the sequestration of guest molecules through solvophobic interactions. For example, in a micellar assembly in water, the interior is hydrophobic and is therefore capable of sequestering lipophilic guest molecules that are otherwise insoluble in water. This binding interaction can be viewed as the micellar assemblies acting as nanocontainers for hydrophobic guest molecules. The reason for the enormous interest in this area is that most therapeutic organic molecules are hydrophobic and thus water insoluble (Kwon and Kataoka 1995; Kwon and Okano 1996; Kwon 1998; Batrakova et al. 2006). The bioavailability of these molecules can be significantly enhanced if they are sequestered inside micelles and subsequently released slowly at concentrations where there is reasonable solubility. In addition to making these molecules bioavailable, such a process could also find use in controlled release or triggered release of drug molecules in biological environments.

Amphiphilic block copolymers can be broadly classified into nonionic and ionic block copolymers. Although both of these block copolymers exhibit the guest sequestration properties mentioned above, nonionic block copolymers have drawn particular interest in combining nanocontainer properties with temperature sensitivity. An example of such a nonionic block copolymer involves poly(ethylene oxide) (PEO) as the hydrophilic block and poly(propylene oxide) (PPO) as the hydrophobic block (Chart 2.1; Cammas-Marion et al. 1999; Chung et al. 2000; Rosler et al. 2001). The PEO block is considered to be hydrophilic, which is due to the formation of hydrogen bonds with water. However, it has been observed that this hydrogen bond based hydration is significantly weakened at higher temperatures.

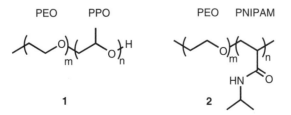

Chart 2.1   Structures of block copolymers: PEO-*b*-PPO (**1**) and PEO-*b*-PNIPAM (**2**).

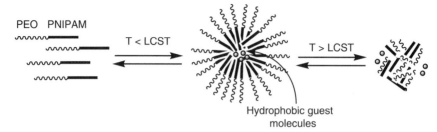

Figure 2.4 Temperature responsive micellar formation in PNIPAM-based polymers.

Thus, the PEO segment actually becomes hydrophobic at higher temperatures. This temperature-dependent change converts the amphiphilic block copolymer to a water-insoluble hydrophobic polymer (Topp et al. 1997; Chung et al. 2000). The temperature at which the polymer exhibits this transition is called its lower critical solution temperature (LCST). In addition to PEO, substituted poly(N-isopropyl acrylamide) (PNIPAM; Chart 2.1) exhibits temperature sensitivity, where the LCST can be tuned by varying the alkyl functionality. The guest encapsulation combined with the temperature-sensitive precipitation of the polymers has been exploited to sequester and separate guest molecules from aqueous solutions (Fig. 2.4).

Despite the interesting combination of temperature sensitivity and micelle forming behavior of nonionic micelles, they have the disadvantage of being less stable mechanically. The lack of mechanical stability is mainly due to the fact that the repeat units in PEO or PNIPAM interact with water through hydrogen bonding, but they are otherwise hydrocarbon based. Therefore, the difference in compatibility between the hydrophilic and hydrophobic blocks is not significant enough. It is because of this that we often find block copolymer micelles that are based on ionic repeat units as the hydrophilic segments to be more stable.

In addition to the hydrophobic interaction mentioned above to encapsulate guest molecules, other types of nonspecific interactions have also been explored to enhance binding. For example, block copolymer micelles based on PEO as hydrophilic segments and poly(β-benzyl L-aspartate) as hydrophobic blocks have used to encapsulate doxorubicin. The encapsulation efficiency of doxorubicin, an aromatic anticancer drug molecule, has been found to be significantly higher. This observation has been attributed to the π–π interaction between the anthracycline moiety of doxorubicin and the benzyl group of poly(β-benzyl L-aspartate) (Cammas-Marion et al. 1999).

Similar to micellar assemblies in water, reverse micelles have also been utilized to bring about nonspecific binding interactions in organic solvents. Akiyoshi et al. (2002) have synthesized an amphiphilic block copolymer containing PEO and an amylase chain as receptor for methyl orange (MO; Chart 2.2). Amylases are insoluble and methoxy-PEO (MPEO) is soluble in chloroform. Hence, an MPEO–amylase block copolymer forms reverse micelles in chloroform. Akiyoshi et al. established the capability of the buried receptors to extract the complementary analyte by studying the ultraviolet visible (UV–vis) spectra. A solution of polymer was shaken

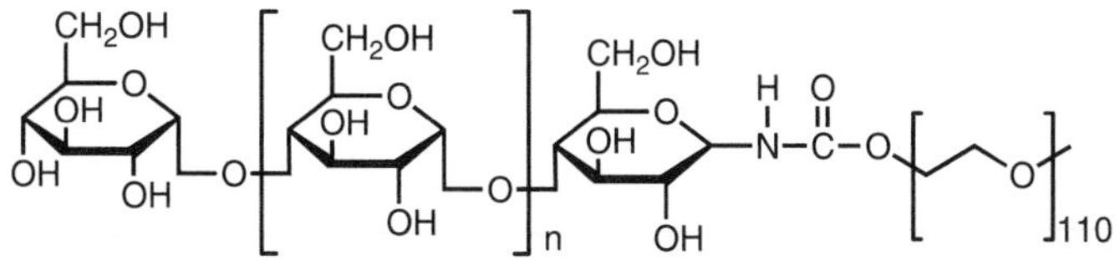

Chart 2.2 Structure of block copolymer containing amylase and PEO.

with MO-containing water. The extraction was confirmed by a peak at 428 nm observed in the UV–vis spectra of the chloroform layer. The 36-nm blueshift observed for the encapsulated MO was attributed to the microenvironment provided by the polymeric assembly.

Most of the discussion above has focused on block copolymers in which the hydrophilic block is made of neutral functionalities, but it is also interesting to consider polymers in which one of the blocks is ionic. The first part of the discussion here will be about block copolymers, where the ionic segment will end up as the core component of an assembly. In order to encapsulate charged molecules in the core of block copolymer micelles, amphiphilic ionic block copolymers consisting of an ionic block and a PEO block have been developed. Because both PEO and polyionic blocks are water soluble, these are also called double-hydrophilic block copolymers (Colfen 2001). Even though the ionic block is water soluble in this type of polymer, it can be converted into a relatively hydrophobic one either by varying the pH of the solution or by the formation of a polyion complex (PIC). The micellization of PEO-*b*-poly(2-vinylpyrine) (P2VP) block copolymer (PEO as shell and P2VP as core) was pH responsive: micellization was observed when increasing the pH of the solution. In contrast, PEO-*b*-poly(methacrylic acid) block copolymers were able to form micelles by decreasing the pH. At high pH, the resultant carboxylate anions of these polymers destabilize the micelles because of the electrostatic repulsion between them (Fig. 2.5; Kabanov et al. 1996).

In ionic block copolymers, the micellar assemblies can also be induced by complexation with oppositely charged molecules, forming a PIC at the core. The electrostatic interaction between PEO–polycation and PEO–polyanion block copolymers is the driving force for the formation of micellar assemblies in which the complex

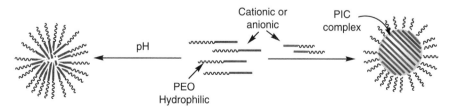

Figure 2.5 Formation of micellar assemblies induced by pH complexation in ionic block copolymers.

of anionic and cationic blocks forms the core. In addition to electrostatic interactions, the increase in entropy attributable to the release of smaller counterions plays a key role in micellar formation. These types of interactions were first observed between PEO-*b*-polylysine and PEO-*b*-polyaspartate (Harada and Kataoka 1995). The combination of PEO-*b*-poly(sodium methacrylate) and poly(*N*-ethyl-4-vinylpyridinium) is also another example of PIC formation. The interest in these types of polymers arises from their ability to deliver charged biomolecules, such as DNA and enzymes at the core. Block copolymer micelles containing a cationic block as the hydrophobic block were capable of binding with charged biological molecules, especially with anionic DNA molecules, forming PIC in the core between the cationic block and anionic DNA molecules. These polymeric micelles are evolving as nonviral DNA delivery vehicles for gene therapy (Katayose and Kataoka 1997; Kakizawa and Kataoka 2002). For example, poly(ethylene glycol)-*b*-poly(L-lysine) block copolymer has been used to trap DNA in its ionic core through electrostatic interactions.

Another class of amphiphilic ionic block copolymers employs an ionic block as the hydrophilic segment linked to a hydrophobic segment. Polystyrene-*b*-poly (4-vinylpyrine) (PS-P4VP) and PS-P2VP were the first reported polymers of this class in which PS is the hydrophobic segment and VP is the hydrophilic segment (Marie et al. 1976; Gauthier and Eisenberg 1987). Gauthier and Eisenberg have established PS-*b*-poly(acrylic acid) (PAA) based block copolymers that form spherical micelles, called "crew-cut" micelles, in cases when the PS chain is long and the PAA chain length is short. PS-*b*-PAA polymers have been used to obtain a variety of other morphologies such as rods and vesicles by tuning the ratio of PS/ PAA by the addition of salt or by mixing organic solvents. They claim that the factors that control these morphological changes include 1) the degree of stretching in the core, 2) intershell (corona) interactions, and 3) core–shell interfacial energy. A change in any of these factors affects the morphology of the assemblies that results in translating the spherical micelles to rods and rods to vesicles (Allen et al. 1999; Choucair and Eisenberg 2003; Chen and Jiang 2005; Rodriguez-Hernandez et al. 2005).

Nolte and others (Cornelissen et al. 1998) reported a new class of amphiphilic ionic block copolymers in which a charged helical polypeptide chain was attached to the hydrophobic PS. The charged helical peptide is poly(isocyanide) based peptide derived from dipeptides, isocyano-L-alanine-L-alanine (IAA) and isocyano-L-alanine-L-histidine (IAH). From circular dichroism (CD) spectra they concluded the formation of a left-handed helix for PS-*b*-PIAA and right-handed helix for PS-*b*-PIAH (Chart 2.3). These polymers have been touted as superamphiphiles forming various structures such as rods, vesicles, and bilayers. When the chain length of PIAA is decreased, the rods change their morphologies to bilayers and vesicles. In another communication, Velonia et al. (2002) reported a protein–polymer hybrid, called "giant amphiphiles," that is based on a lipase enzyme attached to PS. Rodlike micelles were observed for these polymers in tetrahydrofuran.

In addition to the micelle-type assemblies described above, there has been significant interest in developing conditions for forming vesicle-type assemblies from amphiphilic polymers. Polymeric vesicles are formed by bolamphiphilic block

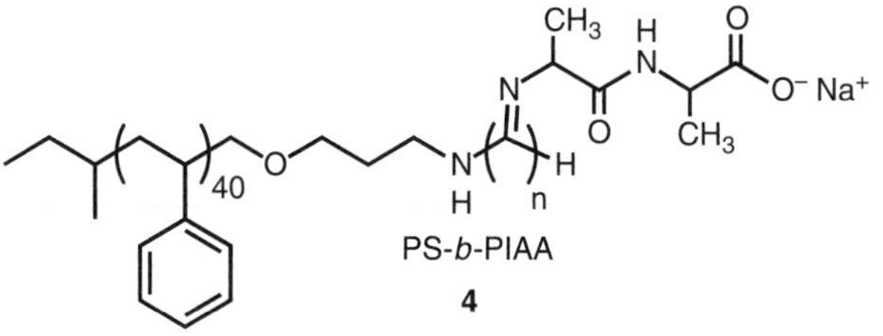

Chart 2.3   Structure of block copolymers PS-*b*-PIAA (**4**) and PS-*b*-PIAH (**5**).

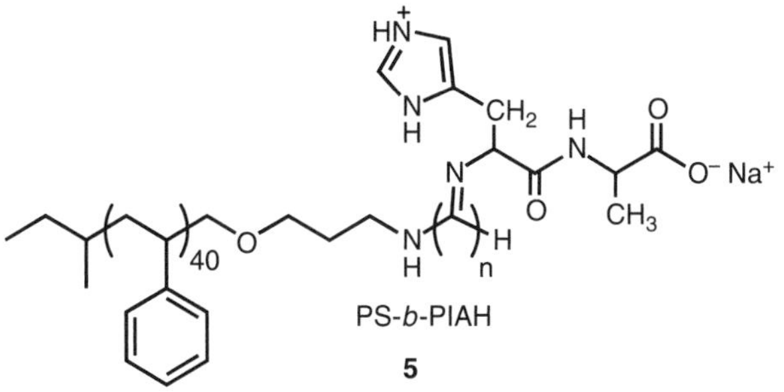

copolymers, in which two terminal blocks are hydrophilic and the middle block is hydrophobic (Discher and Eisenberg 2002; Savariar et al. 2006). Upon formation of the assembly, the hydrophilic terminal blocks are exposed to the solvent as well as entangled in the core, forming a hydrophilic core as well as exterior (Fig. 2.2b). These aggregates are capable of sequestering hydrophilic molecules within the interior and therefore have been touted as potential delivery vehicles for hydrophilic drugs. In addition to the above-mentioned triblock copolymer architectures, vesicular assemblies have been obtained from diblock copolymers by controlling the processing conditions.

2.2.2. Specific Interactions

Specific interactions between receptor and ligand functionality incorporated in the polymer have been popular, because polyvalent interactions significantly enhance binding affinity. The concept of polyvalent interactions to enhance binding affinity to biological molecules and their utility in biomedical applications has been extensively reviewed elsewhere (Mammen et al. 1998; Ercolani 2003; Kitov and Bundle 2003; Badjic et al. 2005; Kiessling et al. 2006). Here, we will focus on the aspects of amphiphilic polymers that exhibit specific binding to targeted receptors. Carbohydrates, peptides, and single-stranded DNA all have been incorporated as components of block copolymers for specific interactions.

The classic example in Nature involving multivalent interactions for specific binding is the double helix formation in DNA. Interesting superstructures have been achieved by hybridizing block copolymer based DNA molecules (Jeong and

Park 2001; Frankamp et al. 2002). Recently, Ding and coworkers (2007) have shown that the properties of DNA-based block copolymers can be modulated by molecular recognition. For this purpose, a block copolymer containing ssDNA (sequence: 5′-CCTCGCTCTGCTAATCCTGTTA-3′) as one block and PPO as another block was synthesized. The polymer was found to be easily soluble in water because of the low glass-transition temperature (T_g) of PPO. In fact, the low T_g was the prime reason for the choice of PPO as one block. A complementary ssDNA (5′5′-TAACAGGATTAGCAGAGCGAGG-3′) having the same number of bases as that of the ssDNA in the block copolymer was added to ssDNA–PPO micelles. The assembly formation was studied before and after the dsDNA formation. In this experiment, the size and shape of the micelles did not change. In the next experiment, a long ssDNA was chosen in such a way that the sequence was the complementary of DNA-*b*-PPO by several times. The long ssDNA strand consists of five [sequence: 5′5′-(TAACAGGATTAGCAGAGCGAGG)-3′] and four ([5′-(TAACAGGATTAG CAGAGCGAGG)-3′], in which five and four DNA-*b*-PPO polymers can be annealed, respectively. Scanning force microscopy images of the hybridized assembly did not show any spherical structures but showed rodlike structures. Figure 2.6 represents the formation of micellar and rodlike structures. The height of the rods was ~2 nm and the length was ~30 nm. The reason for this structural change was the disintegration of the spherical ssDNA–PPO micelle, which was followed by linear organization along the template. Therefore, it is possible to modulate the structural assembly of block copolymer micelles by molecular recognition.

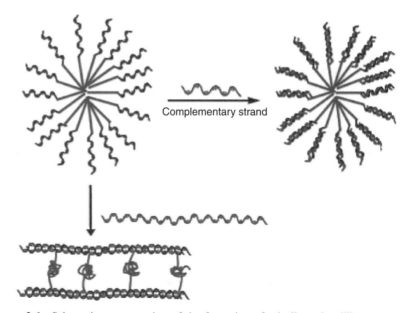

Figure 2.6   Schematic representation of the formation of micelle and rodlike structures of amphiphilic conjugate of oligonucleotide and PPO.

Amphiphilic block copolymers capable of binding to DNA can be used as nonviral delivery vehicles, as mentioned above (Cleek et al. 1997; Jeong and Park 2001). For efficient DNA delivery, the vehicle should have the necessary recognition and encapsulation capabilities. Toward this goal, a block copolymer having biodegradable poly(D,L-lactic-*co*-glycolic acid) (PLGA) as one block and an oligodinucleotide (ODN, CACGTTGAGGGGCAT) as the other block was synthesized. It is expected that the ODN/PLGA conjugate can be more efficiently delivered into the cells by an endocytosis mechanism, in contrast to ODN itself that is transported across the cell membrane by passive diffusion. The ODN/PLGA micelles released ODN in a sustained manner because of the controlled degradation of PLGA.

In addition to utilizing natural ODNs in block copolymers, ODN mimics have also been studied using polymers with the capability of forming hydrogen bonds. Bazzi and colleagues (Bazzi and Sleiman 2002; Bazzi et al. 2003) have reported diblock and triblock amphiphilic polymers with diamidopyridine (DAP) and dicarboximide moieties as molecular recognition units and studied their micelle formation properties (Chart 2.4).

They found the formation of ~40-nm particles in a polymer, where two recognizing blocks are separated by a neutral linker. The demonstration of disassembly is as interesting as assembly formation, because the ease of disassembly is the greatest advantage of noncovalent assemblies. One could imagine how this could be particularly useful in utilizing these types of assemblies as delivery vehicles. Toward this goal, Bazzi and coworkers (Bazzi and Sleiman 2002; Bazzi et al. 2003) synthesized a diblock copolymer containing DAP and a hydrophobic block, which forms ~190-nm aggregates in $CHCl_3$. The aggregation is possibly due to hydrogen bonding between DAP units, which are weakly self-complementary. This weak interaction was proven by the addition of maleimide, which binds more strongly with DAP. This disrupts the hydrogen bonding between DAP units that results in disassembling the micelle. To further prove that micelle disassembly is a result of the specific interaction between maleimide and DAP, *N*-methylmaleimide was added to the micellar solution. *N*-Methylmaleimide has similar structural features to maleimide, except that it does not associate with DAP. The micelle solution remained turbid even after boiling with *N*-methylmaleimide, confirming that the micelle did not disassemble. However, the micelles were disassembled upon the addition of succinimide, *N*-butylthymine, and *N*-hexylthymine.

Similarly, small triblock copolymer **7** containing DAP and thymine (THY) as outer blocks (Chart 2.4) was assembled to form a micelle because of the noncovalent interaction between DAP and THY. These micelles were shown to disassemble upon the addition of THY containing small molecules. The micelles could be efficiently disassembled by small molecule guests, albeit with less selectivity. This is possibly because of the weak interaction between the DAP units. To bring more selectivity, Bazzi and cohorts (Bazzi and Sleiman 2002; Bazzi et al. 2003) synthesized a triblock copolymer containing DAP and THY blocks, which were separated by a hydrophobic block. This polymer formed large micellar aggregates. Because of the strong interaction between DAP and THY, these assemblies were not disassembled by many guest molecules, confirming the strong hydrogen bonding between the DAP and

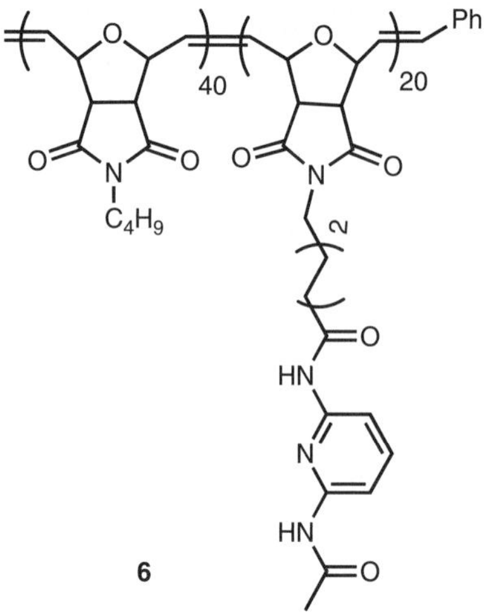

Chart 2.4 Structures of diblock and triblock amphiphilic polymers with diamidopyridine (DAP) and dicarboximide moieties.

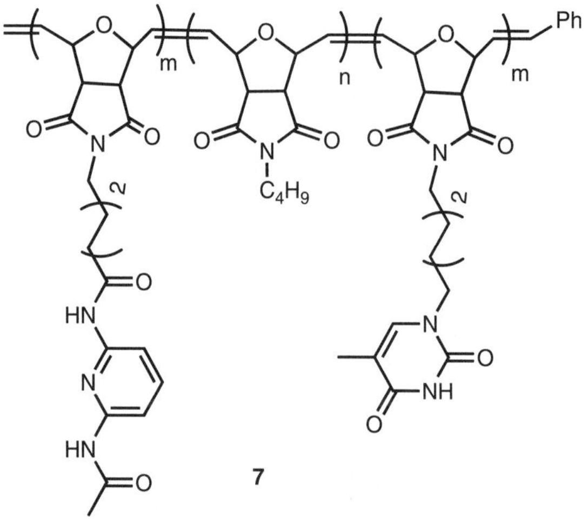

THY units. However, these micelles could be disassembled by THY analogues that indicates the selective disassembly of DAP and THY block copolymer micelles. An interesting aspect of these triblock copolymers is that no assembly was observed from the polymer, where the two complementary blocks are connected directly. No explanations were given about this drastic change in the aggregation behavior. This observation is indeed interesting and rich information might be available, if we look at the molecular origins of this difference.

Carbohydrate–protein interactions have gained particular interest because of implications in areas such as cell-specific drug delivery (Dwek 1996).

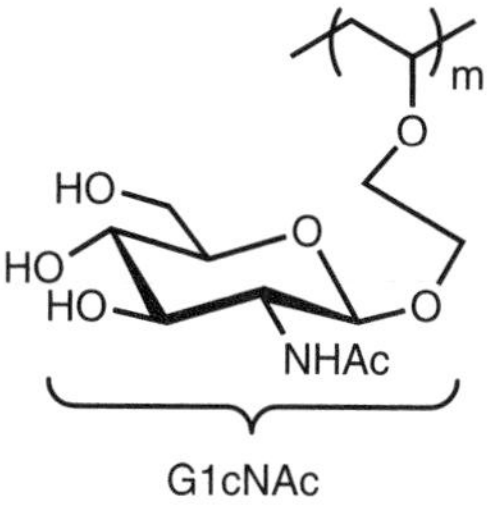

Chart 2.5 Structure of GlcNAc-based homopolymer (**8**) and amphiphilic block copolymer (**9**).

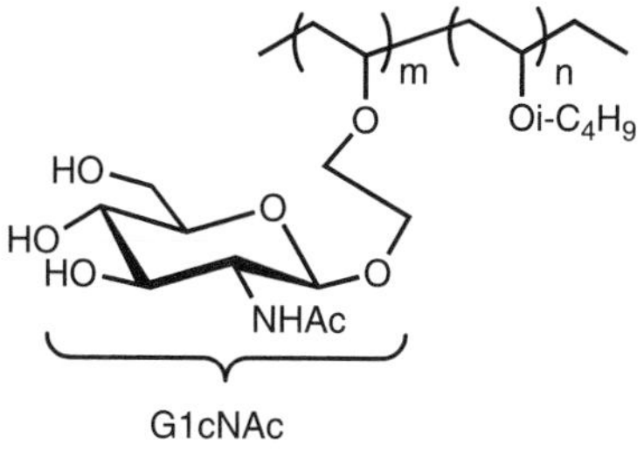

Oligosaccharides on the cell surface play an important role in various biological recognition processes, including intercellular recognition, adhesion, cell growth, and differentiation. For recognition of cell surfaces, polymers containing sugar units are interesting synthetic targets. For example, amphiphilic block copolymer **9** containing isobutyl vinyl ether as the hydrophobic block and N-acetyl-D-glucosamine (GlcNAc) as the hydrophilic block with pendant glucose residues has been synthesized and studied (Chart 2.5; Yamada et al. 1999). These polymers have been found to assemble and disassemble in response to variations in temperature. To study the binding between glucosamine units and wheat germ (WGA) lectins, a natural glycoprotein that specifically recognizes GlcNAc residues was used. Using the tryptophan emission from the WGA lectin as the probe, the interaction between the glucosamines in the block copolymer and the protein was monitored. Considering several water-soluble, carbohydrate-functionalized polymers have been studied to understand polyvalent interactions, it is interesting to question the importance of studying block copolymers for this purpose. This is particularly important considering the fact that block copolymer assemblies, such as micelles, could reduce the accessibility of some of the ligands even within the hydrophilic block for binding to the protein. In this study, the recognition of these polymers was compared with that of homopolymer **8** and corresponding monomers. Both polymer types exhibited higher binding affinity compared to the monomeric GlcNAc (6.8×10^2 M^{-1}). For the polymer containing 20 repeat units of GlcNAc functionality, block copolymer **9** exhibited a binding affinity of 3.3×10^5 M^{-1} compared to 5.9×10^4 M^{-1} for homopolymer **8** with the same number of repeat units. The Hill coefficient for both of these polymers was found to be the same, 2.3. The fivefold enhancement in binding affinity for the block copolymer could be attributed to preorganization. Nonetheless, it is interesting to note that the eventual encapsulation does seem to play a role in the binding affinity, because when the degree of polymerization was increased to 38, the difference in binding affinity between the homopolymer and the block copolymer decreased significantly.

To understand whether the preorganization in a block copolymer assembly indeed provides any advantages in presenting multivalent ligand copies, one could envision the presentation of a single copy of a specific ligand at the hydrophilic chain terminus

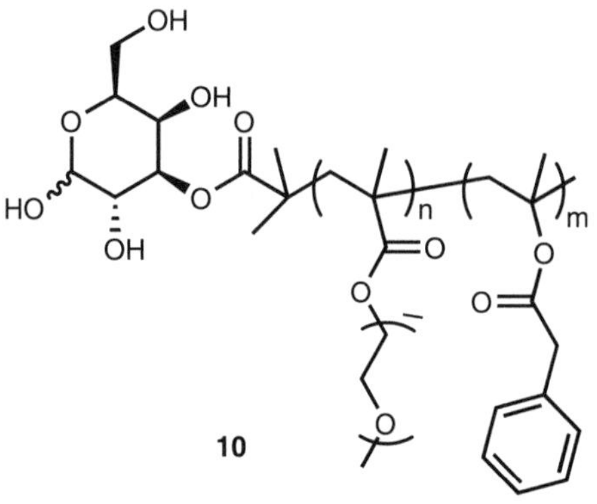

10

Chart 2.6   Structure of block copolymer based on PEO and poly(benzyl methacrylate) (PBMA) containing β-D-galactose unit.

of an amphiphilic block copolymer. Although this has not been addressed specifically in this manner, synthetic methods have been developed to incorporate a single copy of the carbohydrate ligand in a block copolymer. In this case, Bes et al. (2003) prepared glucose and galactose containing block copolymers to recognize lectins containing poly((ethylene glycol)methyl ether methacrylate) as the hydrophilic segment and poly(benzyl methacrylate) as the hydrophobic segment (Chart 2.6). These polymers were synthesized by living radical polymerization utilizing glucose and galactose with 2-bromoisobutyryl bromide as the initiator. To check the binding affinity of the glucose and galactose containing micelles toward lectin, these micelles were passed through a high-performance liquid chromatography column containing RCA-1 lectin. The copolymer containing galactose residues was retained in the column, which was assumed to confirm the binding between the galactose and lectins. Further, a control experiment was done using micelles with protected galactose, which did not bind with lectins.

2.3. AMPHIPHILIC HOMOPOLYMERS

An interesting variation to solve these conflicting properties of block copolymers is to generate amphiphilic homopolymers. Amphiphilic homopolymers have hydrophobic and hydrophilic functionalities in each repeat unit. These polymers have the potential to exhibit lower CACs, while still providing the advantages of preorganization of the ligands. In addition, the polymer architecture does not necessarily reduce the accessibility of ligands, as is necessarily the case with block copolymer assemblies.

Basu et al. (2004) reported a series of amphiphilic homopolymers in which the hydrophilic moiety is a carboxylic acid moiety and the hydrophobic moiety can be based on either an alkyl chain or an aromatic ring (Chart 2.7). These polymers self-assemble into either micelles or reverse micelles, depending on the solvent environment. Figure 2.7 represents the formation of a micellar assembly in a polar

11

12

13

14

15

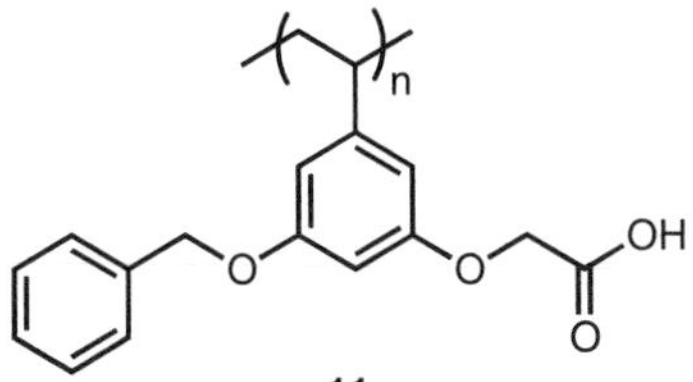

Chart 2.7 Structures of amphiphilic homopolymers.

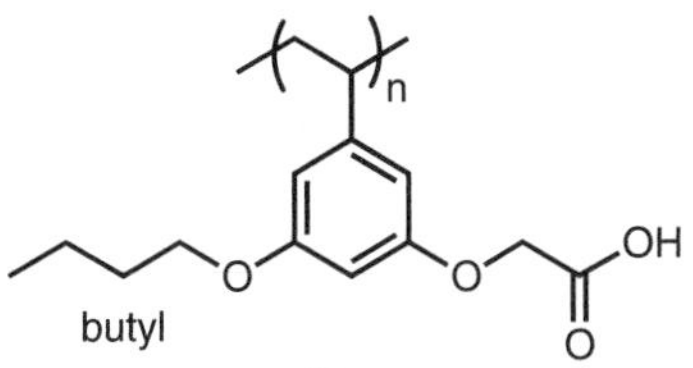

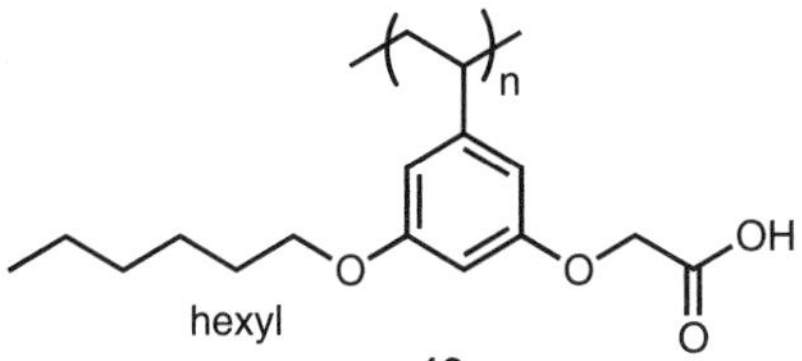

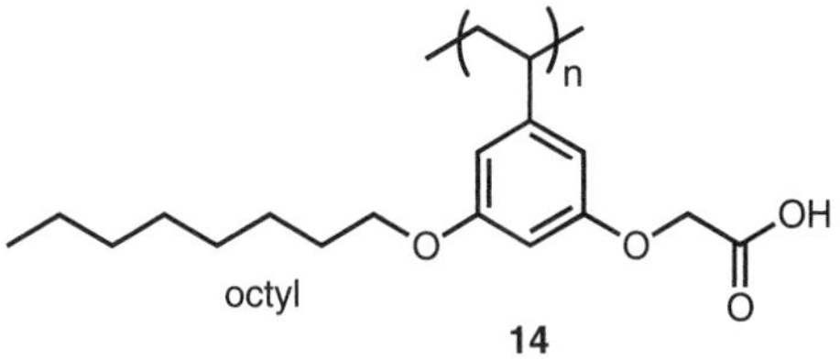

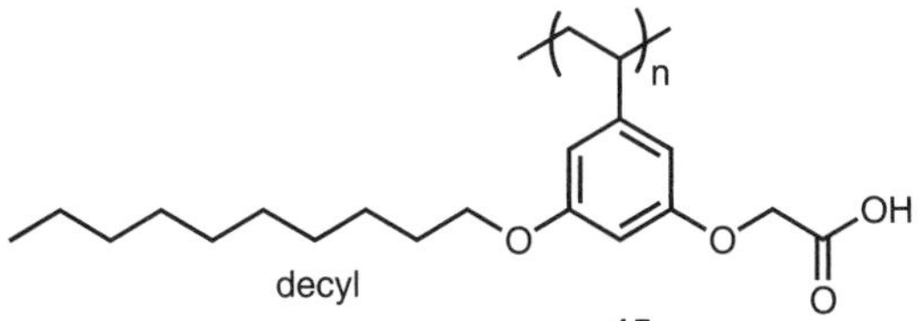

solvent and reverse micellar assemblies in an apolar solvent by the aggregation of amphiphilic homopolymers. The formations of these assemblies were confirmed by a combination of absorption and emission spectroscopy, transmission electron microscopy (TEM), and light scattering experiments.

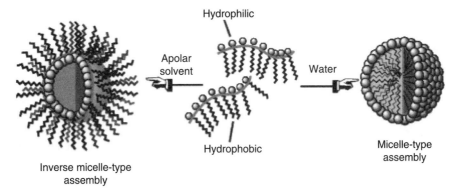

Figure 2.7 Formation of micellar and inverted micellar assemblies from amphiphilic homopolymer.

2.3.1. Container Properties

The host–guest properties (Basu et al. 2005) of amphiphlic homopolymers were studied in polar and apolar solvents. To systematically study the host–guest properties, a series of amphiphilic polymers that differ in lipophilic chain length were synthesized (Chart 2.7). The micellar properties of these polymers were established by solubilizing water-insoluble Reichardt dye (RD) in the aqueous micellar solution. The absorption maxima for RD in the polymer–dye complex observed at 502 nm indicates that the micellar environment is comparable to that of glycerol. The polarity of the interior seems higher, because it looks closer to glycerol, at first glance. However, note that the apolar interiors of these polymer assemblies are comparable to those of the well-characterized micellar interiors formed by small molecular surfactants such as CTAB and SDS. Similarly, inverted micellar properties were studied by solubilizing Rhodamine-6G in toluene in the presence of these polymers. This was confirmed by the absorption spectrum of the polymer–dye complex in toluene. The absorption spectrum confirmed the presence of the dye molecules, and the self-quenching property of Rhodamine-6G was utilized to confirm the encapsulation of the dye within the inverse micelles.

Because these amphiphilic polymers form assemblies in both polar and apolar solvents, we were interested in identifying the fate of the polymer in a heterogeneous mixture of polar and apolar solvents (Basu et al. 2005). We noticed that the guest molecules are released from the micellar or inverse micellar container and end up in the solvent in which they are most soluble. However, this property does not tell us about the fate of the amphiphilic polymer. Three limiting scenarios are possible: 1) the distribution coefficient of the polymer dictates the partition in the aqueous phase and the organic phase as micelle-type and inverted micelle-type assemblies, respectively. The dye molecule gets released during the process of this thermodynamically driven redistribution of the host. 2) The polymer assembly disintegrates by disassembling at the interface of the two solvents and therefore loses the container property. 3) The amphiphilic assemblies are kinetically trapped in the solvent in which they are initially assembled. Using a series of simple experiments, we confirmed that these polymer assemblies are indeed kinetically trapped in the solvent in which they are initially assembled. This unique feature of these polymers has been utilized in an interesting peptide detection and extraction application, which could find use in proteomics (*vide infra*).

Utilizing similar designs, polymers containing bolamphiphilic moieties were synthesized to form vesicle-type assemblies (Chart 2.8; Savariar et al. 2006). Because these polymers contain two hydrophilic carboxylic functionalities connected by alkyl chains of varying lengths, it was expected that such a placement of the functionalities in a homopolymer side chain could provide a more direct pathway to achieve vesicular assemblies. The formation of vesicular assemblies was investigated by TEM, spectroscopy, and light scattering experiments.

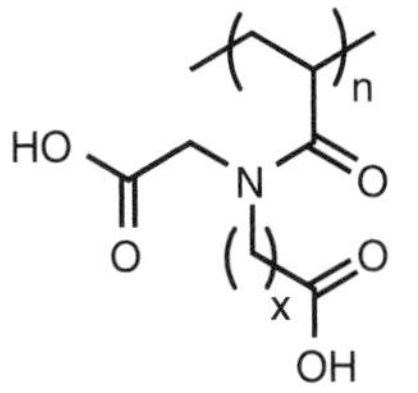

16, x = 5
17, x = 7
18, x = 10
19, x = 15

Chart 2.8   Structures of bolamphiphilic homopolymers.

2.3.2. Recognition of Protein Surfaces

Amphiphilic polymers form flexible assemblies; hence, they can change their conformation upon binding to a surface. This interesting property coupled with the easy modification of charge can be used for recognition of biomolecules like proteins, without necessarily denaturing their structure. This noncovalent binding between proteins and charged polymers could be useful in modulating protein–protein interactions and protein–DNA interactions (Fischer et al. 2002, 2003; Hong et al. 2004). We studied the interaction of anionic polymer micelles with positive patches of chymotrypsin (ChT; Sandanaraj et al. 2005). Amphiphilic polymer micelles are ~40 nm in size with negatively charged carboxylate groups on the surface. Because these polymeric micelles expose a huge amount of negative charges on the surface, we expected them to recognize the positive patch of proteins by electrostatic attraction. Figure 2.8 is a schematic representation of the interaction of proteins with the polymeric micelles.

Polymer–protein binding affinity was studied qualitatively through nondenaturing gel electrophoresis, fluorescence spectroscopy, and enzymatic assay. The formation of the polymer–protein complex and the electrostatic basis of the interaction were confirmed by the inhibition of the enzymatic activity and the recovery of this activity at high ionic strengths, respectively. The dissociation constant of polymer **11** with the protein was determined to be 7×10^{-7} M with a polymer/ChT binding ratio of 1:10. Perhaps the most interesting aspect of this study is the artificial selectivity that this interaction brought to the enzyme toward its substrates. The modulation in activity

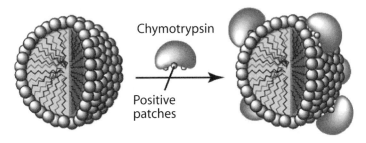

Figure 2.8   Interaction of anionic amphiphilic homopolymeric micelle with the positive patch of protein.

of the ChT–polymer complex was studied toward anionic, neutral, and cationic substrates. Because the polymer–protein interaction is based on electrostatics, we speculated that the negative charge of N-(succinyl)-L-phenylalanyl-p-nitroanilide could influence the inhibition efficiency of ChT (Fischer et al. 2002, 2003). Indeed, there was an enhanced activity of ChT against the cationic substrate, but a decrease in activity in the case of the anionic substrate. These results imply that the negatively charged polymer plays a role in recruiting the substrates to the active site of the protein.

Similarly, this amphiphilic polymer micelle was also used to disrupt the complex between cytochrome c (Cc) and cytochrome c peroxidase (CcP; Sandanaraj, Bayraktar et al. 2007). In this case, we found that the polymer modulates the redox properties of the protein upon binding. The polymer binding exposes the heme cofactor of the protein, which is buried in the protein and alters the coordination environment of the metal. The exposure of heme was confirmed by UV–vis, CD spectroscopy, fluorescence spectroscopy, and electrochemical kinetic studies. The rate constant of electron transfer (k^0) increased by 3 orders of magnitude for the protein–polymer complex compared to protein alone. To establish that the polymer micelle is capable of disrupting the Cc–CcP complex, the polymer micelle was added to the preformed Cc–CcP complex. The k^0 observed for this complex was the same as that of the Cc–polymer complex, which confirms that the polymer micelle is indeed capable of disrupting the Cc–CcP complex.

2.3.3. Protein Sensing

Because these polymeric amphiphiles are capable of binding to proteins, we envisaged the possibility of detecting metalloproteins using fluorescence quenching. Conjugated polymers containing charged functionalities have been utilized as fluorescent sensing elements (Chen et al. 1999; Fan et al. 2002; Wilson et al. 2003; Kim et al. 2005), because the charges can bind to the complementary charges in the proteins, whereas the inherent fluorescence of the polymer could potentially be affected by the cofactors in metalloproteins. Porphyrin-based or heavy metal based cofactors in metalloproteins can accept energy or electrons from the excited state of the conjugated polymer chromophores. Conjugated polymers have been shown to be effective in sensing metalloproteins, but it has also been shown that fluorescence changes can be observed with nonmetalloproteins because of conformational or aggregation changes in conjugated polymers. We hypothesized (Sandanaraj et al. 2006) that utilizing the amphiphilic homopolymers with pendant chromophores (Chart 2.9) would provide all the advantages of conjugated polymers, while being selective only to metalloproteins. This is because even when the protein binding could cause gross conformational changes in the polymer, the emission properties of a pendant chromophore in a nonconjugated polymer should remain unaffected. Therefore, the only feature that this polymer would report is whether there is a metalloprotein cofactor that can accept energy or electrons from its excited state. We have shown that such a strategy indeed results in selective sensors for metalloproteins (Fig. 2.9).

The approach above could distinguish metalloproteins from nonmetalloproteins. However, it is interesting to provide an approach that distinguishes among the

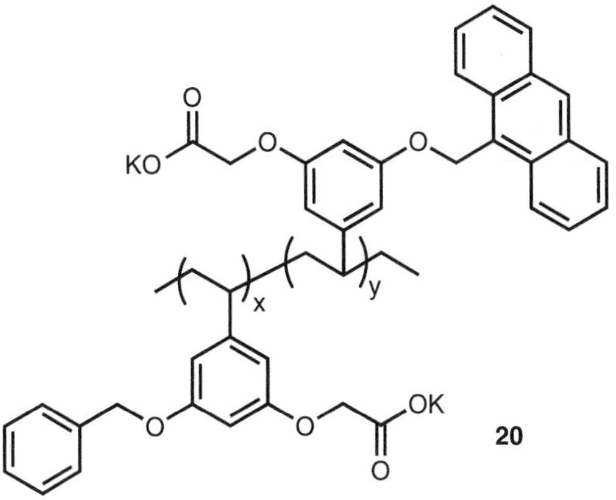

Chart 2.9 Amphiphilic homopolymer containing an anthracene unit as a transducer.

metalloproteins. For this purpose, we adapted a new biomimetic strategy. Nature has evolved a unique mechanism by which a series of less selective bindings result in a pattern, which is then translated to sensing. Examples of such sensing elements in Nature involve our ability to differentiate odor and taste. There have been a variety of approaches in the literature, where multiple receptors that have different binding affinities to a variety of analytes were designed and synthesized. Impressive recognition efficiencies have been demonstrated using this strategy. However, if this strategy were to be broadly applicable, the number of data points needed to attain a reliable pattern for analytes is large. This translates into a significant synthetic enterprise to arrive at a variety of receptors. We conceived that if one develops an orthogonal differential receptor approach, where a variety of data points could be generated from a single receptor, then the synthetic requirements become small. Note that the

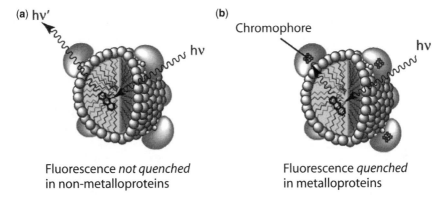

Figure 2.9 Interaction of polymeric micelles that contain anthracene units (a) with non-metalloproteins and (b) with metalloproteins.

polymer micelles have been shown to be noncovalent hosts for hydrophobic guest molecules and these can bind to proteins. If these guest molecules can be varied to be fluorescent dyes, a variety of receptors with different fluorescent transducers can be achieved very rapidly. We hypothesized (Sandanaraj, Demont et al. 2007) that this strategy will work, because factors such as the relative frontier orbital energetics of different fluorescent transducers and the distance between the transducer and the protein will vary for each fluorophore. These variations should provide different responses and therefore result in a fingerprint for different metalloproteins. This was demonstrated to be the case using four different metalloproteins and eight different dye molecules.

2.3.4. Recognition and Detection of Peptides

One of the interesting aspects of amphiphilic homopolymer assemblies is that they are kinetically trapped in the solvents in which they are initially assembled. Therefore, we envisioned the possibility of utilizing inverse micelles in apolar solvents and utilizing electrostatic interactions to selectively remove molecules with complementary charges from aqueous solutions into the organic phase (Combariza et al. 2007). It is particularly interesting to be able to achieve such a separation with peptides based on their isoelectric points (pIs). Peptide detection is capable of ultimately providing information on proteins, especially in mass spectrometry based proteomics (Karas and Hillenkamp 1988; Tanaka et al. 1988; Aebersold and Mann 2003). We hypothesized that 1) the anionic carboxylate interiors are capable of recognizing positively charged peptides and sequestering them through electrostatic interactions and 2) because the interactions are mainly due to electrostatics, the selective separation of proteins can be achieved at different pH values based on the different pI values of protein. This strategy is schematically represented in Figure 2.10.

A toluene solution of polymer **15** was added to an aqueous mixture of peptides. After effective equilibration, the heterogeneous mixture was allowed to separate and the organic layer was analyzed by matrix assisted laser desorption ionization mass spectrometry analysis. Because the pH of the aqueous solution was 7.1, the peptides with pIs above 7.1 were extracted and the peptides with pIs below 7.1

Figure 2.10 Selective extraction of peptides utilizing amphiphilic homopolymeric micelles.

remained in the aqueous solution. These results indicate the positively charged peptides were extracted and encapsulated by negatively charged polymeric interiors, whereas the negatively charged ones were not extracted because of the electrostatic repulsion. In order to study the effect of pH on peptide extraction, experiments were carried out at pH 4.0, 7.1, and 9.0. At these pH values, the extraction cutoff was determined by the pI of the peptide and the pH at which the experiment was carried out. To test the scope of this approach, the experiment was carried out using a mixture of peptides from trypsin digest of myoglobin. In this case also, the positively charged peptides were extracted much better into the apolar organic solvent, compared to the negatively charged ones. These results could have significant implications in applications such as proteomics.

2.4. AMPHIPHILIC DENDRIMERS

Dendrimers are branched macromolecules that can be achieved with well-defined molecular weights. The key interest in dendrimers as candidates for molecular recognition arises because of this molecular weight control and because they attain a globular shape at high molecular weights. Globular shape associated with amphiphilicity in a macromolecule provides the opportunity to achieve unimolecular micelles, that is, the molecule can exhibit the properties of a micelle without the need for aggregation. This opportunity was recognized early on and the synthesis of an amphiphilic dendrimer was reported. These types of amphiphilic dendrimers have been shown to act as unimolecular micelles. More recently, dendrimers that are capable of aggregating have also been reported. It should be noted that there are ample opportunities for molecular recognition in dendrimers, because of the stepwise fashion in which these molecules are assembled. Dendrimers have three different locations in their structure: core, branches, and periphery. Synthetic methodologies have been developed to decorate any of the units mentioned above with the desired functionality in a precise fashion (Newkome et al. 1999; Bharathi et al. 2001; Grayson and Frechet 2001; Newkome et al. 2001; Zhang et al. 2001; Vutukuri et al. 2004; Aathimanikandan et al. 2005; Klaikherd et al. 2006). The topic of amphiphilic dendrimers is further elaborated in a later chapter in this book.

2.5. CONCLUSIONS

Amphiphilic polymers are interesting candidates for recognizing biological and synthetic molecules. Amphiphilic block polymer micelles have been used for the encapsulation of drug molecules for drug delivery applications. These micellar assemblies are capable of releasing the drug molecules to targeted sites in response to external stimuli. In addition to drug delivery, these molecules have also been utilized for the delivery of DNA. Although these constitute examples of nonspecific interactions based on hydrophobic or electrostatic interactions, specific ligand–receptor interactions have also been studied in block copolymers. The preorganized assemblies exhibited by amphiphilic block copolymers seem to provide distinct advantages in studying polyvalent interactions.

Block copolymers have been the subject of intense study and therefore significant information has been obtained about their behavior, but amphiphilic homopolymers have attracted attention only in recent times. Amphiphilic homopolymers could be excellent candidates for polyvalent interactions, where the polyvalent presentation and advantages gained by preorganization of an amphiphilic assembly are folded into a single scaffold. Moreover, these polymers exhibit environment-dependent assemblies that are trapped in the solvents in which they are initially assembled. This property has been exploited to separate and detect peptides, where the separation is dependent on the pI of the peptide and the pH of the solution. Similarly, these polymer micelles provide a unique handle to develop sensor arrays in which the possibility of generating multiple data points from a single receptor has been demonstrated. Overall, the inherent incompatibility of the hydrophilic environment in Nature and the hydrophobic nature of the artificial drug molecules will be among the forces that continue to drive the field of molecular recognition using amphiphilic polymers for a variety of biological and materials applications, including delivery and sensing.

ACKNOWLEDGMENT

We acknowledge the National Institute of General Medical Sciences of the National Institute of Health for their generous support (GM-65255).

REFERENCES

Aathimanikandan SV, Savariar EN, Thayumanavan S. Temperature-sensitive dendritic micelles. J Am Chem Soc 2005;127:14922–14929.

Adams H, Hunter CA, Lawson KR, Perkins J, Spey SE, Urch CJ, Sanderson JM. A supramolecular system for quantifying aromatic stacking interactions. Eur J Chem 2001;7: 4863–4877.

Aebersold R, Mann M. Mass spectrometry-based proteomics. Nature 2003;422:198–207.

Akiyoshi K, Maruichi N, Kohara M, Kitamura S. Amphiphilic block copolymer with a molecular recognition site: induction of a novel binding characteristic of amylose by self-assembly of poly(ethylene oxide)-*block*-amylose in chloroform. Biomacromolecules 2002;3:280–283.

Alexandridis P, Athanassiou V, Fukuda S, Hatton TA. Surface activity of poly(ethylene oxide)-*block*-poly(propylene oxide)-*block*-poly(ethylene oxide) copolymers. Langmuir 1994;10:2604–2612.

Alexandridis P, Lindman B. Amphiphilic block copolymers: self-assembly and applications. Amsterdam: Elsevier Science. 2000.

Allen C, Maysinger D, Eisenberg A. Nano-engineering block copolymer aggregates for drug delivery. Colloids Surf B 1999;16:3–27.

Badjic JD, Nelson A, Cantrill SJ, Turnbull WB, Stoddart JF. Multivalency and cooperativity in supramolecular chemistry. Acc Chem Res 2005;38:723–732.

Basu S, Vutukuri DR, Shyamroy S, Sandanaraj BS, Thayumanavan S. Invertible amphiphilic homopolymers. J Am Chem Soc 2004;126:9890–9891.

Basu S, Vutukuri DR, Thayumanavan S. Homopolymer micelles in heterogeneous solvent mixtures. J Am Chem Soc 2005;127:16794–16795.

Batrakova EV, Bronich TK, Vetro JA, Kabanov AV. Polymer micelles as drug carriers. In: Torchilin VP, editor. Nanoparticulates as drug carriers. London: Imperial College Press. 2006. pp. 57–93.

Bazzi HS, Bouffard J, Sleiman HF. Self-complementary ABC triblock copolymers via ring-opening metathesis polymerization. Macromolecules 2003;36:7899–7902.

Bazzi HS, Sleiman HF. Adenine-containing block copolymers via ring-opening metathesis polymerization: synthesis and self-assembly into rod morphologies. Macromolecules 2002;35:9617–9620.

Bes L, Angot S, Limer A, Haddleton DM. Sugar-coated amphiphilic block copolymer micelles from living radical polymerization: recognition by immobilized lectins. Macromolecules 2003;36:2493–2499.

Bharathi P, Zhao H, Thayumanavan S. Toward globular macromolecules with functionalized interiors: design and synthesis of dendrons with an interesting twist. Org Lett 2001;3:1961–1964.

Cammas-Marion S, Okano T, Kataoka K. Functional and site-specific macromolecular micelles as high potential drug carriers. Colloids Surf B 1999;16:207–215.

Chen D, Jiang M. Strategies for constructing polymeric micelles and hollow spheres in solution via specific intermolecular interactions. Acc Chem Res 2005;38:494–502.

Chen L, McBranch DW, Wang H-L, Helgeson R, Wudl F, Whitten DG. Highly sensitive biological and chemical sensors based on reversible fluorescence quenching in a conjugated polymer. Proc Natl Acad Sci USA 1999;96:12287–12292.

Cheremisinoff NP. Handbook of engineering polymeric materials. New York: Marcel-Dekker; 1997.

Choucair A, Eisenberg A. Control of amphiphilic block copolymer morphologies using solution conditions. Eur Phys J Eng 2003;10:37–44.

Chung JE, Yokoyama M, Okano T. Inner core segment design for drug delivery control of thermo-responsive polymeric micelles. J Controlled Release 2000;65:93–103.

Cleek RL, Rege AA, Denner LA, Eskin SG, Mikos AG. Inhibition of smooth muscle cell growth in vitro by an antisense oligodeoxynucleotide releases from poly(DL-lactic-co-glycolic acid) microparticles. J Biomed Mater Res 1997;35:525–530.

Colfen H. Double-hydrophilic block copolymers: synthesis and application as novel surfactants and crystal growth modifiers. Macromol Rapid Commun 2001;22:219–252.

Combariza MY, Savariar EN, Vutukuri DR, Thayumanavan S, Vachet RW. Polymeric inverse micelles as selective peptide extraction agents for MALDI-MS analysis. Anal Chem 2007;79:7124–7130.

Cornelissen JJLM, Fischer M, Sommerdijk NAJM, Nolte RJM. Helical superstructures from charged poly(styrene)-poly(isocyanodipeptide) block copolymers. Science 1998;280:1427–1430.

Cram DJ. The design of molecular hosts, guests, and their complexes. Science 1988;240: 760–767.

Cubberley MS, Iverson BL. Models of higher-order structure: foldamers and beyond. Curr Opin Chem Biol 2001;5:650–653.

Davidson SMK, Regen SL. Nearest-neighbor recognition in phospholipid membranes. Chem Rev 1997;97:1269–1279.

Davis AP, Wareham RS. Carbohydrate recognition through noncovalent interactions: a challenge for biomimetic and supramolecular chemistry. Angew Chem Int Ed 1999;38: 2979–2996.

De Maeyer L, Trachimow C, Kaatze U. Entropy-driven micellar aggregation. J Phys Chem B 1998;102:8480–8491.

Ding K, Alemduroglu FE, Boersch M, Berger R, Herrmann A. Engineering the structural properties of DNA block copolymer micelles by molecular recognition. Angew Chem Int Ed 2007;46:1172–1175.

Discher DE, Eisenberg A. Polymer vesicles. Science 2002;297:967–973.

Dwek RA. Glycobiology: toward understanding the function of sugars. Chem Rev 1996;96: 683–720.

Ercolani G. Assessment of cooperativity in self-assembly. J Am Chem Soc 2003;125: 16097–16103.

Fan C, Plaxco KW, Heeger AJ. High-efficiency fluorescence quenching of conjugated polymers by proteins. J Am Chem Soc 2002;124:5642–5643.

Fischer NO, McIntosh CM, Simard JM, Rotello VM. Inhibition of chymotrypsin through surface binding using nanoparticle-based receptors. Proc Natl Acad Sci USA 2002;99: 5018–5023.

Fischer NO, Verma A, Goodman CM, Simard JM, Rotello VM. Reversible "irreversible" inhibition of chymotrypsin using nanoparticle receptors. J Am Chem Soc 2003;125: 13387–13391.

Frankamp BL, Uzun O, Ilhan F, Boal AK, Rotello VM. Recognition-mediated assembly of nanoparticles into micellar structures with diblock copolymers. J Am Chem Soc 2002;124:892–893.

Gauthier S, Eisenberg A. Vinylpyridinium ionomers. 2. Styrene-based ABA block copolymers. Macromolecules 1987;20:760–767.

Gellman SH. Foldamers: a manifesto. Acc Chem Res 1998;31:173–180.

Goodman CM, Choi S, Shandler S, DeGrado WF. Foldamers as versatile frameworks for the design and evolution of function. Nature Chem Biol 2007;3:252–262.

Grayson SM, Frechet JMJ. Convergent dendrons and dendrimers: from synthesis to applications. Chem Rev 2001;101:3819–3867.

Hannon MJ. Supramolecular DNA recognition. Chem Soc Rev 2007;36:280–295.

Harada A, Kataoka K. Formation of polyion complex micelles in an aqueous milieu from a pair of oppositely-charged block copolymers with poly(ethylene glycol) segments. Macromolecules 1995;28:5294–5299.

Hill DJ, Mio MJ, Prince RB, Hughes TS, Moore JS. A field guide to foldamers. Chem Rev 2001;101:3893–4011.

Hof F, Craig SL, Nuckolls C, Rebek J Jr. Molecular encapsulation. Angew Chem Int Ed 2002; 41:1488–1508.

Holland RJ, Parker EJ, Guiney K, Zeld FR. Fluorescence probe studies of ethylene oxide/propylene oxide block copolymers in aqueous solution. J Phys Chem 1995;99: 11981–11988.

Hong R, Fischer NO, Verma A, Goodman CM, Emrick T, Rotello VM. Control of protein structure and function through surface recognition by tailored nanoparticle scaffolds. J Am Chem Soc 2004;126:739–743.

Huc I. Aromatic oligoamide foldamers. Eur J Org Chem 2004;17–29.

Israelachvili JN. Intermolecular and surface forces. San Diego: Academic Press. 1991.

Jeong JH, Park TG. Novel polymer–DNA hybrid polymeric micelles composed of hydrophobic poly(DL-lactic acid-*co*-glycolic acid) and hydrophilic oligonucleotides. Bioconjugate Chem 2001;12:917–923.

Kabanov AV, Bronich TK, Kabanov VA, Yu K, Eisenberg A. Soluble stoichiometric complexes from poly(*N*-ethyl-4-vinylpyridinium) cations and poly(ethylene oxide)-*block*-polymethacrylate anions. Macromolecules 1996;29:8999.

Kakizawa Y, Kataoka K. Block copolymer micelles for delivery of gene and related compounds. Adv Drug Deliv Rev 2002;54:203–222.

Karas M, Hillenkamp F. Laser desorption ionization of proteins with molecular masses exceeding 10,000 Daltons. Anal Chem 1988;60:2299–2301.

Katayose S, Kataoka K. Water-soluble polyion complex associates of DNA and poly(ethylene glycol)-poly(L-lysine) block copolymer. Bioconjugate Chem 1997;8:702–707.

Kiessling LL, Gestwicki JE, Strong LE. Synthetic multivalent ligands as probes of signal transduction. Angew Chem Int Ed 2006;45:2348–2368.

Kim I-B, Dunkhorst A, Bunz UHF. Nonspecific interactions of a carboxylate-substituted PPE with proteins. A cautionary tale for biosensor applications. Langmuir 2005;21: 7985–7989.

Kitov PI, Bundle DR. On the nature of the multivalency effect: a thermodynamic model. J Am Chem Soc 2003;125:16271–16284.

Klaikherd A, Sandanaraj BS, Vutukuri DR, Thayumanavan S. Comparison of facially amphiphilic biaryl dendrimers with classical amphiphilic ones using protein surface recognition as the tool. J Am Chem Soc 2006;128:9231–9237.

Kwon GS. Diblock copolymer nanoparticles for drug delivery. Crit Rev Therap Drug Carrier Syst 1998;15:481–512.

Kwon GS, Kataoka K. Block copolymer micelles as long-circulating drug vehicles. Adv Drug Deliv Rev 1995;16:295–309.

Kwon GS, Okano T. Polymeric micelles as new drug carriers. Adv Drug Deliv Rev 1996;21:107–116.

Lazzari M, Lopez-Quintela MA. Block copolymers as a tool for nanomaterial fabrication. Adv Mater 2003;15:1583–1594.

Lehn J-M. Supramolecular polymer chemistry—scope and perspectives. In: Ciferri A, editor. Supramolecular polymers. 2nd ed. New York: Marcel-Dekker; 2005. pp. 3–27.

Li X, Yang D. Peptides of aminoxy acids as foldamers. Chem Commun 2006; 3367–3379.

Li Z-T, Hou J-L, Li C, Yi H-P. Shape-persistent aromatic amide oligomers: new tools for supramolecular chemistry. Chem Asian J 2006;1:766–778.

Licini G, Prins LJ, Scrimin P. Oligopeptide foldamers: from structure to function. Eur J Org Chem 2005;969–977.

Lopes JR, Loh W. Investigation of self-assembly and micelle polarity for a wide range of ethylene oxide-propylene oxide-ethylene oxide block copolymers in water. Langmuir 1998;14:750–756.

Lu JR, Zhao XB, Yaseen M. Biomimetic amphiphiles: biosurfactants. Curr Opin Colloid Interface Sci 2007;12:60–67.

Luk Y-Y, Abbott NL. Applications of functional surfactants. Curr Opin Colloid Interface Sci 2002;7:267–275.

Mammen M, Chio S-K, Whitesides GM. Polyvalent interactions in biological systems: implications for design and use of multivalent ligands and inhibitors. Angew Chem Int Ed 1998;37:2755–2794.

Marie P, Herrenschmidt YL, Gallot Y. Study of the emulsifying power of the block copolymers polystyrene/poly(2-vinylpyridinium chloride) and polyisoprene/poly(2-vinylpyridinium chloride). Makromol Chem 1976;77:2773–2780.

Menger FM, Chlebowski ME, Galloway AL, Lu H, Seredyuk VA, Sorrells JL, Zhang H. A tribute to the phospholipid. Langmuir 2005;21:10336–10341.

Menger FM, Keiper JS. Gemini surfactants. Angew Chem Int Ed 2000;39:1907–1920.

Mueller-Dethlefs K, Hobza P. Noncovalent interactions: a challenge for experiment and theory. Chem Rev 2000;100:143–167.

Nagarajan R. Theory of micelle formation: quantitative approach to predicting micellar properties from surfactant molecular structure. Surface Sci Ser 1997;70:1–81.

Nakashima K, Bahadur P. Aggregation of water-soluble block copolymers in aqueous solutions: recent trends. Adv Colloid Interface Sci 2006;123–126:75–96.

Newkome GR, He E, Moorefield CN. Suprasupermolecules with novel properties: metallodendrimers. Chem Rev 1999;99:1689–1746.

Newkome GR, Moorefield CN, Vogtle F. Dendrimers. 2nd ed. New York: Wiley–VCH. 2001.

Odian G. Principles of polymerization. 4th ed. New York: Wiley; 2004.

Oshovsky GV, Reinhoudt DN, Verboom W. Supramolecular chemistry in water. Angew Chem Int Ed 2007;46:2366–2393.

Philp D, Stoddart JF. Self-assembly in natural and unnatural systems. Angew Chem Int Ed Engl 1996;35:1155–1196.

Prins LJ, Reinhoudt DN, Timmerman P. Noncovalent synthesis using hydrogen bonding. Angew Chem Int Ed 2001;40:2382–2426.

Ravve A. Principles of polymer chemistry. 2nd ed. New York: Springer; 2000.

Rodriguez-Hernandez J, Checot F, Gnanou Y, Lecommandoux S. Toward "smart" nano-objects by self-assembly of block copolymers in solution. Prog Polym Sci 2005;30:691–724.

Rosler A, Vandermeulen GWM, Klok H-A. Advanced drug delivery devices via self-assembly of amphiphilic block copolymers. Adv Drug Deliv Rev 2001;53:95–108.

Ruzette A-V, Leibler L. Block copolymers in tomorrow's plastics. Nature Mater 2005;4:19–31.

Sandanaraj BS, Bayraktar H, Krishnamoorthy K, Knapp MJ, Thayumanavan S. Recognition and modulation of cytochrome c's redox properties using an amphiphilic homopolymer. Langmuir 2007;23:3891–3897.

Sandanaraj BS, Demont R, Aathimanikandan SV, Savariar EN, Thayumanavan S. Selective sensing of metalloproteins from nonselective binding using a fluorogenic amphiphilic polymer. J Am Chem Soc 2006;128:10686–10687.

Sandanaraj BS, Demont R, Thayumanavan S. Generating patterns for sensing using a single receptor scaffold. J Am Chem Soc 2007;129:3506–3507.

Sandanaraj BS, Vutukuri DR, Simard JM, Klaikherd A, Hong R, Rotello VM, Thayumanavan S. Noncovalent modification of chymotrypsin surface using an amphiphilic polymer scaffold: implications in modulating protein function. J Am Chem Soc 2005;127:10693–10698.

Savariar EN, Aathimanikandan SV, Thayumanavan S. Supramolecular assemblies from amphiphilic homopolymers: testing the scope. J Am Chem Soc 2006;128:16224–16230.

Stigers KD, Soth MJ, Nowick JS. Designed molecules that fold to mimic protein secondary structures. Curr Opin Chem Biol 1999;3:714–723.

Sun W-Y, Yoshizawa M, Kusukawa T, Fujita M. Multicomponent metal–ligand self-assembly. Curr Opin Chem Biol 2002;6:757–764.

Tanaka K, Waki H, Ido Y, Akita S, Yoshida Y, Yohida T. Protein and polymer analyses up to m/z 100,000 by laser ionization time-of-flight mass spectrometry. Rapid Commun Mass Spectrom 1988;2:151–153.

Tanford C. The hydrophobic effect: formation of micelles and biological membranes. 2nd ed. New York: Wiley; 1980.

Topp MDC, Dijkstra PJ, Talsma H, Feijen J. Thermosensitive micelle-forming block copolymers of poly(ethylene glycol) and poly(N-isopropylacrylamide). Macromolecules 1997;30:8518–8520.

van Hest JCM, Delnoye DAP, Baars MWPL, van Genderen MHP, Meijer EW. Polystyrene–dendrimer amphiphilic block copolymers with a generation-dependent aggregation. Science 1995;268:1592–1595.

Velonia K, Rowan AE, Nolte RJM. Lipase polystyrene giant amphiphiles. J Am Chem Soc 2002;124:4224–4225.

Vutukuri DR, Basu S, Thayumanavan S. Dendrimers with both polar and apolar nanocontainer characteristics. J Am Chem Soc 2004;126:15636–15637.

Wallimann P, Marti T, Fuerer A, Diederich F. Steroids in molecular recognition. Chem Rev 1997;97:1567–1608.

Wilson JN, Wang Y, Lavigne JJ, Bunz UHF. A biosensing model system: selective interaction of biotinylated PPEs with streptavidin-coated polystyrene microspheres. Chem Commun 2003; 1626–1627.

Won Y-Y, Davis HT, Bates FS. Giant wormlike rubber micelles. Science 1999;283:960–963.

Yamada K, Minoda M, Miyamoto T. Controlled synthesis of amphiphilic block copolymers with pendant N-acetyl-D-glucosamine residues by living cationic polymerization and their interaction with WGA lectin. Macromolecules 1999;32:3553–3558.

Yan D, Zhou Y, Hou J. Supramolecular self-assembly of macroscopic tubes. Science 2004;303:65–67.

Zana R. Dynamics of surfactant self-assemblies: micelles, microemulsions, vesicles, and lyotropic phases. New York: CRC Press; 2005.

Zeng F, Zimmerman SC. Dendrimers in supramolecular chemistry: from molecular recognition to self-assembly. Chem Rev 1997;97:1681–1712.

Zhang L, Bartels C, Yu Y, Shen H, Eisenberg A. Mesosized crystal-like structure of hexagonally packed hollow hoops by solution self-assembly of diblock copolymers. Phys Rev Lett 1997;79:5034–5037.

Zhang L, Yu K, Eisenberg A. Ion-induced morphological changes in "crew-cut" aggregates of amphiphilic block copolymers. Science 1996;272:1777–1779.

Zhang W, Nowlan III DT, Thomson LM, Lackowski WM, Simanek E. Orthogonal, convergent syntheses of dendrimers based on melamine with one or two unique surface sites for manipulation. J Am Chem Soc 2001;123:8914–8922.

CHAPTER 3

SUPRAMOLECULAR CONTROL OF MECHANICAL PROPERTIES IN SINGLE MOLECULES, INTERFACES, AND MACROSCOPIC MATERIALS

DAVID M. LOVELESS, FARRELL R. KERSEY, and STEPHEN L. CRAIG

3.1. INTRODUCTION AND BACKGROUND

Molecular recognition and aggregation, whether highly specific and directional in nature as in some of the examples discussed in Chapter 1, or less so as in the case of solvophobic interactions discussed in Chapter 2, have been demonstrated to be capable of leading to useful, stress-bearing materials with properties that are similar to those of traditional covalent polymers. Rehage and Hoffmann published a series of studies in the 1980s that described the viscoelasticity of aqueous wormlike micellar aggregates (Rehage and Hoffmann 1988, 1991) that were known to form gels at low concentrations (Hayashi and Ikeda 1980; Porte 1980; Candau et al. 1984). These noncovalent aggregates of small, amphiphilic molecules possess mechanical properties typically associated with high molecular weight polymers, properties that are not observed in comparable solutions of nonaggregated small molecules.

Dramatic examples have also been reported for main-chain supramolecular polymers (SPs; Lehn 1993; Ciferri 2005; Fig. 3.1), in which specific and directional molecular recognition events between end groups define the main chain of a linear polymeric assembly. Although main-chain SPs had been created and characterized previously (Broze et al. 1983; Fouquey et al. 1990; Alexander et al. 1993; Bladon and Griffin 1993; St Pourcain and Griffin 1995), it was a groundbreaking paper in 1997 that demonstrated the mechanical potential of supramolecular interactions and catalyzed much of the current interest in the field (Sijbesma et al. 1997).

Molecular Recognition and Polymers: Control of Polymer Structure and Self-Assembly.
Edited by V. Rotello and S. Thayumanavan
Copyright © 2008 John Wiley & Sons, Inc.

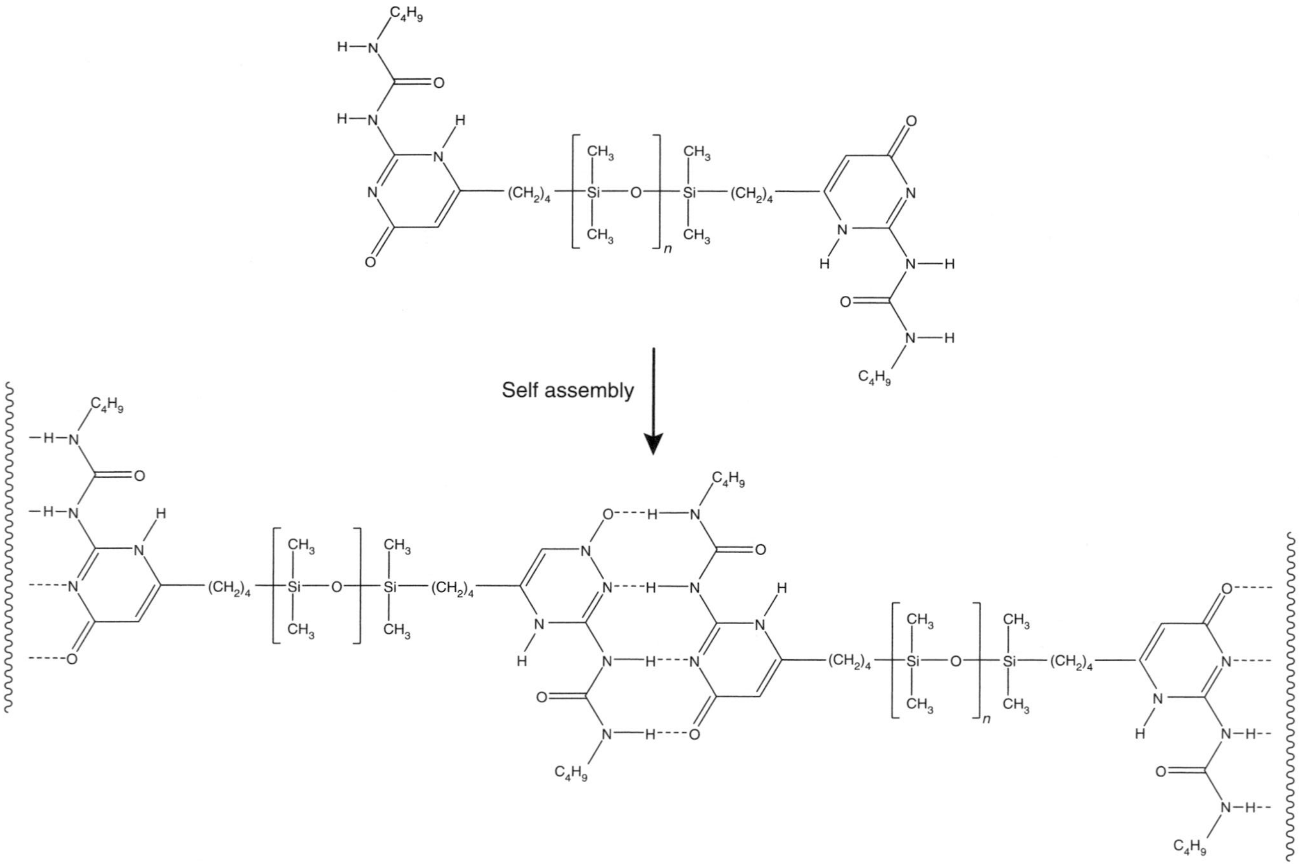

Figure 3.1 Quadruple hydrogen-bonded 2-ureido-4-pyrimidinone units attached to the ends of a polydimethylsiloxane chain assemble to create materials with viscoelastic properties.

In that work, Sijbesma and colleagues covalently appended ureidopyrimidinone (UPy) units to the ends of short hydrocarbons, siloxanes, and poly(ethylene oxide)/poly(propylene oxide) copolymers (Fig. 3.1). The UPy units self-dimerize ($K_{eq} = 6 \times 10^7$ M^{-1} in $CHCl_3$), and when they are attached to the ends of low molecular weight polymers they have a remarkable impact on material properties. The attachment of end-grafted UPy groups to oligomeric siloxanes, for example, converts a thick fluid into a viscoelastic thermoplastic. When extended to three-dimensional networks, supramolecular cross-linking via the UPy units leads to a plateau modulus that is 6 times that found in polymers with the same number of potential covalent cross-linking groups.

A decade later, the UPy functionality remains a paragon of supramolecular utility. The ostensibly weak and transient interactions provided by four hydrogen bonds are strong enough to create polymeric properties. Similar effects have been observed subsequently, and together they demonstrate that interactions that are often regarded as weak do in fact possess sufficient integrity to bear mechanical forces. Supramolecular interactions, like covalent bonds, can form the structural basis for large molecular entities (SPs) that in turn form useful materials.

As we consider the mechanical properties of SPs, it is often useful to consider them in the context of entanglements, which are intermolecular interactions that transfer mechanical forces from one molecule to the next. In this chapter we use the term entanglement in a very general way, so that it includes topological entanglements (one polymer chain is physically wrapped around another), chemical entanglements (attractive intermolecular interactions between polymer chains), and surface adsorption (attractive intermolecular interactions between polymer chains and a particle surface, e.g., from a filler). The important, fundamental characteristic is that it is an interaction that allows a mechanical stress on one molecule to be distributed or transferred to another molecule with which it is entangled.

SPs, in this framework, are reasonably viewed as systems in which specific intermolecular interactions create entanglements that would not exist in the absence of those interactions: the individual molecular constituents are too small, for example, to be physically entangled or to bridge between surfaces. In other words, the transfer of mechanical force from one macromolecule to another must pass through the defining supramolecular interaction. As seen in examples throughout this book, the mechanical properties of SPs are responsive to external stimuli to the same extent that the defining supramolecular interaction is responsive. Polymer properties can be turned "on" and "off," toggling between polymeric and small molecular behavior and creating materials that are more easily processed or recycled. The responsiveness need not be limited to on/off states; it could also be tuned through synthesis and environment to create "smart" materials whose properties adjust along a wide continuum.

The fields of polymer physics and the physical organic chemistry of supramolecular interactions therefore become intrinsically linked. To the extent that chemists are able to relate behavior at the level of polymer physics to characteristics of the isolated small molecular constituents, therefore, the control of SP properties becomes a rational and potentially very precise science. This theme permeates the field, and this book contains examples that span from macroscopic bulk polymers and polymer gels, to interfaces and thin films, to single molecules. As remains the case with covalent polymers, an ongoing challenge is to develop hierarchical pictures that unite material properties to well-defined and specific supramolecular behavior across material regimes.

3.2. MECHANICAL PROPERTIES OF LINEAR SPs

At the most basic level, mechanical properties are necessarily a matter of action and reaction, stimulus and response. The actor—a mechanical stress—is a familiar part of everyday life, but efforts to understand its molecular consequences are only recently coming to the fore in supramolecular chemistry. We next consider examples of how covalent polymers respond to a mechanical stress, and then we extend that examination to the case of SPs. This discussion is not intended to capture all of the details of a thermodynamically rigorous treatment but is instead focused on providing a useful conceptual framework for key concepts related to the mechanics of SPs.

The energy applied through a mechanical strain can either be stored as potential energy or dissipated as heat. In polymers, energy can be stored by the entropic restriction of conformational space available to a given polymer chain. If the polymers are anchored together, or entangled, at various positions, deformation of the material will move those positions apart. The conformational freedom of a given chain is, on average, decreased relative to its equilibrium state; and the collective random, thermal fluctuations of the polymer chain will, on average, pull the entanglements toward that equilibrium state. Thus, if the entanglements are fixed, elastic behavior results; a mechanical stress deforms the material, but the energy is stored in molecular conformational changes that can be used subsequently to perform work, for example, shooting a pebble with a slingshot. This case is essentially that of rubber elasticity for which many excellent reference texts are available.

A complementary case is that in which the entanglements are not fixed but are lost instead. The competition between two mechanisms of entanglement loss (chain slippage vs. chain scission; Fig. 3.2) in covalent polymers provides a useful basis for subsequent discussion of the supramolecular case. Chain slippage typically occurs via reptation, the process by which a polymer chain is constrained to a tube by surrounding polymer chains (Fig. 3.2, process A). Movement within this tube is driven by the diffusion of stored length in a snakelike motion, and the rate of the reptative relaxation drops exponentially as the molecular weight of the polymer increases. The physical consequence of reptation, or other chain slippage mechanisms, is that an entanglement that once anchored an elastic chain is lost, and the energy stored in that elastic chain is lost (dissipated as heat) along with it.

An alternative pathway for entanglement loss is chain scission (Fig. 3.2, process B), in which a covalent bond along the polymer main chain is broken and a stress-bearing, otherwise elastic, chain is lost. Chain scission reactions, for example, homolytic carbon–carbon cleavage, have obviously high activation energies. The stress-free rates of these reactions are therefore typically extremely low,

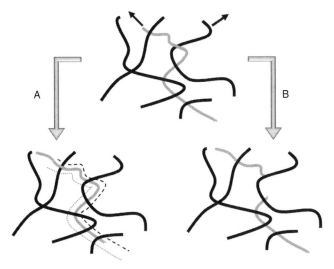

Figure 3.2 A mechanical force causes a loss of an entanglement in a covalent polymeric network by one of two major mechanisms: process A, chain slippage via reptation; process B, chain scission.

and they only occur when high stress is placed on the bonds. Chain scission events lead to an effectively permanent failure in the polymeric material and provide a molecular-level defect at which forces concentrate and macroscopic cracks initiate.

The supramolecular case can be instructively considered in the same context (Fig. 3.3). If the entanglements and the polymer chains remain intact under an applied stress, the SP is elastic. If the SP chain remains intact, but entanglements are lost through chain slippage, then the polymer is viscous and the viscosity will depend on the molecular weight of the SP in the same way that it depends on the molecular weight of the covalent system. In contrast, if entanglements are lost through chain scission, then the stored energy (elasticity) is lost. For SPs, the primary mechanism for chain scission is the simple dissociation of the supramolecular bond. Unlike covalent bonds, supramolecular associations are such that dissociation often occurs readily in the absence of any stress. Although some supramolecular interactions are effectively permanent, many have lifetimes that approach nanoseconds. Chain scission, mediated by supramolecular dissociation, might therefore make significant contributions to material mechanical properties that are not observed in covalent polymers.

Supramolecular chain scission differs further from covalent chain scission, because supramolecular recombination is typically the predominant fate of a ruptured chain; anthropomorphically speaking, the supramolecular moieties are, by their very nature, predisposed to reassociation rather than alternative reaction pathways. This predisposition is not intrinsic to the products of covalent bond rupture, which might lead either to high-energy intermediates with nonspecific reactivity or to species that require catalyst or elevated temperature to recombine. The

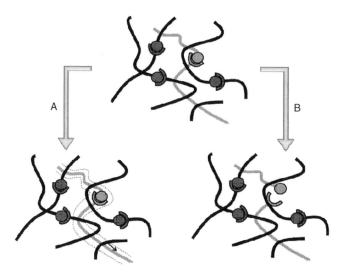

Figure 3.3 Entanglement response to a mechanical force in a supramolecular system: process A, chain slippage via reptation; process B, chain scission via breaking of the supramolecular bond.

tendency to recombine has important consequences for material properties; most notably, it has led to speculation about the potential of SPs to be autonomous self-repairing materials.

These three fates of a mechanically deformed polymer (elastic storage, chain slippage, chain scission) provide a productive context through which to consider the potential applications of SPs. If the supramolecular chain remains intact and energy is either stored or lost through chain slippage, then the supramolecular case reduces to the covalent case. In these situations, SPs can be substituted for covalent polymers in applications while retaining the environmental responsiveness that makes them, for example, potentially more easily processed or recycled. Differences in properties will be found, however, when supramolecular dissociation is a contributing relaxation process.

Sijbesma and Meijer's (1997) extensive work on UPy-terminated monomers beautifully illustrates these concepts. In solution, the average supramolecular degree of polymerization increases with concentration and can be limited by the addition of monofunctional "chain terminators." The viscosity of the solution increases slowly with concentration at first, but it then increases more dramatically ($\sim C^{3.4}$, where C is the concentration of polymer) once a critical concentration is attained. The critical concentration in this case corresponds to the transition from dilute to semidilute conditions, representing the point at which the excluded volumes of the SPs begin to overlap and the diffusion path of individual SPs is restricted by chains in solution. The observed concentration dependence agrees well with scaling laws that assume that the primary mechanism of diffusion is that of the intact SPs rather than being mediated by dissociation/diffusion/reassociation pathways.

However, SP dissociation can be the primary relaxation pathway, and theoretical scaling concepts developed by Cates (1987) to describe the relaxations of wormlike micelles (such as those examined by Rehage and Hoffmann 1988, 1991) are often successfully applied to experimental observations on SPs. The Cates theory provides a useful probe of SP structure and relaxation mechanisms. For example, in their work on reversible neodymium(III) coordination polymers, Vermonden et al. (2004) found that dissociation, rather than reptation, is the fastest route for entanglement relaxation and the wormlike micelle theory fits the data for the supramolecular assemblies well. Even when metals that have trivalent coordination are used, the scaling law behavior is consistent with the Cates model. This result suggests that the trivalent metals form linear chains that include alternating ring structures, although, as discussed in Section 3.3, similar scaling laws are observed in supramolecular networks in which the lifetime of the stress-bearing cross-links determines the relaxation of the network structure.

The overlaps between SPs in semidilute concentrations can be thought of in very similar terms to the entanglements defined above. Supramolecular interactions create large structures that physically interact to determine the mechanical response (in this case, viscous flow). The primary relaxation is the diffusion of an SP that is effectively intact on the timescale of the diffusion process. Thus, at a fixed concentration, the SP properties in dilute solution are therefore quite similar to those of covalent polymers of the same molecular weight and molecular weight distribution.

Similar behavior is observed in many SP solutions, and we highlight only a few of the many examples. Castellano et al. (2000) were able to quantitatively relate the semidilute transition observed in viscosimetry measurements of calixarene-based SP capsules ("polycaps") to that extrapolated from quasielastic light scattering. Xu and coworkers (2004) made a similar analysis on DNA-based SPs. Scaling laws above the critical concentration are similar in all cases. The generality of the behavior supports the expected view that, in the semidilute regime, the SP mechanical response is effectively equivalent to that of covalent systems. The observed viscosity depends on the average SP size, and the dynamic nature of the SP is invisible in the flow properties of the polymer solution.

Yount and colleagues (2003) recently presented a direct and fairly general probe of SP dynamics. Their approach is based on the fact that a direct study of SP dynamics would involve making a change in the dissociation/association kinetics of the defining SP interaction and observing the corresponding changes in the properties of the SP materials. Although this strategy is potentially very informative, a subtle, yet persistent, difficulty exists in distinguishing the contributions of the kinetics of a given molecular interaction (k_{diss}, Fig. 3.4) from those of its thermodynamics (K_{eq}). In most reversibly assembled systems, for example, those based on hydrogen bonding, association occurs at or near the diffusion rate and K_{eq} and k_{diss} are strongly anticorrelated. For SPs, the inverse correlation of k_{diss} and K_{eq} intrinsically frustrates efforts to determine the relative importance of the two contributions. A high K_{eq} leads to increased aggregation, higher SP molecular weights, and slower dynamics within the equilibrium polymer structure. At the same time, a lower k_{diss} leads to slower reversible kinetics along the assembly. Dynamic properties,

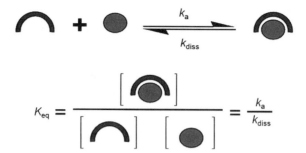

Figure 3.4   Relationship between the equilibrium association constant (K_{eq}) and the rate constants for supramolecular association and dissociation.

whether controlled by the SP equilibrium structure or reversible kinetics, are slowed by both mechanisms.

The independent control of the association kinetics relative to thermodynamics is therefore desirable, and Yount and others showed that N,C,N-pincer metal–ligand coordination motifs such as $1 \cdot 2$ (Fig. 3.5) are well suited to that goal. Pincer compounds **1** and analogs have been synthesized and studied extensively (Rietveld et al. 1997; Albrecht and van Koten 2001; Rodriguez et al. 2002; Slagt et al.

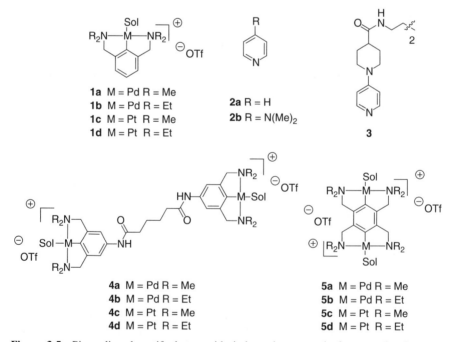

Figure 3.5   Pincer-ligand motifs that provide independent control of supramolecular association/dissociation kinetics relative to thermodynamics.

2004). These and numerous other organometallic coordination system have been used in SPs (Rodriguez et al. 2002). The reversible kinetics of interest are those of ligand exchange. In Pd(II) and Pt(II) complexes, ligand exchange occurs through a sterically congested associative mechanism. Added bulk in the *N*-alkyl substituents R therefore slows the exchange while exerting a lesser effect on the relative energy of the roughly isosteric end points. For example, the association constants for **1a · 2b** and **1b · 2b** in dimethylsulfoxide (DMSO) are (1.6 and 1.3) $\times 10^3$ M^{-1}, respectively, but the rates of ligand exchange differ by nearly 2 orders of magnitude (70–100 s^{-1} for **1a · 2b** and 1.0 s^{-1} for **1b · 2b**; Yount et al. 2003).

The pincer motif can be incorporated into SPs such as linear SPs **3 · 4**. A 1:1 mixture of **3:4a** (4.6 wt%) forms linear SPs, and there is a concomitant increase in solution viscosity. That the increased viscosity of the SP solution is not influenced by the reversibility of the metal–ligand bond is demonstrated by direct comparison: a similar solution of **3 · 4b** has the same viscosity, and so the equilibrium structure of the SP determines the viscosity of the SP solutions; the transience of the main chain does not contribute (Yount et al. 2003).

The situation becomes more complex when considering nonlinear responses such as shear thinning. Under high steady shear, the viscosity of polymer solutions often drops as a result of chain alignment in the flow field. The same behavior has been observed in SPs, but now there are two possible mechanisms for viscosity loss: the alignment of effectively intact SPs or the scission of the relatively weak SP bonds to decrease the average SP molecular weight. The ambiguity is complicated by the typically unknown contributions of an applied mechanical force to the dissociation rate of the supramolecular interaction. For example, Paulusse and Sijbesma (2004) showed that ultrasound-induced flow can rupture coordinative bonds in main-chain SPs, and Kersey et al. (2006) reported the force dependence of ligand exchange reactions in pincer coordination motifs such as **3 · 4**. A discussion of force-dependent supramolecular dynamics is presented in Section 3.5. The elucidation of shear thinning, shear thickening, and other nonlinear viscoelastic responses of SPs under large, rapid deformations constitute one rich area of relatively undeveloped activity in the field.

As solutions become increasingly concentrated and enter the solid state, intact chain slippage mechanisms slow dramatically. In these regimes, the dissociation rates of the SP interaction might change as well, but in general the change will be greatest in the dynamics of the intact SP. The expectation is therefore that at some point a transition will occur, after which SP dissociation provides the principal relaxation mechanism. Recent work on UPy-terminated polycaprolactones in the melt state, for example, reveals ideal Maxwell viscoelastic behavior in the solid state that is described by a single relaxation element with a lifetime of 1.6 ms (van Beek et al. 2007). This lifetime corresponds well to that of the UPy dimer; thus, the mechanism of relaxation appears to involve dissociation of the dimers, subsequent (relatively) rapid rearrangement of the SP subunits, and reassociation of dimers. An understanding of the full structure–activity space (density of SP unit, temperature dependence, yield under high strain rates, etc.) relevant to a given application, and especially under which conditions the SP dissociation is the limiting dynamic event, creates a rational avenue for SP material design. The regions of structure–activity space that are close to transitions between "supramolecular" and "covalent" behavior are especially rich for applications, because we can potentially engineer materials that with modest manipulation could provide the best of both sets of materials in a stimulus-responsive manner.

3.3. MECHANICAL PROPERTIES OF SP NETWORKS

The UPy-terminated polycaprolactone described above constitutes what is effectively a transient polymer network in which individual chains are held in place by entanglements that are rendered transient through supramolecular interaction. Formal SP networks, such as that shown schematically in Figure 3.6, can be created by multiple branching (cross-linking) points from a common linear polymer. Linear polymers aggregate through a supramolecular interaction that constitutes a chemical, rather than topological, entanglement of two chains. With multiple branching sites per chain, aggregation-induced molecular weights increase more rapidly with the addition of the cross-linkers than in the linear case. At early stages of aggregation, the same competition between relaxation mechanisms holds; the aggregates can diffuse or slip past each other as intact structures, or pieces of the aggregated SP can relax by dissociating and reassociating in the applied stress field. At appropriate concentrations of SP interactions (percolation, or the gel point), the polymer network becomes effectively infinite and the diffusion of intact components is negligible. Supramolecular dissociation is expected to dominate the mechanical properties landscape.

These characteristics are illustrated by studies on SP networks formed through the same van Koten type pincer coordination complexes described above. We focus the discussion on SPs formed from poly(4-vinylpyridine) (PVP) that is cross-linked in DMSO by bis(M(II)-pincer) compounds **4a–d** (Fig. 3.5; Yount et al. 2005a, 2005b). The same simple steric effect in the pincer alkylamino ligands is used,

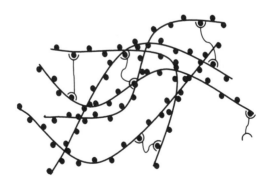

Figure 3.6   Schematic of a supramolecular polymer network.

both within Pd complexes **4a–b** and the slower, more strongly coordinating Pt complexes **4c–d** (Yount et al. 2005a, 2005b). As with the linear SPs, the independent control of kinetics is particularly significant. Cross-linkers **4a** and **4b** are effectively identical structural components within the network, and so their similar thermodynamics ($K_{eq} = \sim 30$ M^{-1} for **1a** · **2a** and **1b** · **2a**) ensure that the extent and nature of crosslinking is essentially the same in the two samples [or between Pt(II) pincer molecules **1c–d**; $K_{eq} = 8 \times 10^3$ M^{-1} for **4c** · **2a** and 4×10^3 M^{-1} for **1d** · **2a**].

The addition of 2% (by functional group) **4b** to a 100 mg mL^{-1} DMSO solution of PVP gives rise to a clear, thick, deep yellow solution whose viscosity is ~ 2000 times greater than that of PVP alone (33 vs. 0.016 Pa s). The viscosity does not increase upon the addition of the same quantity of monomeric **1b**, and the viscosity of **4b** · PVP reverts to that of a free-flowing solution with the addition of the stronger ligand **4b**, which competes the metal away from the PVP side groups. The increased viscosity of **4b** · PVP networks is therefore attributed to some combination of the two broad mechanistic possibilities brought about by interchain cross-linking: the motion of equilibrium structures that are effectively intact on the timescale of the viscous response versus rearrangement within the transient network via the dissocation dynamics of the individual cross-links.

A comparison with cross-linker **4a** proves the underlying dynamics are controlled by metal–ligand dissociation. Ligand exchange kinetics for **4a** are substantially faster than for **4b** but the association thermodynamics are very similar, and the effect of those kinetics is dramatic. At 5% cross-linker, the dynamic viscosity of 100 mg mL^{-1} **4a** · PVP is only 6.7 Pa s, a factor of 80 less than that of the isostructural network **4b** · PVP. Although the association constants are not identical, the effect of the thermodynamics would be to increase the viscosity of **4a** · PVP relative to **4b** · PVP, the opposite direction of that observed. The kinetics dominate even the extent of cross-linking; 5% **4a** · PVP is less viscous by a factor of 5 than is 2% **4b** · PVP.

The frequency-dependent storage (G') and loss moduli (G'') for multiple networks of either **4** · PVP or the related **5** · PVP are reduced to a single master plot when scaled by the corresponding ligand exchange rates, which are measured on model systems (data for **5** · PVP are provided in Fig. 3.7). These scaled plots are similar to linear free energy relationships, in which rate or equilibrium constants have been replaced by material properties. The dynamics at the cross-links determine the dynamic mechanical response of the materials, and subsequent experiments show that the mechanism of ligand/cross-link exchange in the network is the same mechanism observed in model systems: solvent-assisted exchange, in which a solvent molecule of DMSO first displaces the bound pyridine; subsequently, a new pyridine then displaces the DMSO. The solvent-assisted pathway requires that stress relaxation (flow) occur while the crosslink is dissociated from the network. Mechanical properties are determined by the dissociation of the cross-linkers from the network, but it is the rate of dissociation rather than the fraction of time in the dissociated state (equal in **4a** · PVP and **4b** · PVP) that governs the properties (Yount et al. 2005a).

Experiments on SP networks formed from multiple types of cross-linkers show that the response to an applied stress occurs through sequential, individual

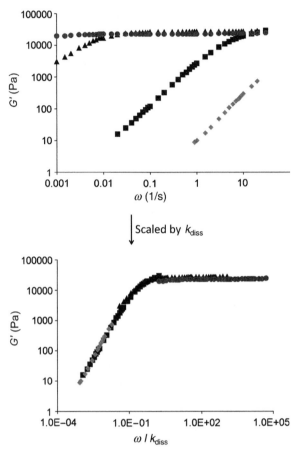

Figure 3.7 (a) Storage modulus versus frequency for the networks **5** · PVP and (b) the same storage modulus scaled by k_{diss} of model complexes **1** · **2a** versus the frequency of oscillation scaled by the same k_{diss}. Each of the networks consists of 5% (by metal functional group per pyridine residue) of (●) **5a**, (▲) **5b**, (■) **5c**, and (♦) **5d** and PVP at 10% by total weight of the network in DMSO at 20°C. Data taken from Loveless et al. (2005).

dissociation and reassociation events (Loveless et al. 2005). Discrete contributions from each type of cross-linker are evident in the mechanical properties, rather than an average of the contributing species. For example, the dynamic viscosity and G' as a function of frequency were compared for five different networks: PVP with 5% of either **5b** or **5c**, 2.5% of either **5b** or **5c**, and 2.5% each of **5b** and **5c**, all at the same concentration of 10% by total weight in DMSO. Below a frequency (ω) of 0.1 s^{-1}, the dynamic viscosity of the mixed network closely mimics that of the network consisting entirely of the slower component **5c** · PVP at a concentration of 2.5% (Fig. 3.8). As ω increases to values greater than 0.1 s^{-1}, the dynamic viscosity of the mixed network increases and eventually plateaus at that of a network comprising PVP and 5% of the faster cross-linking component **5b**.

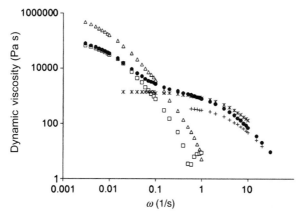

Figure 3.8   Dynamic viscosity versus frequency for (●) 2.5% + 2.5% (**5b** + **5c**) · PVP, (□) 2.5% **5c** · PVP, (△) 5% **5c** · PVP, (+) 2.5% **5b** · PVP, and (✕) 5% **5b** · PVP. All networks are 10% by total weight in DMSO at 20 °C. Data taken from Loveless et al. (2005).

Similar effects are observed in the G' and for different mixtures of networks (including those with three different cross-linkers). In all cases, the frequency onsets and magnitudes of the transitions are anticipated by the behavior of networks with a single cross-linker.

The dominance of the supramolecular dissociation kinetics, however, is only evident above a certain concentration of cross-linker. Relaxation mechanisms below the percolation threshold involve the diffusion of discrete aggregates, and they are therefore independent of the kinetics of the cross-linking interactions. The chemistry of the supramolecular interactions can be exploited in combination with details of the percolation behavior to provide a useful mechanism for engineering dramatic and reversible, chemically induced changes in mechanical properties, because slight shifts in binding equilibria are enough to move from one regime to the other. In the case of pincer-based networks, for example, gel–sol transitions are induced by a range of chemical signals, including competing ligands and acids, some of which can be coupled to thermal processes (Loveless et al. 2007). Because the low viscosity state is independent of the cross-linking kinetics and the high viscosity state depends directly on those kinetics, the magnitude of the change in properties is determined by the kinetics, and several orders of magnitude (or more) changes in viscosity are readily demonstrated. Gel–gel transitions can be similarly engineered by a strategy using mixtures of cross-linkers with different dissociation rates.

The combination of chemically diverse cross-linkers in a single supramolecular material is especially promising in this regard. Pollino and Weck (2005) have explored combinations of interactions (hydrogen bonding, coordination, pseudorotaxanes) for side chain functionalization and cross-linking. The potential benefit of this strategy is in establishing a system that is responsive to a variety of external stimuli independently of each other. SPs containing two sets of cross-links defined by hydrogen bonding and by metal–ligand associations, for example, provide for two accessible handles for addressing responsive materials.

In the case of the PVP/pincer SP networks, the onset of percolation, as defined by the concentration of cross-linkers, depends on the total weight percentage of the SP network and provides another degree of control in the systems. These and other details of the percolation behavior will necessarily be somewhat system specific. Nevertheless, although the present discussion is limited to one representative example, the strategy to exploit sol–gel or other phase transitions as highly sensitive regions of stimulus-responsive behavior in SP systems is general.

The behavior of SP networks is typically that of transient network models, in which the independent relaxations of stress-bearing entanglements determine the dynamic mechanical response of a network (Green and Tobolsky 1946; Lodge 1956; Yamamoto 1956; Cates 1987; Cates and Candau 1990; Tanaka and Edwards 1992; Turner and Cates 1992; Turner et al. 1993; Jongschaap et al. 2001). Without considering the exact structure of the networks, the detailed mechanism of relaxation (Leibler et al. 1991; Tanaka and Edwards 1992), or the extent of cooperativity in the associations, individual dissociation events clearly dominate the mechanical properties: no significant averaging or summation of different components is observed. The independence of the cross-links has significant consequences for the rational, molecular engineering of viscoelastic properties. When entanglements are defined by very specific interactions, the chemical control of properties follows. As long as the strength of the association is great enough to render associated a significant fraction of the crosslinkers, the dynamics of cross-link dissociation, rather than further details of their thermodynamics, are the key design criterion. As a result, quite complex viscoelastic behavior can be engineered given suitable knowledge of the small molecules.

3.4. MECHANICAL PROPERTIES IN SPs AT INTERFACES

Among the areas of materials science in which SPs are likely to bring interesting and important new material properties is the study of interfaces. Here, the nature of stress-bearing interactions must be considered in the context of a nearby surface. Two examples will be considered in this section: main-chain reversible polymer brushes and cross-linked polymer brushes. These two examples are analogs of the main-chain SPs and SP networks described above, but in which the polymer chains are anchored to a surface. In the case of main-chain reversible polymer brushes, we will focus on the transfer of force from one surface to another via SP bridges (Fig. 3.9). In particular, we consider the consequences of a force normal to the surfaces (adhesion), although lateral forces (e.g., friction) might also be influenced. In the case of cross-linked polymer brushes, both normal and lateral forces will be discussed.

Polymer bridging has important consequences in the material and life sciences, including an influence on adhesion, tribology and polymer flow, microtubule formation and function, and cell surface interactions (de Gennes 1979; Fleer et al. 1993). Bridges occur when a polymer chain is either physisorbed or covalently bound to

Figure 3.9   Supramolecular bridges between two surfaces.

two separate surfaces, and covalent polymer bridging has received extensive theoretical (Fleer and Scheutjens 1986) and experimental (Jeppesen et al. 2001) attention. When the potential bridges are linear SPs that can adjust their size and shape in response to the steric constraints imposed by the surfaces (van der Gucht 2002, 2003), the bridging structure and the resulting material properties reflect the chemistry of small-molecule self-assembly. Even sterically induced entropic penalties as small as kT [Boltzmann thermal energy (Boltzmann constant × temperature)] per chain are large enough to shift the chemical potential that drives polymerization and reduce the average SP molecular weight by ∼40%. In situations where mechanical properties scale, for example, as MW$^{3.5}$ (molecular weight raised to the power of 3.5), order of magnitude changes in properties might result from the fairly modest steric energetics.

The theory of SP bridging in this context has begun to receive attention in recent years. Van der Gucht and cohorts (2003) have reported intersurface forces and scaling relationships that differ from covalent analogs. Chen and Dormidontova (2006) have examined the attractive forces of end-adsorbed SPs and their ability to bind to complementary surfaces. These studies show that the molecular weights of supramolecular structures on surfaces vary substantially from those in solution as the concentration of surface absorption sites are increased. The chain orientation relative to the surface is further affected by the concentration of polymer in solution. The dynamic nature of the SP chains allows for surface remodeling in response to changing environment. For example, an increase in solution concentration leads to more dense, rather than taller, surfaces.

The first experimental investigations of SP bridging have also been reported only recently. One series of studies makes use of the DNA-based SPs described in Section 3.2 (Fogleman et al. 2002; Xu et al. 2004). The DNA-based SP monomers comprise

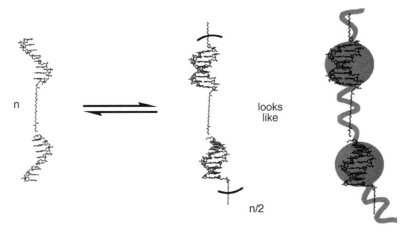

Figure 3.10 Reversible base pairing defines the rigid structure of a reversible polymer system.

oligonucleotide sequences that are covalently linked directly or through a synthetic spacer (Fig. 3.10). Duplex formation creates a linear, polymeric assembly that resembles larger duplex DNA, but in which the main chain is defined by the reversible base pairing. These SPs are intrinsically modular: 1) the thermodynamics and kinetics of the association are determined by the variable base sequence; 2) the linear density of the reversible interactions, and the conformational flexibility along the polymer backbone, are each dependent on a spacer in which much variation is possible; and 3) inter- and intrachain interactions may be tuned by salt concentration in the buffer. Furthermore, enzymatic covalent capture and characterization are possible (Xu et al. 2004).

SP bridging was examined using atomic force microscopy (AFM; Kim et al. 2005). On patterned gold substrates, SP brushes reversibly assembled in solution from different surface densities of a thiol SP anchor. The adhesion between the SP-functionalized tip and four different substrate chemistries was characterized in quasiparallel by AFM. Using a series of comparative control experiments, the contributions of SP-mediated adhesion were quantified and attributed to the formation of multiple surface–tip bridges during contact. The DNA-based monomers enable differentiation between bridges that are formed through molecular recognition of one brush layer with the other and those that are formed by nonmolecular recognition mediated adsorption of the brushes onto opposing surfaces. When similar, but noncomplementary, brushes are displayed on opposing surfaces, for example, the adhesive interaction is much lower than that between the same tip and a complementary substrate, supporting direct SP bridging ($\sim$15–20 bridging contacts present with the polymer that are not present without) as a dominant contributor to adhesion in these systems.

Using the modularity of the DNA-based SPs, several structure–activity relationships have been probed. For example, it was determined that the strength of the supramolecular interaction (hybridization energy) influences the adhesion; weaker base pairing leads to a drop in the amount of force transferred from one surface to the other. In fact, a relatively modest change in average free energy ($\sim 1\,\text{kcal mol}^{-1}$) cuts the adhesion roughly in half for the weaker SP brushes relative to the stronger. The magnitude of the change in adhesion indicates that the association thermodynamics influence the adhesive properties both through the mechanical response of the individual associations (presumably related to their dissociation kinetics) and by influencing the number of adhesive contacts that are formed. Other details of oligonucleotide-based monomer (OM) structure also influence the mechanics of the interface. When a flexible spacer is introduced between the molecular recognition end groups, for example, the adhesion between flexible SP brush surfaces is greater than that between rigid SP brush surfaces of comparable height, even though the individual associations in the flexible brushes are weaker than those in the rigid brush.

The theoretical investigations of van der Gucht and colleagues (2003) predict that the adhesive interaction should have a long-range component, although the contributions should drop off with surface separation. This expectation is met in the experimental systems. When a similarly functionalized AFM tip is held away from (not in contact with) an SP brush surface, bridges spontaneously form across $\sim 5-10\,\text{nm}$ gaps (Kersey et al. 2004). Individual bridging events are observed upon retraction of the AFM tip, and the length of the bridges can be inferred from the force versus distance retraction curves. The length distribution of the SP bridges approximates a Flory distribution (Flory 1969) as expected for linear SPs. The actual distribution of bridge lengths, however, is skewed toward lengths that are much shorter than the equilibrium distributions in solution. Reversible bridging is therefore responsive to the spatial constraints of the intersurface gap, and shorter bridges are preferred to longer ones. The presence of an attractive force provides an interesting mechanism for self-repair, because surfaces that are mechanically disjoined might be slowly pulled into closer contact, where adhesion is greatest. The bridges form across gaps and in surface–surface contact provide snapshots of different points in that process.

An SP interface of greater complexity is the reversibly cross-linked polymer brush, in which the addition of SP cross-links might change the brush conformation and intersurface penetration as well as create entanglements that might contribute, differently, to lateral and normal forces. In this regard, the pincer coordination motif described previously has great utility: the ability to change dynamic response at the level of molecular associations creates the equivalent of a "macromolecular kinetic isotope effect" that can be applied to the investigation. Although actual isotopic substitutions are obviously not involved, the phenomenological similarity with kinetic isotope effects in reaction mechanisms is quite real. In particular, contributions to rate-determining material processes are revealed by kinetic differences in two isostructural systems. Here, structure is preserved at two levels: the molecular structure of the individual SP constituents and the equilibrium association constant (K_{eq}) between molecular recognition partners. The former maintains chemical composition in the material, and the latter maintains structural composition in the extended polymer assembly.

In the case of dense, surface-grafted polymer brush thin films, cross-links are introduced by the simple addition of solutions containing bis(Pd^{II}-pincer) compounds **4a** or **4b** to grafted PVP brushes (Loveless et al. 2006). Because the association constants for pyridine coordination are similar, the uptake of **4a** and **4b** from equimolar solutions into the PVP brushes (at constant grafting density and molecular weight) is effectively equivalent, producing samples with comparable structure (number and placement of cross-links).

AFM studies on the thin (∼50 nm) brush layers demonstrate that the dynamics of supramolecular cross-links contribute to the friction of the soft brush surfaces (Fig. 3.11). When the faster **4a** cross-linker is added, both the absolute friction values and the coefficient of friction (COF) *drop* to ∼30% of those of the uncross-linked PVP control. When the slower cross-linker **4b** is added, however, the absolute friction value and the COF *increase* dramatically: both the COF and the absolute friction values for PVP·**4b** are more than twice that of PVP alone. The absolute lateral force measured is proportional to the normal force applied; this is consistent with the model that the more the tip is pressed into the brush layer the more force it takes to drag it laterally through the brushes.

The combined results demonstrate the complexity of the system. Cross-linking must include kinetic contributions to the lateral resistance that are similar to those observed in the networks, but a combination of structural and dynamic factors is likely responsible for the significant but opposite effects from kinetically dissimilar cross-links. Stimulus-responsive polymer brush layers hold great potential (Minko et al. 2000; Motornov et al. 2003; Granville et al. 2004; Kaholek et al. 2004a,

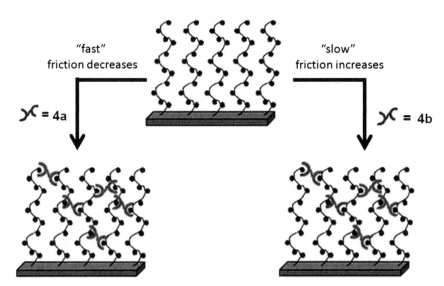

Figure 3.11 Poly(4-vinylpyridine) brushes subsequently exposed to metallic cross-linkers. The addition of **4a** decreases the lateral friction, but the addition of **4b** increases the lateral friction.

2004b; Kizhakkedathu et al. 2004; LeMieux et al. 2004), and the specific modulation of cross-linking kinetics is one method by which to exert control. As in the SP networks discussed above, the cross-linking in the brushes is reversed by chemical competition, but responsiveness to other stimuli, such as temperature, could be engineered.

3.5. MECHANICAL FORCES AND SUPRAMOLECULAR INTERACTIONS

The mechanical properties of SPs described in Sections 3.2–3.4 are, in general, successfully interpreted, often quantitatively, in terms of thermal rate and equilibrium constants, but it is reasonable to expect that the underlying molecular behavior should be perturbed by the application of a mechanical stress. On the whole, the mechanical properties of supramolecular interactions are not well known, and their study constitutes a relatively new but burgeoning research area related to the field of SPs.

The details of the theory of mechanical influence on reaction kinetics are quite complex and, in many respects, still under investigation. A thorough discussion is outside of the scope of this chapter, but a key concept is that a discussion of the mechanical "strength" of a bond is, in reality, a discussion of how mechanical force affects the probability of bond rupture (lifetime). Working from the theories of Arrhenius (1889; Odell and Keller 1986) and Eyring (1935; Kauzmann and Eyring 1940), Kramers (1940) described the link between bond strength and lifetime for weakly bound interactions in solution. His formulation modeled the kinetics of chemical reactions in an overdamped environment, where all states are confined to a deep energy well far from equilibrium; and escape from the well is a function of the diffusion of thermal fluctuations along a reaction coordinate x that crosses the activation barrier. Kramers' model also contains an external force parameter that acts to lower the barrier along the coordinate x as expressed in Eq. (3.1):

$$k_d(F) = \frac{\omega_{min}\omega_{ts}}{2\pi\gamma} \cdot \exp\left(\frac{-E_a + Fx_\beta}{k_B T}\right) \qquad (3.1)$$

where ω_{min} and ω_{ts} represent the angular frequencies of the energy minimum and transition state, respectively; $\pi\gamma$ represents the damping relaxation rate; $k_B T$ represents the thermal energy of the system; and x_β represents the distance between the energy minimum and the transition state along the direction of applied force. For a sharp energy barrier, $x_\beta = \langle x_{ts} \cdot \cos(\theta)\rangle$, where x_{ts} is the position of the transition state along coordinate x. If we subsequently factor out the activation energy component from Eq. (3.1), it can be shown that Kramers' expression sets the scale for stress-free dissociation (k_d^0) shown in Eq. (3.2),

$$k_d(F) = k_d^0 \cdot \exp\left(\frac{Fx_\beta}{k_B T}\right) \qquad (3.2)$$

This expression was derived by Bell (1978), who used Kramers' theory to show that bond lifetime can be shortened by an applied force in processes such as cell adhesion. Although Eq. (3.2) is quite useful, it is in practice limited, most notably by the fact that it assumes that x_β is constant. Typically, measurements of force dependency are made under conditions in which force changes with time, and it is likely that the position of the transition state will move as the shape of the potential surface is perturbed by an applied force (Evans and Ritchie 1997; Hummer and Szabo 2003). Theoretical and empirical treatments of various cases have been put forth in the literature, but they are outside the scope of this chapter and will not be reviewed here.

A recent resurgence in mechanochemical research (Beyer and Clausen–Schaumann 2005) provides technical and intellectual opportunities to correlate the mechanical response of individual molecules to that of SPs, and quantitative, single-molecule force spectroscopy studies have been completed for several supramolecular interactions that are important components of SPs. These interactions have included, but are not limited to, host–guest interactions involving crown ethers, cyclodextrins (CDs), and calixarenes; hydrogen bonding within double-stranded oligonucleotides and between UPy units; hydrophobic interactions between alkanes and fullerenes; and metal–ligand bonds, such as ruthenium(II)-terpyridine $[Ru^{2+}\text{-}(tpy)_2]$. A summary of the forces associated with these interactions is given in Table 3.1.

The experiments summarized in Table 3.1 are done under conditions in which the applied force increases with time. The most-probable rupture force (F^*), which is taken from a distribution of measured values, reflects the competition between force loading and bond rupture and, with the exception of the CD inclusion compounds, rupture force increases with loading rate (Evans and Ritchie 1997). The data emphasize that the mechanical strength of the interactions do not necessarily reflect the thermodynamic strength, or association constant, of the interaction. This point is made by comparison of the DNA duplexes ($K_{eq} > 10^8$ M^{-1}, $F^* < 50$ pN) to the **4b** · pyridine associations ($K_{eq} < 10^2$ M^{-1}, $F^* > 50$ pN) at comparable loading rates. The kinetic nature of the mechanical influence expected from Eq. (3.2) is further shown in a comparison of the rupture forces of the same pyridine ligand from metal complexes **5a** and **4b**. The thermodynamics of ligand coordination are nearly identical for the two Pd complexes, but larger rupture forces are observed for the "slower" **4b** complex.

Such effects are likely to be important. The use of SP interactions to create bio-inspired material properties (e.g., see Chap. 9) implies that the ultimate yield behavior of SP materials could depend on the mechanical response of supramolecular interactions. Paulusse and Sijbesma (2004) have also shown that ultrasound-generated shear stresses can mechanically tear apart coordination SPs, damage that is subsequently repaired during dynamic equilibration once the shear stresses are removed. The mechanical response of supramolecular interactions within materials has potentially important consequences in the context of self-repairing materials, where the rupture of "sacrificial" supramolecular interactions protects a permanent, underlying materials architecture. The dynamic repair of the SP component in

TABLE 3.1 Selected Examples of Supramolecular Interactions, Relevant to Supramolecular Polymers, Studied by Force Spectroscopy

System	F^* (pN)	r_f (nN/s)	Ref.
Metal–Ligand			
Ru^{2+} + (tpy)$_2$[a]	95	1	Kudera et al. (2003)
Dopa + Ti surface[d]	636–847	0.41–250	Lee et al. (2006)
5a + dialkylaminopyridine[a]	50–150	2–120	Kersey et al. (2007)
4b + dialkylaminopyridine[a]	50–200	0.5–150	Kersey et al. (2006)
4b + carbamoylpyridine[a]	50–100	10–100	Kersey et al. (2006)
Host–guest			
18-crown-6 + ammonium[b]	60	–	Kado and Kimura 2003
β-CD + ferrocene[b,c]	55 ± 10	$2 \cdot 10^3$	Schonherr et al. (2000), Zapotoczny et al. (2002)
β-CD + anilylthiol[b,c]	39 ± 15	$2 \cdot 10^3$	Auletta et al. (2004)
β-CD + toluidylthiol[b,c]	45 ± 15	$2 \cdot 10^3$	Auletta et al. (2004)
β-CD + t-butylphenylthiol[b,c]	89 ± 15	$2 \cdot 10^3$	Auletta et al. (2004)
β-CD + adamantylthiol[b,c]	102 ± 15	$2 \cdot 10^3$	Auletta et al. (2004)
Resorc[4]arene + ammonium[d]	45–140	0.15–20	Eckel et al. (2005)
Resorc[4]arene + trimethylammonium[d]	80–125	0.15–20	Eckel et al. (2005)
Hydrogen bonding			
DNA oligonucleotides[a]	20–50	0.016–4	Strunz et al. 1999
Ureido-4[1H]-pyrimidinone dimers[d]	180 ± 21	35	Zou et al. (2005a)
Ureido-4[1H]-pyrimidinone dimers[a]	150–260	5–500	Zou et al. (2005b)
Hydrophobic			
Hexadecane dimers[a]	50–100	0.2–70	Ray et al. (2006)
C60 dimers[a]	50–110	2–70	Gu et al. (2007)

β-CD, β-cyclodextrin.
[a]Polymer tethers used (tip and surface).
[b]Force value determined from autocorrelation of force histogram.
[c]Loading rate-independent system.
[d]Polymer tether used (tip or surface only).

hybrid materials then reverses the damage once the stress is removed. Increasingly sophisticated models of force distributions in materials and along polymer chains and of the force-coupled potential energy surface of a supramolecular dissociation hold great promise in both fundamental and applied areas of polymer science.

3.6. CONCLUSIONS

This chapter highlights recent work that shows the relevance of a small-molecule view of SP mechanical properties, in which reversible and specific intermolecular interactions create entanglements or polymer bridges (Kersey et al. 2004; Kim et al. 2005; Zou et al. 2005a) necessary to transmit force in materials. Many related examples are found throughout this book, and many others are either known or are sure to follow in the years to come. The sophistication available in the field of molecular recognition holds considerable promise as it is applied increasingly to areas of material science. Whether in cross-linked networks or linear supramolecular polymers, the thermodynamics of the intermolecular interaction are a primary design consideration. However, the kinetics of the interactions that create entanglements are particularly important under nonequilibrium conditions such as those imposed by a mechanical stress. Fortunately, the chemist's knowledge and understanding of model systems often translates extremely well to macroscopic and nanoscale materials. The potential for bringing the creativity and diversity inherent in supramolecular chemistry to the rational, molecular design and synthesis of materials and interfaces with very specific and customized properties is such that this book likely provides only an early glimpse of the potential in the field.

REFERENCES

Albrecht M, van Koten G. Platinum group organometallics based on "pincer" complexes: sensors, switches, and catalysts. Angew Chem Int Ed 2001;40:3750–3781.

Alexander C, Jariwala CP, Lee CM, Griffin AC. Hydrogen-bonded main chain liquid crystalline polymers. Polym Prepr (Am Chem Soc Div Polym Chem) 1993;34:168–169.

Arrhenius S. Über die Reaktionsgeschwindigkeit bei der Inversion von Rohrzucker durch Säuren. Z Phys Chem (Leipzig) 1889;4:226–248.

Auletta T, de Jong MR, Mulder A, van Veggel FCJM, Huskens J, Reinhoudt DN, Zou S, Zapotoczny S, Schonherr H, Vancso GJ, Kuipers L. β-Cyclodextrin host–guest complexes probed under thermodynamic equilibrium: thermodynamics and AFM force spectroscopy. J Am Chem Soc 2004;126:1577–1584.

Bell GI. Models for the specific adhesion of cells to cells. Science 1978;200:618–627.

Beyer MK, Clausen-Schaumann H. Mechanochemistry: the mechanical activation of covalent bonds. Chem Rev 2005;105:2921–2948.

Bladon P, Griffin AC. Self-assembly in living nematics. Macromolecules 1993;26:6604–6610.

Broze G, Jerome R, Teyssie P, Marco C. Halato-telechelic polymers. 8. Dependence of the viscoelastic behavior on the prepolymer molecular-weight. Macromolecules 1983;16:1771–1775.

Candau SJ, Hirsch E, Zana R. New aspects of the behavior of alkyltrimethylammonium bromide micelles—light-scattering and viscosimetric studies. J Phys Paris 1984;45:1263–1270.

Castellano RK, Clark R, Craig SL, Nuckolls C, Rebek J Jr. Emergent mechanical properties of self-assembled polymeric capsules. Proc Natl Acad Sci USA 2000;97:12418–12421.

Cates ME. Reptation of living polymers—dynamics of entangled polymers in the presence of reversible chain-scission reactions. Macromolecules 1987;20:2289–2296.

Cates ME, Candau SJ. Statics and dynamics of worm-like surfactant micelles. J Phys Condensed Matter 1990;2:6869–6892.

Chen C-C, Dormidontova EE. Monte Carlo simulations of end-adsoption of head-to-tail reversibly associated polymers. Macromolecules 2006;29:9528–9538.

Ciferri, A. Supramolecular polymers 2nd ed. New York: Taylor & Francis; 2005.

de Gennes PG. Scaling concepts in polymer physics. Ithaca: Cornell University Press. 1979.

Eckel R, Ros R, Decker B, Mattay J, Anselmetti D. Supramolecular chemistry at the single-molecule level. Angew Chem Int Ed 2005;44:484–488.

Evans E, Ritchie K. Dynamic strength of molecular adhesion bonds. Biophys J 1997;72: 1541–1555.

Eyring H. The activated complex in chemical reactions. J Chem Phys 1935;3:107–115.

Fleer GJ, Scheutjens J. Interaction between adsorbed layers of macromolecules. J Colloid Interface Sci 1986;111:504–515.

Fleer GJ, Cohen Stuart MA, Scheutjens JMHM, Cosgrove T, Vincent B. Polymers at Interfaces. London: Chapman and Hall. 1993.

Flory, PJ. Principles of polymer chemistry. Ithaca, NY: Cornell University Press. 1969.

Fogleman EA, Yount WC, Xu J, Craig SL. Modular, well-behaved reversible polymers from DNA-based monomers. Angew Chem Int Ed 2002;41:4026–4028.

Fouquey C, Lehn J-M, Levelut A-M. Molecular recognition directed self-assembly of supra-molecular liquid crystalline polymers from complementary chiral components. Adv Mater 1990;2:254–257.

Granville AM, Boyes SG, Akgun B, Foster MD, Brittain WJ. Synthesis and characterization of stimuli-responsive semifluorinated polymer brushes prepared by atom transfer radical polymerization. Macromolecules 2004;37:2790–2796.

Green MS, Tobolsky AV. A new approach to the theory of relaxing polymeric media. J Chem Phys 1946;14:80–92.

Gu C, Ray C, Guo S, Akhremitchev BB. Single-molecule force spectroscopy measurements of interactions between C60 fullerene molecules. J Phys Chem C 2007;111:12898–12905.

Hayashi S, Ikeda S. Micelle size and shape of sodium dodecyl-sulfate in concentrated NaCl solutions. J Phys Chem 1980;84:744–751.

Hummer G, Szabo A. Kinetics from nonequilibrium single-molecule pulling experiments. Biophys J 2003;85:5–15.

Jeppesen C, Wong JY, Kuhl TL, Israelachvili JN, Mullah N, Zalipsky S, Marques CM. Impact of polymer tether length on multiple ligand–receptor bond formation. Science 2001;293: 465–468.

Jongschaap RJJ, Wientjes RHW, Duits MHG, Mellema J. A generalized transient network model for associative polymer networks. Macromolecules 2001;34:1031–1038.

Kado S, Kimura K. Single complexation force of 18-crown-6 with ammonium ion evaluated by atomic force microscopy. J Am Chem Soc 2003;125:4560–4564.

Kaholek M, Lee W-K, LaMattina B, Caster KC, Zauscher S. Preparation and characterization of stimulus-responsive poly(n-isopropylacrylamide) brushes and nanopatterns. Polym Mater Sci Eng 2004a;90:226–227.

Kaholek M, Lee WK, LaMattina B, Caster KC, Zauscher S. Fabrication of stimulus-responsive nanopatterned polymer brushes by scanning-probe lithography. Nano Lett 2004b;4: 373–376.

Kauzmann W, Eyring H. The viscous flow of large molecules. J Am Chem Soc 1940;62: 3133–3135.

Kersey FR, Lee G, Marszalek P, Craig SL. Surface-to-surface bridges formed by reversibly assembled polymers. J Am Chem Soc 2004;126:3038–3039.

Kersey FR, Loveless DM, Craig SL. A hybrid polymer gel with controlled rates of cross-link rupture and self-repair. J R Soc Interface 2007;4:373–380.

Kersey FR, Yount WC, Craig SL. Single-molecule force spectroscopy of bimolecular reactions: system homology in the mechanical activation of ligand substitution reactions. J Am Chem Soc 2006;128:3886–3887.

Kim J, Liu Y, Ahn SJ, Zauscher S, Karty JM, Yamanaka Y, Craig SL. Self-assembly and properties of main-chain reversible polymer brushes. Adv Mater 2005;17:1749–1753.

Kizhakkedathu JN, Norris-Jones R, Brooks DE. Synthesis of well-defined environmentally responsive polymer brushes by aqueous ATRP. Macromolecules 2004;37:734–743.

Kramers, HA. Brownian motion in a field of force and the diffusion model of chemical reactions. Physica 1940;4:284–304.

Kudera M, Eschbaumer C, Gaub HE, Schubert US. Analysis of metallo-supramolecular systems using single-molecule force spectroscopy. Adv Funct Mater 2003;13:615–620.

Lee H, Scherer NF, Messersmith PB. Single-molecule mechanics of mussel adhesion. Proc Natl Acad Sci USA 2006;103:12999–13003.

Lehn JM. Supramolecular chemistry—molecular information and the design of supramolecular materials. Makromol Chem Macromol Symp 1993;69:1–17.

Leibler L, Rubinstein M, Colby RH. Dynamics of reversible networks. Macromolecules 1991; 24:4701–4707.

LeMieux MC, Minko S, Usov D, Shulha H, Stamm M, Tsukruk VV. Manipulating polymer interfaces: adaptive morphology of mixed brushes in a selective and non-selective surrounding. Polym Mater Sci Eng 2004;90:372–373.

Lodge AS. A network theory of flow birefringence and stress in concentrated polymer solutions. Trans Faraday Soc 1956;52:120–130.

Loveless DM, Abu-Lail NI, Kaholek M, Zauscher S, Craig SL. Molecular contributions to the mechanical properties of reversibly cross-linked surface grafted polymer brushes. Angew Chem Int Ed 2006;45:7812–7814.

Loveless DM, Jeon SL, Craig SL. Rational control of viscoelastic properties in multicomponent associative polymer networks. Macromolecules 2005;38:10171–10177.

Loveless DM, Jeon SL, Craig SL. Chemoresponsive viscosity switching of a metallo-supramolecular polymer network near the percolation threshold. J Mater Chem 2007;17:56–61.

Minko S, Stamm M, Goreshnik E, Usov D, Sidorenko A. Environmentally responsive polymer brush layers for switchable surface properties. Polym Mater Sci Eng 2000;83:533–534.

Motornov M, Minko S, Eichhorn KJ, Nitschke M, Simon F, Stamm M. Reversible tuning of wetting behavior of polymer surface with responsive polymer brushes. Langmuir 2003; 19:8077–8085.

Odell JA, Keller A. Flow-induced chain fracture of isolated linear macromolecules in solution. J Polym Sci B 1986;24:1889–1916.

Paulusse JMJ, Sijbesma RP. Reversible mechanochemistry of a PdII coordination polymer. Angew Chem Int Ed 2004;43:4460–4462.

Pollino JM, Weck M. Non-covalent side-chain polymers: design principles, functionalization strategies and perspectives. Chem Soc Rev 2005;34:193–207.

Porte G, Appell J, Poggi Y. Experimental investigations on the flexibility of elongated cetylpyridinium bromide micelles. J Phys Chem 1980;84:3105–3110.

Ray C, Brown JR, Akhremitchev BB. Single-molecule force spectroscopy measurements of "hydrophobic bond" between tethered hexadecane molecules. J Phys Chem B 2006;110: 17578–17583.

Rehage H, Hoffmann H. Rheological properties of viscoelastic surfactant systems. J Phys Chem 1988;92:4712–4719.

Rehage H, Hoffmann H. Viscoelastic surfactant solutions—model systems for rheological research. Mol Phys 1991;74:933–973.

Rietveld MHP, Grove DM, van Koten G. Recent advances in the organometallic chemistry of aryldiamine anions that can function as N,C,N'- and C,N,N'-chelating terdentate "pincer" ligands: an overview. New J Chem 1997;21:751–771.

Rodriguez G, Albrecht M, Schoenmaker J, Ford A, Lutz M, Spek AL, van Koten G. Bifunctional pincer-type organometallics as substrates for organic transformations and as novel building blocks for polymetallic materials. J Am Chem Soc 2002;124:5127–5138.

Schonherr H, Beulen MWJ, Bugler J, Huskens J, van Veggel FCJM, Reinhoudt DN, Vancso GJ. Individual supramolecular host–guest interactions studied by dynamic single molecule force spectroscopy. J Am Chem Soc 2000;122:4963–4967.

Sijbesma RP, Beijer FH, Brunsveld L, Folmer BJB, Hirschberg J, Lange RFM, Lowe JKL, Meijer EW. Reversible polymers formed from self-complementary monomers using quadruple hydrogen bonding. Science 1997;278:1601–1604.

Slagt MQ, van Zwieten DAP, Moerkerk A, Gebbink R, van Koten G. NCN–pincer palladium complexes with multiple anchoring points for functional groups. Coord Chem Rev 2004; 248:2275–2282.

St Pourcain CB, Griffin AC. Thermoreversible supramolecular networks with polymeric properties. Macromolecules 1995;28:4116–4121.

Strunz T, Oroszlan K, Schafer R, Guntherodt H-J. Dynamic force spectroscopy of single DNA molecules. Proc Natl Acad Sci USA 1999;96:11277–11282.

Tanaka F, Edwards SF. Viscoelastic properties of physically cross-linked networks—transient network theory. Macromolecules 1992;25:1516–1523.

Turner MS, Cates ME. Linear viscoelasticity of wormlike micelles—a comparison of micellar reaction-kinetics. J Phys II 1992;2:503–519.

Turner MS, Marques C, Cates ME. Dynamics of wormlike micelles—the bond-interchange reaction scheme. Langmuir 1993;9:695–701.

van Beek DJM, Spiering AJH, Peters GWM, te Nijenhius K, Sijbesma RP. Unidirectional dimerization and stacking of ureidopyrimidinone end groups in polycaprolactone supramolecular polymers. Submitted 2007.

van der Gucht J, Besseling NAM, Stuart MAC. Surface forces, supramolecular polymers, and inversion symmetry. J Am Chem Soc 2002;124:6202–6205.

van der Gucht J, Besseling NAM, Fleer GJ. Surface forces induced by ideal equilibrium polymers. J Chem Phys 2003;119:8175–8188.

Vermonden T, van Steenbergen MJ, Besseling NAM, Marcelis ATM, Hennink WE, Sudholter EJR, Stuart MAC. Linear rheology of water-soluble reversible neodymium(III) coordination polymers. J Am Chem Soc 2004;126:15802–15808.

Xu J, Fogleman EA, Craig SL. Structure and properties of DNA-based reversible polymers. Macromolecules 2004;37:1863–1870.

Yamamoto M. The visco-elastic properties of network structure. 1. General formalism. J Phys Soc Jpn 1956;11:413–421.

Yount WC, Juwarker H, Craig SL. Orthogonal control of dissociation dynamics relative to thermodynamics in a main-chain reversible polymer. J Am Chem Soc 2003;125: 15302–15303.

Yount WC, Loveless DM, Craig SL. Small-molecule dynamics and mechanisms underlying the macroscopic mechanical properties of coordinatively cross-linked polymer networks. J Am Chem Soc 2005a;127:14488–14496.

Yount WC, Loveless DM, Craig SL. Strong means slow: dynamic contributions to the mechanical properties of supramolecular networks. Angew Chem Int Ed 2005b;44:2746–2748.

Zapotoczny S, Auletta T, de Jong MR, Schonherr H, Huskens J, van Veggel FCJM, Reinhoudt DN, Vancso GJ. Chain length and concentration dependence of β-cyclodextrin-ferrocene host–guest complex rupture forces probed by dynamic force spectroscopy. Langmuir 2002;18:6988–6994.

Zou S, Schonherr H, Vancso GJ. Stretching and rupturing individual supramolecular polymer chains by AFM. Angew Chem Int Ed 2005a;44:956–959.

Zou S, Schonherr H, Vancso GJ. Force spectroscopy of quadruple H-bonded dimers by AFM: dynamic bond rupture and molecular time–temperature superposition. J Am Chem Soc 2005b;127:11230–11231.

PART II

POLYMER FORMATION AND SELF-ASSEMBLY

CHAPTER 4

HYDROGEN BOND FUNCTIONALIZED BLOCK COPOLYMERS AND TELECHELIC OLIGOMERS

BRIAN D. MATHER and TIMOTHY E. LONG

4.1. SCIENTIFIC RATIONALE AND PERSPECTIVE

Recently, increasing attention has been devoted to the study of noncovalent interactions, particularly hydrogen bonding, in the construction and design of supramolecular materials (Ilhan et al. 2000; Brunsveld et al. 2001). This chapter highlights recent advances in the development of block and telechelic polymers containing hydrogen bonds, with a specific emphasis on the tunability of physical properties and morphology through the introduction of hydrogen bonding groups. Hydrogen bonding enables the introduction of thermoreversible properties into macromolecules through the creation of specific noncovalent intermolecular interactions (Yamauchi, Lizotte, Hercules, et al. 2002). The strength of these interactions is a strong function of temperature, solvent, humidity, and pH, thus allowing control of properties through a number of environmental parameters. The strength of hydrogen bonding associations is further tunable via structural parameters and molecular design of the hydrogen bonding sites. Association strengths range from DNA nucleobases, which have association constants near $100 \, \text{M}^{-1}$ (Kyogoku et al. 1967b), to self-complementary ureidopyrimidone (UPy) hydrogen bonding groups, which possess association constants near $10^7 \, \text{M}^{-1}$.

Hydrogen bonding interactions are important for the development of self-assembling supramolecular materials, which are defined as materials in which monomeric units are reversibly bound via secondary interactions to form polymer-like structures that exhibit polymeric properties in solution as well as in bulk (Brunsveld et al. 2001). Rotello used hydrogen bond functional polymers to direct the formation of large vesicles (Ilhan et al. 2000), reversibly attach polymers on

Molecular Recognition and Polymers: Control of Polymer Structure and Self-Assembly.
Edited by V. Rotello and S. Thayumanavan
Copyright © 2008 John Wiley & Sons, Inc.

surfaces (Norsten et al. 2003), and reversibly attach functional small molecules on polymers (Ilhan et al. 2001). Yamauchi and colleagues incorporated quadruple hydrogen bonding self-complementary UPy multiple hydrogen bonding groups into random copolymers (Yamauchi et al. 2003) and onto the chain end of homopolymers (Yamauchi, Lizotte, Hercules, et al. 2002) to introduce controlled rheological performance and thermoreversible properties to novel supramolecular structures.

Hydrogen bonding interactions have played a major role in polymer science. From the hydrogen bonding in polyurethanes (McKeirnan et al. 2002) to the structures of polypeptides (Aggeli et al. 1997) we can observe the cohesive effect that hydrogen bonding interactions introduce to various macroscopic properties. More recently, elegant hydrogen bonding arrays were introduced into synthetic polymers and yielded novel polymers containing reversible linkages, which are tuned with such environmental variables as temperature, moisture, and solvent polarity. The resultant polymers often exhibit a stronger temperature dependence of the melt viscosity, suggesting possible processing advantages.

The advantages of materials that contain hydrogen bonds are numerous (Brunsveld et al. 2001). Beyond the novel rheological properties of hydrogen bonding systems, the introduction of hydrogen bonds often results in increased moduli and tensile strengths (Reith et al. 2001; Sheth, Wilkes, et al. 2005), as well as increased polarity and adhesion (Vezenov et al. 2002; Viswanathan et al. 2006; Eriksson et al. 2007). Hydrogen bonding interactions enable such diametrically opposed effects as the compatibilization of blends (Coleman et al. 1991) and the induction of microphase separation (Yuan et al. 2006). They can also be used to create dynamic micelles in solution, and the structure is tunable through the introduction of guest molecules, which alter solvent polarity or introduce covalent cross-links. In the case of block copolymers, the introduction of hydrogen bonding groups leads to a synergistic combination of microphase separation and hydrogen bonding associations (Lillya et al. 1992; Brunsveld et al. 2001; Mather et al. 2007a). Furthermore, guest molecules may be sequestered in specific domains of the block copolymer, allowing the reversible attachment of functional groups or biological molecules (Thibault et al. 2003; Bondzic et al. 2004).

4.2. HYDROGEN BONDING INTERACTIONS IN MACROMOLECULAR DESIGN

4.2.1. Fundamentals of Hydrogen Bonding

Hydrogen bonds are formed between molecules containing an electronegative atom that possesses lone electron pairs (usually O, N, or F), called acceptors (A), and molecules containing covalent bonds between hydrogen and an electronegative atom (usually O-H, N-H, S-H), called donors (D). The polarized nature of the X-H bond (X = O, N) results in a highly electropositive hydrogen that is attracted toward bond formation with the electron-rich electronegative acceptor atoms.

The preferred geometry of the hydrogen bond is typically close to a linear conformation, with the hydrogen atom positioned along the line connecting the heteroatoms, although there are many examples of nonlinear hydrogen bonding (Lommerse et al. 1997). A dominant property of hydrogen bonds is their directionality and requirement for closeness. The strength of individual hydrogen bonds ranges from 1 to 10 kcal/mol (Coleman et al. 1991), which leads to longer, weaker bonds compared to covalent bonds (70–110 kcal/mol). Thus, when hydrogen bonds serve as the primary secondary interaction, the resultant structures are typically closer to thermodynamic equilibrium and are quite dynamic with respect to changes in the environment. This is due to the simple fact that the thermal energy (kT) more closely matches the bond strengths in these systems. For instance, an association energy of 3 kcal/mol leads to a 1% level of dissociation among units in a single component bulk liquid bound with this energy at ambient temperature (Coleman et al. 1991). This figure obviously represents a time average; the typical lifetime of a hydrogen bond in water, for example, is roughly 10^{-11} s (Conde and Teixerira 1983).

The acidity of the donor and the Lewis basicity of the acceptor play important roles in the strength of the interaction (Beijer et al. 1996; Kawakami and Kato 1998). Phenols, which possess stronger acidity than aliphatic alcohols, produce stronger hydrogen bonds with acceptors. These properties, which are measured in terms of pK_a or pK_b, are founded in the electronic structure of the donors and acceptors and their ability to stabilize negative or positive charges, respectively. Electron withdrawing substituents such as fluorine increase the acidity of the hydrogen bond donors, as in the case of the hexafluoroisopropanol group (Liu et al. 2000). In the extreme case, mixing a strongly acidic hydrogen bond donor with a strongly basic hydrogen bond acceptor results in an acid–base neutralization reaction and the formation of an ion pair. Sedlak et al. (2003) discussed the fact that hydrogen bonding interactions are indeed founded in electrostatic attractive and repulsive interactions.

$$\text{A-H} + \text{B} \xrightleftharpoons{\text{hydrogen bonding interaction}} \text{A-H}\!-\!\text{B}$$

$$\text{A-H} + \text{B} \xrightleftharpoons{\text{formation of ion pairs}} \text{A}^- + \text{BH}^+ \tag{4.1}$$

The common measure of the strength of association or complexation is the association constant (K_a), defined in terms of the equilibrium between associated and dissociated units [Eq. (4.2)]:

$$\text{A} + \text{B} \rightleftharpoons \text{A} - \text{B} \quad K_a = \frac{[\text{A} - \text{B}]}{[\text{A}][\text{B}]} \tag{4.2}$$

The K_a value is a key measurement of the strength of the hydrogen bonding association and determines the dynamics of systems incorporating the hydrogen bonding groups. Thus, for weaker hydrogen bonding interactions, the lifetimes of associated species are shorter than for systems with higher association constants. This has a dramatic effect on material properties, such as relaxation rates, creep, modulus, melt viscosity, and so forth. In order to tune the strength of the association constant, numerous molecular parameters are accessible. A common approach for increasing the association strength is the incorporation of additional hydrogen bonding sites within a single hydrogen bonding array, which is called "multiple hydrogen bonding." The association of two species is unfavorable entropically, but this loss of entropy is not greatly increased from the association of additional pairs of donors and acceptors on the same two molecules. Single to quadruple hydrogen bonding units exist, and some larger arrays containing six (Zeng et al. 2000) and eight (Folmer et al. 1999) hydrogen bonds were synthesized and studied. In addition, the strength of the hydrogen bonding interaction is tuned via the pattern of hydrogen bonding groups. Research has shown that a number of factors are important in the use of multiple hydrogen bonded systems (Beijer et al. 1996; Brunsveld et al. 2001). Alternation of donor and acceptor groups in a multiple hydrogen bonding array reduces the strength of the association via repulsive interactions of opposing adjacent donors or acceptors (Jorgensen and Pranata 1990; Beijer et al. 1998). Thus, Meijer's quadruple hydrogen bonding (DDAA) UPy group is designed with minimum alternation of donor and acceptor units (Beijer et al. 1998). When multiple hydrogen bonding groups are considered, the possibility of self-complementarity arises. The simplest case of self-complementarity is probably the carboxylic acid dimer, which contains two hydrogen bonds. In contrast, complementary hydrogen bonding arrays are found in DNA, which possesses both adenine–thymine (A-T) and cytosine–guanine (C-G) pairs. Another factor to consider in heterocylic bases is the ability of the base to tautomerize into different forms (Lee and Chan 1972). This can lead to complexity in the behavior of the hydrogen bonding group, and different hydrogen bonding guests can shift the equilibrium between tautomers (Ligthart et al. 2005).

The reversible nature of the hydrogen bond is not limited to thermoreversibility, although that is the most commonly exploited and studied property of these systems. Hydrogen bonds are also sensitive to other environmental factors, such as the polarity of the medium and the presence of competitive solvents and water. Deans et al. (1999) observed that solvent polarity has a strong effect on the hydrodynamic volume for self-associating polymers. Thus, humidity and exposure to polar media are two other factors that are considerations in the application of hydrogen bonded materials. Another factor that strongly affects hydrogen bonding, as suggested in Eq. (4.1), is concentration. Lower concentrations of monomeric units lead to lower equilibrium concentrations of complexed units. Solution pH can also affect hydrogen bonding systems, leading to pH controlled thickening (Sotiropoulou et al. 2003) in cases of interpolymer complexes between poly(acrylic acid) and polyacrylamide. Complexes of poly(acrylic acid) and poly(*N*-vinylpyrrolidone) exhibit molecular weight dependent critical pHs below which stable complexes are formed because of a transition to polyelectrolyte species at higher pH (Nurkeeva et al. 2003).

The hydrogen bond is extensively used in nature, particularly in the construction of proteins (Aggeli et al. 1997), DNA (Voet and Voet 1995), and RNA. In all cases, it performs the role of establishing a reversible structure that allows such processes as replication and transcription in DNA and reactions in enzymes and it is integral to many molecular recognition events in living organisms. One goal of modern science is to incorporate such functionality and dynamics into synthetic materials. In the case of DNA, hydrogen bonding performs in concert with numerous other noncovalent interactions such as electrostatic interactions and $\pi-\pi$ stacking (Vanommeslaeghe et al. 2006) to yield the double helix structure and to give DNA its dynamic properties.

Numerous analytical tools are useful in the study of hydrogen bonding phenomenon. Fourier transform IR (FTIR; Kyogoku et al. 1967b) and NMR (Fielding 2000) techniques as well as UV visible (Lutz, Thunemann, Rurack 2005) are useful for determining association constants or observing association phenomena. Advanced techniques are necessary for the measurement of very high association constant systems (Söntjens et al. 2000). Rheology (Müller et al. 1996; Yamauchi, Lizotte, Hercules, et al. 2002) and mechanical analysis (Reith et al. 2001) are useful for determining the influence of hydrogen bonding interactions on thermal and mechanical properties, whereas microscopy [transmission electron microscopy (TEM), atomic force microscopy (AFM)] is primarily useful for studying changes in morphology. Solution rheology is a particularly useful tool for hydrogen bonded systems because of the ability to introduce solvents of varying dielectric constants and to study the hydrogen bonding interaction as a function of concentration (Sijbesma et al. 1997; McKee et al. 2004). Light scattering techniques (dynamic light scattering, static light scattering) are often used to characterize micellar or aggregate structures resulting from hydrogen bonding associations in solution (Liu and Jiang 1995; Liu and Zhou 2003).

4.2.2. Performance Advantages of Hydrogen Bond Containing Polymers

Reversibility of the hydrogen bond confers unique properties to polymeric and supramolecular materials. In supramolecular science, reversible bonding is important because it allows the self-assembly process to occur. Molecular recognition and self-organization are hailed as the mechanisms of supramolecular growth (Kato 1996). Controlled geometric placement of hydrogen bonding groups leads to efficient molecular recognition, which is optimized through self-organization. Furthermore, the directionality of the hydrogen bond assists in the construction of supramolecular structures because random orientations of hydrogen bonding associations leading to disordered networks are not favored. In the field of polymer processing, thermoreversibility is of interest because of the promise for lower melt viscosities of hydrogen bond containing polymers. Thus, a lower molecular weight hydrogen bonding polymer could potentially afford mechanical properties approaching those of a higher molecular weight nonfunctional polymer over a short time scale and yet exhibit lower melt viscosity when heated above the dissociation temperature of the hydrogen bonding groups (Yamauchi, Lizotte, Hercules, et al. 2002; Yamauchi et al. 2004). The hydrogen bonding polymer would exhibit a higher apparent molecular weight at room temperature than it actually possessed.

In the case of ditopic molecules with two hydrogen bonding groups, linear noncovalent polymers can form in either solution or the solid state (Fig. 4.1). In this case, the degree of noncovalent polymerization (DP) depends directly on the association constant (K_a) in the medium and the concentration of the molecule (c) as shown in Eq. (4.3) (Xu et al. 2004):

$$DP = 4K_a c/(-1 + (1 + 8K_a c)^{0.5}) \sim (2K_a c)^{0.5} \tag{4.3}$$

Another feature of hydrogen bonding systems is that they are dynamic, and this often allows them to achieve the most thermodynamically favored state. This dynamic property of hydrogen bonding allows self-assembly to occur at ambient conditions, unlike many covalent polymerizations (Saadeh et al. 2000). Deans and colleagues (1999) were able to create a thermodynamic cycle to examine the competitive processes of self-association of a hydrogen bonding polymer and association with a small molecule complementary guest in solution. In the case of hydrogen bonded networks, the dynamic character leads to the presence of fewer network defects, such as dangling ends and unincorporated chains (sol). In support of this hypothesis, Lange et al. (1999) observed a higher plateau modulus for a hydrogen bonded network compared to a covalent network. Karikari et al. (2007) studied nucleobase-functionalized poly(D,L-lactide) star polymers and observed higher viscosity for the solution blends of complimentarily functionalized polymers.

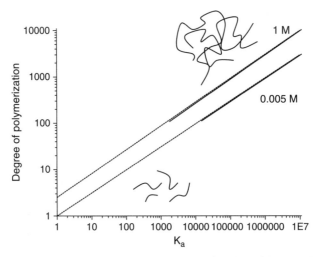

Figure 4.1 Degree of noncovalent polymerization as a function of the association constant (K_a). The effect of concentration is illustrated in the two parallel lines. Reprinted from Brunsveld et al. (2001). Copyright 2001 American Chemical Society.

Hydrogen bonding has improved mechanical properties in a number of systems and is often grouped with other noncovalent cohesive intermolecular interactions, performing as the "glue" that sticks chains. Hydrogen bonding supramolecular assembly in conjunction with covalent polymer synthesis often leads to a positive impact on the mechanical properties such as the stress at break and percentage of elongation (Reith et al. 2001) as well as the elastic modulus (Shandryuk et al. 2003). Nissan (1976) theorized that for a hydrogen bonded solid the Young's modulus scales with the number of hydrogen bonds per unit volume to the $1/3$ power. Hydrogen bonding results in an increased glass-transition temperature (T_g), which is due to restricted molecular mobility and temporary "cross-links" (Yamauchi, Lizotte, Hercules, et al. 2002). Hydrogen bonding interactions are credited, in part (along with microphase separation), with the outstanding mechanical properties of polyurethanes. Sheth and coworkers recently demonstrated that hydrogen bonding in model single unit hard segment poly(urethane urea)s leads to the development of ribbonlike hard phases containing stacks of hydrogen bonding urethanes or ureas that are disrupted upon the addition of branching units in the hard phase or addition of hydrogen bond screening agents such as lithium chloride (Sheth, Wilkes, et al. 2005). Further studies of single hard segment polyurethanes and polyureas revealed that the symmetry of the hard segment precursor, which affects the packing ability of hydrogen bonding groups, has a dramatic influence on the thermal integrity of the hard phases. Model polyureas based on *p*-phenylene diisocyanate and poly(tetramethylene oxide) possess rubbery plateaus extending to 250 °C, whereas meta substitution leads to rubbery plateaus extending only 100 °C (Sheth, Klinedinst, et al. 2005). McKeirnan et al. (2002) utilized variable temperature FTIR to show that the hydrogen bonds in polyurethanes persist up to temperatures beyond 100 °C and that, even in the melt, $\sim 75\%$ of the urethane linkages were involved in hydrogen bonding.

Hydrogen bonding interactions confer unique melt (Yamauchi et al. 2003) and solution (Lele and Mashelkar 1998) rheological behavior. The temperature sensitivity of the hydrogen bond holds promise for increasing the temperature dependence of the melt viscosity while providing an effectively higher molecular weight at room temperature. Thus, lower molecular weight hydrogen bond containing polymers may be employed. However, increased melt and solution viscosities are typically observed at temperatures where hydrogen bonds still exist. Lillya and colleagues (1992) showed that the zero shear melt viscosity of a 650 g/mol carboxylic acid terminated poly(tetrahydrofuran) [poly(THF)] was 2.5 times that of the corresponding protected (ester) version telechelic carboxylic acid functionalized poly(THF). This high melt viscosity dropped above 62 °C for the 650 g/mol polymer and above 50 °C for the 1000 g/mol polymer. Sivakova et al. (2005) also observed sudden decreases in viscosity in hydrogen bonding systems. Based on the sticky reptation theory that Leibler and colleagues (1991) developed, the terminal relaxation time (t_D) for a thermoreversible, hydrogen bonded network is longer than for a corresponding nonfunctionalized polymer. Rogovina et al. (1995) observed reversible network formation in self-associating carboxyl functionalized poly(dimethylcarbosiloxane) with 0.5–2.5 mol% of the carboxylic acid group. Müller et al. (1995) conducted some of the first major rheological studies of multiple hydrogen bond functionalized polymers. They studied randomly functionalized, urazole and phenylurazole containing, anionically synthesized polybutadienes. Increased plateau shear moduli, melt viscosity, and a shifting of relaxation times to lower frequencies (longer times) were observed in randomly substituted urazole containing polybutadienes (de Lucca Freitas et al. 1987). Zero shear viscosities, which are sensitive to molecular weight, but not to molecular weight distribution (MWD), were shown to increase with the mole percentage of phenylurazole substitution that gave an increasing "effective" molecular weight. The real part of the creep compliance at low shear rates (J_c^0), which is sensitive to changes in the MWD but not to molecular weight, was found to increase with increasing phenylurazole content because of the effective broadening in the MWD in the thermoreversible network. Similar results were obtained from relaxation time spectra, where a broadening of the relaxation time distribution occurred for phenylurazole-substituted polymers. Increased flow activation energies were also observed through fitting of Williams–Landel–Ferry parameters for the shift factors of the storage modulus. Comparing plateau moduli above and below a frequency corresponding to the dissociation rate of the phenylurazoles led to the determination of the number of reversible contacts, indicating that 60–70% of the phenyurazole stickers were complexed at any one point in time.

The solution behavior of hydrogen bond containing polymers is equally interesting compared to the solid state. In the dissolved state, polymers exhibit greater freedom to assume conformations that are thermodynamically favored. Furthermore, the dynamic nature of solutions allows rapid response to changes in the environment. In solution, factors such as the concentration and solvent dielectric constant are used to control the equilibrium between associated and dissociated states. For instance, Zhao et al. (1995) established that self-association of the 4-hydroxystyrene residues in symmetric poly(styrene-b-4-hydroxystyrene) diblock copolymers led to the formation of rodlike micelles in low polarity toluene, which was not observed in the more polar THF solvent. Reversible noncovalent cross-linking of similar polymers with 1,4-butanediamine induced micelle formation (Yoshida and Kunugi 2002). Placement of specific molecules within micelles was also achieved using hydrogen bonding (Zhang et al. 2004). The solvent dielectric constant strongly affects the solution rheological behavior of hydrogen bond containing polymers. Solution rheological studies of UPy-functionalized poly(methyl methacrylate) indicated that the solution viscosity increased with decreasing dielectric constant (polarity) of the solvent and that the critical concentration for entanglement (C_e) decreased, presumably because of the higher effective molecular weight of the associated structures (McKee et al. 2004). Furthermore, the slope of the viscosity versus concentration above C_e increased with decreasing solvent polarity, illustrating the effect of the concentration on the equilibrium between associated and dissociated states. Hydrogen bonding was utilized to induce micelle formation. Chen and Jiang (2005) studied micelle formation via hydrogen bonding between homopolymers of poly(4-vinylpyridine) and carboxylic acid terminated polystyrene or polyisoprene. They found that assemblies of these homopolymers formed micelles when selective solvents were added for either block. The size of such micelles was controlled via the mass ratio of the two homopolymers (Wang et al. 2001). Moreover, the polydispersity of the micelles changed with the solvent composition and narrowed for less selective solvent mixtures.

Other potential applications for hydrogen bond containing polymers exist, such as reversible attachment of guest molecules (Ilhan et al. 2001), reversible cross-linking (Thibault et al. 2003), improved melt processing behavior (Yamauchi, Lizotte, Hercules, et al. 2002), self-healing materials (Lange et al. 1999), shape-memory polymers (Liu et al. 2006), recyclable thermosets (Lange et al. 1999), induction of liquid crystallinity (Gulik-Krzywicki et al. 1993), induction of phase mixing (Kuo et al. 2003) and demixing (de Lucca Freitas et al. 1998), templated polymerization (Khan et al. 1999), drug-selective chromatographic media (Kugimiya et al. 2001), hydrogen bonded layer by layer assemblies (Wang et al. 1999), and reinforcement of the orientation of nonlinear optic materials (Huggins et al. 1997).

4.3. HYDROGEN BOND CONTAINING BLOCK COPOLYMERS

Hydrogen bonding telechelic and block copolymers both differ from randomly functionalized copolymers in that the placement of the hydrogen bonding groups is localized to specific region(s) in the polymer backbone. This typically leads to different behavior compared to random functionalization because of the presence of both phase separation and hydrogen bonding interactions. Thus, in the bulk state, many of these systems consist of both separated domains or blocks of hydrogen bonding groups and nonhydrogen bonded regions.

One particular advantage of introducing hydrogen bonding groups in a block copolymer structure is the multiplicative and cooperative effects of the hydrogen bonding interactions that are attributable to the multitude of adjacent hydrogen bonding groups (Pan et al. 2000). Pan and colleagues observed association constants of $(4.5 \text{ to } 7.9) \times 10^5 \text{ M}^{-1}$ for poly(4-vinylpyridine-b-NLO) with poly(4-hydroxystyrene-b-styrene) (a relatively good hydrogen bonding donor) and $4.4 \times 10^3 \text{ M}^{-1}$ for poly(2-hydroxyethyl methacrylate-b-methyl methacrylate) (a relatively poor hydrogen bonding donor) in toluene/CH_2Cl_2 (99:1) mixtures. These values are much higher than those expected for a single vinylpyridine–hydroxystyrene hydrogen bonding interaction. Kriz et al. (2006) found that the cooperativity of the association of poly(4-vinylpyridine) and poly(4-hydroxystyrene) increased with the degree of polymerization and reached a maximum at a 1:1 functional group molar ratio.

In addition, only short hydrogen bonding blocks are necessary to achieve phase separation, because of the influence of the associations on the tendency of block copolymers to microphase separate. Although this tendency is often measured through the Flory–Huggins interaction parameter χ (Leibler 1980; Lee and Han 2002b), hydrogen bonding interactions must be accounted for through additions of other parameters into thermodynamic treatments, because χ refers to nonspecific interactions and is repulsive whereas hydrogen bonding is attractive. Coleman and Painter (1995, 2006) thoroughly discussed thermodynamic treatments of hydrogen bonding effects in polymer blend miscibility. Considerations must be made for screening and accessibility of hydrogen bonding interactions (because they are directional) as well as the competition between self-association and complementary association. In some cases, single hydrogen bonding groups at the chain ends phase separate from the bulk polymer (Lillya et al. 1992; Sivakova et al. 2005). This is particularly possible when hydrogen bonding interactions occur in multiple directions, facilitating the organization of the chain ends (Kautz et al. 2006).

4.3.1. Block Copolymers Involving Single Hydrogen Bonding Groups

Hydrogen bonding block copolymers containing single hydrogen bonding groups have attracted attention because of the simplicity of their preparation via anionic polymerization techniques of commonly available monomers. One specific template is poly(vinylpyridine), which is often blended with complementary poly(methacrylic acid) (Jiang et al. 2003) or poly(4-hydroxystyrene). Anionic polymerization possesses a limitation in these cases that is due to the sensitivity of anionic polymerization to protic functionality. Thus, protecting group strategies are often employed. In the case of 4-hydroxystyrene, silyl protecting groups are employed; in the case of methacrylic acid residues, *tert*-butyl ester protecting groups are employed. Another strategy involves postpolymerization modification. Liu and Jiang (1995) introduced carboxylic acid groups and hydroxyl groups into styrene–ethylene–butylene–styrene polymers via Friedel–Crafts acylation and subsequent oxidation or reduction of the ketone functional groups. Other strategies involve protecting group chemistry. In the case of 4-hydroxystyrene based blocks, a silyl protected monomer is polymerized and deprotected after completion of the polymerization.

Block copolymers containing complementary monomers such as 2-vinylpyridine and 4-hydroxystyrene or methacrylic acid were shown to produce novel morphologies in which the complementary segments are mixed in domains, which are phase separated from the bulk polymer (Asari et al. 2005). Jiang et al. (2003) blended poly (styrene-*b*-butadiene-*b*-*tert*-butyl methacrylate-*co*-methacrylic acid) (SBT/A) with poly(styrene-*b*-2-vinylpyridine) (SV) to obtain a compatibilized blend. The degree of hydrogen bonding interaction was varied through controlling the level of hydrolysis of the *tert*-butyl methacrylate residues to methacrylic acid residues. For partially hydrolyzed SBT polymers (18–51% hydrolyzed), blending with SV led to a transition from a double gyroid or lamellar morphology (SBT/A–A/TBS) to hcp V-T/A cylinders in B lamella alternating with S lamella. Asari and colleagues (2006) conducted a similar study where poly(2-vinylpyridine-*b*-isoprene-*b*-2-vinylpyridine) triblocks with a mole fraction of vinylpyridine of 0.07 and poly(styrene-*b*-4-hydroxystyrene) with a 4-hydroxystyrene mole fraction of 0.14 were blended in either a 1:1 or 2:1 2-vinylpyridine/4-hydroxystyrene ratio. Archimedean tile morphologies resulted, which consisted of rows of vinylpyridine with 4-hydroxystyrene cylinders staggered between polyisoprene and polystyrene lamella for the 1:1 ratio or hexagons of polystyrene in an polyisoprene matrix with poly(2-vinylpyridine) with poly(4-hydroxystyrene) cylinders at the corners of the styrene hexagons (Fig. 4.2). Asari et al. established the 4-hydroxystyrene/2-vinylpyridine hydrogen bonding

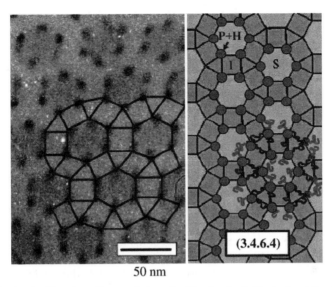

Figure 4.2 An Archimedean tile morphology for blends of poly(2-vinylpyridine-*b*-isoprene-*b*-vinylpyridine) with poly(styrene-*b*-4-hydroxystyrene) in a 2:1 vinylpyridine/hydroxystyrene blend. The vinylpyridine/hydroxystyrene domains are the cylinders at the vertices of polystyrene hexagons within a polyisoprene continuous phase. Reprinted from Asari et al. (2006). Copyright 2006 American Chemical Society.

interaction from ^{1}H-NMR, where a shifting of the OH proton was visible with the addition of vinylpyridine containing copolymers.

Coleman, Painter, and coworkers developed theoretical treatments to describe hydrogen bonding in polymer blends (Coleman et al. 1991; Painter et al. 1997). In general, the compatibilization of blends reduces the sizes of the domains present and increases the mechanical properties while reducing the melt viscosity as Xu, Chen, et al. (1999) observed in studies of blending syndiotactic polystyrene with a thermoplastic polyurethane using a poly(styrene-*b*-4-vinylpyridine) block copolymer. The reduction of interfacial tension through the use of hydrogen bonding block copolymers is the primary mechanism to improve blend miscibility. Blends of poly(2,5-dimethyl-1,4-phenylene oxide) (PPO) and poly(ethylene-*co*-acrylic acid) were compatibilized using poly(styrene-*b*-2-ethyl-2-oxazoline) because of the hydrogen bonding interactions of the amine and the carboxylic acid and the compatibility of the polystyrene block with PPO (Xu, Zhao, et al. 1999). The tensile strength and elongation of blends increased as the domain size decreased with the addition of a small amount (~5 wt %) of poly(styrene-*b*-2-ethyl-2-oxazoline). Edgecombe et al. (1998) pursued living anionic and atom transfer radical polymerization (ATRP) synthesis of block and graft copolymers of polystyrene and poly(4-hydroxystyrene) for strengthening the interface between polystyrene and poly(4-vinylpyridine). The hydrogen bonding between the poly(4-hydroxystyrene) block and poly(4-vinylpyridine) led to strong interactions that allowed the use of poly(4-hydroxystyrene) blocks that were below the entanglement molecular weight (M_c). In contrast, the polystyrene blocks possessed molecular weights above M_c, which allowed some degree of entanglement with the polystyrene side of the interface.

Self-association and multiple modes of association of hydrogen bonding groups frequently leads to complexity in the behavior of hydrogen bonded systems in which complementary hydrogen bonding groups are involved. Han et al. (2000) synthesized poly(styrene-b-vinylphenyldimethylsilanol) polymers via anionic polymerization of silane (Si-H) containing monomers and subsequent postpolymerization oxidation with dimethyldioxirane The stronger acidity of the silanol group compared to the hydroxyl group was expected to result in greater hydrogen bonding donating capabilities. However, self-association of the silanol residues was found to play a large role in the behavior of these copolymers. Block copolymers with a range of 11–33% conversion to silanol groups were miscible with the weak hydrogen bond accepting homopolymer poly(n-butyl methacrylate) whereas higher or lower conversions led to macrophase separation. Blends of the 21% silanol block copolymer with the stronger hydrogen bond acceptor poly(4-vinylpyridine) or poly(N-vinylpyrrolidone) formed clear films in all cases and exhibited hydrogen bonding through changes in the FTIR spectrum.

4.3.2. Nucleobase Containing Hydrogen Bonding Block Copolymers

Nucleobase containing hydrogen bonding block copolymers are typically synthesized via techniques that are amenable to protic functionality of the hydrogen bonding groups such as living radical or methathesis polymerization techniques. In addition, nucleobase-functional polymers are synthesized via postpolymerization techniques, although these are typically hampered by incomplete conversion and side reactions (Pollino and Weck 2005). Nucleobase associations possess strengths in the range of $100 \, M^{-1}$ (A-T, two hydrogen bonds) to $10^4 \, M^{-1}$ (C-G, three hydrogen bonds), which places them between simple single hydrogen bonds and synthetic quadruple or higher multiple hydrogen bonding interactions (Kyogoku et al. 1967b; Thomas and Kyogoku 1967). Recent computational studies on the energy of hydrogen bonding interactions in DNA duplexes revealed values of $\sim 14 \, \text{kcal/mol}$ for the A-T pair and $\sim 27 \, \text{kcal/mol}$ for the C-G pair (Sponer et al. 2004). For adenine–uracil (A-U) hydrogen bonds in chloroform, the association energy was determined experimentally from an Arrhenius plot of the association constants, yielding a value of $6.2 \, \text{kcal/mol}$ (Kyogoku et al. 1967a). Nucleobases are complementary in nature, so that specific pairs such as adenine and thymine or uracil (A-T, A-U) or cytosine and guanine (C-G) are typically employed. Sivakova et al. (2005) noted that the behavior of the isolated nucleobases in synthetic polymers is quite different from their behavior in DNA where they bond to a complementary base. Multiple association modes are also possible for these nucleobases, including the primary Watson–Crick mode that is present in DNA and the less commonly observed Hoogsteen association mode (Ghosal and Muniyappa 2006), as well as several self-association modes. The self-association modes of typical nucleobases are extremely weak in nature and exhibit association constants of less than $10 \, M^{-1}$

(Kyogoku et al., 1967b). As in DNA, other noncovalent interactions such as $\pi-\pi$ stacking are often present with these heterocyclic bases.

In a similar fashion to DNA, synthetic nucleobase containing polymers exhibited "melting" in solution that was attributable to the unraveling of cooperatively associated chains (Lutz, Thunemann, Rurack 2005). Note, however, that the structure of the associated species was not elucidated and likely possessed little resemblance to the double helix of DNA. Lutz and colleagues observed this melting phenomenon in mixtures of random copolymers of adenine and thymine functionalized styrenic monomers, 9-(4-vinylbenzyl)adenine and 1-(vinylbenzyl) thymine, with dodecylmethacrylate (Lutz, Thunemann, Nehring 2005; Lutz, Thunemann, Rurack 2005). These copolymers organized in relatively low dielectric constant organic solvents (chloroform and dioxane) at concentrations near $3 \times 10^{-5} \, M^{-1}$. A decrease in the UV absorbance near 260 nm for the complex compared to the individual copolymers suggests the formation of stacked nucleotide base pairs. The hypochromic effect, which is also observed in DNA, increased with solution concentration because of a shift to a more associated state. The hypochromic effect diminished with heating, as a result of thermal disruption of the hydrogen bonded complex, and a characteristic "melting" temperature was measured at 325 K, which is similar to the melting of A-T DNA oligonucleotides at 325 K.

Marsh and colleagues (1999) reported on the use of ATRP for the synthesis of uridine- and adenosine-functionalized methacrylate homopolymers as well as end functionalization using uridine and adenosine functional initiators. These uridine and adenonsine monomers consisted of nucleotide bases uracil and adenine connected to the five membered deoxyribose sugars present in DNA and subsequently bonded to methacrylate groups. In their study, silyl protecting groups were employed on the hydroxyl groups of the uridine and adenosine monomers to improve their solubility in organic solvents. Narrow polydispersities were obtained; however, the molecular weights of the hydrogen bonding homopolymers were typically low ($<10\,000 \, g/mol$). The T_g values for the hydrogen bonding methacrylate homopolymers were near 140 °C, even at these low molecular weights.

Mather et al. (2007a) recently utilized nitroxide mediated polymerization from a novel difunctional initiator to synthesize adenine- and thymine-functionalized triblock copolymers. Hydrogen bonding interactions were observed for blends of the complementary nucleobase-functionalized block copolymers in terms of increased specific viscosity as well as higher scaling exponents for specific viscosity as a function of solution concentration. In the solid state, the blends exhibited evidence of a complementary A-T hard phase, which formed upon annealing, and dynamic mechanical analysis revealed higher softening temperatures. Morphological development of the block copolymers was studied using small-angle X-ray scattering (SAXS) and AFM, which revealed intermediate interdomain spacings and surface textures for the blends compared to the individual precursors.

Spijker and colleagues (2005) synthesized nucleobase-functionalized block copolymers containing thymine via ATRP of a thymine methacrylate monomer from a poly(ethylene glycol) (PEG) macroinitiator. This polymer was introduced into the polymerization of an adenine containing an alkyl methacrylate monomer in order to attempt to template the polymerization of the adenine containing monomer. However, a rate enhancement of the adenine monomer polymerization reaction was observed. Similar rate enhancements were observed with the addition of a thymine low molar mass molecule, suggesting that competitive coordination of the adenine groups to the copper ATRP catalyst occurred, which slowed the reaction rates and was disrupted because of the hydrogen bonding association to thymine.

Bazzi and Sleiman (2002) synthesized adenine-functionalized block copolymers using ring-opening methathesis polymerization (ROMP) of substituted oxonorbornene monomers (Fig. 4.3). The adenine containing block copolymers self-assembled during evaporation from dilute solutions into cylindrical shaped structures (Fig. 4.4). The ability to form these structures was attributed to the ability of adenine to hydrogen bond in two directions with neighboring adenine molecules. Surprisingly, low degrees of crystallinity in the adenine homopolymers were observed from wide-angle X-ray scattering and differential scanning calorimetry. Novel, noncovalent protecting chemistry involving the addition of complementary succinimide was utilized to improve the conversion of the adenine containing monomer.

Inaki (1992) synthesized a wide range of nucleobase-functionalized random and homopolymers. In addition, Inaki et al. (1980) synthesized block copolymers containing thymine and uracil groups in the main chain through ring-opening cationic and anionic polymerization of cyclic derivatives of the nucleobases.

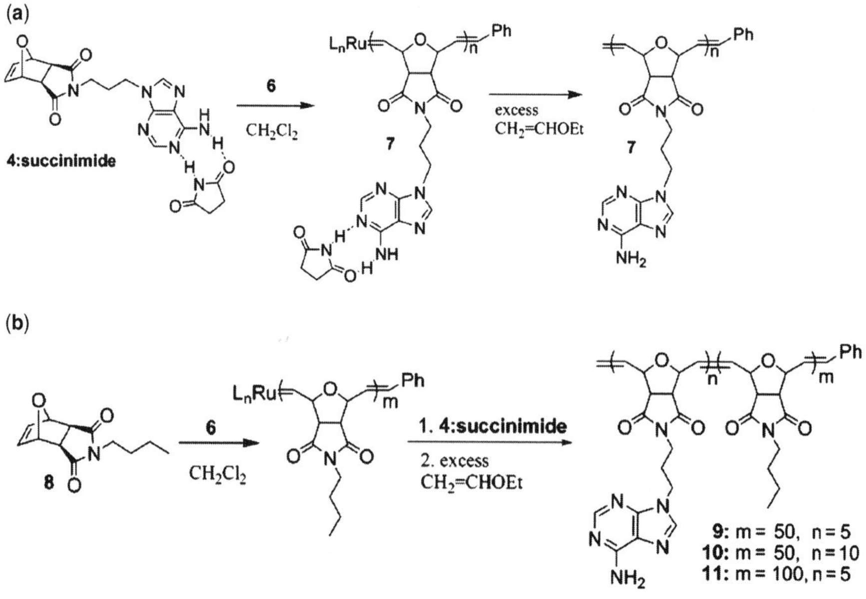

Figure 4.3 Synthesis of adenine containing block copolymers via a ROMP methodology. Reprinted from Bazzi and Sleiman (2002). Copyright 2002 American Chemical Society.

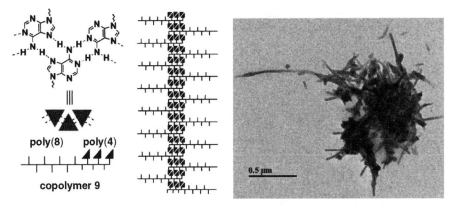

Figure 4.4 Formation of cylindrical aggregates via multidirectional self-association bonding of adenine functional block copolymers. Reprinted from Bazzi and Sleiman (2002). Copyright 2002 American Chemical Society.

Hydrogen bonding in nucleobase-functionalized block copolymers was utilized to template nanostructured materials in solution for potential drug delivery applications. Zhou et al. (2002) introduced cross-linkable and hydrogen bonding functionality in the form of poly(2-cinnamoyloxyethyl methacrylate-*co*-2-thyminylacetoxyethyl methacrylate-*b*-*tert*-butyl acrylate). Complementary, noncross-linkable poly(2-hydrocinnamoyloxyethyl methacrylate-*co*-9-adeninylacetoxyethyl methacrylate) was also synthesized. These complementary polymers were solution blended in chloroform/cyclohexane mixtures to produce micelles containing the nucleotide pairs with soluble *tert*-butyl acrylate blocks extending into solution (Zhou et al. 2002). Photocross-linking of these micelles and subsequent extraction of the adenine-functionalized polymers resulted in porous nanospheres, which absorbed adenine from solution more rapidly than nonextracted or noncross-linked micelles. Kinetics varying from first order for low molar mass porogens to zero order for higher molar mass porogens were observed, which suggested potential applications in drug delivery (Liu and Zhou 2003). In a related study, Hu and Liu produced mixed micelles from poly(2-cinnamoyloxyethyl methacrylate-*co*-2-thyminylacetox-yethyl methacrylate-*b*-*tert*-butyl acrylate) and poly(2-cinnamoyloxyethyl methacry-late-*co*-9-adeninylacetoxyethyl methacrylate-*b*-styrene) (Hu and Lui 2005; Yan et al. 2006). These mixed micelles were photocross-linked and then the *tert*-butyl acrylate blocks were hydrolyzed to produce poly(acrylic acid) residues, which phase separated from the styrene-functionalized block copolymer tails in the micellar corona, as observed with TEM. The phase separation resulted in either a two-faced appearance or flowerlike patterns in the corona.

Lehn and Rotello's groups extensively studied diacyldiaminopyridines that form triple hydrogen bonded associated structures with uracil and thymine groups (Gulik-Krzywicki et al. 1993; Carroll et al. 2003; Stubbs and Weck 2003; Sanyal et al. 2004). Ilhan et al. (2000) studied the formation of vesicles from mixtures of randomly functionalized complementary copolymers containing thymine groups and another with diacyldiaminopyridine units. The complementary flavin guest molecule was incorporated into the ~ 3 μm spherical vesicles. Bis(thymine) functional small molecules served as noncovalent cross-linkers when mixed with randomly functionalized diacyldiaminopyridine-functionalized polystyrene (Thibault et al. 2003). In chloroform solutions, mixing of these complementary units led to the formation of ~ 1 μm microgels. Temperature-dependent turbidity measurements at 700 nm showed that clearing of the solutions was obtained upon heating to 50 °C, which was due to thermal disruption of hydrogen bonding, and turbidity was reproducibly obtained upon cooling to room temperature. Sanyal et al. (2004) further demonstrated reversible attachment of diacyldiaminopyridine containing styrenic copolymers to thymine modified surfaces using X-ray photoelectron spectroscopy and a quartz crystal microbalance to measure the adsorption. The adsorbed polymer was easily removed via washing with hydrogen bond screening solvents such as chloroform/ ethanol mixtures but was retained upon washing with chloroform (Sanyal et al. 2004). Noncovalent attachment of diacyldiaminopyridine-functionalized polyoligosilsequioxanes to thymine functionalized polystyrene was also studied (Carroll et al. 2003). Stubbs and Weck (2003) also examined ROMP homopolymers of diacyldiaminopyridine- and diacyldiaminotriazine-functionalized norbornenes. Because of the propensity for self-association of these hydrogen bonding units, homopolymers precipitated from solution during polymerization. The polymers were redissolved slowly through the addition of 1-butylthymine in chloroform. To avoid precipitation during polymerization, the authors utilized noncovalent protecting chemistry with the addition of 1-butylthymine.

Block copolymers containing the diacyldiaminopyridine group were also synthesized. Starting with poly(*tert*-butyl acrylate-*b*-hydroxyethylmethacrylate), block copolymers synthesized from ATRP, Li et al. (2006) carried out postpolymerization esterification of the hydroxyl groups with a carboxylic acid containing acrylamide-functionalized diacyldiaminopyridine hydrogen bonding cross-linker. The unreacted hydroxyl groups were then reacted with methacrylic anhydride to introduce further cross-linkable groups. The block copolymers hydrogen bonded with alkyl functional thymines and uracils in nonselective solvents as observed in FTIR spectra. The introduction of a selective solvent (cyclohexane) produced micelles that were then cross-linked in solution. Extraction of the alkyl functional thymines or uracils through dialysis led to hollow nanospheres. Binding isotherms of the nanospheres with the thymine or uracil containing molecules were obtained.

4.3.3. Block Copolymers Containing DNA Oligonucleotides

Conjugates of synthetic polymers with DNA oligonucleotides were also synthesized. Noro and coworkers (2005) utilized the phosphoramidite route to incorporate nucleotide bases in a stepwise fashion from a hydroxyl containing polymer precursor. Coupling of thymidine phosphoramidite to polymeric hydroxyl groups was achieved with benzimidazoleum triflate (Fig. 4.5). Oxidation with *tert*-butyl hydroperoxide led to the formation of phosphodiester bonds. The addition of acid allowed deprotection of the remaining hydroxyl groups of the deoxyribose sugar linkage, enabling

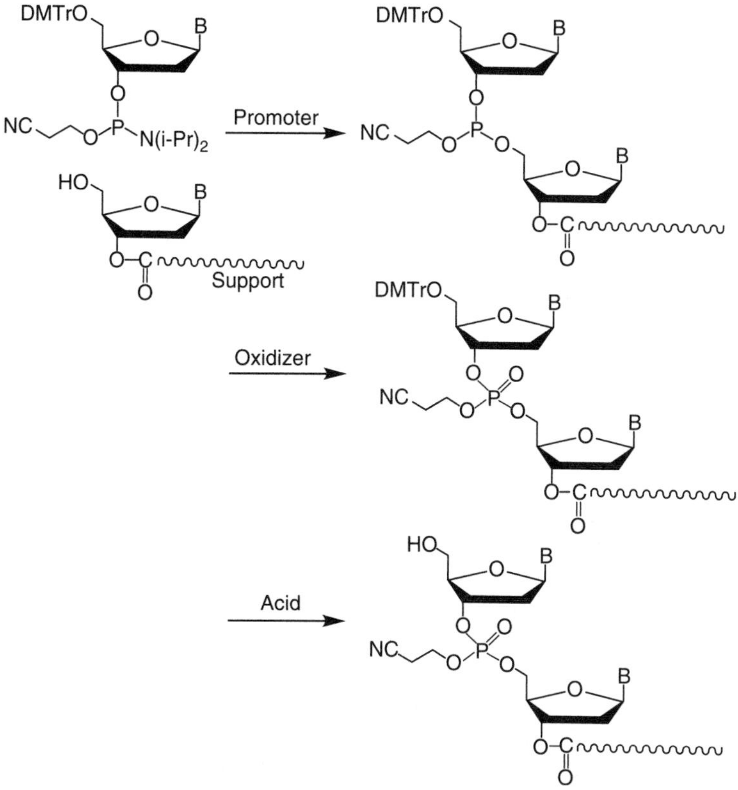

Figure 4.5 Scheme for stepwise synthesis of oligonucleotides via coupling of phosphoramidite-functionalized nucleosides. Reprinted from Noro et al. (2005). Copyright 2005 American Chemical Society.

repetition of the chemistry to generate multiple thymidine residues. Noro et al. (2005) synthesized anionic deuterated polystyrenes with five terminal thymidine units that formed a cylindrical morphology. Microphase separation was also visible for a single terminal thymidine unit; however, the exact morphological structure was not determined.

Xu and colleagues (2004) synthesized DNA oligonucleotides consisting of 8-base self-complementary sequences repeated twice in the same molecule to engender ditopic linear self-assembly. The association of these molecules resulted in a degree of polymerization of roughly 10, corresponding to a molecular weight of 182 000 g/mol at a 0.54 mM concentration in water. Injections of varying volumes of the associated polymer onto a size exclusion chromatograph resulted in decreasing apparent molecular weight for smaller injection volumes, which was consistent with dilution leading to a lower effective molecular weight. The solution viscosity exhibited a slope discontinuity from 1.8 to 3.7, indicating a transition from dilute to semidilute unentangled behavior, which is characteristic of high molecular

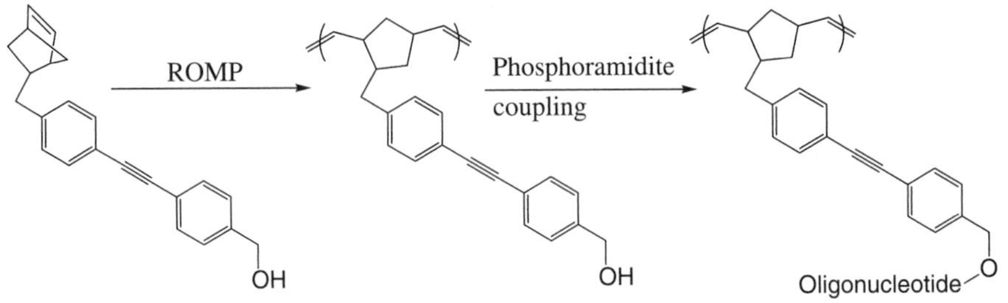

Figure 4.6 Phosphoramidite coupling onto hydroxyl-functionalized ROMP polymers (Watson et al. 2001).

weight polymers. The radius of gyration (R_g) of the associated structures scaled with $M_w^{0.7}$, indicating rodlike behavior typical of DNA. The introduction of flexible PEG spacers between the self-complementary 8-base sequences led to cyclic formation and slow reorganization into linear, high molecular weight species.

Watson et al. (2001) synthesized ROMP homopolymers and block copolymers from hydroxyl containing norbornene monomers (Fig. 4.6). The hydroxyl functional groups were phosphoramidated and attached to DNA oligonucleotides. Complementary DNA oligonucleotides containing ROMP polymers were mixed in solution and resulted in precipitation of the hydrogen bonded complex. Heating of this complex produced a melting point similar to that observed in pure DNA oligonucleotides near 75 °C. Gold nanoparticles containing DNA strands were complexed with complementary difunctional polymeric oligonucleotides, resulting in aggregation of the gold nanoparticles into a network structure as observed with TEM and surface plasmon resonance.

Peptide nucleic acids (PNAs) are a class of polymers with peptide or pseudopeptide backbones and pendant nucleotide bases. PNAs typically lack the helicity of DNA but benefit from stronger association and solubility in organic

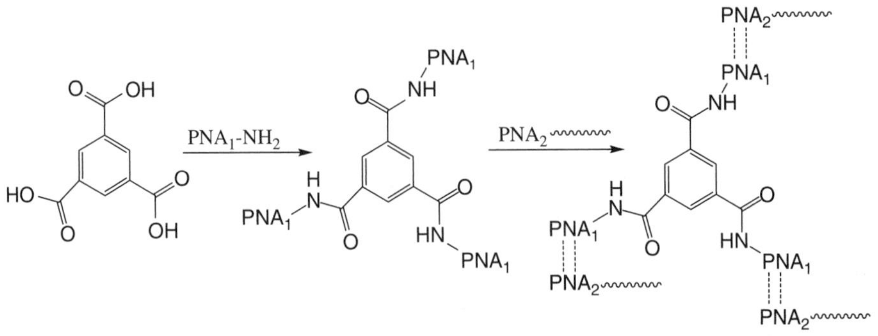

Figure 4.7 Synthesis of PNA-functionalized crosslinker molecule and association of a complementary PNA-terminated polymer synthesized by ATRP (Wang et al. 2005).

solvents. Wang and colleagues (2005) polymerized hydroxyethylacrylate from a bromofunctional PNA hexamer via ATRP (Fig. 4.7). This poly(hydroxyethyl acrylate-*b*-PNA) associated with a PNA with three complementary arms, resulting in an apparent molecular weight that was 3 times higher than the precursor. Solution phase "melting" of these PNA assemblies occurred at 30 °C. Biologically inspired hydrogen bonding block copolymers containing peptide sequences (Arimura et al. 2005) or polysaccharide blocks (Bosker et al. 2003) were also studied recently. Klok et al. (2000) studied the synthesis of poly(styrene-*b*-γ-benzyl-L-glutamate) copolymers.

4.3.4. Block Copolymers Containing Other Hydrogen Bonding Arrays

Dalphond et al. (2002) also studied hydrogen bonding ROMP block copolymers containing dicarboximide hydrogen bonding groups as the polar and hydrogen bonding block and alkylated dicarboximide residues as the hydrophobic block (Fig. 4.8). Upon the addition of water to a THF solution of the block copolymers, large (100–300 nm) spherical aggregates formed that were stained with CsOH, which deprotonated the dicarboximide groups. These polymers were capable of hydrogen bonding with adenine because of the similarity of the dicarboximide group to thymine and uracil. In another paper, Bazzi et al. (2003) described the ROMP of ABC and BAC triblock copolymers containing dicarboximide groups (A) and complementary diacyldiaminopyridine groups (C) as well as a blocked, alkylated dicarboximide monomer (B). The placement of the complementary groups on opposite ends of the block copolymer (ABC) resulted in large aggregates in chloroform solution as well as micelles, whereas placement of these complementary blocks adjacent to one another did not result in aggregate formation, suggesting a tendency toward intramolecular association.

Deans and coworkers (1999) studied diaminotriazine randomly functionalized polystyrenes, which exhibited strong intramolecular self-association leading to five-fold decreased R_g values compared to nonfunctional polystyrenes in low polarity

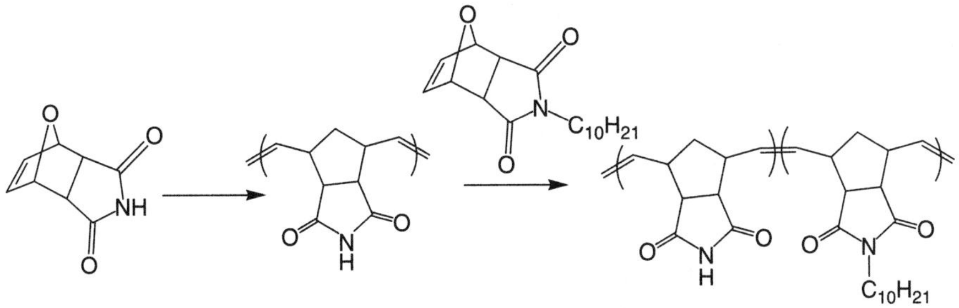

Figure 4.8 Synthesis of dicarboximide functional hydrogen bonding block copolymers via ROMP (Dalphond et al. 2002).

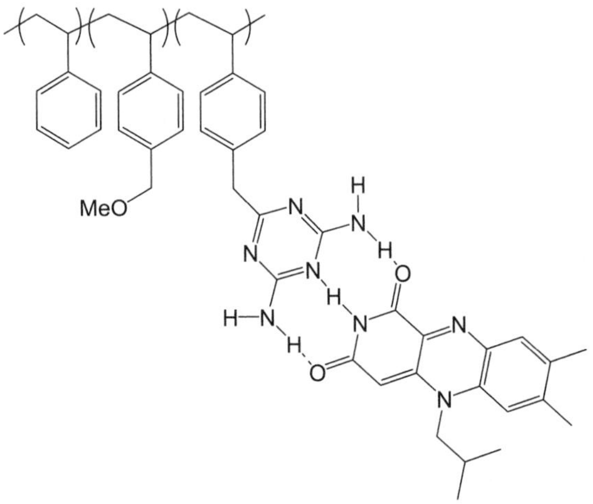

Figure 4.9   Hydrogen bonding attachment of flavin to diaminotriazine-functional polystyrene (Deans et al. 1999).

solvents such as chloroform (Fig. 4.9). The absence of acyl groups on diaminotriazine led to stronger self-association and opened more modes of self-association because of the presence of additional N-H protons. The self-association of the diaminotriazine functional polymers led to lower association constants with complementary flavin guests. Through a study of the temperature dependence of the association constants between polymeric diaminotriazine and flavin as well as monomeric diaminotriazine and flavin, entropic and enthalpic thermodynamic parameters were deduced for the polymer unfolding process required to allow the association with flavin. Based on a comparison of the enthalpy of unfolding ($-4.66\,\text{kcal/mol}$) with that of association ($7.00\,\text{kcal/mol}$), it was proposed that two hydrogen bonds were involved in the self-associated folded chain whereas three were involved in the associated complex.

4.3.5. Order–Disorder Transitions (ODTs) in Hydrogen Bonding Block Copolymers

ODTs occur when a microphase separated block copolymer melt is heated to the point that the phase separated domains begin mixing. In block copolymers containing hydrogen bonding functionality, hydrogen bonding interactions often control miscibility. Thus, the thermoreversibility of the hydrogen bonding interactions plays a role in the ODT. Yuan and colleagues (2006) studied ODTs in hydrogen bonding hydroxyl-functionalized diblock dendrons. In their work, poly(benzyl ether) and polyester containing dendrons were coupled and peripheral hydroxyl groups on the polyester dendron were subsequently deprotected (Fig. 4.10). SAXS indicated a lamellar morphology that disordered at 50 °C, which corresponded to a transition from bound to free hydroxyl absorbances in FTIR measurements (Yuan et al. 2006).

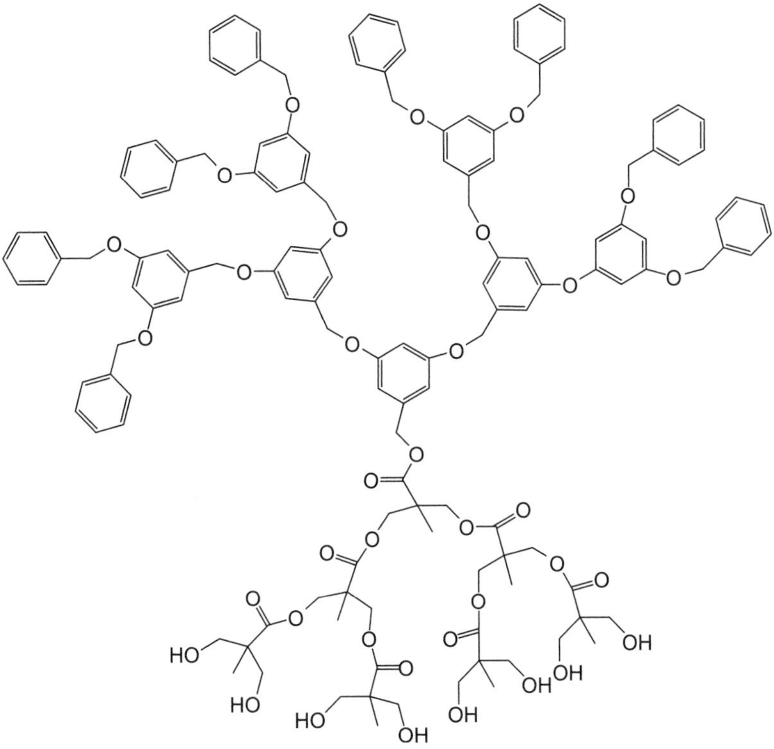

Figure 4.10 Hydrogen bonding block codendrimer synthesized by Yuan et al. (2006).

Lee and Han (2002a) studied low molecular weight poly(styrene-*b*-isoprene) diblocks synthesized anionically in THF. The high 1,2 and 3,4 enchainment produced from the polar polymerization solvent resulted in the absence of microphase separation between the blocks. Upon hydroboration–oxidation of the isoprene double bonds, hydroxyl groups were selectively introduced to the isoprene units and microphase separation was obtained. An ODT temperature near 195 °C was obtained for a poly(styrene-*b*-hydroxylated isoprene) with block molecular weights of 14 000 to 3000, and the ODT was much higher for longer hydroxylated isoprene blocks. In situ FTIR spectroscopy indicated the loss of hydrogen bonds between hydroxyl groups near the ODT that was measured using melt rheology.

4.4. TELECHELIC HYDROGEN BOND FUNCTIONAL POLYMERS

Telechelic hydrogen bonding is an alternate strategy for localized hydrogen bonding functionality. Although the ability to increase the association strength via cooperative hydrogen bonding of neighboring groups is sacrificed, more well-defined supramolecular structures are attainable. The effect of telechelic hydrogen bonding groups diminishes with increasing molecular weight; however, in low molecular weight systems even weak hydrogen bonding groups lead to significant changes in properties (Lillya et al. 1992). Recently, attention has centered on strong telechelic hydrogen bonding, because the degree of association of telechelic polymers depends on the association constant raised to the one-half power (Xu et al. 2004). Meijer (Cates 1987; Hirschberg et al. 1999; Brunsveld et al. 2001; Ligthart et al. 2005) and Long (Yamauchi, Lizotte, Hercules, et al. (2002); McKee et al. 2004) have extensively studied the self-complementary quadruple hydrogen bonding unit 2-uriedo-4[1H]-pyrimidone (UPy). This DDAA quadruple hydrogen array forms very strong self-associated dimers in solution, with association constants near $6 \times 10^7 \, M^{-1}$ in chloroform and dimer lifetimes of 170 ms in chloroform (Söntjens et al. 2000). Telechelic functionalization of molecules with UPy was shown to result in the formation of hydrogen bonded extended chain structures, where the degree of polymerization (as determined from solution viscosity) depended on the solution concentration (Cates 1987). Much like a step-growth condensation polymerization, this leads to sensitivity of the noncovalent degree of polymerization on monofunctional end cappers and the formation of cycles are observed in cases where the spacer between the UPy groups is bent (Sijbesma et al. 1997).

Some of the earliest telechelic UPy containing polymers were based on oligomeric (DP = 2, 100) poly(dimethylsiloxane) (Hirschberg et al. 1999). The degrees of association observed through melt rheological measurements ranged from 100 to 20 and decreased with increasing molecular weight. The activation energy also decreased with increasing molecular weight and was lower for a benzyl protected precursor (37 kJ/mol). A rubbery plateau was observed in melt rheological experiments for the short (DP = 2) telechelic polymer, indicating entanglements in the melt and associated structures with molecular weight above M_c. Solution rheological measurements indicated that the solution viscosity scaled with the concentration at a value of 3.9, whereas a benzyl protected analogue produced a scaling value of 1.06 (Fig. 4.11).

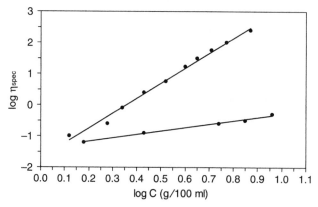

Figure 4.11   Solution viscosity of (a) telechelic UPy functional poly(dimethylsiloxane) and (b) a benzyl protected analog in chloroform at 20 °C. Reprinted from Cates (1987). Copyright 1987 American Chemical Society.

The higher exponent for the hydrogen bonded compound was consistent with the Cates model for concentration-dependent association (Cates 1987).

Lange et al. (1999) functionalized poly(ethylene oxide-*co*-propylene oxide) three-arm star oligomers with a number-average molecular weight (M_n) of 6000 g/mol (2000 g/mol arms) with UPy and urethane hydrogen bonding groups. Because of the presence of three terminal UPy units, noncovalent networks could form. Thus, Lange and cohorts observed the formation of elastic solids from the viscous liquid precursors, as well as increased solution viscosity in chloroform with increased concentration dependence due to the concentration dependence of the hydrogen bonding association. In melt rheological studies, plateaus in the storage modulus were observed, which were not present in the oligomeric precursor and resembled those expected for high molecular weight polymers. A urea-terminated analog exhibited a higher plateau modulus than the UPy containing polymer, which was attributed to the formation of stacked arrays of hydrogen bonding groups instead of simple dimers. In contrast, a covalent polyurethane network prepared with the same oligomeric precursor exhibited a lower plateau modulus than the UPy functional oligomers, which was attributed to the formation of a more thermodynamically perfect network for the UPy polymer, because of the reversibility of the hydrogen bonding interactions in comparison to the irreversible covalent bonds.

Because the UPy group preferentially associates into dimers in both the solution and solid states, the introduction of only two telechelic UPy groups leads to increased effective molecular weight without evidence of phase separation or elastic character for the low T_g precursor. In recent work, Kautz et al. (2006) placed urea and urethane groups adjacent to a telechelic UPy group, resulting in elastic behavior due to phase separation. These urethane and urea groups were responsible for lateral stacking of the chain ends, much like what Sheth and colleagues observed in model single hard segment polyurethanes and polyureas (Sheth, Klinedinst, et al. 2005; Sheth, Wilkes, et al. 2005). The lateral stacking of these telechelic hydrogen bonding groups gives rise to fibril-like structures in the surface morphology as observed from AFM, and more pronounced stacking is observed for the stronger hydrogen bonding urea functional copolymers. The presence of the UPy group increased the strain at break fourfold compared to simple telechelic urea functionality because of the noncovalent chain extension and effectively more difficult chain pullout from the hydrogen bonded fibrils.

Li et al. (2004) synthesized napthyridine (Napy) complementary hydrogen bonding arrays for UPy. These Napy groups associate to one particular tautomer of UPy with K_a values near those of the UPy homodimerization ($5 \times 10^6 \, M^{-1}$). Solution rheological studies involving mixtures of bis(UPy) and bis(Napy) low molar mass compounds indicated that solution viscosity remained high as the bis(Napy) was added to the bis(UPy), until more than one equivalent was added, upon which time the solution viscosity decreased (Ligthart et al. 2005). Recently, Scherman and coworkers (2006) functionalized a semicrystalline polyoctenamer synthesized using ROMP with telechelic Napy functionality and blended it with telechelic UPy functionalized poly(ethylene-*co*-propylene) (Fig. 4.12). This resulted in the microphase separated structures observed via AFM, whereas mixtures of

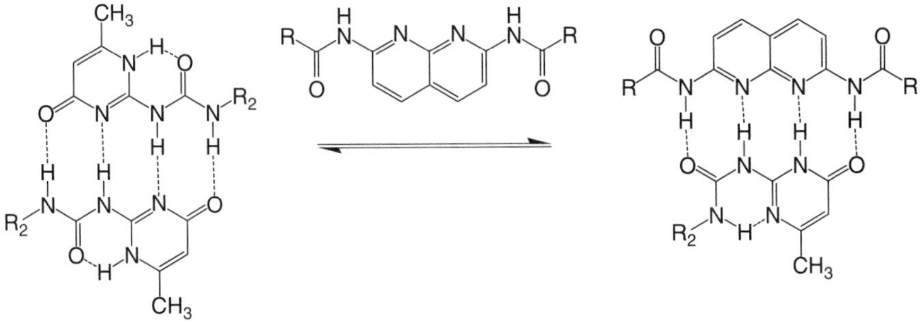

Figure 4.12 Tautomeric equilibrium of UPy dimer with Napy-UPy heterodimer (Scherman et al. 2006).

nonfunctional analogs resulted in macrophase separation. Thus, hydrogen bonding was utilized to effectively connect the blocks of a multiblock polymer.

Yamauchi, Lizotte, Hercules, et al. (2002) studied the end functionalization of polystyrene, polyisoprene, and poly(isoprene-b-styrene) with UPy groups. The introduction of the hydrogen bonding groups lead to increased T_g and melt viscosities, with a decrease in the melt viscosity at 80 °C, which was attributed to hydrogen bonding dissociation. Elkins and cohorts (2006) also studied the terminal functionalization of star-shaped and linear poly(ethylene-co-propylene)s. This resulted in dramatic increases in the melt viscosity, particularly in the case of the star-shaped polymer, and a shift of the terminal region to longer times. Further SAXS and AFM studies of telechelic and star-shaped UPy functional poly(ethylene-co-propylene)s indicated the presence of microphase separated domains of UPy groups, suggesting association beyond simple dimerization (Mather, Elkins, et al. 2007). Recently, Yamauchi et al. (2004) telechelically end-functionalized polyesters with UPy via a postpolymerization reaction with an isocyanate containing UPy. Melt rheological analysis indicated higher viscosity for the UPy containing polymer with a dramatic decrease near 80 °C to levels comparable to a nonfunctional analog. The tensile strength and elongation of the UPy containing polyester matched those of a nonfunctional polyester with twice the molecular weight; however, the melt viscosity was lower than for this high molecular weight polyester. Elkins and colleagues also studied random UPy functionalization of linear and branched methacrylates (Elkins et al. 2005; McKee et al. 2005).

Telechelic nucleobase functional polymers have also gathered attention recently. One group synthesized complementary nucleobase (thymine, adenine, and purine) end-functional polystyrenes via Michael addition reactions to terminal acrylate groups, introduced via postfunctionalization of anionic hydroxyl terminated polystyrene (Yamauchi, Lizotte, Long 2002). The hydrogen bonding groups were found to associate according to ^{1}H-NMR studies, in which chemical shift changes of hydrogen bonded protons diminished on heating to 95 °C, indicating thermoreversible association. Karikari et al. (2007) applied a similar synthetic strategy to star-shaped poly(D,L-lactide). Sivakova et al. (2005) functionalized poly(tetramethylene oxide) with acylated adenine and cytosine nucleobases. The film forming, solidlike properties of the films were attributed to the weak self-association of the adenine and cytosine groups ($<5 \, M^{-1}$), in concert with phase separation, which increased the hydrogen bonding association via concentration of the hydrogen bonding units. Rheological and X-ray scattering data studies on the adenine-functionalized polymer indicated the formation of stacks of the benzyl protected adenine units showing the presence of π stacking, which diminished on heating above 70 °C. Blockage of the hydrogen bonding properties via alkyl substitution at the hydrogen bonding (NH) protons resulted in reversion of the functional polymer to a liquid state. The protected cytosine containing polymers did not exhibit π-stacking interactions, which were likely due to the bulky aliphatic *tert*-butyl containing acyl group, but they exhibited gel-like behavior in rheological studies.

Binder et al. (2005) compatibilized blends of polyisobutylene and a bisphenol A containing poly(ether ketone) using telechelic complementary hydrogen bonding. A thymine and diaminopyridine hydrogen bonding pair with an association constant near $800 \, M^{-1}$ was employed. This resulted in microphase separation as observed from TEM and SAXS. From variable temperature SAXS studies, the blend was stable up to 250 °C, at which point the poly(ether ketone) phase reached the T_g and macrophase separation occurred. A stronger Janus wedge and barbituric acid hydrogen bonding pair with an association constant near $10^4 - 10^6 \, M^{-1}$ resulted in stability of the blend to nearly 80 °C above the poly(ether ketone) glass transition (Fig. 4.13).

Castellano and coworkers (2000) developed a calixarene with four terminal urea groups in a radial, cone-shaped geometry (Fig. 4.14). The self-complementarity of this hydrogen bonding calixarene exhibited association constants of $10^6 - 10^8$ M^{-1} and required a central solvent molecule as a space filling guest (either *o*-dichlorobenzene or chloroform). The strong hydrogen bonding of ditopic calixarenes

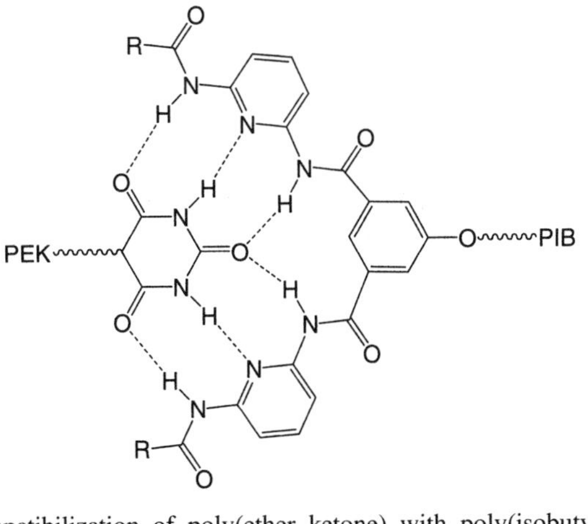

Figure 4.13 Compatibilization of poly(ether ketone) with poly(isobutylene) using Janus wedge hydrogen bonding (Binder et al. 2005).

Figure 4.14   Calixarene synthesized by Castellano et al. (2000). Reprinted from Castellano et al. (2000). Copyright 2000 National Academy of Sciences USA.

led to the observation of a C* in the solution viscosity versus concentration profile as well as the ability to draw fibers from solution. Furthermore, the introduction of a tetrafunctional calixarene resulted in noncovalent networks in which greater than 99% of the material was associated at only 5 wt% concentration. The gels behaved as solids on short time scales but flowed over longer time scales.

4.5. COMBINING HYDROGEN BONDING WITH OTHER NONCOVALENT INTERACTIONS

The combining of multiple noncovalent interactions is currently a topic of intense study. DNA, which is an exquisite example of supramolecular assembly, is composed of interactions including hydrogen bonding, electrostatic interactions (Herbert et al. 2006) due to the charged phosphodiester linkages, hydrophobic interactions, and $\pi-\pi$ stacking of the nucleobases (Vanommeslaeghe et al. 2006). This combination of noncovalent interactions enables DNA to perform functions such as replication and protein synthesis (Voet and Voet 1995). Combining multiple noncovalent assembly mechanisms results in versatility and tunable properties, particularly when the interaction mechanisms respond to different stimuli. Recently, Nair and coworkers (2006) introduced both metal–ligand and hydrogen bonding interactions into ROMP block copolymers (Fig. 4.15). The polymers consisted of blocks of palladium sulfur–carbon–sulfur (SCS) pincer complexes (which bind nitrile groups) and diacyldiaminopyridine (which binds to thymine) functionalized ROMP monomers. The presence of the metal coordinating groups did not affect the hydrogen bonding capabilities of the diacyldiaminopyridine containing block copolymers, suggesting orthogonality of the associative mechanisms.

Recently, Hofmeier et al. (2005) reported the use of both metal–ligand and UPy hydrogen bonding interactions in the main chain using a heterotelechelic poly(ε-caprolactone). Highly associated noncovalent polymers were formed upon the addition of iron salts to a poly(ε-caprolactone) containing UPy and terpyridine end groups. Sudden decreases in melt viscosity at temperatures ranging from 100 to 127 °C were attributed to the dynamics of the metal–ligand associations. Replacing iron with zinc salt decreased the temperature of this transition to 70–80 °C.

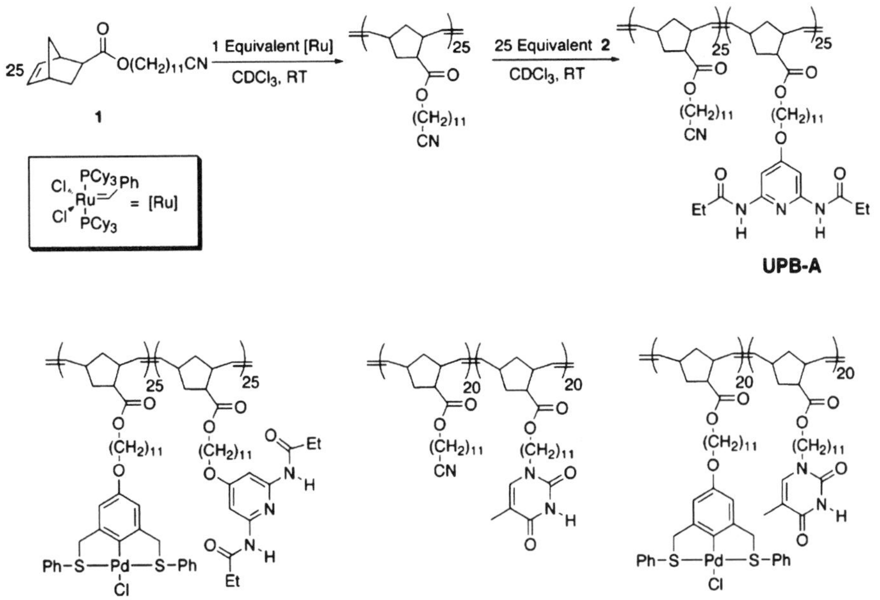

Figure 4.15 Dual side chain noncovalent interaction modes including both hydrogen bonding and metal–ligand interactions. Reprinted from Nair et al. (2006). Copyright 2006 American Chemical Society.

4.6. REVERSIBLE ATTACHMENT OF GUEST MOLECULES VIA HYDROGEN BONDING

Reversible attachment of small molecules through recognition processes in hydrogen bond containing polymers is an area of current interest. For a homopolymer blended with low molar mass hydrogen bonding molecules, novel morphologies and behaviors were observed. Ruokolainen et al. (1997) blended poly(4-vinylpyridine) with pentadecylphenol. At a 1:1 ratio of pyridine/phenol, a microphase separated texture composed of lamella was observed by SAXS in which a second diffraction maxima occurred at $2q^*$. Heating this blend resulted in a marked decrease and broadening of the SAXS maxima at 65 °C, which was attributed to a transition from a microphase separated morphology to a homogeneous system. This ODT was confirmed with rheological measurements and the storage modulus (G') exhibited a sudden decrease at 65 °C, and G' and the loss modulus (G'') began to follow scaling relationships expected for homopolymer melts ($G' \sim \omega^2$, $G'' \sim \omega$).

In the case of hydrogen bonding block copolymers, reversible attachment of guests holds promise to place guest molecules in specific domains and to influence the size and morphology of those domains. The morphology of block copolymers is classically controlled using changes in the volume fractions of various blocks, which are related to the relative molecular weights of each block. In AB diblock copolymers, this ranges from spherical dispersed phases at low volume fractions to cylinders and gyroid structures with increasing volume fraction and finally lamella at near equal volume fractions for the two monomers (Leibler 1980). The introduction of small molecule hydrogen bonding groups serves to control this volume fraction through selective incorporation into particular domains, thereby influencing the morphology. For example, Polushkin and colleagues studied blends of pentadecyl-phenol and poly(styrene-b-4-vinylpyridine) (Polushkin et al. 2001; Albderda van Ekenstein et al. 2003). In this case, two length scales were observed: a longer length scale (20–90 nm) corresponding to microphase separation between poly-styrene and poly(4-vinylpyridine) blocks and a smaller length scale (3.5 nm) corre-sponding to the lamellar structure in the poly(4-vinylpyridine)-pentadecylphenol comb domains. Mathers and colleagues (2007b) recently introduced phosphonium ionic groups with hydrogen bonding functionality to complementarily functionalized block copolymers. The block copolymers shifted from a cylindrical morpho-logy to a lamellar morphology upon introduction of the hydrogen bonding phos-phonium salt.

In block lamellar diblock copolymers, the molecular weight dependence of the long period typically follows a dependence on the molecular weight to the 0.67 power in the case of strong phase separation ($D \sim N^{0.67}$). However, in the case of poly(styrene-b-4-vinylpyridine) with pentadecylphenol an exponent of 0.8 was observed, suggesting weaker phase separation, which was attributed to the partial miscibility of the pentadecylphenol with the styrenic lamella. High shear stability of hydrogen bonded pentadecylphenol and poly(4-vinylpyridine) groups below their ODT was demonstrated at strains of 100% and 0.5 Hz frequency. In the case of poly(styrene-b-4-vinylpyridine) blends with pentadecylphenol, shearing was a useful strategy for orienting the lamella parallel to the shear direction and the hydro-gen bonded comb structure redeveloped upon cooling (Makinen et al. 2000). Shearing of blends with cylindrical morphology resulted in the orientation of the hexa-gonally packed cylinders along the shear direction. The soluble pentadecylphenol was extracted from the oriented cylinders of poly(4-vinylpyridine) using methanol, resulting in a nanoporous structure (Makki-Ontto et al. 2001).

Ten Brinke's group also studied the hydrogen bonding attachment of octyl gallate (an octyl-functionalized trihydroxybenzene) to poly(isoprene-b-vinylpyridine) block copolymers containing small 2- or 4-vinylpyridine blocks of 10–28 units (Bondzic et al. 2004). For 4-vinylpyridine block copolymers, the morphologies consisted of cylinders of 4-vinylpyridine and octyl gallate in a polyisoprene matrix, and cylinder diameters increased with octyl gallate content. This was not surprising given that the hard phase weight fraction ranged from 15.8 to 25.2 wt%. For the poly(2-vinylpyridine) blocks, the attachment led to different morphologies, depending on the amount of octyl gallate present. The morphology shifted to lamellar with increasing octyl gallate content, and the morphologies exhibited temperature dependence such that the lamellar morphology showed a decrease in lamellar spacing with temperature and a transition to a cylindrical morphology at 150 °C (Fig. 4.16). Shifting of the mor-phology and the changing sizes of the morphological features was attributed to dis-sociation of hydrogen bonding guests, and the poly(2-vinylpyridine) blocks were thought to relax from a stretched state upon the dissociation of the octyl gallate.

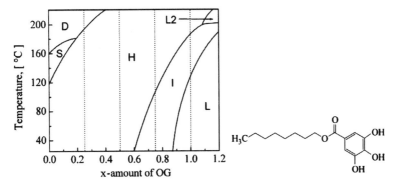

Figure 4.16   Phase diagram of poly(isoprene-*b*-2-vinylpyridine) with octyl gallate, indicating the transition between different morphologies with octyl gallate content and temperature (D, disordered; S, spherical; H, hexagonal; L, lamellar; L2, lamellar with reduced spacing; I, intermediate state). Reprinted from Bondzic et al. (2004). Copyright 2004 American Chemical Society.

FTIR revealed that hydrogen bonding decreased with heating above 100 °C, so at 190 °C almost all of the pyridine groups were not hydrogen bonded.

Chao et al. (2004) studied the reversible attachment of liquid crystalline side chains to poly(acrylic acid-*b*-styrene) block copolymers via hydrogen bonding of imidazole groups to mesogens. The reversibility of the attachment of the mesogens, which tended to form smectic phases, was proposed to play an integral role in the ability of these block copolymers to undergo reorientation of the lamella relative to external electrodes with an alternating electric field above the T_g. Comparisons with covalently bound chromophores showed that reorientation of the lamella did not occur. Further studies involving poly(methacrylic acid-*b*-styrene) with imidazole containing mesogens allowed the fabrication of temperature-dependent photonic bandgap materials that exhibited reflectivity maxima from 560 to 569 nm that reversibly changed as a function of temperature from 40 to 70 °C (Osuji et al. 2002).

Recently, ten Brinke and Ikkala's group demonstrated the attachment of ionic molecules (zinc dodecylbenzene sulfonate) to poly(styrene-*b*-4-vinylpyridine) via coordination of the zinc ions to the pyridine rings (Valkama et al. 2003). These blends possessed lamellar morphology, even at low weight fractions (0.23) of the 4-vinylpyridine and zinc dodecylbenzene sulfonate, possibly due to the strength of the ionic interactions. Strong segregation attributable to ionic interactions often leads to altered morphology in ionomers (Mani et al. 1999). Although the crystallinity of the zinc dodecylbenzene sulfonate was suppressed in the blends, lamellar packing present in the pure salt was observed in the blends, indicating a lamella within lamella structure. The zinc salt was selectively removed using methanol washes.

4.7. CONCLUSIONS AND SUMMARY

Hydrogen bonding interactions have gained significant attention recently as a means of introducing novel, thermoreversible interactions to polymers, influencing both physical property performance as well as morphology and processing behavior. The recent development of multiple hydrogen bonding groups that associate more strongly than single hydrogen bonds has led to greater control of physical and mechanical properties.

The introduction of hydrogen bonding groups in block copolymers leads to increased hydrogen bonding interactions through locally increased hydrogen bonding group concentration and cooperativity of the hydrogen bonding groups. The microphase separation in hydrogen bond containing block copolymers often exhibits greater temperature dependence, and ODTs are linked directly to the hydrogen bonding phenomenon. In the case of complementary hydrogen bonding groups, such as nucleobases, the introduction of guest molecules and complexation of complementary polymer blocks is achievable. Interesting morphologies are observed in both the solution and solid state for hydrogen bonding block copolymers.

Telechelic hydrogen bonding functionality has been the focus of great attention in recent years. Telechelic functionality results in a simpler system for studying hydrogen bonding, although microphase separation can still complicate the system in the solid state. Self-complementary quadruple hydrogen bonding UPy groups have been a strong focus in this area. Solution rheology is a particularly powerful tool for studying these systems, and both concentration and the present of "chain blocker" monofunctional hydrogen bonding groups control the effective molecular weight in solution.

REFERENCES

Aggeli A, Bell M, Boden N, Keen JN, Knowles PF, McLeish TCB, Pitkeathly M, Radford SE. Responsive gels formed by the spontaneous self-assembly of peptides into polymeric B-sheet tapes. Nature 1997;386:259–262.

Albderda van Ekenstein G, Polushkin E, Nijland H, Ikkala O, ten Brinke G. Shear alignment at two length scales: comb-shaped supramolecules self-organized as cylinders-within-lamellar hierarchy. Macromolecules 2003;36:3684–3688.

Arimura H, Ohya Y, Ouchi T. Formation of core–shell type biodegradable polymeric micelles from amphiphilic poly(aspartic acid)-*block*-polylactide diblock copolymer. Biomacromolecules 2005;6:720–725.

Asari T, Arai S, Takano A, Matsushita Y. Archimedean tiling structures from ABA/CD block copolymer blends having intermolecular association with hydrogen bonding. Macromolecules 2006;39:2232–2237.

Asari T, Matsuo S, Takano A, Matsushita Y. Three-phase hierarchical structures from AB/CD diblock copolymer blends with complemental hydrogen bonding interaction. Macromolecules 2005;38:8811–8815.

Bazzi H, Bouffard J, Sleiman H. Self-complementary ABC triblock copolymers via ring-opening metathesis polymerization. Macromolecules 2003;36:7899–7902.

Bazzi H, Sleiman H. 2002.Adenine-containing block copolymers via ring-opening metathesis polymerization: synthesis and self-assembly into rod morphologies. Macromolecules 35:9617–9620.

Beijer FH, Sijbesma RP, Kooijman H, Spek AL, Meijer EW. Strong dimerization of ureidopyrimidones via quadruple hydrogen bonding. J Am Chem Soc 1998;120:6761–6769.

Beijer FH, Sijbesma RP, Vekemans JAJM, Meijer EW, Kooijman H, Spek AL. Hydrogen-bonded complexes of diaminopyridines and diaminotriazines: opposite effect of acylation on complex stabilities. J Org Chem 1996;61:6371–6380.

Binder WH, Bernstorff S, Kluger C, Petraru L, Kunz MJ. Tunable materials from hydrogen-bonded pseudo block copolymers. Adv Mater 2005;17:2824–2828.

Bondzic S, DeWit J, Polushkin E, Schouten AJ, Ten Brinke G, Ruokolainen J, Ikkala O, Dolbnya I, Bras W. Self-assembly of supramolecules consisting of octyl gallate hydrogen bonded to polyisoprene-*block*-poly(vinylpyridine) diblock copolymers. Macromolecules 2004;37:9517–9524.

Bosker WTE, Agoston K, Cohen Stuart MA, Norde W, Timmermans JW, Slaghek TM. Synthesis and interfacial behavior of polystyrene–polysaccharide diblock copolymers. Macromolecules 2003;36:1982–1987.

Brunsveld L, Folmer BJB, Meijer EW, Sijbesma RP. Supramolecular polymers. Chem Rev 2001;101:4071–4098.

Carroll JB, Waddon AJ, Nakade H, Rotello VM. "Plug and play" polymers. Thermal and X-ray characterizations of noncovalently grafted polyhedral oligomeric silsesquioxane (POSS)–polystyrene nanocomposites. Macromolecules 2003;36:6289–6291.

Castellano RK, Clark R, Craig SL, Nuckolls C, Rebek J. Emergent mechanical properties of self-assembled polymeric capsules. Proc Natl Acad Sci USA 2000;97:12418–12421.

Cates ME. Reptation of living polymers: dynamics of entangled polymers in the presence of reversible chain-scission reactions. Macromolecules 1987;20:2289–2296.

Chao CY, Li X, Ober CK, Osuji C, Thomas EL. Orientational switching of mesogens and microdomains in hydrogen bonded side-chain liquid-crystalline block copolymers using AC electric fields. Adv Funct Mater 2004;14:364–370.

Chen D, Jiang M. Strategies for constructing polymeric micelles and hollow spheres in solution via specific intermolecular interactions. Acc Chem Res 2005;38:494–502.

Coleman MM, Graf JF, Painter PC. Specific interactions and the miscibility of polymer blends. Lancaster (PA): Technomic; 1991.

Coleman MM, Painter PC. Hydrogen bonded polymer blends. Prog Polym Sci 1995;20:1–59.

Coleman MM, Painter PC. Prediction of the phase behaviour of hydrogen-bonded polymer blends. Austral J Chem 2006;59:499–507.

Conde O, Teixerira JJ. Hydrogen bond dynamics in water studied by depolarized Rayleigh scattering. Physics 1983;44:525–529.

Dalphond J, Bazzi H, Kahrim K, Sleiman H. Synthesis and self-assembly of polymers containing dicarboximide groups by living ring-opening metathesis polymerization. Macromol Chem Phys 2002;203:1988–1994.

Deans R, Ilhan F, Rotello VM. Recognition-mediated unfolding of a self-assembled polymeric globule. Macromolecules 1999;32:4956–4960.

de Lucca Freitas L, Jacobi MM, Goncalves G. Microphase separation induced by hydrogen bonding in poly(1,4-butadiene)-*block*-poly(1,4-isoprene) diblock copolymers: an example of supramolecular organization via tandem interactions. Macromolecules 1998;31:3379–3382.

de Lucca Freitas L, Stadler R. Thermoplastic elastomers by hydrogen bonding. 3. Interrelations between molecular parameters and rheological properties. Macromolecules 1987;20: 2478–2485.

Edgecombe BD, Stein JA, Frechet JMJ, Xu Z, Kramer EJ. The role of polymer architecture in strengthening polymer—polymer interfaces: a comparison of graft, block, and random copolymers containing hydrogen-bonding moieties. Macromolecules 1998;31:1292–1304.

Elkins C, Park T, McKee MG, Long TE. Synthesis and characterization of poly(2-ethylhexyl methacrylate) copolymers containing pendant, self-complementary multiple-hydrogen-bonding sites. J Polym Sci Part A Polym Chem 2005;43:4618–4631.

Elkins C, Viswanathan K, Long TE. Synthesis and characterization of star-shaped poly(ethylene-co-propylene) polymers bearing terminal self-complementary multiple hydrogen-bonding sites. Macromolecules 2006;39:3132–3139.

Eriksson M, Notley SM, Pelton R, Wagberg L. The role of polymer compatibility in the adhesion between surfaces saturated with modified dextrans. J Colloid Interface Sci 2007;310:312–320.

Fielding L. Determination of association constants (Ka) from solution NMR data. Tetrahedron 2000;56:6151–6170.

Folmer BJB, Sijbesma RP, Kooijman H, Spek AL, Meijer EW. Cooperative dynamics in duplexes of stacked hydrogen-bonded moieties. J Am Chem Soc 1999;121:9001–9007.

Ghosal G, Muniyappa K. Hoogsteen base-pairing revisited: resolving a role in normal biological processes and human diseases. Biochem Biophys Res Commun 2006;343:1–7.

Gulik-Krzywicki T, Fouquey C, Lehn JM. Electron microscopic study of supramolecular liquid crystalline polymers formed by molecular-recognition-directed self-assembly from complementary chiral components. Proc Natl Acad Sci USA 1993;90:163–167.

Han YK, Pearce EM, Kwei TK. Poly(styrene-b-vinylphenyldimethylsilanol) and its blends with homopolymers. Macromolecules 2000;33:1321–1329.

Herbert HE, Halls MD, Hratchian HP, Raghavachari K. Hydrogen-bonding interactions in peptide nucleic acid and deoxyribonucleic acid: a comparative study. J Phys Chem B 2006;110:3336–3343.

Hirschberg JHKK, Beijer FH, van Aert HA, Magusin PCMM, Sijbesma RP, Meijer EW. Supramolecular polymers from linear telechelic siloxanes with quadruple-hydrogen-bonded units. Macromolecules 1999;32:2696–2705.

Hofmeier H, Hoogenboom R, Wouters MEL, Schubert US. High molecular weight supramolecular polymers containing both terpyridine metal complexes and ureidopyrimidinone quadruple hydrogen-bonding units in the main chain. J Am Chem Soc 2005;127: 2913–2921.

Hu J, Lui G. Chain mixing and segregation in B-C and C-D diblock copolymer micelles. Macromolecules 2005;38:8058–8065.

Huggins KE, Son S, Stupp SI. Two-dimensional supramolecular assemblies of a polydiacetylene. 1. Synthesis, structure, and third-order nonlinear optical properties. Macromolecules 1997;30:5305–5312.

Ilhan F, Galow TH, Gray M, Clavier G, Rotello VM. Giant vesicle formation through self-assembly of complementary random copolymers. J Am Chem Soc 2000;122:5895–5896.

Ilhan F, Gray M, Rotello VM. Reversible side chain modification through noncovalent interactions. "Plug and play" polymers. Macromolecules 2001;34:2597–2601.

Inaki Y. Synthetic nucleotide analogs. Prog Polym Sci 1992;17:515–570.

Inaki Y, Futagawa H, Takemoto K. Polymerization of cyclic derivatives of uracil and thymine. J Polym Sci Polym Chem Ed 1980;18:2959–2969.

Jiang S, Gopfert A, Abetz V. Novel morphologies of block copolymer blends via hydrogen bonding. Macromolecules 2003;31:6171–6177.

Jorgensen WL, Pranata J. Importance of secondary interactions in triply hydrogen bonded complexes: guanine-cytosine vs uracil-2,6-diaminopyridine. J Am Chem Soc 1990;112:2008–2010.

Karikari A, Mather BD, Long TE. Association of star-shaped poly(D,L-lactide)s containing nucleobase multiple hydrogen bonding. Biomacromolecules 2007;8:302–308.

Kato T. Supramolecular liquid-crystalline materials: molecular self-assembly and self-organization through intermolecular hydrogen bonding. Supramol Sci 1996;3:53–59.

Kautz H, van Beek DJM, Sijbesma RP, Meijer EW. Cooperative end-to-end and lateral hydrogen-bonding motifs in supramolecular thermoplastic elastomers. Macromolecules 2006;39:4265–4267.

Kawakami T, Kato T. Use of intermolecular hydrogen bonding between imidazolyl moieties and carboxylic acids for the supramolecular self-association of liquid-crystalline side-chain polymers and networks. Macromolecules 1998;31:4475–4479.

Khan A, Haddleton DM, Hannon MJ, Kukulj D, Marsh A. Hydrogen bond template-directed polymerization of protected 5′-acryloylnucleosides. Macromolecules 1999;32:6560–6564.

Klok HA, Langenwalter JF, Lecommandoux S. Self-assembly of peptide-based diblock oligomers. Macromolecules 2000;33:7819–7826.

Kriz J, Dybal J, Brus J. Cooperative hydrogen bonds of macromolecules. 2. Two-dimensional cooperativity in the binding of poly(4-vinylpyridine) to poly(4-vinylphenol). J Phys Chem B 2006;110:18338–18346.

Kugimiya A, Mukawa T, Takeuchi T. Synthesis of 5-fluorouracil-imprinted polymers with multiple hydrogen bonding interactions. Analyst 2001;126:772–774.

Kuo SW, Chan SC, Chang FC. Correlation of material and processing time scales with structure development in isotactic polypropylene crystallization. Macromolecules 2003;36:6653–6661.

Kyogoku Y, Lord RC, Rich A. An infrared study of hydrogen bonding between adenine and uracil derivatives in chloroform solution. J Am Chem Soc 1967a;89:496–504.

Kyogoku Y, Lord RC, Rich A. The effect of substituents on the hydrogen bonding of adenine and uracil derivatives. Proc Natl Acad Sci USA 1967b;57:250–257.

Lange RFM, van Gurp M, Meijer EW. Hydrogen-bonded supramolecular polymer networks. J Polym Sci Part A Polym Chem 1999;37:3657–3670.

Lee GYC, Chan SI. Tautomerism of nucleic acid bases. 11. Guanine. J Am Chem Soc 1972;94:3218–3229.

Lee KM, Han CD. Order–disorder transition induced by the hydroxylation of homogeneous polystyrene-*block*-polyisoprene copolymer. Macromolecules 2002a;35:760–769.

Lee KM, Han CD. Microphase separation transition and rheology of side-chain liquid-crystalline block copolymers. Macromolecules 2002b;35:3145–3156.

Leibler L. Theory of microphase separation in block copolymers. Macromolecules 1980;13:1602–1617.

Leibler L, Rubinstein M, Colby RH. Dynamics of reversible networks. Macromolecules 1991;24:4701–4707.

Lele AK, Mashelkar RA. Energetically crosslinked transient network (ECTN) model: implications in transient shear and elongation flows. J Non-Newtonian Fluid Mech 1998;75:99–115.

Li XQK, Jiang X, Wang XZ, Li ZT. Novel multiply hydrogen-bonded heterodimers based on heterocyclic ureas. Folding and stability. Tetrahedron 2004;60:2063–2069.

Li Z, Ding J, Day M, Tao Y. Molecularly imprinted polymeric nanospheres by diblock copolymer self-assembly. Macromolecules 2006;39:2629–2636.

Ligthart GBWL, Ohkawa H, Sijbesma RP, Meijer EW. Complementary quadruple hydrogen bonding in supramolecular copolymers. J Am Chem Soc 2005;127:810–811.

Lillya CP, Baker RJ, Huette S, Winter HH, Lin H-G, Shi J, Dickinson LC, Chien JCW. Linear chain extension through associative termini. Macromolecules 1992;25:2076–2080.

Liu G, Guan C, Xia H, Guo F, Ding X, Peng Y. Novel shape-memory polymer based on hydrogen bonding. Macromol Rapid Commun 2006;27:1100–1104.

Liu G, Zhou J. First- and zero-order kinetics of porogen release from the cross-linked cores of diblock nanospheres. Macromolecules 2003;36:5279–5284.

Liu L, Jiang M. Synthesis of novel triblock copolymers containing hydrogen-bond interaction groups via chemical modification of hydrogenated poly(styrene-*block*-butadiene-*block*-styrene). Macromolecules 1995;28:8702–8704.

Liu S, Zhu H, Zhao H, Jiang M, Wu C. Interpolymer hydrogen-bonding complexation induced micellization from polystyrene-*b*-poly(methyl methacrylate) and PS(OH) in toluene. Langmuir 2000;16:3712–3717.

Lommerse JPM, Price SL, Taylor R. Hydrogen bonding of carbonyl, ether, and ester oxygen atoms with alkanol hydroxyl groups. J Comput Chem 1997;18:757–774.

Lutz JF, Thunemann AF, Nehring R. Preparation by controlled radical polymerization and self-assembly via base-recognition of synthetic polymers bearing complementary nucleobases. J Polym Sci Part A Polym Chem 2005;43:4805–4818.

Lutz JF, Thunemann AF, Rurack K. DNA-like "melting" of adenine- and thymine-functionalized synthetic copolymers. Macromolecules 2005;38:8124–8126.

Makinen R, Ruokolainen J, Ikkala O, De Moel K, ten Brinke G, De Odorico W, Stamm M. Orientation of supramolecular self-organized polymeric nanostructures by oscillatory shear flow. Macromolecules 2000;33:3441–3446.

Makki-Ontto R, de Moel K, de Odorico W, Ruokolainen J, Stamm M, ten Brinke G, Ikkala O. "Hairy tubes": mesoporous materials containing hollow self-organized cylinders with polymer brushes at the walls. Adv Mater 2001;13:117–121.

Mani S, Weiss RA, Williams CE, Hahn SF. Microstructure of ionomers based on sulfonated block copolymers of polystyrene and poly(ethylene-*alt*-propylene). Macromolecules 1999;32:3663–3670.

Marsh A, Khan A, Haddleton DH, Hannon MJ. Atom transfer polymerization: use of uridine and adenosine derivatized monomers and initiators. Macromolecules 1999;32:8725–8731.

Mather BD, Baker MB, Beyer FL, Green MD, Berg MAG, Long TE. Supramolecular triblock copolymers containing complementary nucleobase molecular recognition. Macromolecules 2007a;40:6834–6845.

Mather BD, Baker MB, Beyer FL, Green MD, Berg MAG, Long TE. Multiple hydrogen bonding for the noncovalent attachment of ionic functionality in block copolymers. Macromolecules 2007b;40:4396–4398.

Mather BD, Elkins CL, Beyer FL, Long TE. Morphological analysis of telechelic ureidopyrimidone functional hydrogen bonding linear and star-shaped poly(ethylene-*co*-propylene)s. Macromol Rapid Commun 2007;28:1601–1606.

McKee MG, Elkins C, Long TE. Influence of self-complementary hydrogen bonding on solution rheology/electrospinning relationships. Polymer 2004;45:8705–8715.

McKee MG, Elkins CL, Park T, Long TE. Influence of random branching on multiple hydrogen bonding in poly(alkyl methacrylate)s. Macromolecules 2005;38:6015–6023.

McKeirnan RL, Heintz AM, Hsu SL, Atkins EDT, Penelle J, Gido SP. Influence of hydrogen bonding on the crystallization behavior of semicrystalline polyurethanes. Macromolecules 2002;35:6970–6974.

Müller M, Dardin A, Seidel U, Balsamo V, Ivan B, Spiess HW, Stadler R. Junction dynamics in telechelic hydrogen bonded polyisobutylene networks. Macromolecules 1996;29:2577–2583.

Müller M, Seidel U, Stadler R. Influence of hydrogen bonding on the viscoelastic properties of thermoreversible networks: analysis of the local complex dynamics. Polymer 1995;36:3143–3150.

Nair KP, Pollino JM, Weck M. Noncovalently functionalized block copolymers possessing both hydrogen bonding and metal coordination centers. Macromolecules 2006;39:931–940.

Nissan AH. H-bond dissociation in hydrogen bond dominated solids. Macromolecules 1976;9:840–850.

Noro A, Nagata Y, Tsukamoto M, Hayakawa Y, Takano A, Matsushita Y. Novel synthesis and characterization of bioconjugate block copolymers having oligonucleotides. Biomacromolecules 2005;6:2328–2333.

Norsten TB, Jeoung E, Thibault RJ, Rotello VM. Specific hydrogen-bond-mediated recognition and modification of surfaces using complementary functionalized polymers. Langmuir 2003;19:7089–7093.

Nurkeeva ZS, Mun GA, Khutoryanskiy VV, Bitekenova AB, Dubolazov AV, Esirkegenova SZ. pH Effects in the formation of interpolymer complexes between poly(*N*-vinylpyrrolidone) and poly(acrylic acid) in aqueous solutions. Eur Physical J E Soft Matter 2003;10:65–68.

Osuji C, Chao CY, Bita I, Ober CK, Thomas EL. Temperature-dependent photonic bandgap in a self-assembled hydrogen bonded liquid-crystalline block copolymer. Adv Funct Mater 2002;12:753–758.

Painter PC, Veytsman B, Kumar S, Shenoy S, Graf JF, Xu Y, Coleman MM. Intramolecular screening effects in polymer mixtures. 1. Hydrogen-bonded polymer blends. Macromolecules 1997;30:932–942.

Pan J, Chen M, Warner W, He M, Dalton L, Hogen-Esch TE. Synthesis and self-assembly of diblock copolymers through hydrogen bonding. Semiquantitative determination of binding constants. Macromolecules 2000;33:7835–7841.

Pollino JM, Weck M. Non-covalent side-chain polymers: design principles, functionalization strategies, and perspectives. Chem Soc Rev 2005;34:193–207.

Polushkin E, Alberda van Ekenstein GOR, Knaapila M, Ruokolainen J, Torkkeli M, Serimaa R, Bras W, Dolbnya I, Ikkala O, ten Brinke G. Intermediate segregation type chain length dependence of the long period of lamellar microdomain structures of supramolecular comb–coil diblocks. Macromolecules 2001;34:4917–4922.

Reith RL, Eaton FR, Coates WG. Polymerization of ureidopyrimidinone-functionalized olefins by using late-transition metal Ziegler–Natta catalysts: synthesis of thermoplastic elastomeric polyolefins. Angew Chem Int Ed 2001;40:2153–2156.

Rogovina LZ, Vasilev VG, Papkov VS, Shchegolikhina OI, Slonimskii GL, Zhdanov AA. The peculiarities of physical network formation in carboxyl-containing poly(dimethylcarbosiloxane). Macromol Symp 1995;93:135–142.

Ruokolainen J, Torkkeli M, Serimaa R, Komanschek E, ten Brinke G, Ikkala O. Order–disorder transition in comblike block copolymers obtained by hydrogen bonding between homopolymers and end-functionalized oligomers: poly(4-vinylpyridine)-pentadecylphenol. Macromolecules 1997;30:2002–2007.

Saadeh H, Wang L, Yu L. Supramolecular solid-state assemblies exhibiting electrooptic effects. J Am Chem Soc 2000;122:546–547.

Sanyal A, Norsten TB, Uzun O, Rotello VM. Adsorption/desorption of mono- and diblock copolymers on surfaces using specific hydrogen bonding interactions. Langmuir 2004;20:5958–5964.

Scherman OA, Ligthart GBWL, Ohkawa H, Sijbesma RP, Meijer EW. Olefin metathesis and quadruple hydrogen bonding: a powerful combination in multistep supramolecular synthesis. Proc Natl Acad Sci USA 2006;103:11850–11855.

Sedlak M, Simunek P, Antonietti M. Synthesis and 15N NMR characterization of 4-vinylbenzyl substituted bases of nucleic acids. J Heterocyclic Chem 2003;40:671–675.

Shandryuk GA, Kuptsov SA, Shatalova AM, Plate NA, Talroze RV. Liquid crystal H-bonded polymer networks under mechanical stress. Macromolecules 2003;36:3417–3423.

Sheth JP, Klinedinst DB, Wilkes GL, Yilgor E, Yilgor I. Role of chain symmetry and hydrogen bonding in segmented copolymers with monodisperse hard segments. Polymer 2005;46:7317–7322.

Sheth JP, Wilkes GL, Fornof AR, Long TE, Yilgor I. Probing the hard segment phase connectivity and percolation in model segmented poly(urethane urea) copolymers. Macromolecules 2005;38:5681–5685.

Sijbesma RP, Beijer FH, Brunsveld L, Folmer BJB, Hirschberg JHKK, Lange RFM, Lowe JKL, Meijer EW. Reversible polymers formed from self-complementary monomers using quadruple hydrogen bonding. Science 1997;278:1601–1604.

Sivakova S, Bohnsak DA, Mackay ME, Suwanmala P, Rowan SJ. Utilization of a combination of weak hydrogen-bonding interactions and phase segregation to yield highly thermosensitive supramolecular polymers. J Am Chem Soc 2005;127:18202–18211.

Söntjens SHM, Sijbesma RP, van Genderen MHP, Meijer EW. Stability and lifetime of quadruply hydrogen bonded 2-ureido-4[1H]-pyrimidinone dimers. J Am Chem Soc 2000;122:7487–7493.

Sotiropoulou M, Bokias G, Staikos G. Soluble hydrogen-bonding interpolymer complexes and pH-controlled thickening phenomena in water. Macromolecules 2003;36:1349–1354.

Spijker HJ, Dirks AJ, van Hest JCM. Unusual rate enhancement in the thymine assisted ATRP process of adenine monomers. Polymer 2005;46:8528–8535.

Sponer J, Jurecka P, Hobza P. Accurate interaction energies of hydrogen-bonded nucleic acid base pairs. J Am Chem Soc 2004;126:10142–10151.

Stubbs LP, Weck M. Towards a universal polymer backbone: design and synthesis of polymeric scaffolds containing terminal hydrogen-bonding recognition motifs at each repeating unit. Chem Eur J 2003;9:992–999.

Thibault RJ, Hotchkiss PJ, Gray M, Rotello VM. Thermally reversible formation of microspheres through non-covalent polymer cross-linking. J Am Chem Soc 2003;125: 11249–11252.

Thomas GJ, Kyogoku Y. Hypochromism accompanying purine–pyrimidine association interactions. J Am Chem Soc 1967;89:4170–4175.

Valkama S, Ruotsalainen T, Kosonen H, Ruokolainen J, Torkkeli M, Serimaa R, ten Brinke G, Ikkala O. Amphiphiles coordinated to block copolymers as a template for mesoporous materials. Macromolecules 2003;36:3986–3991.

Vanommeslaeghe K, Mignon P, Loverix S, Tourwe D, Geerlings P. Influence of stacking on the hydrogen bond donating potential of nucleic bases. J Chem Theory Comput 2006;2:1444–1452.

Vezenov D, Zhuk A, Whitesides G, Lieber C. Chemical force spectroscopy in heterogeneous systems: intermolecular interactions involving epoxy polymer, mixed monolayers, and polar solvents. J Am Chem Soc 2002;124:10578–10588.

Viswanathan K, Ozhalici H, Elkins CL, Heisey C, Ward TC, Long TE. Multiple hydrogen bonding for reversible polymer surface adhesion. Langmuir 2006;22:1099–1105.

Voet D, Voet JG. Biochemistry. New York: Wiley; 1995.

Wang L, Fu Y, Wang Z, Fan Y, Zhang X. Investigation into an alternating multilayer film of poly(4-vinylpyridine) and poly(acrylic acid) based on hydrogen bonding. Langmuir 1999;15:1360–1363.

Wang M, Zhang G, Chen D, Jiang M, Liu S. Noncovalently connected polymeric micelles based on a homopolymer pair in solutions. Macromolecules 2001;34:7172–7178.

Wang Y, Armitage BA, Berry GC. Reversible association of PNA-terminated poly(2-hydroxyethyl acrylate) from ATRP. Macromolecules 2005;38:5846–5848.

Watson KJ, Park SJ, Im JH, Nguyen SBT, Mirkin CA. DNA-block copolymer conjugates. J Am Chem Soc 2001;123:5592–5593.

Xu J, Fogleman EA, Craig SL. Structure and properties of DNA-based reversible polymers. Macromolecules 2004;37:1863–1870.

Xu S, Chen B, Tang T, Huang B. Syndiotactic polystyrene/thermoplastic polyurethane blends using poly(styrene-*b*-4-vinylpyridine) diblock copolymer as a compatibilizer. Polymer 1999;40:3399–3406.

Xu S, Zhao H, Tang T, Dong L, Huang B. Effect and mechanism in compatibilization of poly(styrene-*b*-2-ethyl-2-oxazoline) diblock copolymer in poly(2,6-dimethyl-1,4-phenylene oxide)/poly(ethylene-*ran*-acrylic acid) blends. Polymer 1999;40:1537–1545.

Yamauchi K, Lizotte JR, Hercules DM, Vergne MJ, Long TE. Combinations of microphase separation and terminal multiple hydrogen bonding in novel macromolecules. J Am Chem Soc 2002;124:8599–8604.

Yamauchi K, Lizotte JR, Long TE. Synthesis and characterization of novel complementary multiple-hydrogen bonded (CMHB) macromolecules via a Michael addition. Macromolecules 2002;35:8745–8750.

Yamauchi K, Lizotte JR, Long TE. Thermoreversible poly(alkyl acrylates) consisting of self-complementary multiple hydrogen bonding. Macromolecules 2003;36:1083–1088.

Yamauchi K, Kanomata A, Inoue T, Long TE. Thermoreversible polyesters consisting of multiple hydrogen bonding (MHB). Macromolecules 2004;37:3519–3522.

Yan X, Liu G, Hu J, Willson CG. Coaggregation of B-C and D-C diblock copolymers with H-bonding C blocks in block-selective solvents. Macromolecules 2006;39:1906–1912.

Yoshida E, Kunugi S. Micelle formation of nonamphiphilic diblock copolymers through non-covalent bond cross-linking. Macromolecules 2002;35:6665–6669.

Yuan F, Wang W, Yang M, Zhang X, Li J, Li H, He B, Minch B, Lieser G, Wegner G. Layered structure and order-to-disorder transition in a block codendrimer caused by intermolecular hydrogen bonds. Macromolecules 2006;39:3982–3985.

Zeng H, Miller RS, Flowers RA, Gong B. A highly stable, six-hydrogen-bonded molecular duplex. J Am Chem Soc 2000;122:2635–2644.

Zhang W, Shi L, An Y, Wu K, Gao L, Liu Z, Ma R, Meng Q, Zhao C, He B. Adsorption of poly(4-vinyl pyridine) unimers into polystyrene-*block*-poly(acrylic acid) micelles in ethanol due to hydrogen bonding. Macromolecules 2004;37:2924–2929.

Zhao JQ, Pearce EM, Kwei TK, Jeon HS, Kesani PK, Balsara NP. Micelles formed by a model hydrogen-bonding block copolymer. Macromolecules 1995;28:1972–1978.

Zhou J, Li Z, Liu G. Diblock copolymer nanospheres with porous cores. Macromolecules 2002;35:3690–3696.

CHAPTER 5

NONCOVALENT SIDE CHAIN MODIFICATION

KAMLESH P. NAIR and MARCUS WECK

5.1. INTRODUCTION

The combination of noncovalent chemistry and polymer science has led to the emergence of the relatively new field of supramolecular polymer chemistry. In the past two decades this new area of science has seen phenomenal growth that is mainly attributable to the significant advances in supramolecular and synthetic polymer chemistry. Clearly, the synergy of these two areas of science has produced a quantum leap in supramolecular polymer science. Almost endless permutations and combinations of noncovalent interactions such as hydrogen bonding, metal coordination, and coulombic bonding with a wide variety of polymeric scaffolds have been investigated with the goal to form highly functionalized complex nanoscale architectures and materials. By using single or multiple noncovalent interactions, polymer properties can be tailored precisely and easily.

In this chapter we will focus on side chain functionalized supramolecular polymers as well as main chain noncovalent functionalized polymers, which are the two main areas of supramolecular polymers. We will initially discuss the design principles and methodology of side chain functionalization, in particular, multifunctionalization. In the later part of the chapter, we will discuss in detail two important applications of side chain functionalized supramolecular polymers. The first application involves the use of noncovalent interactions to yield highly functionalized materials, whereas the second application involves the reversible noncovalent cross-linking of polymers to yield responsive materials.

Molecular Recognition and Polymers: Control of Polymer Structure and Self-Assembly.
Edited by V. Rotello and S. Thayumanavan
Copyright © 2008 John Wiley & Sons, Inc.

5.1.1. Supramolecular Polymers

Supramolecular polymers can be defined as "the formation of polymeric materials via noncovalent interactions using self-assembly" and can be categorized into main and side chain supramolecular polymers (Bosman et al. 2003). The vast majority of reports in the literature focus on main chain supramolecular polymers, materials that are held together by noncovalent interactions such as hydrogen bonding, metal coordination, or coulombic interactions (Brunsveld et al. 2001). Because the degree of polymerization and ultimately the stability of the polymer backbone are dependent upon the strength of the noncovalent interaction, main chain supramolecular polymers often utilize recognition motifs that have very strong binding efficiencies.

In contrast, side chain supramolecular polymers are polymers in which the polymer backbone is based on covalent bonds but the side chains of the polymers are noncovalently functionalized (Pollino and Weck 2005; Weck 2007). The unique advantage of noncovalent side chain functionalization is that it combines the robustness of the covalent main chain polymer with the reversibility and flexibility of noncovalent interactions; hence, it has been used in synthesizing "tailor-made" materials with controlled architectures and properties. Furthermore, because the degree of polymerization and the stability of the polymer backbone are independent of the strength of the noncovalent interaction, a vast variety of interactions ranging from weak to the strongest noncovalent interactions can be utilized in these systems. Hydrogen bonding, metal coordination, coulombic interactions, and dipole–dipole interactions are some of the most extensively employed noncovalent interactions used for side chain functionalization of polymers.

By choosing the appropriate noncovalent interaction, quantitative functionalization of polymer scaffolds can be obtained. Such a strategy offers an important advantage in the field of material design, because by simply varying the desired noncovalent functionality self-assembled along a polymeric scaffold, a single parent polymer scaffold can be transformed into a family of functionalized polymers with very different and tunable physical and chemical properties. Therefore, this strategy is capable of circumventing lengthy sequential synthetic steps based on covalent chemistry and thus has the potential to allow for easier, faster, and more efficient materials optimization.

Depending on the desired application, appropriate polymeric scaffolds can be synthesized depending upon the synthetic feasibility of the covalent scaffolds. For example, copolymers such as random, alternating, diblock, and triblock copolymers, as well as cross-linked polymeric networks, hyperbranched polymers, dendrimers, and graft polymers have been employed as polymeric scaffolds for side chain functionalization. Furthermore, functional polymer backbones such as liquid crystalline polymers (Kato et al. 1996) and biodegradable polymers (Dankers et al. 2006), in addition to polymeric macrostructures such as vesicles, aggregates, networks, and inorganic materials (such as nanoparticles) have been utilized (Carroll et al. 2002).

5.2. STRATEGIES TOWARD NONCOVALENT SIDE CHAIN FUNCTIONALIZATION OF POLYMERIC SCAFFOLDS

Side chain functionalization of covalently functionalized polymers can take place either by "prepolymerization functionalization," which is the functionality that has been bestowed to the monomer, or by "postpolymerization functionalization," which is after the polymerization, when the polymer backbone is subsequently functionalized with the desired moiety (Pollino and Weck 2005). Although both approaches have been employed successfully in covalent polymer chemistry, the first approach can be synthetically more demanding but always yields 100% functionalization, which is not the case for most postpolymerization functionalization strategies.

In supramolecular side chain functionalized polymers, the prepolymerization functionalization strategy is not available. However, using appropriate design principles and noncovalent functionalization strategies, we can tune the postpolymerization functionalization strategy from fully functionalized polymers to very weak and nonperfect functionalized polymers, depending on the application in mind. Clearly, the strength of the noncovalent interaction in the desired medium is key to this strategy. Noncovalent interactions with the full spectrum of association constants from very weak (e.g., hydrogen bonding, dipole–dipole, $\pi-\pi$ stacking, or hydrophilic interactions) to fairly strong (e.g., metal coordination or coulombic interactions) have been used to functionalize polymeric scaffolds. It is important to note that the strength of these association constants is dependent on external factors such as the temperature and solvent. When we describe the strength of a noncovalent interaction, unless otherwise noted, we report it at room temperature in a noncompetitive solvent such as a nonpolar halogenated solvent. In the following subsections we describe the most common side chain polymer functionalization strategies that are based on hydrogen bonding, metal coordination, and coulombic charge interactions. These examples of noncovalent interactions to functionalized polymeric scaffolds along the side chains are not meant to be conclusive; instead, they demonstrate the basic design principles behind the functionalization strategies and give some selected examples from the literature. Wide varieties of recognition motifs for the functionalization step are imaginable and have been used often. Thus, it is not possible to mention all of them here.

5.2.1. Side Chain Functionalization Using Hydrogen Bonding

Hydrogen bonding interactions are the most widely employed noncovalent interactions for the functionalization of polymeric scaffolds (Armstrong and Buggy 2005; Wilson 2007). Over the past 20 years, a wide variety of hydrogen bonding motifs ranging from one-, two-, three-, four-, and six-point recognition motifs to higher order systems have been developed. Figure 5.1 describes some of the more common recognition motifs described in the literature. The popularity of hydrogen bonding is because the strength of the hydrogen bonded complexes can be tuned easily by using these different hydrogen bonding motifs and by altering the acidity

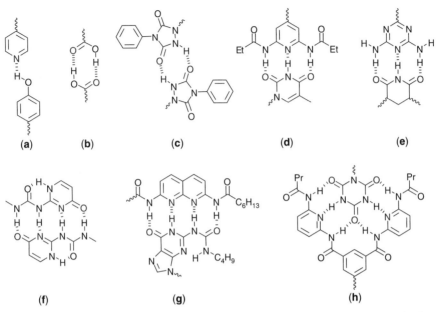

Figure 5.1 Examples of hydrogen bonding motifs used in side chain functionalizations of polymers: (a) one-point complementary, (b, c) two-point dimerizing, (d, e) three-point complementary, (f) four-point dimerizing, (g) four-point complementary, and (h) six-point complementary hydrogen bonding motifs.

and basicity of the donor (D) and acceptor (A) moieties, respectively (Cooke and Rotello 2002). By choosing the appropriate hydrogen bonding motif, binding constants as low as 1 M^{-1} for a single hydrogen bonding interaction to binding constants of more than 10^6 M^{-1} for multiple hydrogen bonded interactions can be "programmed" (Gong 2007).

Typically the hydrogen bonding interactions have a strength in solution of 10–120 kJ/mol (depending on the solvent) with an approximate range of length of 0.15–10.0 nm (Binder 2005). All hydrogen bonding motifs can be categorized into two distinct classes: 1) self-dimerizing interactions in which the hydrogen bonding motifs have a high tendency of self-complexation (e.g., self-dimerizing carboxylic acid groups, urazole groups) and 2) complementary hydrogen bonding interactions in which the recognition motifs have a complementary recognition partner with which a stable complex formation is preferentially formed over self-complexation. Examples of such complementary motifs are the recognition pair 2,6-diaminopyridine and thymine (Fig. 5.1d) as well as cyanuric acid and the Hamilton wedge receptor (Fig. 5.1 h).

In general, self-assembly using hydrogen bonding interactions is especially facile compared to other interactions requiring no prior activation as in the case of metal coordination. It consists of simply "add and stir" chemistry of the recognition unit involved. Because hydrogen bonding interactions are thermally reversible, this strategy allows the strength of the interactions as well as the thermal responsiveness of the system to be controlled to a great extent. Figure 5.2 describes some examples of side chain functionalized supramolecular scaffolds using hydrogen bonding. The importance of hydrogen bonding interactions can be judged from the vast number of examples reported in the literature ranging from small molecule self-assembly (Sijbesma and Meijer 1999) to self-assembled supramolecular polymers (Shimizu 2007; Weck 2007). The versatility and ease of self-assembly of hydrogen bonding interactions have made it a popular choice in both main chain (Shimizu 2007) and side chain supramolecular polymers (Pollino and Weck 2005).

The groups of Weck (South et al. 2007) and Sleiman (Bazzi et al. 2003; Chen et al. 2005) synthesized a variety of copolymers functionalized by either thymine or

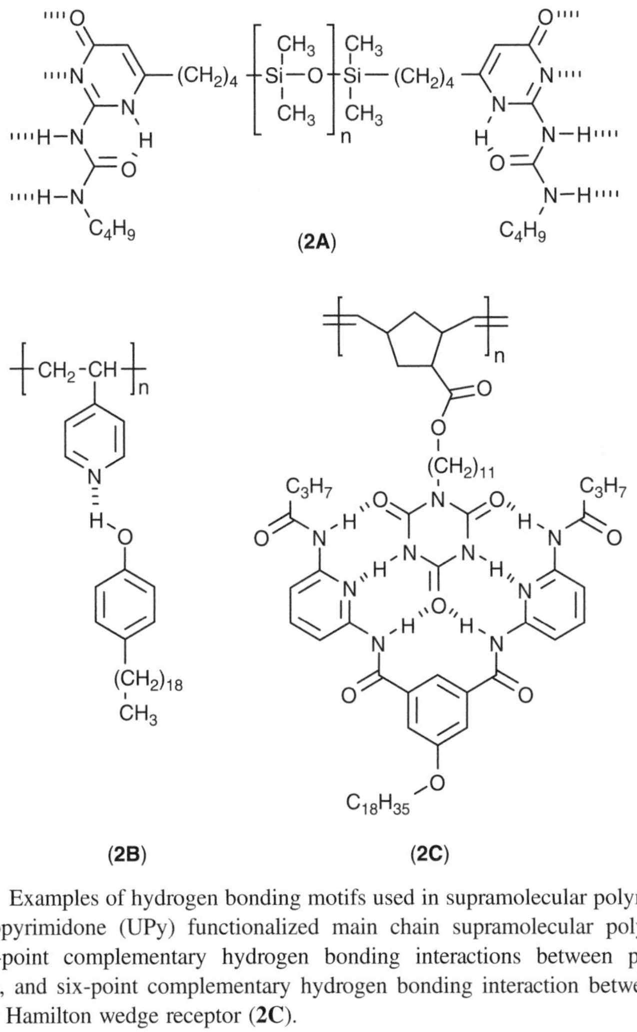

Figure 5.2 Examples of hydrogen bonding motifs used in supramolecular polymers: dimerizing ureidopyrimidone (UPy) functionalized main chain supramolecular polymers (**2A**), simple one-point complementary hydrogen bonding interactions between pyridine and phenol (**2B**), and six-point complementary hydrogen bonding interaction between cyanuric acid and the Hamilton wedge receptor (**2C**).

diaminopyridine derivatives or biologically important groups such as biotin, adenine, and so forth using ring-opening methathesis polymerization (ROMP). Sleiman's group then investigated their self-assembly into nanostructures (such as star micelles) using hydrogen bonding moieties to control the formation of the three-dimensional structures (Bazzi et al. 2003; Chen et al. 2005).

Carroll and colleagues (2003) carried out postpolymerization functionalization of polystyrene copolymers containing randomly dispersed chloromethylstyrene functional groups with 2,4-diaminotriazine or 2,6-diaminopyridine derivatives, yielding polystyrene copolymers with terminal hydrogen bonding recognition moieties along the side chains. These scaffolds were then functionalized noncovalently using hydrogen bonding with a variety of small molecules, including sesquiloxane, to form inorganic–organic hybrid materials (Carroll et al. 2003). Such a strategy of obtaining a host of different materials from the same parent polymeric scaffold just by varying the anchoring moieties was named "plug and play" polymers by the Rotello group (Carroll et al. 2003).

5.2.2. Side Chain Functionalization Using Metal Coordination

The second major class of noncovalent interactions that has been employed extensively for polymeric functionalization is metal coordination. Side chain metal functionalized polymers possess the characteristic properties of both the metal and the polymer components, giving rise to a variety of hybrid materials thereby exhibiting metal-specific properties such as conductivity and magnetism while maintaining the benefit of solubility and processability because of the polymer backbone. These metal–ligand interactions are fairly temperature insensitive compared to hydrogen bonding interactions. However, they are highly sensitive to ligand displacement reactions and are therefore considered to be chemoresponsive. Advantages of metal complexes include their highly controlled synthesis; the formation of strong noncovalent bonds in noncompeting solvents; and the potential application of metal containing polymers in areas such as supported catalysis (Yu et al. 2005), electro-optical materials (Long 1995), and chemically responsive gels (Hofmeier and Schubert 2003).

Side chain metal functionalized polymers fall into two classes according to the position of the metal complex with respect to the polymer backbone. In the first class, the metal is covalently tethered to the polymer backbone, whereas the complementary component, the ligand, is coordinated to the "polymeric scaffold." In the second class of side chain metal functionalized polymers, the ligand is covalently attached to the polymer backbone to form a "polymeric ligand species," often called a macroligand, whereas the metal center is then subsequently complexed onto the polymer. In both cases, the resultant polymers may possess identical structures and the choice of the synthetic strategy is dependent on the ease of the synthetic method: the synthesis of polymeric ligand scaffolds is more synthetically accessible compared to the polymerization of a metal containing monomer because of the limited number of polymerization methods as a result of the metal intolerance of most polymerization techniques. Although many examples of side chain metal coordinated polymers exist, only a handful are designed to serve as recognition motifs for controlled side chain functionalization. The two most widely encountered metal coordination recognition units in side chain supramolecular polymers are palladated sulfur–carbon–sulfur (SCS) pincer complexes and metal–terpyridine/bipyridine complexes.

Palladated SCS pincer complexes are an important class in coordination chemistry and have been widely used in applications ranging from catalysis (Yu et al. 2005) to anchoring units in functional materials (van Manen 2000). They consist of a metal-lated tridentate pincer ligand having a square planar coordination sphere with only one chemically accessible coordination site for self-assembly with monodentate ligands such as nitriles, pyridines, thiocyanates, or phosphines. The stability of the metal–ligand complex is in the order phosphine > pyridine > thiocyanate ~ nitrile. Pollino and Weck (2002) reported the preparation of polynorbornenes bearing palladated SCS pincer complexes at every repeat unit that could then be functionalized with pyridines or nitriles (Fig. 5.3a). They also reported the synthesis of a polynorbornene macroligand bearing a nitrile group at every repeat unit that could then be functionalized with a small molecule of SCS Pd pincer center (Fig. 5.3b; Nair et al. 2006). In all cases, these polymers were formed via living polymerization methods that allow low polydispersities and full stoichiometric control over the molecular weight.

Metal–terpyridine/bipyridine complexes are the second class of common metal–ligand interactions used in side chain functionalized polymers. The importance of these metal complexes lies in their electro-/photoluminescent properties as well as the possibility to direct self-assembly in supramolecular systems. Schubert and Eschbaumer (2002) extensively used terpyridine-based metal coordination for both side chain and main chain functionalizations. Shunmugam and Tew (2005) utilized a postpolymerization functionalization approach to covalently attach terpyridine functional groups on both random and block copolymers based on methacrylates that were copolymerized using atom transfer radical polymerization (ATRP; Fig. 5.3e). Subsequent metal complexation of the terpyridine groups resulted in copolymers generating emissive materials in the blue, green, and red regions (Shunmugam and Tew 2005). In contrast to the postpolymerization approach described by Tew and coworkers (2005), Carlise and Weck (2004) employed prepolymerization functionalization. They reported the synthesis of a norbornene monomer functiona-lized with a ruthenium–bipyridine complex that was subsequently polymerized using Grubbs' third-generation catalyst, leading to well-defined polymers with 100% functionalization (Fig. 5.3c).

5.2.3. Side Chain Functionalization Using Coulombic Interactions

Coulombic interactions are among the most widely encountered noncovalent inter-actions in polymeric systems that are rivaled only by hydrogen bonding and van der Waals interactions in their frequency (Faul and Antonietti 2003). The most important example of using coulombic interactions in materials science is in the field of "ionomers," which are tailor-made materials. Ionomers are widely used

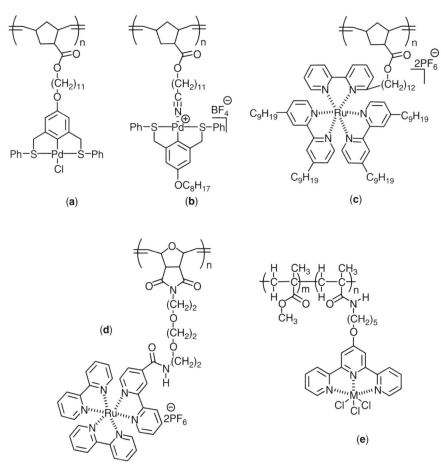

Figure 5.3 Examples of metal coordination motifs employed for side chain functionalization of polymers.

commercially because of their unique physical properties such as enhanced impact strength, toughness, and thermal reversibility (Eisenberg and Kim 1998). Other examples of coulombic interactions in side chain functionalized polymeric systems include charged block copolymers (Schaedler et al. 1998), cross-linked polymers using charge interactions (Ghosh et al. 1998), self-assembled dendrimers (Gittins and Twyman 2003), coulombically linked side chain liquid crystalline polymers, and coulombic polyamphiphiles, among others (Vuillaume and Bazuin 2003). The polymeric scaffold can either be positively or negatively charged; hence, it can be considered to be a "polyelectrolyte." The majority of the coulombically functionalized side chain polymers are based on "postpolymerization" functionalizations because of the often severe interference of the charged coulombic centers during the synthesis. More recently, with the advent of highly functional group tolerant yet mild polymerization techniques such as ROMP, there have been reports of

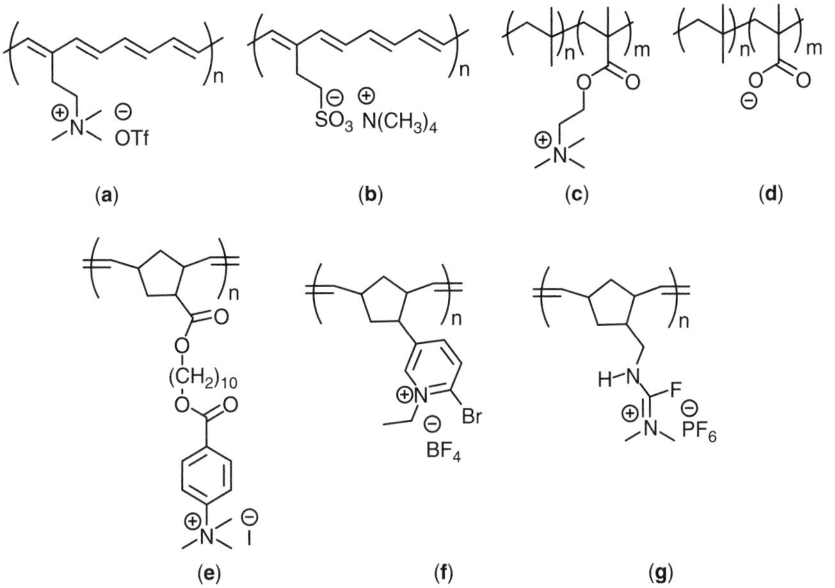

Figure 5.4 Examples of charged coulombic moieties employed for side chain functionalization of polymers.

polymerization of charged monomers. For example, Langsdorf et al. (1999, 2001) reported the ROMP of both positively and negatively charged cyclooctatetraenes using a tungsten-based Schrock catalyst (Fig. 5.4a and b).

Barrett and colleagues (2005) polymerized a positively charged norbornene monomer using Grubbs' second-generation ruthenium catalyst (Fig. 5.4f and g). Nair and Weck (2007) also polymerized positively charged norbornenes using Grubbs' third-generation catalyst (Fig. 5.4e). They reported 100% conversions in less than 1 h with the polymers exhibiting a low polydispersity index of 1.2–1.3.

Unfortunately, the polymerizations of charged monomers are mostly uncontrolled, preventing the easy formation of block copolymers. Fang and Kennedy (2002) circumvented this problem by using a postpolymerization method to introduce the charge species onto one or more blocks of block copolymers that were synthesized using ATRP (Fig. 5.4c and d).

5.3. NONCOVALENT MULTIFUNCTIONALIZATION OF THE SIDE CHAINS OF POLYMERIC SCAFFOLDS

The previous section described the supramolecular side chain functionalization of polymers based on a single recognition motif. However, biological systems use a wide variety of noncovalent interactions such as hydrogen bonding, metal coordination, and hydrophobic interactions in an orthogonal fashion to introduce function, diversity, and complexity. Although most of the concepts of supramolecular polymer science are inspired by the natural self-assembly process, the vast majority of supramolecular polymers are based solely on one molecular recognition motif. In the future, supramolecular side chain functionalized polymers for advanced applications will require multiple functionalities combined with controlled architectures. Noncovalent multifunctionalized materials will provide unique advantages such as rapid optimization via reversible functionalization to give highly advanced materials whose responsiveness to external stimuli can be tuned. Such systems will open new possibilities for the preparation of dynamic and rapidly optimized "smart" materials (Service 2005).

One strategy for the noncovalent multifunctionalization of side chain supramolecular polymers would be the controlled employment of multiple noncovalent interactions within the same polymeric system. Such a strategy should allow for the tailoring of material properties by exploiting the differences in the nature of these reversible interactions as well as multifunctionalization. An important prerequisite for the use of multiple interactions in such a system is that all noncovalent interactions have to be orthogonal to each other, or at least the effects of one interaction in the presence of another one must be clearly understood. The Weck group developed a modular polymer multifunctionalization strategy using prefabricated polymeric scaffolds possessing pendant units for hydrogen bonding, metal coordination, and coulombic interactions (South et al. 2007). They demonstrated that a family of materials derived from a single polymer backbone can be prepared rapidly and quantitatively with desirable properties via an orthogonal self-assembly approach. We now describe some of these studies in detail.

5.3.1. Combination of Hydrogen Bonding and Metal Coordination Interactions

The Weck group was the first to establish the orthogonality of hydrogen bonding interactions and metal coordination in side chain polymers (Pollino et al. 2004). The hydrogen bonding interactions they employed were based on the three-point hydrogen bonding complex between 2,6-diaminopyridine and *N*-butylthymine whereas the metal coordination was based on a palladated SCS pincer metal complex that was functionalized via coordination chemistry with pyridine or nitriles. To measure the strength of the hydrogen bonding interactions and to demonstrate the orthogonality of the two noncovalent interactions, they characterized the systems extensively using either ^{1}H-NMR spectroscopy or isothermal calorimetric titrations. The initial studies involved random polymers based on polynorbornenes. They found that the strength of the hydrogen bonding interactions was independent of the presence or absence of the metal coordinated sites. Furthermore, the studied interactions were also independent of the functionalization route used, that is, metal coordination followed by hydrogen bonding, hydrogen bonding followed by metal coordination, or all in a single step (Fig. 5.5).

In a subsequent study, Weck's group investigated the effect of the architecture of the polymer backbone on the orthogonality of these noncovalent interactions

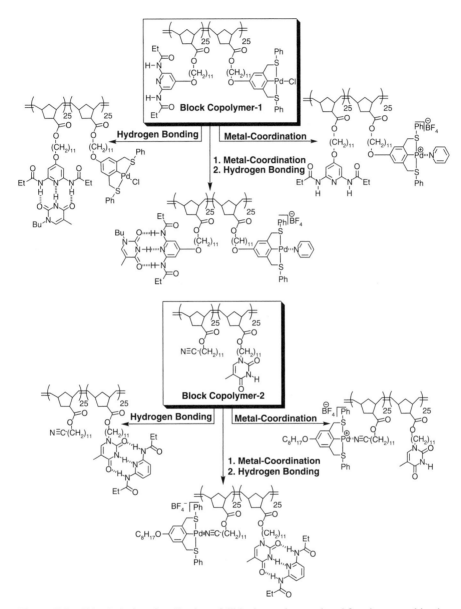

Figure 5.5 Side chain functionalization of diblock copolymers **1** and **2**, using a combination of hydrogen bonding and metal coordination.

(Nair et al. 2006). They synthesized diblock copolymers using ROMP, in which each block was functionalized by either hydrogen bonding or metal coordination motifs. For example, block copolymer **1** contains 2,6-diaminopyridine and palladated SCS pincer recognition units along the side chains of the polymeric scaffolds with *N*-butylthymine and pyridine being the complementary recognition units, respectively. Block copolymer **2** utilizes a thymine and a nitrile functional unit covalently linked to the polymer side chains with 2,6-diaminopyridine with palladated SCS pincer recognition units being used as the complementary recognition units. The Weck group demonstrated for these systems that the degree of polymerization, block copolymerization, block copolymer composition, and metal coordination have no profound effect on the strength of the hydrogen bonding interactions along the polymer backbones. This suggests that the metal coordination and hydrogen bonding are also orthogonal to each other in block copolymers.

Next, Nair et al. (2006) investigated the effect of the nature and placement of the hydrogen bonding group either covalently attached to the polymer backbone or used as a complementary recognition unit. They found that the placement of the hydrogen bonding group had a limited impact on both the strength of the hydrogen bonding interactions and the solubility of the copolymers. For example, the hydrogen bonding unit attached to the polymer backbone is switched in block copolymers **1** and **2** from 2,6-diaminopyridine for copolymer **1** to thymine for copolymer **2**. Nevertheless, both terminal hydrogen bonding units are complementary to each other and consist of identical ADA-DAD units; hence, it is expected that the association constant (K_a) values for all polymers will be similar. However, the K_a values of all block copolymers based on thymine are significantly lower (about 10–30%) than those based on 2,6-diaminopyridine, which can be attributed to the dimerization of the thymine groups attached to the polymer backbone. In contrast, block copolymers based on 2,6-diaminopyridine have a low propensity to dimerize, resulting in higher association constants.

The effect of metal placement along the copolymers was also investigated. Diblock copolymer **1** is based on the SCS palladated pincer complex whereas diblock copolymer **2** is based on nitrile groups. The nitrile functionalized polymers in essence act as a "macroligand" that can coordinate with the small molecule Pd pincer, whereas the covalent functionalization of the pincer complex onto the polymer can be seen as a "polymeric metal center" that needs to be activated prior to functionalization. When using an appropriate solvent, quantitative metal coordination took place and no interference of the hydrogen bonding moieties during the metal coordination steps were observed for either case. These studies proved that the hydrogen bonding and metal coordination are orthogonal and independent of the composition and the macromolecular architecture, whereas the placement of the functional groups covalently attached to the polymer or as a small molecule affects the strength of the hydrogen bonding interactions in a limited manner.

5.3.2. Combination of Hydrogen Bonding and Coulombic Charges

The Weck research group reported the combination of hydrogen bonding and coulombic interactions in the same system (Nair et al. 2008). Random polynorbornene-based copolymer **3** containing both terminal hydrogen bonding and coulombic sites along the polymer side chains was synthesized via ROMP. The charged coulombic sites on the polymer are quaternary ammonium iodides, whereas 2,6-diaminopyridines serve as the hydrogen bonding sites. All copolymers

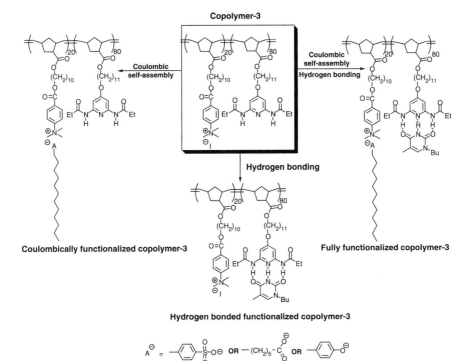

Figure 5.6 Functionalization strategies of random copolymer **3** based on a combination of hydrogen bonding and coulombic interactions.

were functionalized subsequently via self-assembly using hydrogen bonding and coulombic interactions, as shown in Figure 5.6.

The hydrogen bonding interactions between 2,6-diaminopyridine and *N*-butylthymine were studied before and after the functionalization of the quaternary ammonium groups with sulfonates, carboxylates, and phenolates. These coulombic species have one, two, or three oxygen atoms that have the potential to act as hydrogen bonding donor moieties and thus are capable of disrupting the hydrogen bonding self-assembly between 2,6-diaminopyridine and *N*-butylthymine. It was found that hydrogen bonding was independent of the nature and presence of the coulombic interactions. These results proved that the studied hydrogen bonding interactions are orthogonal to the coulombic interactions, and that both interactions can be used independently of each other in the same system for noncovalent functionalization.

5.3.3. Multiple Hydrogen Bonding Interactions: Self-Sorting on Polymers

Weck group also investigated the incorporation of multiple hydrogen bonding interactions along a single polymer backbone for both random and block copolymers.

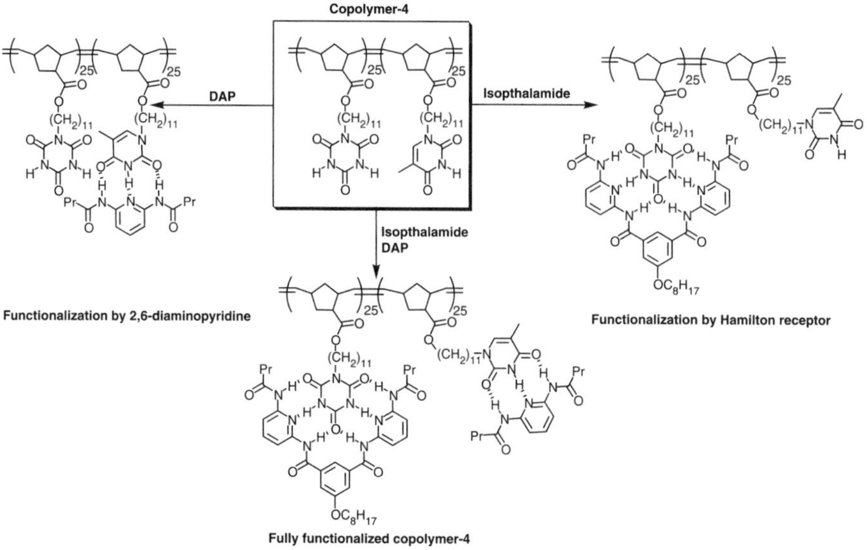

Figure 5.7   Functionalization strategies of copolymer **4** based on multiple hydrogen bonding interactions.

Copolymer **4** (Fig. 5.7) based on polynorbornene was functionalized with two different hydrogen bonding side chains based on thymine and cyanuric acid recognition groups (Burd and Weck 2005). They demonstrated that the thymine and cyanuric acid units were able to self-assemble with their complementary 2,6-diamimopyridine and Hamilton wedge moieties, respectively, even in the presence of competitive recognition sites. Thus, selective functionalization of the copolymers can be accomplished by a one-step orthogonal self-assembly approach.

The selective self-assembly of a receptor molecule with its complementary recognition unit in the presence of a competitive recognition unit has been described as self-sorting in the literature. Using the described system containing two hydrogen bonding units, Burd and Weck (2005) were able to prove the concept of self-sorting in synthetic polymers for the first time and suggest the design of complex polymeric materials containing competitive noncovalent interactions.

5.3.4. Terpolymer Functionalization Strategies: Combing Hydrogen Bonding, Metal Coordination, and Pseudorotaxane Formation

Random polynorbornene-based terpolymer **5** (Fig. 5.8) functionalized with SCS palladated pincer complexes, dibenzo[24]*crown*-8 (DB24C8) rings, and 2,6-diaminopyridine units was synthesized by ROMP (South et al. 2006). The palladium complex serves as the anchoring unit for metal coordination, 2,6-diaminopyridine as the DAD hydrogen bonding moiety, and DB24C8 as the precursor for pseudorotaxane formation, which has been studied extensively by Stoddart and his group

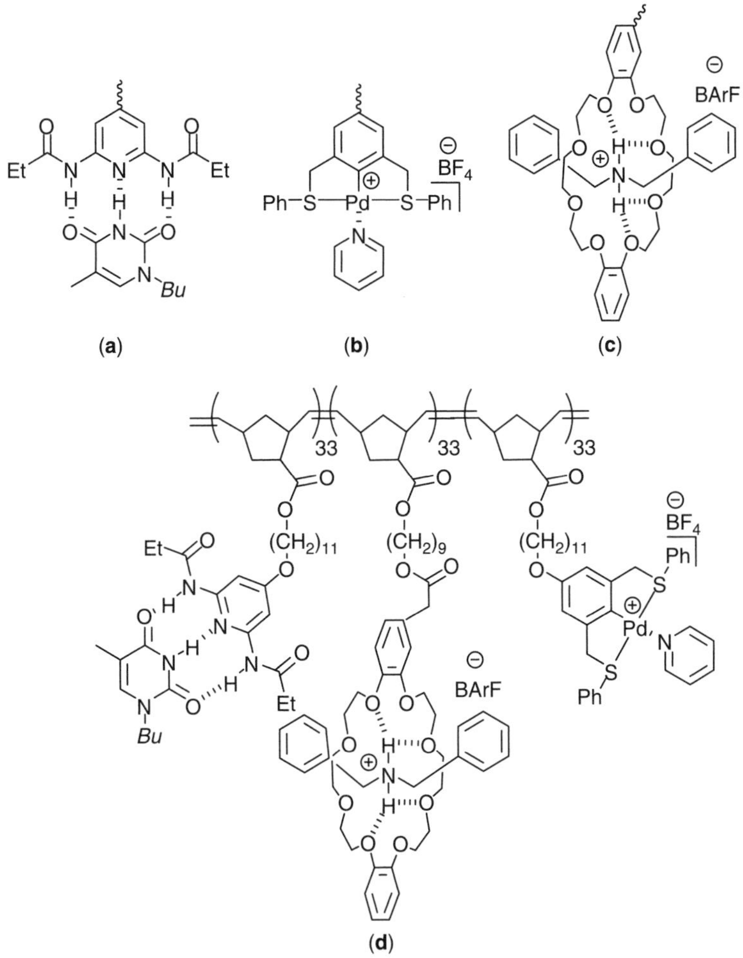

Figure 5.8   Three recognition motifs based on (a) hydrogen bonding interactions between 2,6-diaminopyridine and thymine, (b) metal coordination of sulfur–carbon–sulfur (SCS) Pd pincer with pyridine, (c) pseudorotaxane formation between dibenzo[24]*crown*-8 (DB24C8) and dibenzylammonium ions, and (d) fully functionalized terpolymer **5**.

(South et al. 2006). Side chain functionalization of these terpolymers was achieved by self-assembling 1) pyridines to the palladated pincer complexes, 2) dibenzylammonium ions to the DB24C8 rings, and 3) thymine to the 2,6-diaminopyridine receptors. By following the hydrogen bonding as well as the pseudorotaxane formation by ^{1}H-NMR spectroscopy and isothermal titration calorimetry, the Weck and Stoddart groups were able to show that the association constants were unaffected by neighboring functionalities on the polymer backbone, demonstrating for the first time orthogonality in the recognition expressed by three well-defined and discrete recognition sites.

5.4. APPLICATIONS OF NONCOVALENTLY FUNCTIONALIZED SIDE CHAIN COPOLYMERS

Having discussed self-assembly strategies toward noncovalently functionalized side chain supramolecular polymers as well as studies toward the orthogonality of using multiple noncovalent interactions in the same system, this section presents some of the potential applications of these systems as reported in the literature. The applications based on these systems can be broadly classified into two areas: 1) self-assembled functional materials and 2) functionalized reversible network formation.

5.4.1. Self-Assembled Functional Materials

The majority of applications of side chain functionalized supramolecular polymers reported in the literature can be classified as self-assembled functional materials. This strategy involves the noncovalent anchoring of functional moieties such as mesogens, fluorescent tags, bioactive molecules, metal centers, and so forth to a polymeric scaffold, resulting in the introduction of function along the polymer and the formation of self-assembled functionalized materials. Highly recognition group specific and well-defined assembly is key to this strategy, because nonspecific side chain functionalization would result in the formation of ill-defined materials. We will discuss two widely encountered self-assembled functional materials: supramolecular side chain liquid crystalline polymers (SSCLCPs) and self-assembled macrostructures.

SSCLCPs. Over a decade ago, Kumar and coworkers (1992) reported the first application of hydrogen bonded side chain functionalized polymers. Hydrogen bonding interactions were used to noncovalently attach mesogens onto polysiloxanes and polyacrylate's thereby forming thermally reversible side chain liquid crystalline polymers (Fig. 5.9, SSCLCPs **6** and **7**). They reported a variety of liquid crystalline mesophases including smectic and nematic ones through small variations of the mesogen and/or the polymeric backbone. These studies were key to the field of side chain supramolecular polymers because they demonstrated for the first time the applicability of the noncovalent functionalization strategy in materials science.

Although the majority of the SSCLCPs reported in the literature are based on a single type of noncovalent interaction, Bazuin and Sallenave (2007) reported the application of multiple noncovalent interactions in side chain supramolecular polymers to form functional materials. They used poly(pyridylpyridinium dodecyl methacrylate) bromide with terminal pyridyl groups along the side chains as hydrogen bond acceptors that were proximal to an ion pair at the polymeric scaffold to anchor a series of phenolic mesogens to the scaffold by using single-point hydrogen bonding interactions (Fig. 5.9, SSCLCP **8**). Bazuin and Sallenave showed that the hydrogen bond complexation of the mesogens onto the polymer scaffold was successful, leading to the formation of SSLCPs that exhibited well-defined thermotropic

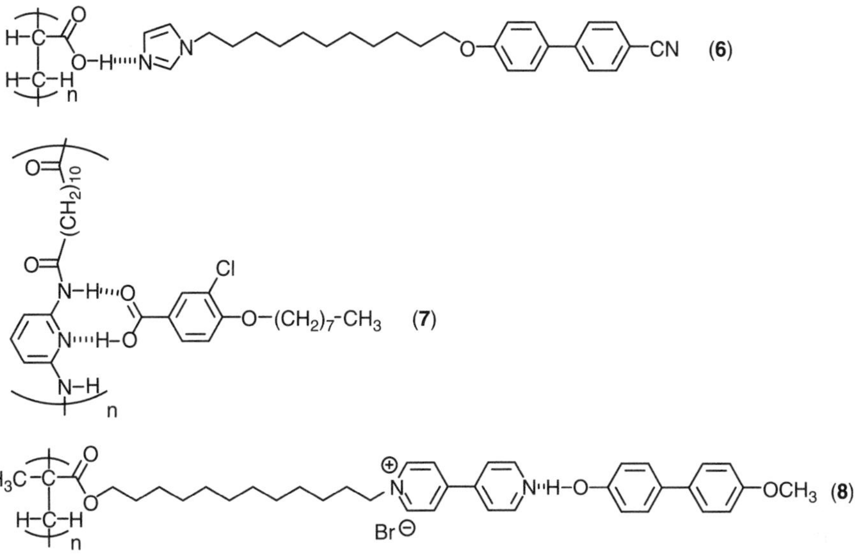

Figure 5.9 Examples of supramolecular side chain liquid crystalline polymers (SSCLCPs) based on hydrogen bonding (**6** and **7**) and coulombic interactions and hydrogen bonding (**8**).

liquid crystalline characteristics. The hydrogen bonding between the scaffold and the phenol functionalized mesogens took place selectively and completely despite the presence of potentially interfering coulombic groups. The presence of the coulombic groups resulted in relatively high glass-transition temperatures of around 80 °C in the presence of the lengthy side chains, whereas the mesogen that was created by the hydrogen bonding complex formation promoted the liquid crystalline character.

Self-Assembled Nanostructures. SSCLCPs are generally based on linear uncross-linked polymeric scaffolds, but there have been many reported examples of the utilization of side chain supramolecular polymers to serve as "scaffolds" for nanostructure formations by assembling microspheres or gold nanoparticles to supramolecular polymer side chains, such as dendronized polymers. Following this strategy, hybrid materials can be obtained by using an organic-based polymer and functionalizing it with suitable inorganic materials such as gold- or silicon-based nanoparticles. In an extension to the concept of plug and play polymers, Rotello and his group (Arumugan et al. 2007) functionalized nanoparticles with hydrogen bonding moieties and noncovalently cross-linked them with polystyrenes containing complementary hydrogen bonding sites along the side chains (Fig. 5.10). Rotello named this concept a "brick and mortar" assembly in which the nanoparticles serve as bricks and the polystyrene cross-linkers as the mortar that connects all bricks and keeps them in place (Arumugan et al. 2007).

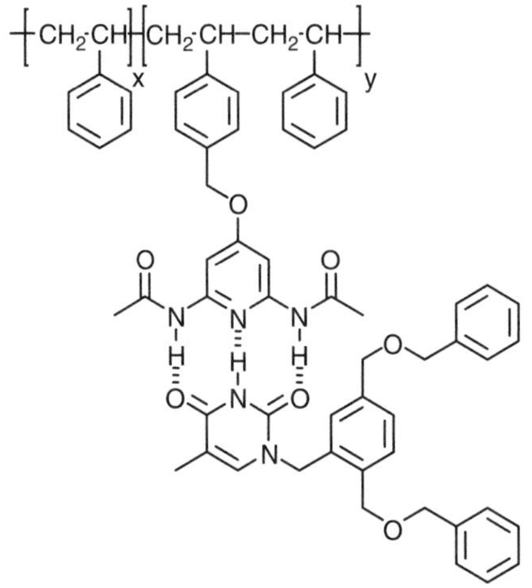

(9A) 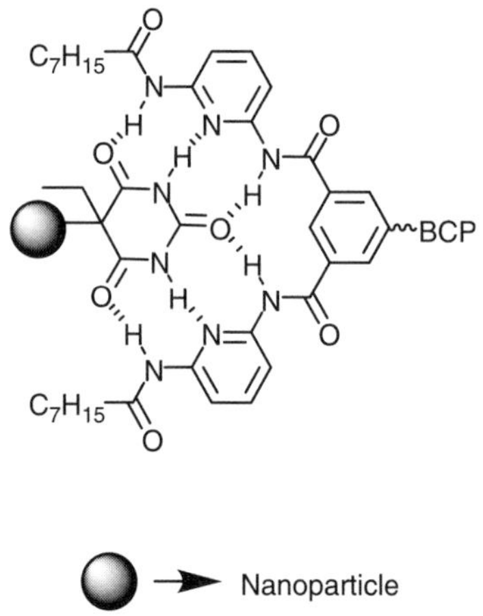   (9B)

Figure 5.10 Diverse applications using hydrogen bonding interactions employed by Rotello and coworkers (Arumugam et al. 2007): (a) self-assembled dendronized polymer and (b) inorganic–organic hybrid material using barbiturate functionalized nanoparticles and Hamilton wedge functionalized block copolymers.

5.4.2. Network Formation Using Side Chain Supramolecular Polymers

Cross-linking of polymers and the formation of elastomers is one of the most important commercial processes in polymer science. The discovery of the vulcanization of rubber by Goodyear over a century ago represented a quantum leap for polymer and material science. Covalent cross-linking of polymers is an inherently irreversible approach and poses severe disadvantages such as limited control over the final network architecture and potential side reactions such as chain degradation. Furthermore, cross-linked polymers are difficult to process and recycle, an important goal in today's society. As a result, scientists have attempted to create reversible or bidirectional cross-linking processes. Strategies that fulfill this goal are thermally reversible transformations such as the Diels–Alder reaction (Gheneim et al. 2002; Liu and Chen 2007), photochemical reversible cross-linking (Trenor et al. 2004), the use of morphological variations such as in thermoplastic elastomers, as well as topological cross-linking (Okumura and Ito 2001).

Noncovalent cross-linking of side chain supramolecular polymers using hydrogen bonding, metal coordination, coulombic interactions, or dipole–dipole interactions among others have also been employed to reversibly cross-link side chain functionalized copolymers. Noncovalent cross-linking has the advantage that it can be used to cross-link reversible polymer chains without chain degradation or other side reactions. Ionomers represent one successful class of materials that employs noncovalent interactions (Eisenberg and Kim 1998). In addition to the inherent advantages of noncovalent cross-links such as reversibility and greater control over the network architecture, noncovalent strategies allow for the tuning of materials properties because of the responsiveness of these interactions and ultimately the material toward external stimuli. For example, the employment of metal coordination for noncovalent cross-linking results in materials that are sensitive toward redox reactions or metal–ligand displacement agents, whereas the use of hydrogen bonding interactions yields thermally sensitive materials. Clearly, noncovalent cross-linking strategies offer a route toward responsive materials with tunable properties that are otherwise not accessible.

Network Formation Using Hydrogen Bonding. A variety of hydrogen bonding interactions have been employed by scientists to cross-link polymeric materials including the described one-point, two-point, three-point, four-point, and six-point hydrogen bonding donor/acceptor recognition moieties. Polymer networks based on hydrogen bonding can be broadly classified into three classes: 1) polymer networks based on self-associating hydrogen bonds, 2) polymer networks synthesized via the use of complementary linker molecules, and 3) polymer networks based on complementary polymer blends (Fig. 5.11). This chapter follows this classification to describe the polymer networks based on hydrogen bonding that are reported in the literature.

Self-Associative Polymer Network. In self-associative polymer networks (often called one component systems), the hydrogen bonding recognition units that are covalently attached to the polymer backbone have an appreciable tendency for self-association, that is, self-dimerize, which leads to interchain cross-linking of the polymers. As a result, the system is inherently cross-linked and does not require any external cross-linking agents for network formation (Fig. 5.11a). Because the cross-linking is based on dimerization phenomena, to achieve effective

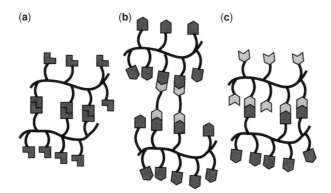

Figure 5.11 The three distinct classes of hydrogen bonded polymer networks: (a) self-associative polymer network, (b) polymer network synthesized via the use of complementary linker molecules, and (c) complementary polymer blends.

cross-linking the functional groups attached to the polymer chains must exhibit a very high tendency toward dimerization, that is, they must have a high dimerization constant. If weak interactions are employed, additional stabilizing effects such as phase separation are needed to form cross-linked three-dimensional networks (Freitas et al. 1998). Self-associative polymer networks based on the dimerization of urazole units (two-point hydrogen bonding; Freitas and Stadler 1987), 2-ureido-4[1H]-pyrimidone (UPy; four-point hydrogen bonding; Lange et al. 1999), simple carboxylic acid groups (two-point hydrogen bonding; Shandryuk et al. 2003), sulfonamide (two-point hydrogen bonding; Peng and Abetz 2005), and 1,2,4-triazole (three-point hydrogen bonding) have been reported (Fig. 5.12; Chino and Ashiura 2001).

Examples of self-associative polymer networks using hydrogen bonding include the work by Hilger and Stadler (1990) who reported reversible polymer network formation via intermolecular hydrogen bonding of (4-carboxyphenyl)urazole groups attached to the side chains of polyisobutylene. Effective cross-linking results from the two-point hydrogen bonding dimerization of the urazole moieties as well as the carboxylic acid groups. Although the individual two-point hydrogen bonding interactions are fairly weak ($K_{dimer} < 100$ M^{-1} in chloroform), the system is also based on highly ordered two-dimensional urazole clusters that phase separate from the amorphous nonpolar polymer, resulting in additional stabilization of the cross-linked polymeric network. Lange et al. (1999) reported the use of UPy, a hydrogen bonding array that dimerizes through an array of four hydrogen bonds in a self-complementary (DDAA) manner with a dimerization constant as high as 6×10^7 M^{-1} in chloroform, to cross-link polymers. Similar approaches using UPy as the cross-linking directing group have been reported by Elkins and colleagues (2005) who functionalized poly(methyl methacrylate) using UPy to form a self-associative polymer network (Fig. 5.13, **10**). Furthermore, Rieth et al. (2001) copolymerized UPy functionalized monomers with 1-hexene to form polyolefin-based elastomers.

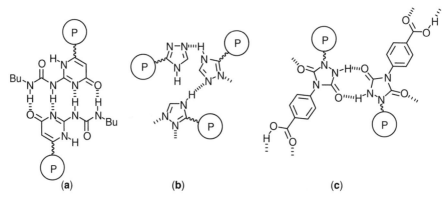

Figure 5.12 Self-associative hydrogen bonded polymer networks based on (a) 2-ureido-4[1H]-pyrimidone (UPy), (b) 1,2,4-triazine, and (c) urazole carboxylic acid.

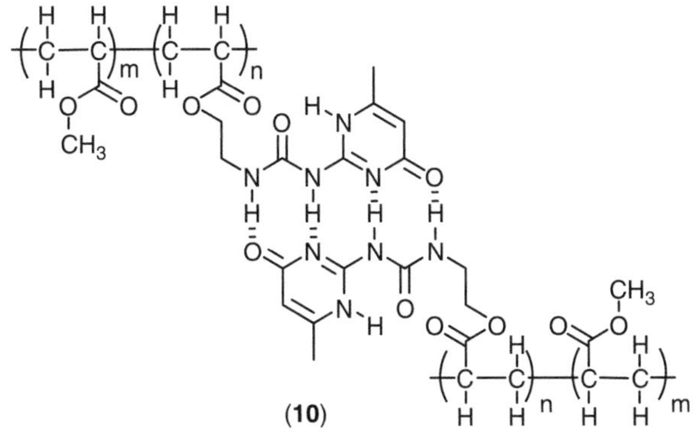

Figure 5.13 Self-associative polymer network formed via the dimerization of 2-ureido-4[1H]-pyrimidone groups attached to the polymer backbone.

Although self-complementary polymer networks clearly show distinct advantages over conventional covalent cross-linked polymers, a limitation of this system is that the system always remains "cross-linked" and would exhibit "uncross-linked" behavior only at temperatures above the hydrogen bond dissociation temperature. Because this system relies on the dimerization of the functional groups that are covalently attached to the polymer chains, the system displays some of the same limitations as covalent ones. As a result, to tune network properties, such as the degree of cross-linking or the cross-linking density, new generations of optimized polymeric material have to be redesigned and resynthesized. Therefore, this strategy does not allow the full exploitation of the advantages of supramolecular self-assembly.

Polymer Networks Based on the Use of Complementary Linkers. In polymeric networks that are based on the employment of complementary linkers (a two component system), the hydrogen bonding recognition units attached to the polymer chains undergo little or no self-association and hence cannot effectively "cross-link" the polymer chains. Such a system represents an "open" system. An external chemical agent or a "cross-linking agent" has to be added that is able to undergo hydrogen bonding with the recognition groups attached to the polymer side chains, resulting in the effective cross-linking of the polymer chains through interchain hydrogen bonding (Fig. 5.14b). In such a system, a polymer can be converted from being a completely "uncross-linked material," that is, no cross-linking agent present, to a completely "cross-linked system" with the addition of exactly 1 equiv of cross-linking agent based on the recognition motifs along the polymer backbone. Changes of the stoichiometry of the recognition site/linker ratio can be used to tailor the physical properties of the resulting polymeric network and therefore obtain a range of materials ranging from highly viscous fluids to highly viscoelastic solids, all resulting from the same parent polymer.

The network strength of these systems can also be altered by varying the stability of the hydrogen bonded complex formation between the cross-linking agent and the

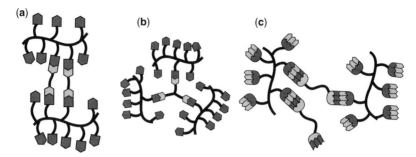

Figure 5.14 Schematic representation of polymer network formation based on the addition of complementary linkers using (a) single hydrogen bond formation and bifunctional cross-linking agents, (b) single hydrogen bond formation and trifunctional cross-linking agents, and (c) three-point hydrogen bond formation and bifunctional cross-linking agents.

functional groups attached to the polymer chains. For example, the Weck research group demonstrated that networks based on three-point hydrogen bonding complex formation between 2,6-diaminopyridine and thymine ($K_a \sim 10^3$ M^{-1}) exhibit lower solution viscosities in chloroform compared to similar systems in which the network is based on a stronger six-point complex formation between cyanuric acid and isophthalamide wedge receptors ($K_a \sim 10^6$ M^{-1}; Pollino et al. 2004).

Another important factor in tuning the network properties is the molecular architecture of the cross-linking agent; by varying the functionality of the cross-linking we can control the cross-linking density of the network. As a result, the addition of 1 equiv (based on recognition sites NOT the linker to polymer ratio) of a tetrafunctional cross-linking agent will result in greater cross-linking efficiency compared to the addition of 2 equiv of a difunctional cross-linking agent. These advantages of the two component cross-linking systems allow for tunability of network properties when compared to the one component system just described. However, the number of reports in the literature using such a two component system is limited when compared to the one component systems.

Kawakami and Kato cross-linked an SSCLCP with side chain carboxylic acid groups by using bis-pyridine (Kato 1996) or bis-imidazoyl compound as the cross-linking agent (Fig. 5.15; Kawakami and Kato 1998). The addition of the bis-pyridine caused interchain cross-linking through the one-point hydrogen bonding of the side chain carboxylic groups with the bis-pyridine, resulting in cross-linking and increased mesophase stability of the liquid crystalline state.

As explained earlier, the role of the cross-linking agent is key in two component systems and affects the properties of the final network. The work of Thibault et al. (2003) illustrates the importance of the cross-linking agent. They employed bis-thymine based cross-linking agents with different linker lengths to reversibly cross-link 2,6-diaminopyridine functionalized copolymers (Fig. 5.16) that formed discrete micron-sized spherical polymeric aggregates (Thibault et al. 2003).

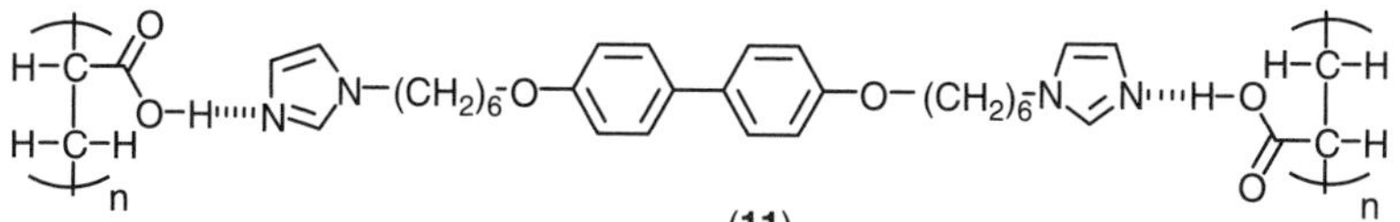

(11)

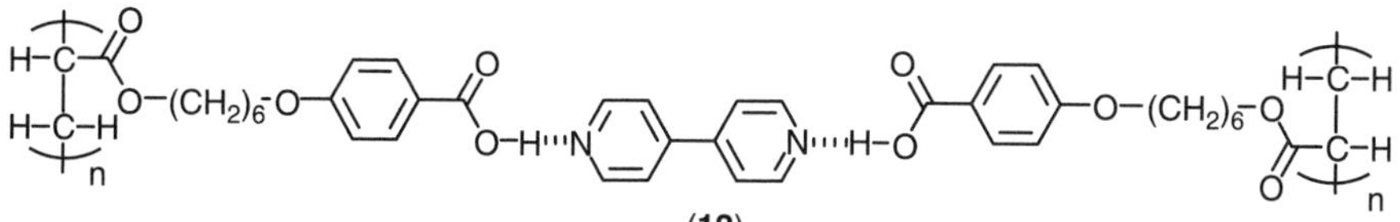

(12)

Figure 5.15 Examples of linker cross-linked polymer networks exhibiting liquid crystalline polymer characteristics.

They demonstrated that the linker length influenced the median diameter of the spherical aggregate that was formed, resulting in good control over the aggregate dimension.

Nair and Weck (2007) extended this approach by using a recognition motif attached to the polymer that is able to form two different and potentially competing hydrogen bonding motifs. This strategy has the potential to tailor material properties as well as the degree and strength of the cross-linking in a highly controllable fashion. The Weck system is based on a polynorbornene backbone that was functionalized with cyanuric acid groups (poly-CA) along the side chains. The cyanuric acid side chains then could be noncovalently cross-linked with linker molecules by hydrogen bonding interactions in two distinct ways: with three-point cyanuric acid–diaminotriazine or six-point cyanuric acid–Hamilton wedge interactions as shown in Figure 5.17.

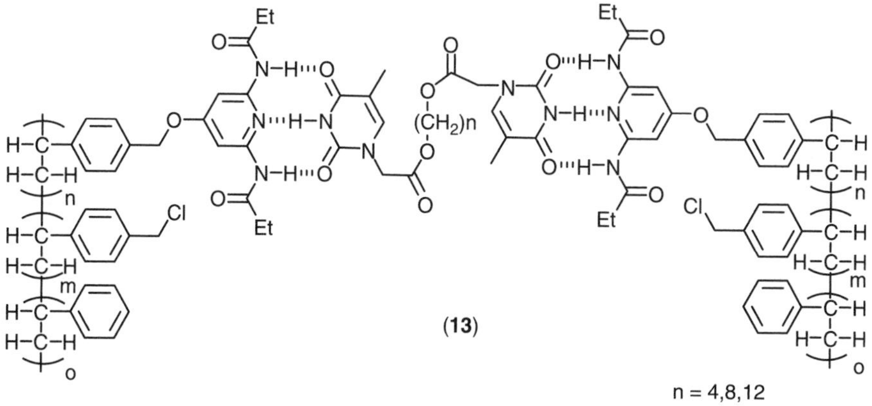

(13)

n = 4,8,12

Figure 5.16 2,6-Diaminopyridine side chain functionalized polystyrene cross-linked via bis-thymine cross-linking agents.

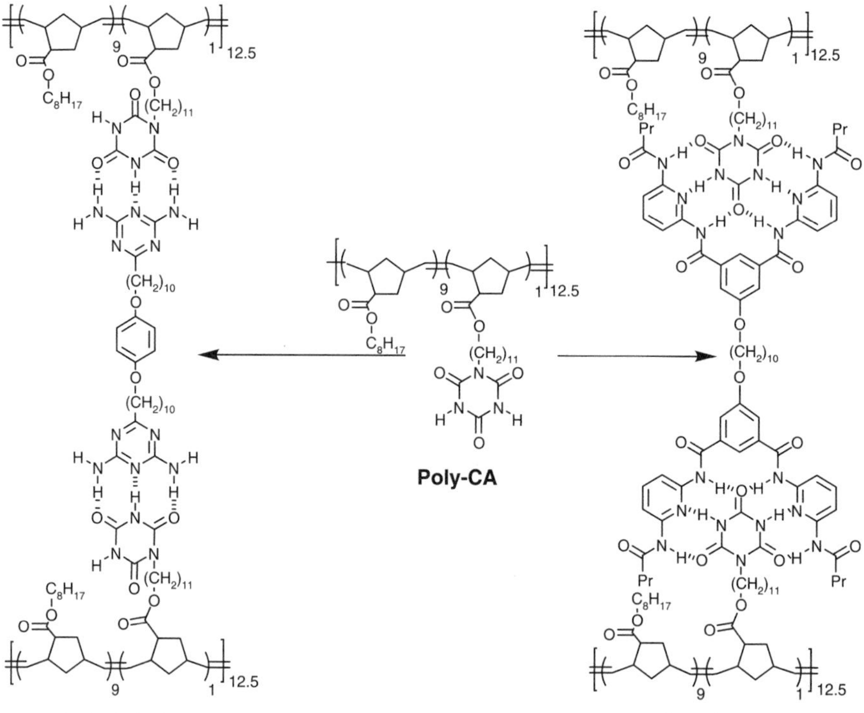

Poly-CA-triazine   **Poly-CA-wedge**

Figure 5.17   Noncovalent cross-linking strategies of cyanuric acid functionalized poly-CA to form (a) poly-CA-triazine by using a diaminotriazine cross-linking agent or (b) poly-CA-wedge by using an isophthalamide cross-linking agent.

These two cross-linking strategies of poly-CA with the two cross-linking agents have drastically different properties. The elastic and loss moduli of the network poly-CA-triazine as a function of cross-linking agent concentration are provided in Figure 5.18. The graph depicting the cross-linking profile can be divided into two regions: a viscous liquid below 60% linker concentration and a predominantly highly viscoelastic solid behavior above 60% linker concentration. The regime at 60% linker concentration can be termed as the cross-over region where the elastic modulus becomes greater than the loss modulus and highly elastic solids are obtained.

In contrast to the triazine-based cross-linking agent, the addition of the wedge cross-linking agent to poly-CA did not result in the formation of a viscoelastic solid. Although the wedge-based cross-linking agent is an efficient cross-linking agent, increasing the loss modulus of the network even more than the triazine-based cross-linking agent, it never creates a viscoelastic network. Clearly, the markedly different natures of these two cross-linked networks have to be attributed to the difference in the microstructures of the networks. The triazine-based cross-linking agent is able to form a true three-dimensional network via multiple hydrogen

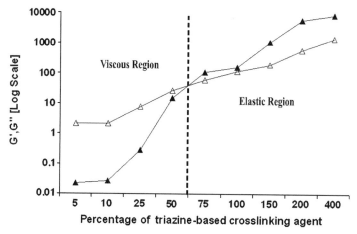

Figure 5.18 Cross-linking profile of poly-CA using the triazine-based cross-linking agent. Filled symbols denote the elastic modulus (G') whereas empty symbols denote the loss modulus (G'') at a strain value of 0.1. The percentage of the triazine-based cross-linking agent is based on the cyanuric acid groups attached to the polymer.

bonding interactions with the cyanuric acid groups; thus, each cross-linking agent is able to bind more than two cyanuric acid groups. However, the wedge-based cross-linking agent is able to connect only two cyanuric acid groups and forms no true three-dimensional network, having a continuum. As a result, the network microstructures of the two polymer networks are markedly different. Hence, when using complementary hydrogen bonding networks, by choosing the appropriate cross-linking agent we can obtain a high degree of control over important network parameters such as the cross-linking density, strength of the cross-links, network microstructure, and extent of polymer cross-linking.

Polymer Networks Based on Polymer Blends. The physical combination of two or more chemically different polymers to yield a hybrid material possessing the desirable properties of all the combined polymers is a well-established part of materials science. However, serious challenges, such as the inherent immiscibility of different polymers leading to phase separated materials, have to be overcome for materials applications. Many strategies have been used to minimize interfacial energy and to reduce the propensity for phase separation, including the use of compatibilizers and introduction of reactive groups to covalently connect individual polymers within the blend. One strategy to overcome microphase separation is the use of hydrogen bonding interactions. The polymers that need to be blended can be functionalized with complementary hydrogen bonding functional groups and, when blended either in solution or the melt, can undergo interchain hydrogen bonding interactions between the two inherently immiscible polymers, thereby suppressing phase separation and forming a homogeneous polymer blend (Fig. 5.14c).

Most systems described in the literature rely on fairly weak hydrogen bonding complexes based mainly on one- or two-point interactions between functional

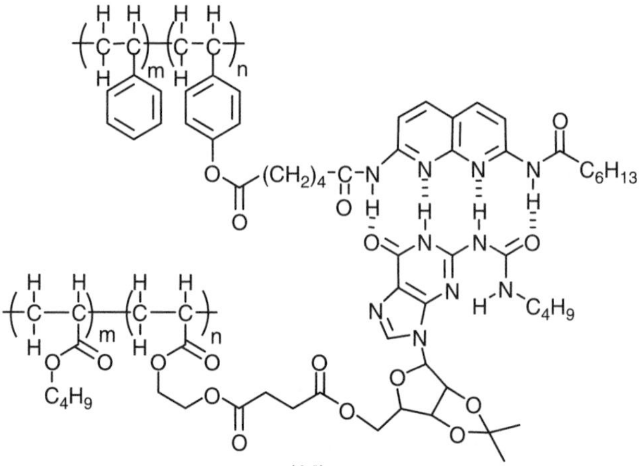

(14)

Figure 5.19 Homogenous polymer blend of 2,7-diamido-1,8-naphthyridine (DAN) functionalized polystyrene and urea of guanosine (UG) functionalized poly(butyl methacrylate) (**14**), based on the four-point complementary complex formation between DAN and UG.

groups such as hydroxyl, carboxyl, pyridyl, and amino groups. As a result of the weakness (low association strength) of the resulting complexes, to achieve homogenous blend formation high mole percentages of the hydrogen bonding functional groups are required. Unfortunately, this often results in materials with undesirable properties such as hygroscopisity or high frictional coefficients. Park and Zimmerman (2006) reported a system that has the potential to overcome these shortcomings. They employed a four-point hydrogen bonding system between urea of guanosine (UG) and 2,7-diamido-1,8-naphthyridine (DAN), which has a $K_a \sim 5 \times 10^7 \text{ M}^{-1}$, to blend two immiscible polymers such as polystyrene and poly(butyl methacrylate) (Fig. 5.19). They demonstrated that a mixture of polystyrene and poly(butyl methacrylate) functionalized with DAN and UG, respectively, formed a homogeneous blend (Fig. 5.19, **14**) with no evidence of phase separation, even at a low concentration of the hydrogen bonding functional groups.

Network Formation Using Metal Coordination. The second class of noncovalent interactions that has been employed in polymer cross-linking is metal coordination. Metal coordination has a number of advantages over hydrogen bonding. First, metal coordination is among the strongest noncovalent interaction used in self-assembly in a variety of solvents (such as halogenated and aprotic ones) and the solid state. Second, hydrogen bonding is generally thermoreversible, but metal coordination is essentially chemoreversible, that is, cross-linking by metal coordination can be reversed by a chemical species. Third, the introduction of metal centers into a cross-linked matrix also potentially confers distinct function to the material such as phosphorescence or fluorescence. Although not belonging to the class of side chain functionalized polymers, Rowan and cohorts report a class of metallosupramolecular gels that are multiresponsive and multistimuli as well as having the capability of being photoelectroluminescent materials (Beck and Rowan 2003; Iyer et al. 2005).

Cross-linked polymers based on metal coordination can be broadly classified into two classes. The first consists of systems in which the metal center is covalently attached to the polymer backbone essentially consisting of "macromolecular metallic centers," and cross-linking is achieved through bi- or multifunctional small molecule ligands (Fig. 5.20, **15**; Pollino et al. 2004). The second class consists of "macroligands" that are then cross-linked by the addition of bi- (or multi-) functionalized metal complexes. Loveless and colleagues (2005) conducted extensive studies of metal cross-linked networks using poly(vinylpyridine) and bis-functionalized pincer complexes based on Pd and Pt (Fig. 5.20, **16**). Because they employed a polymeric macroligand based on poly(vinylpyridine), they were able to noncovalently cross-link the same polymer chain by different bis-metal complexes; they used single as well as multiple metal complex formation for polymer cross-linking

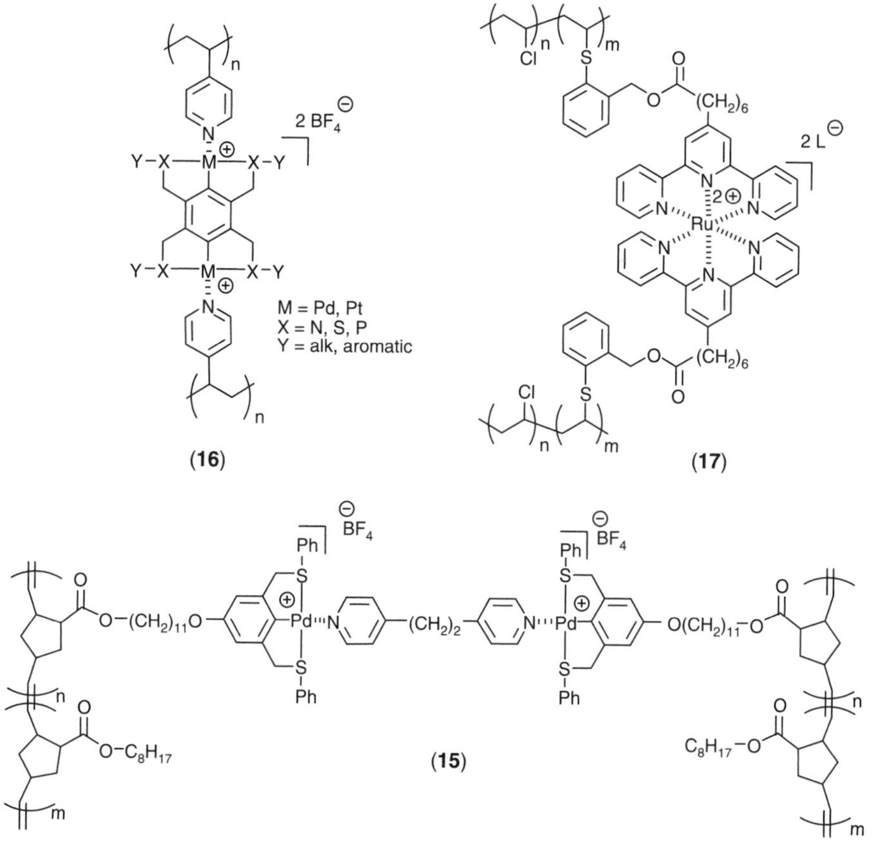

Figure 5.20 Examples of metal cross-linked polymer networks.

without any significant interference of the different metal centers (Loveless et al. 2005).

Meier and Schubert (2003) used terpyridine-based metal coordination for polymer cross-linking. They postpolymerized a commodity plastic such as poly(vinyl chloride) to introduce terpyridine moieties in the side chains. The functionalized polymer was then cross-linked by complexation of the terpyridine groups with ruthenium to form a metal cross-linked polymer (Fig. 5.20, **17**). This example illustrates the importance of using postpolymerization functionalization in converting easily available commodity plastics into high value materials. Furthermore, Hofmeier and Schubert (2003) cross-linked a terpyridine functionalized poly(methacrylate) polymer using Fe(II) and Zn(II) and demonstrated that the addition of Fe(II) resulted in more efficient cross-linking than the addition of Zn(II). They also completely decross-linked the Zn(II)-based network by the addition of hydroethyl-(ethylenediaminetetraacetic acid), thus demonstrating the complete reversibility of metal cross-linked polymers.

Multifunctional Polymer Networks: Combining Cross-linking and Polymer Functionalization. Side chain functionalized polymers offer a strategy to use multiple noncovalent interactions that can be used to reversibly and simultaneously cross-link as well as functionalize the polymeric scaffolds to form highly functionalized cross-linked polymers with unprecedented complexity. This strategy involves the employment of an orthogonal functionalization and cross-linking strategy. The functionalization is achieved by using monofunctionalized moieties that are noncovalently anchored to the scaffold whereas cross-linking is achieved by using bifunctional cross-linking agents to cause interchain cross-linking. When orthogonal noncovalent interactions for cross-linking and functionalization results are employed, both processes should be mutually independent and noninterfering. The Weck group synthesized highly functionalized noncovalently cross-linked polymers prepared from a single "universal polymer backbone" via directional self-assembly processes using a combination of metal coordination and hydrogen bonding (Pollino et al. 2004). They report a functionalization/cross-linking strategy in which the polymeric scaffold can be noncovalently cross-linked by employing one of the self-assembly motifs and the second one is used for the noncovalent functionalization. The Weck system is based on a terpolymer functionalized with 2,6-diaminopyridine as the hydrogen bonding receptor moieties, palladated SCS pincer complexes for metal coordination, and a third inert spacer monomer to increase the polymer solubility and to dilute the recognition units (Fig. 5.21).

Two distinct cross-linking/functionalization schemes were investigated: 1) cross-linking via hydrogen bonding interaction employing either a bis-thymine or bis-perylene unit with the palladated SCS pincer centers used for polymer functionalization by metal coordination or 2) cross-linking via metal coordination using a bis-pyridine cross-linking agent with the 2,6-diaminopyridine along the polymer scaffold functionalized with thymine derivatives. Extensive polymer cross-linking was observed in all cases as investigated by viscometry. However, the metal coordinated cross-linked scaffold exhibited significantly higher viscosity than its hydrogen

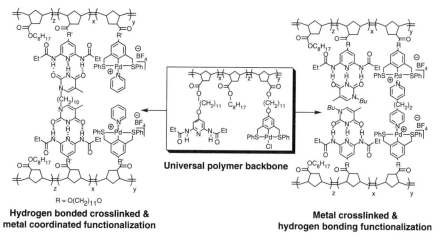

Figure 5.21 Orthogonal noncovalent cross-linking as well as functionalization strategy of terpolymer using hydrogen bonding and metal coordination interactions.

bonded analogue. The independent, noninteracting behaviors of these two modes of self-assembly allowed for the creation of a self-assembled, multifunctional, and cross-linked material in one self-assembly step.

The reversibility of this system was also studied. The polymer scaffold could be fully defunctionalized and decross-linked by 1) heating to disrupt the hydrogen bonds (thus exhibiting thermoresponsiveness) and 2) addition of PPh$_3$ to break the metal–pyridine complex via competitive ligand interaction (thus exhibiting chemoresponsiveness). Hence, by employing different noncovalent interactions, we can design a system that is responsive to multiple stimuli, opening the potential for the easy fabrication of "smart materials."

5.5. CONCLUSIONS AND OUTLOOK

In conclusion, we can say with certainty that supramolecular side chain functionalized polymers combine the best of both worlds of covalent polymer formation and side chain functionalization through noncovalent interactions. This synergistic combination of two very successful areas of science confers the distinct advantage of using such a strategy for materials design and optimization. The ongoing development of synthetic polymer chemistry and supramolecular science will greatly benefit the synthesis of more complex controlled polymer architectures using side chain functionalization. The self-assembly strategies presented in this chapter represent only a small selection from the now vast amount of research in this still rapidly expanding area of multifunctional materials based on noncovalent interactions. As the field of functional materials design and synthesis continues to demand smaller and smaller size ranges for increasingly complex devices and applications, self-assembly will have to emerge as the most important tool available to scientists for the development of these materials. The production of materials in the nanoscale size range becomes much more efficient as the amount of direct manual manipulation of the material on a nanoscale is decreased, that is, by the creation of smart materials that can manipulate themselves on that small scale. The examples described in this chapter demonstrate that scientists are well on their way to this goal. As research continues in this field and scientists continue to take clues from Nature, our understanding and mastery of both materials synthesis and design as well as small-scale noncovalent biological processes will continue to improve.

ACKNOWLEDGMENTS

We gratefully acknowledge the support and enthusiasm of former and current group members, collaborators, and colleagues. Financial support from the Weck group for the research described in this chapter has been provided by the National Science Foundation (ChE-0239385) and the Office of Naval Research (MURI Award N00014-03-1-0793).

REFERENCES

Armstrong G, Buggy M. Hydrogen-bonded supramolecular polymers: a literature review. J Mater Sci 2005;40:547–559.

Arumugam P, Xu H, Srivastava S, Rotello VM. "Bricks and mortar" nanoparticle self-assembly using polymers. Polym Int 2007;56:461–466.

Barrett AGM, Bibal B, Hopkins BT, Köebberling J, Love AC, Tedeschi L. Facile and purification free synthesis of peptides utilizing ROMP gel- and ROMP sphere-supported coupling reagents. Tetrahedron 2005;61:12033–12041.

Bazzi HS, Bouffard J, Sleiman HF. Self-complementary ABC triblock copolymers via ring-opening metathesis polymerization. Macromolecules 2003;36:7899–7902.

Beck JB, Rowan SJ. Multistimuli, multiresponsive metallo-supramolecular polymers. J Am Chem Soc 2003;125:13922–13923.

Binder WH. Polymeric ordering by H-bonds. Mimicking nature by smart building blocks. Monatsh Chem 2005;136:1–19.

Bosman AW, Brunsveld L, Folmer BJB, Sijbesma RP, Meijer EW. Supramolecular polymers: from scientific curiosity to technological reality. Macromol Symp 2003;201:143–154.

Brunsveld L, Folmer BJB, Meijer EW, Sijbesma RP. Supramolecular polymers. Chem Rev 2001;101:4071–4097.

Burd C, Weck M. Self-sorting in polymers. Macromolecules 2005;38:7225.

Carlise JR, Weck M. Side-chain functionalized polymers containing bipyridine coordination sites: polymerization and metal-coordination studies. J Polym Sci Part A Polym Chem 2004;42:2973–2984.

Carroll JB, Frankamp BL, Rotello VM. Self-assembly of gold nanoparticles through tandem hydrogen bonding and polyoligosilsesquioxane (POSS)–POSS recognition processes. Chem Commun 2002;(17):1892–1893.

Carroll JB, Waddon AJ, Nakade H, Rotello VM. "Plug and play" polymers. Thermal and x-ray characterizations of noncovalently grafted polyhedral oligomeric silsesquioxane (POSS)–polystyrene nanocomposites. Macromolecules 2003;36:6289–6291.

Chen B, Metera K, Sleiman HF. Biotin-terminated ruthenium bipyridine ring-opening metathesis polymerization copolymers: synthesis and self-assembly with streptavidin. Macromolecules 2005;38:1084–1090.

Chino K, Ashiura M. Thermoreversible cross-linking rubber using supramolecular hydrogen-bonding networks. Macromolecules 2001;34:9201–9204.

Cooke G, Rotello VM. Methods of modulating hydrogen bonded interactions in synthetic host–guest systems. Chem Soc Rev 2002;31:275–286.

Dankers PYW, Zhang Z, Wisse E, Grijpma DW, Sijbesma RP, Feijen J, Meijer EW. Oligo(trimethylene carbonate)-based supramolecular biomaterials. Macromolecules 2006;39:8763–8771.

Eisenberg A, Kim J-S. Introduction to ionomers. New York: Wiley Interscience; 1998.

Elkins CL, Park T, McKee MG, Long TE. Synthesis and characterization of poly(2-ethylhexyl methacrylate) copolymers containing pendant, self-complementary multiple-hydrogen-bonding sites. J Polym Sci Part A Polym Chem 2005;43:4618–4631.

Fang Z, Kennedy JP. Novel block ionomers. I. Synthesis and characterization of polyisobutylene-based block anionomers. J Polym Sci Part A Polym Chem 2002;40:3662–3678.

Faul CFJ, Antonietti M. Ionic self-assembly: facile synthesis of supramolecular materials. Adv Mater 2003;15:673–683.

Freitas LDL, Jacobi MM, Goncalves G, Stadler R. Microphase separation induced by hydrogen bonding in a poly(1,4-butadiene)-*block*-poly(1,4-isoprene) diblock copolymer—an example of supramolecular organization via tandem interactions. Macromolecules 1998;31:3379–3382.

Freitas LDL, Stadler R. Thermoplastic elastomers by hydrogen bonding. 3. Interrelations between molecular parameters and rheological properties. Macromolecules 1987;20:2478–2485.

Gheneim R, Perez-Berumen C, Gandini A. Diels–Alder reactions with novel polymeric dienes and dienophiles: synthesis of reversibly cross-linked elastomers. Macromolecules 2002;35:7246–7253.

Ghosh S, Rasmusson J, Inganaes O. Supramolecular self-assembly for enhanced conductivity in conjugated polymer blends. Ionic crosslinking in blends of poly(3,4-ethylenedioxythiophene)-poly(styrene sulfonate) and poly(vinylpyrrolidone). Adv Mater 1998;10:1097–1099.

Gittins PJ, Twyman LJ. Dendrimers and supramolecular chemistry. Supramol Chem 2003;15:5–23.

Gong B. Engineering hydrogen-bonded duplexes. Polym Int 2007;56:436–443.

Hilger C, Stadler R. New multiphase architecture from statistical copolymers by cooperative hydrogen bond formation. Macromolecules 1990;23:2095–2097.

Hofmeier H, Schubert US. Supramolecular branching and crosslinking of terpyridine-modified copolymers: complexation and decomplexation studies in diluted solution. Macromol Chem Phys 2003;204:1391–1397.

Ghosh S, Rasmusson J, Inganäs O. Supramolecular self-assembly for enhanced conductivity in conjugated polymer blends. Ionic crosslinking in blends of poly(3,4-ethylenedioxythiophene)-poly(styrene sulfonate) and poly(vinylpyrrolidone). Adv Mater 1998;10:1097–1099.

Iyer PK, Beck JB, Weder C, Rowan SJ. Synthesis and optical properties of metallo-supramolecular polymers. Chem Commun 2005;(3):319–321.

Kato T. Supramolecular liquid-crystalline materials: molecular self-assembly and self-organization through intermolecular hydrogen bonding. Supramol Sci 1996;3: 53–59.

Kato T, Kihara H, Ujiie S, Uryu T, Fréchet JMJ. Structures and properties of supramolecular liquid-crystalline side-chain polymers built through intermolecular hydrogen bonds. Macromolecules 1996;29:8734–8739.

Kawakami T, Kato T. Use of intermolecular hydrogen bonding between imidazolyl moieties and carboxylic acids for the supramolecular self-association of liquid-crystalline side-chain polymers and networks. Macromolecules 1998;31:4475–4479.

Kumar U, Kato T, Fréchet JMJ. Use of intermolecular hydrogen bonding for the induction of liquid crystallinity in the side chain of polysiloxanes. J Am Chem Soc 1992;114:6630–6639.

Lange RFM, Van Gurp M, Meijer EW. Hydrogen-bonded supramolecular polymer networks. J Polym Sci Part A Polym Chem 1999;37:3657–3670.

Langsdorf BL, Zhou X, Adler DH, Lonergan MC. Synthesis and characterization of soluble, ionically functionalized polyacetylenes. Macromolecules 1999;32:2796–2798.

Langsdorf BL, Zhou X, Lonergan MC. Kinetic study of the ring-opening metathesis polymerization of ionically functionalized cyclooctatetraenes. Macromolecules 2001;34: 2450–2458.

Liu Y-L, Chen Y-W. Thermally reversible cross-linked polyamides with high toughness and self-repairing ability from maleimide- and furan-functionalized aromatic polyamides. Macromol Chem Phys 2007;208:224–232.

Long NJ. Organometallic compounds for nonlinear optics—the search for en-light-enment! Angew Chem Int Ed Engl 1995;34:21–38.

Loveless DM, Jeon SL, Craig SL. Rational control of viscoelastic properties in multicomponent associative polymer networks. Macromolecules 2005;38:10171–10177.

Meier MAR, Schubert US. Terpyridine-modified poly(vinyl chloride): possibilities for supramolecular grafting and crosslinking. J Polym Sci Part A Polym Chem 2003;41:2964–2973.

Nair KP, Pollino JM, Weck M. Noncovalently functionalized block copolymers possessing both hydrogen bonding and metal coordination centers. Macromolecules 2006;39: 931–940.

Nair KP, Weck M. Noncovalently functionalized poly(norbornene)s possessing both hydrogen bonding and coulombic interactions. Macromolecules 2007;40:211–219.

Nair KP, Breedveld V, Weck M. Complementary hydrogen bonded thermoreversible polymer networks with tunable properties. Macromolecules 2008; in press.

Okumura Y, Ito K. The polyrotaxane gel: a topological gel by figure-of-eight cross-links. Adv Mater 2001;13:485–487.

Park T, Zimmerman SC. Formation of a miscible supramolecular polymer blend through self-assembly mediated by a quadruply hydrogen-bonded heterocomplex. J Am Chem Soc 2006;128:11582–11590.

Peng C-C, Abetz V. A simple pathway toward quantitative modification of polybutadiene: a new approach to thermoreversible cross-linking rubber comprising supramolecular hydrogen-bonding networks. Macromolecules 2005;38:5575–5580.

Pollino JM, Nair KP, Stubbs LP, Adams J, Weck M. Crosslinked and functionalized universal polymer backbones via simple, rapid, and orthogonal multi-site self-assembly. Tetrahedron 2004;60:7205–7215.

Pollino JM, Stubbs LP, Weck M. One-step multifunctionalization of random copolymers via self-assembly. J Am Chem Soc 2004;26:563–567.

Pollino JM, Weck M. Supramolecular side-chain functionalized polymers: synthesis and self-assembly behavior of polynorbornenes bearing Pd^{II} SCS pincer complexes. Synthesis 2002;9:1277–1285.

Pollino JM, Weck M. Non-covalent side-chain polymers: design principles, functionalization strategies, and perspectives. Chem Soc Rev 2005;34:193–207.

Rieth LR, Eaton RF, Coates GW. Polymerization of ureidopyrimidinone-functionalized olefins by using late-transition metal Ziegler–Natta catalysts: synthesis of thermoplastic elastomeric polyolefins. Angew Chem Int Ed 2001;40:2153–2156.

Sallenave X, Bazuin CG. Interplay of ionic, hydrogen-bonding, and polar interactions in liquid crystalline complexes of a pyridylpyridinium polyamphiphile with (azo)phenol-functionalized molecules. Macromolecules 2007;40:5326–5336.

Schädler V, Kniese V, Thurn-Albrecht T, Wiesner U, Spiess HW. Self-assembly of ionically end-capped diblock copolymers. Macromolecules 1998;31:4828–4837.

Schubert US, Eschbaumer C. Macromolecules containing bipyridine and terpyridine metal complexes: towards metallosupramolecular polymers. Angew Chem Int Ed 2002;41:2892–2926.

Service RF. How far can we push chemical self-assembly? Science 2005;309:95.

Shandryuk GA, Kuptsov SA, Shatalova AM, Plate NA, Talroze RV. Liquid crystal H-bonded polymer networks under mechanical stress. Macromolecules 2003;36:3417–3423.

Shimizu LS. Perspectives on main-chain hydrogen bonded supramolecular polymers. Polym Int 2007;56:444–452.

Shunmugam R, Tew GN. Efficient route to well-characterized homo, block, and statistical polymers containing terpyridine in the side chain. J Polym Sci Part A Polym Chem 2005;43:5831–5843.

Sijbesma RP, Meijer EW. Self-assembly of well-defined structures by hydrogen bonding. Curr Opin Colloid Interface Sci 1999;4:24–32.

South CR, Leung KCF, Lanari D, Stoddart JF, Weck M. Noncovalent side-chain functionalization of terpolymers. Macromolecules 2006;39:3738–3744.

Tew GN, Aamer KA, Shunmugam R. Incorporation of terpyridine into the side chain of copolymers to create multi-functional materials. Polymer 2005;46:8440–8447.

Thibault RJ, Hotchkiss PJ, Gray M, Rotello VM. Thermally reversible formation of microspheres through non-covalent polymer cross-linking. J Am Chem Soc 2003;125:11249–11252.

Trenor SR, Shultz AR, Love BJ, Long TE. Coumarins in polymers: from light harvesting to photo-cross-linkable tissue scaffolds. Chem Rev 2004;104:3059–3077.

van Manen H-J, van Veggel FCJM, Reinhoudt DN. Non-covalent synthesis of metallodendrimers. In: Dendrimers IV, 2001, Volume 217 (Topics in Current Chemistry Series), Springer, Berlin/Heidelberg, pp. 121–162.

Vuillaume PY, Bazuin CG. Self-assembly of a tail-end pyridinium polyamphiphile complexed with *n*-alkyl sulfonates of variable chain length. Macromolecules 2003;36:6378–6388.

Weck M. Side-chain functionalized supramolecular polymers. Polym Int 2007;56:453–460.

Wilson AJ. Non-covalent polymer assembly using arrays of hydrogen-bonds. Soft Mater 2007;3:409–425.

Yu K, Sommer W, Richardson JM, Weck M, Jones CW. Evidence that SCS pincer Pd(II) complexes are only precatalysts in Heck catalysis and the implications for catalyst recovery and reuse. Adv Synth Catal 2005;347:161–171.

CHAPTER 6

POLYMER-MEDIATED ASSEMBLY OF NANOPARTICLES USING ENGINEERED INTERACTIONS

HUNG-TING CHEN, YUVAL OFIR, and VINCENT M. ROTELLO

6.1. INTRODUCTION

The fabrication of advanced materials and devices on the nanometer scale is the subject of intensive research and has many potential applications in areas including electronics, catalysis, and biosensing (Fendler 1996; Giannelis 1996; Sanchez et al. 2001; Bhat et al. 2003; Shenhar and Rotello 2003; Binder 2005; Shenhar, Jeoung, et al. 2005; Shenhar, Norsten, et al. 2005; Haryono and Binder 2006; Arumugam et al. 2007; Xu et al. 2007). The traditional "top-down" approach that utilizes photo- or electron beam lithography processes to construct two-dimensional (2-D) or three-dimensional (3-D) structures not only reaches its inherent resolution limitation because of diffraction problems but also is inefficient in terms of energy and materials. The alternative "bottom-up" approach utilizes the self-assembly processes of diverse building blocks by noncovalent interactions, providing a potentially low cost and facile way to provide sophisticated 3-D nanostructures.

Nanoparticles are extensively used as versatile building blocks in the bottom-up approach because of their tunable nanoscale sizes and unique physical properties (Murray et al. 2000; El-Sayed 2001; Brock et al. 2004; Masala and Seshadri 2004). Assembly of nanoparticles into nanostructures can be achieved by applying the "bricks and mortar" strategy, in which the particles serve as bricks and the polymer functions as mortar (Fig. 6.1). In such an approach, a polymer with complementary functionality integrates each nanoparticle to form a polymer-mediated assembly through molecular interactions (Ozin and Arsenault 2005; Haryono and Binder 2006). Among those intermolecular forces, electrostatic interaction and hydrogen bonding are the most utilized interactions in supramolecular chemistry.

Molecular Recognition and Polymers: Control of Polymer Structure and Self-Assembly.
Edited by V. Rotello and S. Thayumanavan
Copyright © 2008 John Wiley & Sons, Inc.

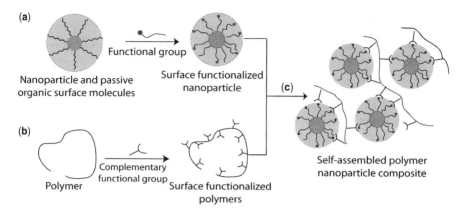

Figure 6.1 General schematic representation of polymer-mediated assembly of nanoparticles: (a) functionalization of nanoparticles through place-exchange method, (b) incorporation of complementary functional group to polymers, and (c) self-assembly of nanoparticles through electrostatic or hydrogen bonding interactions.

In general, electrostatic interaction provides a straightforward method for assembling nanoparticles but with limited control of the resulting structure because of kinetic trapping in the assembly process. Hydrogen bonding interactions in self-assembly processes tend to be easier to modulate, giving rise to the formation of morphologically controlled and highly well-ordered structures that are attributable to their thermodynamically driven equilibrium nature (Beijer et al. 1998; Prins et al. 2001; Lehn, 2002; Sivakova and Rowan, 2005; Xu et al. 2005).

6.2. DESIGN OF NANOPARTICLES AND POLYMERS

Effective functionalization of nanoparticle surfaces and complementary polymers is a prerequisite for realization of polymer-mediated nanoparticle assembly. The metal or semiconductor nanoparticles can be easily functionalized through a two-step process. Nanoparticles are synthesized by photo- or chemical-reduction of metal salt in the presence of capping agents (Brust et al. 1994, 1995). The desired functionality can be attached to the particle at this stage or introduced on the nanoparticle surface by place exchange with the appropriate functional ligand (Templeton et al. 1998, 2000). The self-assembled organic monolayer of nanoparticles not only passivates the surface to prevent particle agglomeration but also modulates the solubility (Hostetler et al. 1999; Shenhar and Rotello 2003; Drechsler et al. 2004). The ability to fine-tune the solubility of the nanoparticles allows them to be characterized by standard solution-phase techniques, such as nuclear magnetic resonance, infrared, and ultraviolet (UV) visible spectroscopy.

A well-controlled polymerization method in terms of polymer length, polydispersity, and spatial distribution of functional groups is crucial for effective nanoparticle assembly (Chiefari et al. 1998; Bielawski and Grubbs 2000; Coessens et al. 2001;

Hawker et al. 2001; Hawker and Wooley 2005). Recent developments in living radical polymerization allow the preparation of structurally well-defined block copolymers with low polydispersity. These polymerization methods include atom transfer free radical polymerization (Coessens et al. 2001), nitroxide-mediated polymerization (Hawker et al. 2001), and reversible addition fragmentation chain transfer polymerization (Chiefari et al. 1998). In addition to their ease of use, these approaches are generally more tolerant of various functionalities than anionic polymerization. However, direct polymerization of functional monomers is still problematic because of changes in the polymerization parameters upon monomer modification. As an alternative, functionalities can be incorporated into well-defined polymer backbones after polymerization by coupling a side chain modifier with tethered reactive sites (Shenhar et al. 2004; Carroll et al. 2005; Malkoch et al. 2005). The modification step requires a clean (i.e., free from side products) and quantitative reaction so that each site has the desired chemical structures. Otherwise it affords poor reproducibility of performance between different batches.

6.3. SELF-ASSEMBLY OF POLYMER–PARTICLE NANOCOMPOSITES

In this section, we focus on the strategies of controlling nanoparticle assemblies through functionalized polymer scaffolds, starting from interparticle spacing in bulk aggregates to 3-D morphologically controlled hierarchical nanostructures.

6.3.1. Control of Interparticle Spacing

The overall optical and electronic properties of nanoparticle assembly are affected by neighboring particles in a strongly distance-dependent fashion (Hovel et al. 1993; Schmitt et al. 1997; Sandrock and Foss 1999; Taton et al. 2000; Ung et al. 2001). Although the control of interparticle spacing on the assembly can be achieved by manipulating the nanoparticle size and shapes, the use of separate entities as spacers provides a more modular method to regulate the interparticle distances. Structurally, regular dendrimers can provide useful "spacers" because of their globular geometry, and they allow systematic control the interparticle spacing through the choice of dendrimer generations.

Rotello's research group first demonstrated the direct control of interparticle spacing through self-assembly of gold nanoparticles with poly(amido-amine) (PAMAM) dendrimers (Fig. 6.2; Frankamp, Boal, et al. 2002). The assembled nanocomposite was obtained by the formation of salt bridges between carboxylic acid-functionalized gold nanoparticles and the peripheral amine group of PAMAM. Small angle X-ray scattering (SAXS) revealed a monotonic increase of interparticle distance from 4.1 to 6.1 nm with an increase of dendrimer generations from G0 to G6. The same approach was applied to dendrimer-mediated assembly of gold nanoparticles with a larger diameter (6.0 nm; Srivastava, Frankamp, et al. 2005). The interparticle spacing calculated from the SAXS spectra in the same PAMAM

(a)

(b)

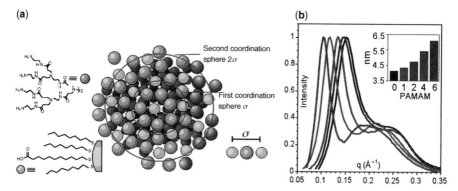

Figure 6.2 (a) Gold nanoparticles assembled using PAMAM dendrimers and (b) small angle X-ray scattering (SAXS) plots of nanoparticle assembly indicating systematic control of interparticle distance by dendrimer generations. Reprinted with permission from Frankamp, Boal, et al. (2002). Copyright 2002 American Chemical Society.

generation confirmed previous studies. Moreover, the surface plasma resonance (SPR) of resulting thin films exhibited a gradual blueshift over an 84-nm range with the increase of PAMAM generations (Fig. 6.3). The observed shift of SPR bands was attributed to a decrease of nanoparticle dipolar coupling with increasing interparticle spacing. Thus, the manipulation of dipolar coupling between nanoparticles provides a new way to fine-tune the optical properties of nanoparticle assemblies.

By applying the same principle, Rotello's group also showed that dendrimer-mediated assembly of magnetic nanoparticles (γ-Fe$_2$O$_3$) can directly modulate

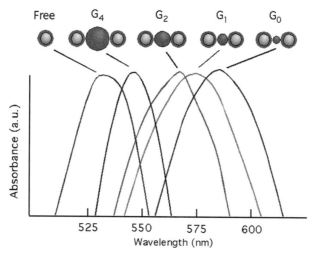

Figure 6.3 UV visible spectra of assembled gold nanoparticle thin films by dendrimers of different generations ranging from G0 to G4. Reprinted with permission from Srivastava, Frankamp, et al. (2005). Copyright 2005 American Chemical Society.

their collective magnetic properties based on precise control of interparticle spacing (Frankamp et al. 2005). The transition from superparamagnetic to ferromagnetic (i.e., the blocking temperature) for iron oxide (Fe_2O_3) nanoparticle assemblies shifted over a substantial range with increasing dendrimer generations. This result can be explained by an increase of the activation energy for spin flipping associated with an increase of interparticle decoupling. The ability to substantially reduce dipolar coupling between particles could have a significant impact on applications such as magnetic storage, which requires the placement of densely packed magnetic nanoparticles in a minimum space.

In addition to dendrimers, charged biomacromolecules such as DNA or proteins have also been utilized as "mortar" to assemble nanoparticles. DNA has been demonstrated to be a particularly versatile construction material because of its flexible length scale, rigidity, and duplex helical structures (Storhoff and Mirkin 1999; Fu et al. 2004; Becerril et al. 2005; Ongaro et al. 2005). Nanoparticle–DNA nanocomposites can be formed because of the affinity of the anionic phosphate linkage of DNA strands and cationic modified nanoparticles. The DNA-mediated assembly of iron platinum (FePt) nanoparticles exhibited an increase in spacing from 7.8 nm for the particle alone up to 8.9 nm for the DNA–particle assembly (Fig. 6.4; Srivastava, Arumugam, et al. 2007). Although a certain degree of denaturation in the bonded duplex DNA was observed through circular dichroism, the reduced dipolar coupling between each particle due to the insertion of DNA molecules still allowed fine-tuning of the magnetic behavior of the material.

In a similar fashion, the Rotello group reported several protein-mediated assemblies of nanoparticles (Srivastava, Verma, et al. 2005; Verma et al. 2005). Unstable proteins such as chymotrypsin are readily denatured upon prolonged interaction with functionalized nanoparticles because of the exposure of proteins to the hydrophobic layer (Srivastava, Verma, et al. 2005). Addition of a hydrophilic portion to the monolayer by inserting a short tetraethylene glycol between the charged terminal group and hydrophobic aliphatic chain circumvented this denaturation problem (Hong et al. 2004).

A ferritin–FePt nanoparticle bionanocomposite was fabricated in later studies, which shows the integrated magnetic behavior of the synthetic and biological components (Fig. 6.5; Srivastava, Samanta, Jordan, et al. 2007). A transmission electron

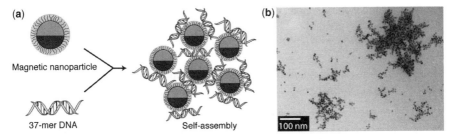

Figure 6.4   (a) Self-assembly of magnetic FePt nanoparticles by DNAs and (b) a TEM micrograph of networklike FePt–DNA aggregates; scale bar = 100 nm. Reprinted with permission from Srivastava, Samanta, Arumugam, et al. (2007). Copyright 2007 RSC Publishing.

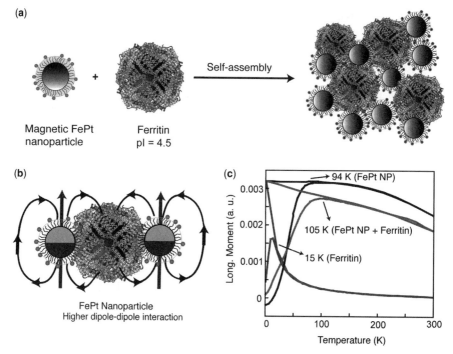

Figure 6.5 (a) The formation of ferritin-mediated self-assembly of FePt nanoparticles via electrostatic interactions, (b) magnetic dipole–dipole interaction of ferritins assembled with FePt nanoparticles, and (c) zero field cooling and field cooling results for the ferritin–FePt nanoparticle composite film and individual components. Reprinted with permission from Srivastava, Samanta, Jordan, et al. (2007). Copyright 2007 American Chemical Society.

microscopy (TEM) micrograph of the nanoparticle–ferritin nanocomposite exhibited a densely packed spherical entity with an essentially monodisperse diameter of ∼130 nm. This ensemble also expressed unique magnetic properties in terms of blocking temperature, net magnetic moment, coercivity, and magnetic remnance that differ from either of their constituent particles. These integrated magnetic properties could be attributable to the interparticle magnetic interaction by a dipolar or exchange mechanism (Zheng et al. 2002; Farrell, Cheng, et al. 2004; Farrell, Ding, et al. 2004).

6.3.2. Self-Assembly of Nanoparticles Mediated by Polymers on the Planar Substrates

Self-assembly of nanoparticles in well-ordered 2-D arrays represents a major goal in the fabrication of microelectronics devices (Sun et al. 2002, 2003). Different strategies have been developed to tackle the challenge of well-organized nanoparticles in a 2-D plate surface (Andres et al. 1996; Spatz et al. 2000). Schmid and coworkers (2000) reported a long-range ordered sulfonic acid functionalized nanoparticle array

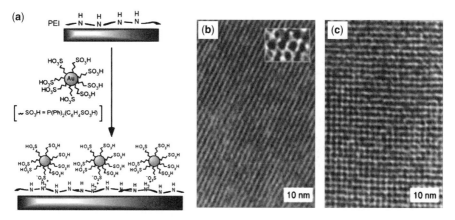

Figure 6.6 (a) Schematic representation of PEI-mediated assembly of gold nanoparticles. Transmission electron micrographs of (b) hexagonal and (c) cubic packing arrangements of nanoparticles. Reprinted with permission from Schmid et al. (2000). Copyright 2000 Wiley InterScience.

in a polyethyleneimine (PEI) coated surface through electrostatic interactions (Fig. 6.6). The low molecular weight PEI ($M_W = 60\,000$) permitted the nanoparticles to have enough mobility to rearrange on the surface. The final well-ordered array exhibited either hexagonal or cubic packing, depending on the crystalline phase of the underlying PEI support. This study suggested that the PEI surface acts as a template to direct the arrangement of nanoparticles on the substrate.

In addition to electrostatic interaction, multipoint hydrogen bonding has been utilized for selective deposition of nanoparticles on polymer template surfaces to generate ordered functional arrays (Murray et al. 2000). Incorporation of such recognition units provides a new direction in the assembly based on the reversible and specific nature of hydrogen bonds. Binder et al. (2005) reported selective binding of gold nanoparticles onto a microphase separated diblock copolymer film (Fig. 6.7). Gold nanoparticles (5.0 nm) were functionalized by ligands containing a barbituric acid moiety, which features a strong affinity with the Hamilton receptor through six-point hydrogen bonding ($K_a = 1.2 \times 10^5 \text{ M}^{-1}$). One block of the copolymer was postfunctionalized with complementary Hamilton receptors through Huisgen 1,3-dipolar cycloaddition ("click chemistry"), whereas the other block bore a fluorinated side chain to enhance microseparation. Nanoparticles selectively deposited on a specific domain with Hamilton receptors of the block copolymer. This intelligent approach takes advantages of both microseparated block copolymer film and specific hydrogen bonding to directly create a patterned surface in the microscale range, which has great potential applications in multifunctional biosensors and novel electronic, mechanical, and photonic devices.

The formation of a patterned surface through selective segregation of nanoparticles onto one domain of a block copolymer film can be used to produce complex 2-D nanoparticle arrays in a controlled fashion. However, the resulting

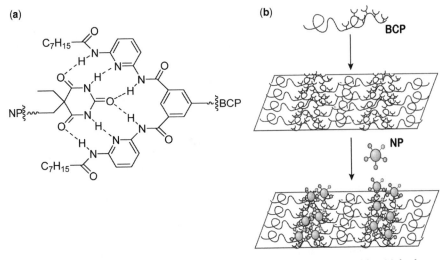

Figure 6.7 Illustration of multipoint hydrogen bonding based self-assembly: (a) hydrogen bond formation between barbituric acid functionalized gold nanoparticles and Hamilton receptor functionalized block copolymers and (b) selective deposition of nanoparticles on a microphase-separated block copolymer film. Reprinted with permission from Binder et al. (2005). Copyright 2005 American Chemical Society.

nanoparticle assembly is usually fragile and loses its order. Rotello and his group used an efficient and mild cross-linking protocol based on an orthogonal self-assembly strategy to provide stable nanoparticle structures, which was applicable to a large variety of nanoparticle core and periphery functionalities (Fig. 6.8; Shenhar et al. 2005). The protocol involved creating a block copolymer mediated nanoparticle assembly by binding of terpyridinyl (Terpy)-functionalized gold nanoparticles on the annealed polystyrene-*b*-poly(methyl methacrylate) (PS-*b*-PMMA) film. Similar to the findings of Zehner and coworkers (Zehner et al. 1998; Zehner and Sita 1999), these hydrophobic nanoparticles preferentially bind to the relatively hydrophobic PS domains to form a patterned surface based on the polarity differences of the PS and PMMA domains. The cross-linking step of the assembled nanoparticle film was performed by immersing the sample into a 10 mg mL^{-1} ethanol solution of [Fe(H$_2$O)$_6$(BF$_4$)$_2$], followed by washing with ethanol and drying with argon. The reinforced nanoparticle superstructure was accomplished by the formation of an iron diterpyridine (Fe(Terpy)$_2^{2+}$) complex through coordination chemistry between nanoparticles. The robustness of the cross-linked film was demonstrated through solvent swelling experiments. As expected, the sample dipped in the iron solution still maintained an organized superstructure upon the swelling of the polymer film by chloroform vapor whereas the nanoparticle structure of the control sample disassociated because of the increased mobility of the underlying polymer film swollen by the chloroform vapor.

An orthogonally self-assembling method can be achieved by tailoring different functionalities on predefined regions of surfaces. Xu et al. (2006) demonstrated

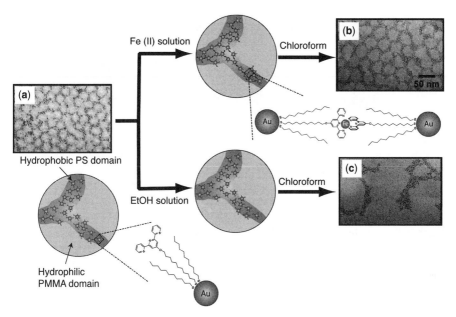

Figure 6.8 TEM micrographs of (a) "dots" patterned samples formed through affinity of Terpy-functionalized gold nanoparticles on PS domain, (b) Fe-treated cross-linked samples, and (c) ethanol-treated samples after swelling in chloroform vapor. Reprinted with permission from Shenhar et al. (2005). Copyright 2005 Wiley InterScience.

this concept by taking advantage of both electrostatic and hydrogen bonding interactions (Fig. 6.9). Initially, the orthogonally functionalized surface was fabricated by cross-linking poly(vinyl *N*-methyl pyridine) (PVMP) on a silicon wafer to serve as a positively charged surface. Next, a thymine-functionalized PS (Thy-PS) was spin cast on the top of the PVMP layer and exposed to UV light. The generated pattern surface contained Thy-PS squares with positive PVMP channels, whose areas featured an ability to perform hydrogen bonding and electrostatic interaction. The orthogonal self-assembly process was then examined by the use of diaminopyridine-functionalized PS (DAP-PS) and carboxylate-derivatized cadmium selenide/zinc sulfide core–shell nanoparticles (CdSe/ZnS-COOH). The three-point hydrogen bonding of DAP-Thy pairs and electrostatic interaction of methylated pyridinium–carboxylate led to the spontaneous fabrication of self-assembled orthogonal patterns. The preferential deposition of nanoparticles on specific sites of the polymer templated surfaces displayed the possibility of generating complex self-assembled patterns and also provided a new horizon for the hierarchical organization of nanostructure materials on surfaces.

A new methodology, which combines traditional (top-down) lithography and self-assembly (bottom-up) approaches, has been developed for fabrication of complex nanoscale colloidal structures on a surface over a large domain. The principle of this method is to create a chemically patterned surface by well-established soft lithography, followed by self-assembling nanoparticles selectively deposited

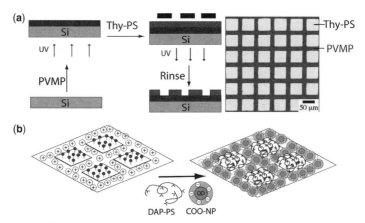

Figure 6.9 A schematic representation of orthogonal process for nanoparticles self-assembly: (a) a patterned silicon wafer with Thy-PS and PVMP polymers fabricated through photolithography and (b) orthogonal surface functionalization through Thy-PS/DP-PS recognition and PVMP/acid–nanoparticle electrostatic interaction. Reprinted with permission from Xu et al. (2006). Copyright 2006 American Chemical Society.

on specific domains. Zheng and colleagues (2002) reported the formation of two different sets of colloid particle arrays based on this method. The gold surface was first chemically patterned by stamping of $HS(CH_2)_{15}COOH$ through a microcontact printing technique (Fig. 6.10). The remaining bare gold regions absorbed a second alkanthiol, $HS(CH_2)_{11}(OCH_2CH_2)_3OH)$ (EG), which resists polyelectrolyte deposition in solutions. Using the layer-by-layer (LBL) alternating adsorption process, 5.5 bilayers of poly(diallyldimethylammonium chloride) (PDAC) and sulfonated PS (SPS) with an outmost positive layer were adsorbed on the COOH self-assembled monolayer (SAM) surface with high selectivity. The LBL polyelectrolyte thin films were intentionally introduced to provide affinity toward negatively charged nanoparticles through electrostatic attraction and to provide opportunities to enhance the adhesion between the multilayer surface and the colloid. The patterned LBL films selectively templated the adsorption of negatively charged PS sphere (PSS) particles, further modified with one PDAC layer on top of the PSS to prevent the deposition of positively charged nanoparticles on the LBL films in the next step. Finally, a second set of dodecyltrimethylammonium bromide modified PSS or positively charged amidine-terminated PS particles were directed toward EG SAM to obtained a surface patterned with two component nanoparticle arrays.

Another technique was introduced by Hua et al. (2002), who used photolithography in combination with a lift-off approach to direct growth of nanoparticle films (Fig. 6.11). At the beginning, a photoresist was patterned through a mask by the standard UV irradiation process. Then this substrate was entirely covered with polyelectrolyte through the alternate LBL method to provide an adhesive layer for deposition of oppositely charged nanoparticles. After forming a nanoparticle film over the entire substrate, the underlying photoresist was dissolved by

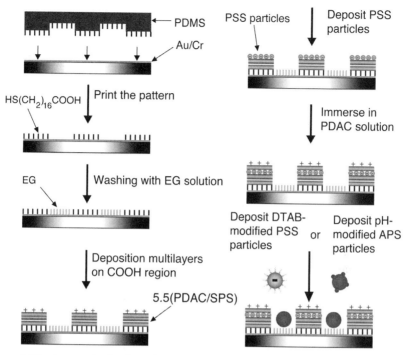

Figure 6.10 A procedural schematic representation for deposition of two kinds of particle arrays on the patterned template. Reprinted with permission from Zheng et al. (2002). Copyright 2002 Wiley InterScience.

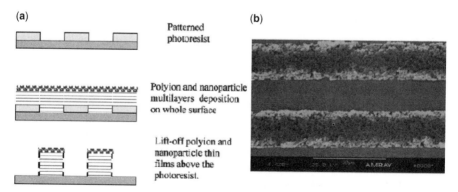

Figure 6.11 (a) Scheme of patterning nanoparticle thin films and (b) scanning electron micrograph of self-assembly patterns after lift-off process. Reprinted with permission from Hua et al. (2002). Copyright 2002 American Chemical Society.

immersion and transonication in the solution. During the dissolution, the resist layer was lifted off, resulting in removing the LBL film on top of the photoresist at the selected area.

Crespo-Biel and coworkers developed various patterning strategies to create polymer-mediated nanoparticle films on active cyclodextrin (CD) SAMs through host–guest CD–adamante (Ad) interactions (Crespo-Biel, Dordi, et al. 2006; Crespo-Biel, Ravoo, et al. 2006). They utilized CD modified gold nanoparticles (CD-Au nanoparticles) and admantyl-terminated poly(propylene imine) dendrimers (Ad-PPI) as the basis for a supramolecular LBL assembly process. The first approach is based on nanotransfer printing introduced by Park and Hammond (2004). Initially, the poly(dimethylsiloxane) (PDMS) stamps were pretreated by a UV/ozone treatment, resulting in a slightly negatively charged surface (Fig. 6.12). This weakly oxidized surface was employed as the substrate for the formation of the LBL supramolecular assembly, which was composed of alternative CD-Au nanoparticles and Ad-PPIs. Because of the stronger interaction of Ad-PPI and CD-SAM, adsorbed LBL assemblies on the PDMS stamps were completely transferred onto a full CD-SAM by the microcontact printing method. Moreover, the thickness of the patterned thin films could be manipulated by an LBL sorption process. Integration of nano-imprinting lithography and the lift-off approach led to a similar patterned LBL assembly templated on the surface. The prepatterned PMMA structure obtained through nanoimprinting lithography served as a physical barrier for the formation of CD-SAMs. The CD-SAMs were placed on the native silicon oxide areas through three steps of functionalization. The LBL assemblies of CD-Au nanoparticles and Ad-PPIs were performed on the entire area, followed by the lift-off process in the acetone that led to patterned LBL films on the silicon substrate.

6.3.3. Control of 3-D Hierarchical Organization in the Solution

Highly structured, 3-D nanoparticle–polymer nanocomposites possess unique magnetic, electronic, and optical properties that differ from individual entities, providing new systems for the creation of nanodevices and biosensors (Murray et al. 2000; Shipway et al. 2000). The choice of assembly interactions is a key issue in order to obtain complete control over the thermodynamics of the assembled system. The introduction of reversible hydrogen bonding and flexible linear polymers into the bricks and mortar concept gave rise to system formation in near-equilibrium conditions, providing well-defined structures.

A random diaminotriazine-immobilized PS (Triaz-PS) was employed as the mortar, whereas a complementary thymine-functionalized gold nanoparticle (Thy-Au) functioned as bricks (Fig. 6.13; Boal et al. 2000, 2002). A black solid was formed immediately after fixing of Triaz-PS and Thy-Au. In contrast, no precipitate was observed when gold nanoparticles functionalized with structurally analogous N-methyl thymine ligands (Me-Thy-Au) were used. This strongly indicated the essentiality of the assembly process by three-point hydrogen bonding. The diameter of these resulting spherical nanoparticle aggregates was strongly dependent on the temperature at which the assembly was performed. Sizes of 100 nm to 0.5–1 μm

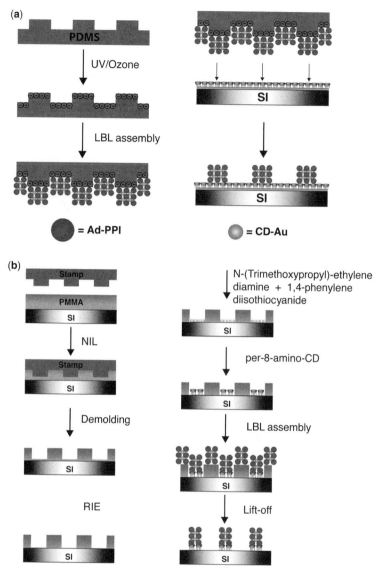

Figure 6.12 Preparation of patterned LBL assemblies (a) by nanotransfer printing and (b) by sequentially using nanoimprinting lithography, CD SAM formation, and lift-off process. Reprinted with permission from Crespo-Biel et al. (2006).

were achieved by varying the assembly temperature from ambient temperature to −20 °C. This temperature-dependent aggregation size demonstrated the thermodynamic control of such systems using the bricks and mortar methodology. In a follow-up investigation, the use of Triaz-functionalized PS-*b*-PS diblock copolymers (PS-*b*-Triaz-PS) with three different chain lengths provided a greater degree of

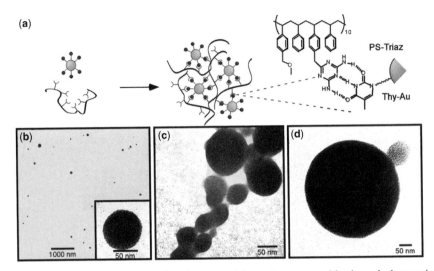

Figure 6.13 (a) Recognition-mediated nanoparticle–polymer assembly through three-point hydrogen bonding between diaminotriazine–polystyrene (Triaz-PS) and thymine-gold nanoparticles (Thy-Au). TEM micrographs of Thy-Au/Triaz-PS aggregates formed at (b) 23 °C, (c) 10 °C, and (d) −20 °C. Reprinted with permission from Boal et al. (2000). Copyright 2000 Nature Publishing Group.

control over the aggregation sizes (Fig. 6.14; Frankamp, Uzun, et al. 2002). The average core sizes calculated from TEM micrographs are 50–70% of the effective hydrodynamic radii (R_h) determined by dynamic light scattering, which suggested the polymer chains within the core are somewhat extended relative to the PS corona.

Vesicles and liposomes are versatile supermolecular systems and possess numerous potential applications in targeting agents, microreactors, encapsulations, and drug delivery. The Rotello group developed novel recognition-mediated polymersomes

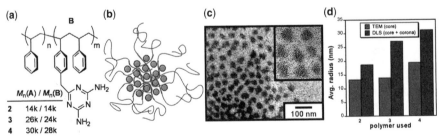

Figure 6.14 (a) Polystyrene-*b*-diaminotriazine-polystyrene diblock copolymers used to control the aggregate size. (b) Schematic structure of gold nanoparticles aggregated in diblock polymer micelles. (c) TEM micrograph, showing the control of aggregate size. (d) Comparison between average core size measured by TEM images and hydrodynamic radii calculated by dynamic light scattering. Reprinted with permission from Frankamp, Uzun, et al. (2002). Copyright 2002 American Chemical Society.

(RIPs) that feature a 3–5 μm diameter and a ~50-nm membrane thickness (Ilhan et al. 2000; Uzun et al. 2004, 2005). These RIPs are synthesized through hydrogen bonding of two complementary DAP and Thy-PS formed in the nonpolar media. The vesicular structure can be disrupted by small molecules, containing thymine functionality through competitive hydrogen binding. This provides a mechanism for controlled release of materials incorporated in the vesicles. However, the stability of vesicles can be further imparted to this RIPs by photoirradiation under UV light (200–300 nm), where the double bond of thymine moieties are cross-linked through $[2\pi_s + 2\pi_s]$ cyclization (Thibault et al. 2006). The cross-linked RIP expressed a larger size after irradiation that was perhaps due to the ripening process (Capek 2002), and the size distribution could be tuned efficiently according to the duration of photoirradiation. Thymine-functionalized cadmium selenide nanoparticles can be easily tethered into the irradiated vesicles without destroying the structures. Again, N-methylated thymine-functionalized cadmium selenides are not taken up by the RIPs because of the lack of hydrogen bond formation. This tailored stability of RIPs provides control of the uptake and release of guest molecules into or out of vesicular polymeric structures.

6.4. CONCLUSIONS AND OUTLOOK

Polymer-mediated self-assembly of nanoparticles provides a versatile and effective approach for the fabrication of new materials. This bottom-up strategy builds up nanocomposite materials from diverse nanosized building blocks by incorporation of molecular-level recognition sites. The flexibility and reversibility of self-assembly processes imparted by specific molecular interactions facilitates the formation of defect-free superstructures, and it can be further explored in fields ranging from electronics to molecular biology.

The examples given in this chapter represent a demonstration of the proof of the principle of using the brick and mortar approach to create hierarchical structures. Although great progress has been made in terms of controlling morphologies and the resultant physical properties of nanoparticle assemblies, deeper understanding of the fundamental principles in self-assembly and advances in polymer and nanoparticle synthesis are necessary to obtain the desired complex nanoarchitecture and to bring promising nanotechnology into practical devices.

REFERENCES

Andres RP, Bielefeld JD, Henderson JI, Janes DB, Kolagunta VR, Kubiak CP, Mahoney WJ, Osifchin RG. Self-assembly of a two-dimensional superlattice of molecularly linked metal clusters. Science 1996;273:1690–1693.

Arumugam P, Xu H, Srivastava S, Rotello VM. "Bricks and mortar" nanoparticle self-assembly using polymers. Polym Int 2007;56:461–466.

Becerril HA, Stoltenberg RM, Wheeler DR, Davis RC, Harb JN, Woolley AT. DNA-templated three-branched nanostructures for nanoelectronic devices. J Am Chem Soc 2005;127: 2828–2829.

Beijer FH, Kooijman H, Spek AL, Sijbesma RP, Meijer EW. Self-complementarity achieved through quadruple hydrogen bonding. Angew Chem Int Ed 1998;37:75–78.

Bhat RR, Genzer J, Chaney BN, Sugg HW, Liebmann-Vinson A. Controlling the assembly of nanoparticles using surface grafted molecular and macromolecular gradients. Nanotechnology 2003;14:1145–1152.

Bielawski CW, Grubbs RH. Highly efficient ring-opening metathesis polymerization (ROMP) using new ruthenium catalysts containing N-heterocyclic carbene ligands. Angew Chem Int Ed 2000;39:2903–2906.

Binder WH. Supramolecular assembly of nanoparticles at liquid–liquid interfaces. Angew Chem Int Ed 2005;44:5172–5175.

Binder WH, Kluger C, Straif CJ, Friedbacher G. Directed nanoparticle binding onto microphase-separated block copolymer thin films. Macromolecules 2005;38:9405–9410.

Boal AK, Ilhan F, DeRouchey JE, Thurn-Albrecht T, Russell TP, Rotello VM. Self-assembly of nanoparticles into structured spherical and network aggregates. Nature 2000;404: 746–748.

Boal AK, Gray M, Ilhan F, Clavier GM, Kapitzky L, Rotello VM. Bricks and mortar self-assembly of nanoparticles. Tetrahedron 2002;58:765–770.

Brock SL, Perera SC, Stamm KL. Chemical routes for production of transition-metal phosphides on the nanoscale: implications for advanced magnetic and catalytic materials. Chem Eur J 2004;10:3364–3371.

Brust M, Walker M, Bethell D, Schiffrin DJ, Whyman R. Synthesis of thiol-derivatized gold nanoparticles in a 2-phase liquid–liquid system. J Chem Soc Chem Commun 1994;801–802.

Brust M, Fink J, Bethell D, Schiffrin DJ, Kiely C. Synthesis and reactions of functionalized gold nanoparticles. J Chem Soc Chem Commun 1995;1655–1656.

Capek I. Sterically and electrosterically stabilized emulsion polymerization. Kinetics and preparation. Adv Colloid Interface Sci 2002;99:77–162.

Carroll JB, Jordan BJ, Xu H, Erdogan B, Lee L, Cheng L, Tiernan C, Cooke G, Rotello VM. Model systems for flavoenzyme activity: site-isolated redox behavior in flavin-functionalized random polystyrene copolymers. Org Lett 2005;7:2551–2554.

Chiefari J, Chong YK, Ercole F, Krstina J, Jeffery J, Le TPT, Mayadunne RTA, Meijs GF, Moad CL, Moad G, Rizzardo E, Thang SH. Living free-radical polymerization by reversible addition–fragmentation chain transfer: the RAFT process. Macromolecules 1998;31:5559–5562.

Coessens V, Pintauer T, Matyjaszewski K. Functional polymers by atom transfer radical polymerization. Prog Polym Sci 2001;26:337–377.

Crespo-Biel O, Dordi B, Maury P, Peter M, Reinhoudt DN, Huskens J. Patterned, hybrid, multilayer nanostructures based on multivalent supramolecular interactions. Chem Mater 2006;18:2545–2551.

Crespo-Biel O, Ravoo BJ, Reinhoudt DN, Huskens J. Noncovalent nanoarchitectures on surfaces: from 2D to 3D nanostructures. J Mater Chem 2006;16:3997–4021.

Drechsler U, Erdogan B, Rotello VM. Nanoparticles: scaffolds for molecular recognition. Chem Eur J 2004;10:5570–5579.

El-Sayed MA. Some interesting properties of metals confined in time and nanometer space of different shapes. Acc Chem Res 2001;34:257–264.

Farrell D, Cheng Y, Ding Y, Yamamuro S, Sanchez-Hanke C, Kao C-C, Majetich SA. Dipolar interactions and structural coherence in iron nanoparticle arrays. J Magn Magn Mater 2004;282:1–5.

Farrell D, Ding Y, Majetich SA, Sanchez-Hanke C, Kao C-C. Structural ordering effects in Fe nanoparticle two- and three-dimensional arrays. J Appl Phys 2004;95:6636–6638.

Fendler JH. Self-assembled nanostructured materials. Chem Mater 1996;8:1616–1624.

Frankamp BL, Boal AK, Rotello VM. Controlled interparticle spacing through self-assembly of Au nanoparticles and poly(amidoamine) dendrimers. J Am Chem Soc 2002;124:15146–15147.

Frankamp BL, Boal AK, Tuominen MT, Rotello VM. Direct control of the magnetic interaction between iron oxide nanoparticles through dendrimer-mediated self-assembly. J Am Chem Soc 2005;127:9731–9735.

Frankamp BL, Uzun O, Ilhan F, Boal AK, Rotello VM. Recognition-mediated assembly of nanoparticles into micellar structures with diblock copolymers. J Am Chem Soc 2002;124:892–893.

Fu AH, Micheel CM, Cha J, Chang H, Yang H, Alivisatos AP. Discrete nanostructures of quantum dots/Au with DNA. J Am Chem Soc 2004;126:10832–10833.

Giannelis EP. Polymer layered silicate nanocomposites. Adv Mater 1996;8:29–35.

Haryono A, Binder WH. Controlled arrangement of nanoparticle arrays in block-copolymer domains. Small 2006;2:600–611.

Hawker CJ, Bosman AW, Harth E. New polymer synthesis by nitroxide mediated living radical polymerizations. Chem Rev 2001;101:3661–3688.

Hawker CJ, Wooley KL. The convergence of synthetic organic and polymer chemistries. Science 2005;309:1200–1205.

Hong R, Fischer NO, Verma A, Goodman CM, Emrick T, Rotello VM. Control of protein structure and function through surface recognition by tailored nanoparticle scaffolds. J Am Chem Soc 2004;126:739–743.

Hostetler MJ, Templeton AC, Murray RW. Dynamics of place-exchange reactions on monolayer-protected gold cluster molecules. Langmuir 1999;15:3782–3789.

Hovel H, Fritz S, Hilger A, Kreibig U, Vollmer M. Width of cluster plasmon resonances—bulk dielectric functions and chemical interface damping. Phys Rev B 1993;48:18178–18188.

Hua F, Cui TH, Lvov Y. Lithographic approach to pattern self-assembled nanoparticle multilayers. Langmuir 2002;18:6712–6715.

Ilhan F, Galow TH, Gray M, Clavier G, Rotello VM. Giant vesicle formation through self-assembly of complementary random copolymers. J Am Chem Soc 2000;122:5895–5896.

Lehn JM. Toward self-organization and complex matter. Science 2002;295:2400–2403.

Malkoch M, Thibault RJ, Drockenmuller E, Messerschmidt M, Voit B, Russell TP, Hawker CJ. Orthogonal approaches to the simultaneous and cascade functionalization of macromolecules using click chemistry. J Am Chem Soc 2005;127:14942–14949.

Masala O, Seshadri R. Synthesis routes for large volumes of nanoparticles. Annu Rev Mater Sci 2004;34:41–81.

Murray CB, Kagan CR, Bawendi MG. Synthesis and characterization of monodisperse nanocrystals and close-packed nanocrystal assemblies. Annu Rev Mater Sci 2000;30:545–610.

Ongaro A, Griffin F, Beeeher P, Nagle L, Iacopino D, Quinn A, Redmond G, Fitzmaurice D. DNA-templated assembly of conducting gold nanowires between gold electrodes on a silicon oxide substrate. Chem Mater 2005;17:1959–1964.

Ozin GA, Arsenault AC. Nanochemistry: a chemical approach to nanomaterials. Cambridge: Royal Society of Chemistry; 2005.

Park J, Hammond PT. Multilayer transfer printing for polyelectrolyte multilayer patterning: direct transfer of layer-by-layer assembled micropatterned thin films. Adv Mater 2004;16: 520–525.

Prins LJ, Reinhoudt DN, Timmerman P. Noncovalent synthesis using hydrogen bonding. Angew Chem Int Ed 2001;40:2383–2426.

Sanchez C, Soler-Illia G, Ribot F, Lalot T, Mayer CR, Cabuil V. Designed hybrid organic–inorganic nanocomposites from functional nanobuilding blocks. Chem Mater 2001;13: 3061–3083.

Sandrock ML, Foss CA. Synthesis and linear optical properties of nanoscopic gold particle pair structures. J Phys Chem B 1999;103:11398–11406.

Schmid G, Baumle M, Beyer N. Ordered two-dimensional monolayers of Au-55 clusters. Angew Chem Int Ed 2000;39:181–183.

Schmitt J, Decher G, Dressick WJ, Brandow SL, Geer RE, Shashidhar R, Calvert JM. Metal nanoparticle/polymer superlattice films: fabrication and control of layer structure. Adv Mater 1997;9:61–65.

Shenhar R, Rotello VM. Nanoparticles: scaffolds and building blocks. Acc Chem Res 2003;36:549–561.

Shenhar R, Sanyal A, Uzun O, Nakade H, Rotello VM. Integration of recognition elements with macromolecular scaffolds: effects on polymer self-assembly in the solid state. Macromolecules 2004;37:4931–4939.

Shenhar R, Norsten TB, Rotello VM. Polymer-mediated nanoparticle assembly: structural control and applications. Adv Mater 2005;17:657–669.

Shenhar R, Jeoung E, Srivastava S, Norsten TB, Rotello VM. Crosslinked nanoparticle stripes and hexagonal networks obtained via selective patterning of block copolymer thin films. Adv Mater 2005;17:2206–2210.

Shipway AN, Katz E, Willner I. Nanoparticle arrays on surfaces for electronic, optical, and sensor applications. Chem Phys Chem 2000;1:18–52.

Sivakova S, Rowan SJ. Nucleobases as supramolecular motifs. Chem Soc Rev 2005;34:9–21.

Spatz JP, Mossmer S, Hartmann C, Moller M, Herzog T, Krieger M, Boyen HG, Ziemann P, Kabius B. Ordered deposition of inorganic clusters from micellar block copolymer films. Langmuir 2000;16:407–415.

Srivastava S, Frankamp BL, Rotello VM. Controlled plasmon resonance of gold nanoparticles self-assembled with PAMAM dendrimers. Chem Mater 2005;17:487–490.

Srivastava S, Samanta B, Arumugam P, Han G, Rotello VM. DNA-mediated assembly of iron platinum (FePt) nanoparticles. J Mater Chem 2007;17:52–55.

Srivastava S, Samanta B, Jordan BJ, Hong R, Xiao Q, Tuominen MT, Rotello VM. Integrated magnetic bionanocomposites through nanoparticle-mediated assembly of ferritin. J Am Chem Soc 2007;129:11776–11780.

Srivastava S, Verma A, Frankamp BL, Rotello VM. Controlled assembly of protein–nanoparticle composites through protein surface recognition. Adv Mater 2005;17:617–621.

Storhoff JJ, Mirkin CA. Programmed materials synthesis with DNA. Chem Rev 1999;99:1849–1862.

Sun SH, Anders S, Hamann HF, Thiele JU, Baglin JEE, Thomson T, Fullerton EE, Murray CB, Terris BD. Polymer mediated self-assembly of magnetic nanoparticles. J Am Chem Soc 2002;124:2884–2885.

Sun SH, Anders S, Thomson T, Baglin JEE, Toney MF, Hamann HF, Murray CB, Terris BD. Controlled synthesis and assembly of FePt nanoparticles. J Phys Chem B 2003;107: 5419–5425.

Taton TA, Mirkin CA, Letsinger RL. Scanometric DNA array detection with nanoparticle probes. Science 2000;289:1757–1760.

Templeton AC, Hostetler MJ, Warmoth EK, Chen SW, Hartshorn CM, Krishnamurthy VM, Forbes MDE, Murray RW. Gateway reactions to diverse, polyfunctional monolayer-protected gold clusters. J Am Chem Soc 1998;120:4845–4849.

Templeton AC, Wuelfing MP, Murray RW. Monolayer protected cluster molecules. Acc Chem Res 2000;33:27–36.

Thibault RJ, Uzun O, Hong R, Rotello VM. Recognition-controlled assembly of nanoparticles using photochemically crosslinked recognition-induced polymersomes. Adv Mater 2006;18:2179–2183.

Ung T, Liz-Marzan LM, Mulvaney P. Optical properties of thin films of Au/SiO$_2$ particles. J Phys Chem B 2001;105:3441–3452.

Uzun O, Sanyal A, Nakade H, Thibault RJ, Rotello VM. Recognition-induced transformation of microspheres into vesicles: morphology and size control. J Am Chem Soc 2004;126: 14773–14777.

Uzun O, Xu H, Jeoung E, Thibault RJ, Rotello VM. Recognition-induced polymersomes: structure and mechanism of formation. Chem Eur J 2005;11:6916–6920.

Verma A, Srivastava S, Rotello VM. Modulation of the interparticle spacing and optical behavior of nanoparticle ensembles using a single protein spacer. Chem Mater 2005;17:6317–6322.

Xu H, Hong R, Lu TX, Uzun O, Rotello VM. Recognition-directed orthogonal self-assembly of polymers and nanoparticles on patterned surfaces. J Am Chem Soc 2006;128: 3162–3163.

Xu H, Norsten TB, Uzun O, Jeoung E, Rotello VM. Stimuli responsive surfaces through recognition-mediated polymer modification. Chem Commun 2005;5157–5159.

Xu H, Srivastava S, Rotello VM. Hydrogen bonded polymers. Berlin: Springer-Verlag; 2007. pp 179–198.

Zehner RW, Lopes WA, Morkved TL, Jaeger H, Sita LR. Selective decoration of a phase-separated diblock copolymer with thiol-passivated gold nanocrystals. Langmuir 1998; 14:241–244.

Zehner RW, Sita LR. Electroless deposition of nanoscale copper patterns via microphase-separated diblock copolymer templated self-assembly. Langmuir 1999;15:6139–6141.

Zeng H, Li J, Liu JP, Wang ZL, Sun S. Exchange-coupled nanocomposite magnets by nano-particle self-assembly. Nature 2002;420:395–398.

Zheng HP, Lee I, Rubner MF, Hammond PT. Two component particle arrays on patterned polyelectrolyte multilayer templates. Adv Mater 2002;14:569–572.

CHAPTER 7

METALLOSUPRAMOLECULAR POLYMERS, NETWORKS, AND GELS

BLAYNE M. MCKENZIE and STUART J. ROWAN

7.1. INTRODUCTION

Metallosupramolecular polymers (MSPs) can be classified as a subset of metal-containing polymers (Adb-El-Aziz et al. 2006) in which the metal–ligand coordination is reversible and dynamic. Thus, the supramolecular bond in these systems is the metal–ligand coordination bond. In such systems, addition of an appropriate metal ion to a ligand-containing (macro)monomer or polymer (Fig. 7.1) results in either self-assembly of polymeric structures (Dobrawa and Würthner 2005; Schubert et al. 2006) or a change in the structural nature and/or properties of the polymer. Through judicious design and/or selection of the (macro) monomer, the resulting polymeric "aggregate" can have the metal ion placed within the polymeric backbone, as part of a side chain, and/or as part of a branching/cross-linking site. These organic/inorganic hybrid materials potentially offer the attractive combination of the functionality of metal ions, the mechanical properties and processability of polymers, and the self-assembly characteristics and dynamic nature of supramolecular chemistry.

The exact classification of MSPs can be difficult to define because not all metal-containing polymers exhibit reversibility at the metal–ligand coordination site or display polymer-like properties. For example, Ru^{II} complexes with multidentate aza-ligands can offer facile routes to metalloblock copolymers and amphiphilic metallopolymer micelles (Lohmeijer and Schubert 2003a; Fustin et al. 2007), but the slow binding kinetics of such coordination motifs can render the systems nondynamic on a "reasonable" timescale. Crystal-engineered structures, the so-called metal–organic frameworks (Yaghi et al. 2003; Kitagawa et al. 2004), are another class of metal-containing materials that offer interesting and innovative technologies.

Molecular Recognition and Polymers: Control of Polymer Structure and Self-Assembly.
Edited by V. Rotello and S. Thayumanavan
Copyright © 2008 John Wiley & Sons, Inc.

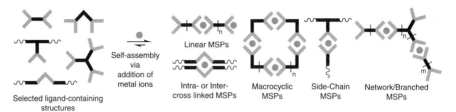

Figure 7.1 Schematic representation of a selection of different possible MSP architectures.

However, such systems are not discussed here given their lack of traditional polymer-like properties.

MSPs differ in a number of ways from more conventional polymers in that their formation (or change in structure) occurs though supramolecular assembly and as such does not require the use of an initiator or catalyst. Furthermore, the presence of supramolecular motifs as an integral part of the polymer can impart a dynamic nature onto these materials. The reversibility of the kinetically labile metal–ligand coordination bonds establishes a continual binding equilibrium and potentially allows the development of stimuli-responsive systems. For example, a change in environmental conditions, such as temperature, concentration, pH, and so forth, can alter the amount of metal ions complexed to the ligand, which in turn can drastically affect the properties of the material. This is in contrast to most conventional covalent polymer systems, where environmental conditions have distinctly less effect on polymer properties. Another feature of these systems is that the resulting material can also utilize the functionality inherent in the metal ions. Especially of interest are those functionalities such as fluorescence, catalysis, conductivity, molecular binding/sensing, and so on, where coordination can alter the properties of the metal ion.

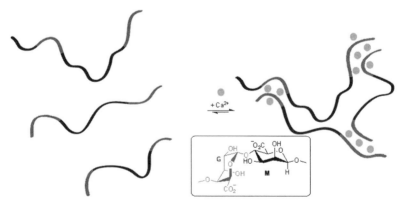

Figure 7.2 Schematic representation of alginate cross-linking with divalent cations and the chemical structure of the constituent repeat units, guluronate (**G**) and mannuronate (**M**). Carboxylate groups present along the backbone (largely from the **G** residues) interact with multivalent cations to yield metal ion cross-linked gels.

MSPs are not just an academic curiosity. Nature uses reversible metal–ligand interactions in a variety of ways, and inspiration for synthetic MSPs arises from a variety of natural metal-containing biopolymers. [For example, this has led to the field of artificial metalloenzymes, which merges protein chemistry and structure with synthetic organometallic chemistry to access hybrid catalysts (Creus and Ward 2007).] Metalloproteins often utilize metal binding cofactors as centers for catalysis, electron transport, or substrate binding. An example of other biomaterials that have attracted significant industrial interest and can be classified as MSPs are alginate polysaccharides (Ertesvåg and Valla 1998). These biopolymers are found in brown seaweeds and are linear blocky copolymers of guluronic acid (G) and mannuronic acid (M). In the presence of certain metal ions, which can cross-link the polymers through the carboxylate functionality on the sugar repeat units, these biopolymers form gels (Fig. 7.2). Such biomaterials hold promise for a variety of applications including biomedical, sequestering agents, and viscosity modifiers.

7.2. METAL–LIGAND BINDING MOTIFS

With access to a deep catalog of ligands and nearly half the periodic table offering applicable metal ions, a vast range of metal–ligand complexes can be envisaged. The characteristics of a particular metal–ligand supramolecular motif can change dramatically, depending on the combination of metal ion and ligand utilized. Significant variations in motif structure (geometry, metal–ligand stoichiometry), dynamics (kinetic lability), and thermodynamic stability can all easily be achieved. With such a wide variety of potential metal–ligand supramolecular motifs available, the researcher can choose the appropriate motif to impart the desired properties onto the targeted material.

Some typical examples of metal–ligand supramolecular motifs and ligands utilized in MSPs are provided in Figure 7.3a and b. The coordination number and geometry (e.g., tetrahedral, square planar, octahedral, etc.) of the complex depends on the choice of metal ion as well as ligand. The combination of these two components dictates the stoichiometry of the motif. For example, 2:1 ligand/metal complexes can be obtained with bidentate ligands bound to metal ions with a coordination number of 4 or alternatively with terdentate ligands binding to hexacoordinate metal ions. It is also possible to design other metal–ligand stoichiometries; for example, 3:1 ligand/metal complexes can be obtained through the use of bidentate ligands bound to hexacoordinate metal ions or terdentate ligands with lanthanide ions, which prefer higher coordination numbers (8–10). Such systems will generally be homocoordinate species, in that only one type of ligand is bound to the metal ion. An additional level of structural diversity can be envisioned if metal–ligand motifs can be designed that allow exclusively the formation of a single heterocoordinate complex. In dynamic systems, this level of control is generally not easy to attain and a combination of homo- and heterocoordination species results. There are however classes of metal–ligand species that have been utilized as heterocoordinate motifs, for

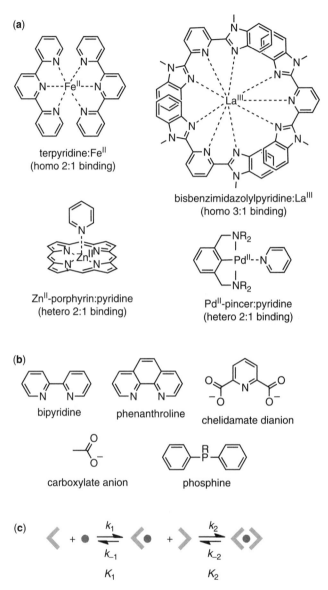

(a)

terpyridine:Fe^{II}
(homo 2:1 binding)

bisbenzimidazolylpyridine:La^{III}
(homo 3:1 binding)

Zn^{II}-porphyrin:pyridine
(hetero 2:1 binding)

Pd^{II}-pincer:pyridine
(hetero 2:1 binding)

(b)

bipyridine   phenanthroline   chelidamate dianion

carboxylate anion   phosphine

(c)

k_1 k_{-1} K_1 k_2 k_{-2} K_2

Figure 7.3 (a) Selected metallosupramolecular motifs and (b) selected ligands, which have been used in the construction of MSPs. (c) Schematic representation of the equilibria (K) and rate constants (k) present in a 2 : 1 ligand–metal supramolecular motif. Dotted lines in the diagram represent the metallosupramolecular reversible bonds in the motif.

example, the palladium pincer ligand/pyridine complex or the metalloporphyrin/ pyridine motif. In such compounds, the pincer ligand–metal bond or the porphyrin metal bond are generally kinetically inert, but reversibility is bestowed on the motif through the dynamic nature of the metal–pyridine coordination.

Other important considerations when selecting a metal–ligand supramolecular motif are its thermodynamic (e.g., K_1 and K_2) and kinetic (e.g., k_1, k_{-1}, k_2, k_{-2}) parameters (Fig. 7.3c). Thermodynamic factors primarily control the size of the MSP and depend on both the metal ion and choice of ligand. For example, oligodentate ligands can be expected to bind orders of magnitude stronger than monodentate ligands. Kinetic factors prove to be just as important for the behavior of an MSP (vide infra) and metal–ligand motifs exhibit a wide range of complexation/ decomplexation rates. For example, studies on the thermodynamic and kinetic data of the binding of terpyridine (terpy) with a series of metal ions (Holyer et al. 1966) show that the overall binding constants ($K_1 K_2$) for the 2:1 complexes can vary from 10^{18} to 10^{21} M^{-1} for Co^{II} and Fe^{II}, respectively, for example, and that the dissociation rates can also vary significantly, for example, the k_{-2} (Fig. 7.3c) value of terpy varies from about 10^{-3} to 10^{-7} s^{-1} for $Co^{II}(terpy)_2$ and $Fe^{II}(terpy)_2$, respectively.

Such facets of dynamic metal–ligand binding offer a range of tunable properties in MSPs. We now discuss a selection of examples that reveals how these systems differ from typical covalent polymers and small molecule metal complexes yet combine the properties of both to form a new class of materials.

7.3. LINEAR AND MACROCYCLIC MAIN CHAIN MSPs

In linear main chain MSPs, metal–ligand coordination drives the formation of the polymer. Such systems can be accessed easily through the addition of appropriate metal ions to a ditopic ligand end-capped monomer unit in a 1:1 ratio. The resulting polymer will then contain multiple metal complexes along its backbone (Fig. 7.1). Alternatively, if the macromonomers only contain one ligand, then addition of a metal ion will result in an MSP that has only one supramolecular motif within its backbone (Lohmeijer and Schubert 2003b). In recent years, a wide variety of main chain MSPs have been prepared and studied. One aspect of these MSPs that differentiates them from covalent polymers is their dynamic behavior, namely, their ability to polymerize/depolymerize in response to environmental factors (concentration, temperature, etc.). A number of dynamic processes are possible of course from polymerization/depolymerization to ring-chain equilibria and chain scission and mixing, analogous to those observed in certain reversible covalent polymerizations (Rowan et al. 2002).

One of the earliest examples of a soluble main chain MSP was reported by Lahn and Rehahn (2001), who utilized phenanthroline end-capped monomers with Cu^I or Ag^I ions (Fig. 7.4). Their studies demonstrated how the concentration and solvent could influence not only the nature of the MSP but also the kinetic lability of the metal–ligand interaction. MSPs formed in noncoordinating solvents [e.g., 1,1,2,2-tetrachloroethane-d_2 (TCE-d_2)] did not exhibit any dynamic behavior. However, the behavior of the polymer changed dramatically in the presence of coordinating solvents, such as acetonitrile. In these solvents large amounts of cyclic species are formed upon dilution, consistent with the presence of ring chain equilibria. It was suggested that the solvent acts as a chain scission agent, aiding the decomplexation

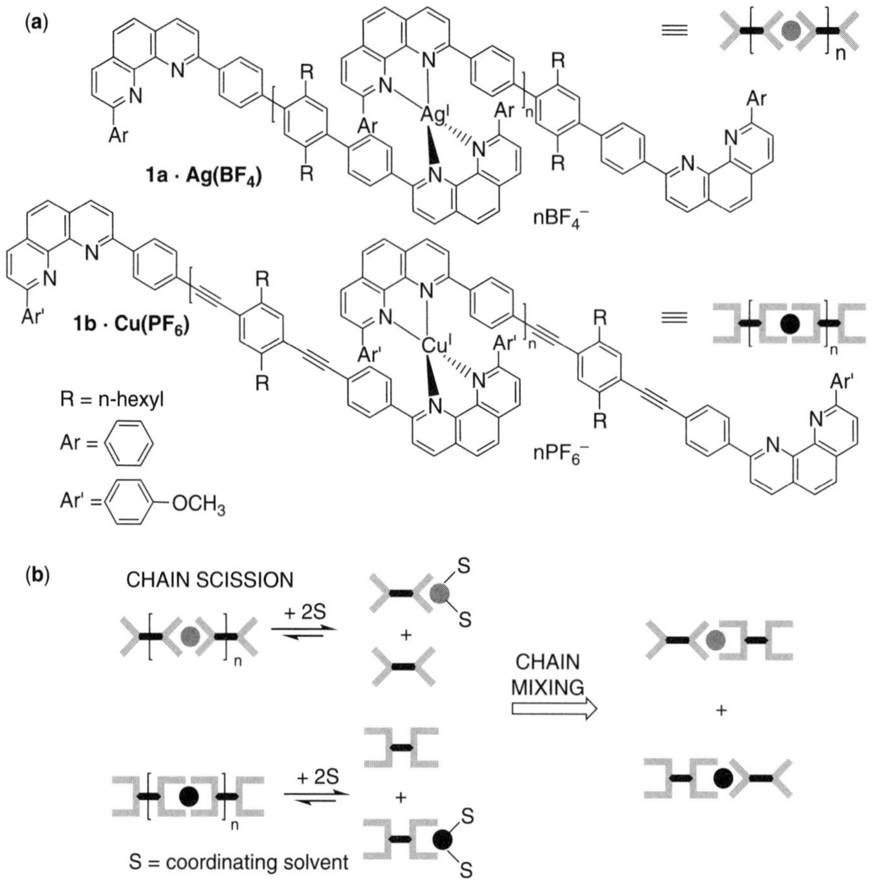

Figure 7.4 (a) MSPs composed of ditopic phenanthroline ligands (**1a** and **1b**) complexed to Ag^I or Cu^I ions. (b) Schematic representation of the solvent displacement mechanism of MSP dissociation and chain mixing (Lahn and Rehahn 2001).

of the ligand from the metal ion. In an elegant demonstration of the effects of solvent on the dynamic behavior of these systems, two different polymers were mixed together: **1a** · AgBF$_4$ and **1b** · CuPF$_6$. This system appeared to be stable in TCE-d_2 for at least 3 days, showing no significant interchange of components. However, upon the addition of acetonitrile, exchange processes occurred, resulting in the formation of mixed copolymers. These experiments highlight the important role of the solvent in the dissociation/exchange mechanism of MSPs, adding a further complication in determining if a metallopolymer exhibits dynamic, supramolecular behavior. For example, this system could be considered an MSP in acetonitrile; however, it does not show supramolecular behavior in the noncoordinating solvent.

The dynamic nature of the metal–ligand bond is of course one of the key features of MSPs. What makes these systems even more versatile from this point of view is that different metal ions will exhibit different binding kinetics and thermodynamics. A pertinent demonstration of this was recently provided by Chiper et al. (2007), who investigated the possibility of using gel permeation chromatography (GPC) to analyze different metal-containing MSPs. Prior investigations into the GPC analysis of labile MSPs, for example, Zn^{II}-terpy systems, revealed that fragmentation of the polymeric aggregate occurs as it passes down the column. Metallopolymers that contain more kinetically inert metal complexes (e.g., Ru^{II}-terpy) within the backbone can be successfully analyzed by this technique because their kinetically inert nature makes them less like MSPs and more like covalent polymers. Chiper and colleagues prepared a range of metal–terpy MSPs (Co^{II}, Fe^{II}, and Ni^{II}) that, based on available binding constant data, should fall between the Zn^{II} and Ru^{II} systems in terms of binding stability. Initial GPC studies were carried out on mono-end-capped terpy–poly(ethylene oxide) (PEO) metal complexes and using a range of sample concentrations. In the Co^{II} and Fe^{II} systems, lowering the concentration resulted in a detectable shift to macromonomer in the GPC whereas the Ni^{II} complexes were stable across the range of concentrations investigated (0.0011–0.0221 mM). This is consistent with the stronger binding strength of terpy to Ni^{II} than Co^{II} or Fe^{II}. Of course, binding kinetics may also play a role in determining the stability of a complex as it passes down the column. Further GPC studies on Ni^{II} complexes of a ditopic terpy–PEO macromonomer revealed an average degree of polymerization of 28 units, suggesting that this MSP was robust enough to survive GPC analysis.

From a supramolecular viewpoint, one possible advantage of using metal–ligand interactions (e.g., vs. hydrogen bonding) is their relative insensitivity to the presence of water. This has allowed a number of studies of MSPs in aqueous environments. For example, Vermonden and coworkers (2003) utilized short ethylene oxide spacers between pyridine-2,6-dicarboxylate end groups (**2**) to prepare MSPs in aqueous solutions and studied the effects of concentration, temperature, and metal–ligand stoichiometry in these systems (Fig. 7.5). As was also shown in the Rehahn system, decreasing the concentration of a 1:1 complex of **2**·$Zn(ClO_4)_2$ increased the fraction of rings. In addition, they showed that increasing temperature increased the fraction

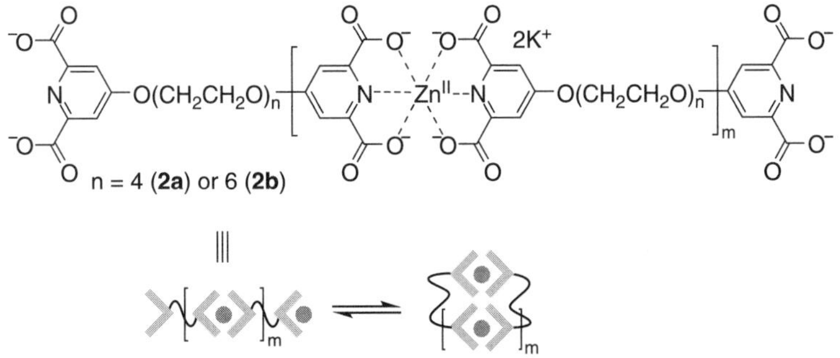

Figure 7.5 Water-soluble MSPs, based on a 2,6-pyridine dicarboxylate end-capped oligo(ethylene oxide) ditopic monomer (**2a** and **2b**) and Zn^{II} ions prepared by Vermonden and coworkers (2003).

of polymer chains, suggesting an entropically controlled ring-opening polymerization is occurring.

One key aspect of any MSP is the metal ion–ligand stoichiometry (Chen and Dormidontova 2003). In Vermonden et al.'s (2003) system, two ligands can bind to one Zn^{II} ion; thus, a 1:1 ratio of **2** to $Zn(ClO_4)_2$ is required to obtain high molecular weight aggregates. Deviation from this stoichiometry would result in an excess of one of the components, yielding chain ends and a significant decrease in the molecular weight. Using a combination of viscosity as a probe for molecular weight and theoretical models, Vermonden's group investigated how the length of the ethylene oxide core ($n = 4$ or 6) and metal–ligand stoichiometry influenced the ring chain equilibria. It would be expected that a maximum in the reduced viscosity would be observed at a 1:1 metal–ligand ratio. For $2a \cdot Zn(ClO_4)_2$ this is indeed what is observed at monomer concentrations of 19.2 and 37.8 mM. For $2b \cdot Zn(ClO_4)_2$, such a maximum is observed at a higher monomer concentration (34.5 mM). However, at lower concentrations (17.8 mM) a dip in the reduced viscosity occurs at a 1:1 metal–ligand ratio. Theoretical studies suggest that this occurs as a consequence of the longer ethylene oxide core, which allows a significant portion of the monomer units to form rings and reduces the viscosity of the system. This behavior is not observed in $2a \cdot Zn(ClO_4)_2$ as the shorter core hinders the formation of small macrocycles at the lower concentration.

Ring formation can also be induced by careful monomer design. Various groups have synthesized terpy-containing multifunctional ligands with geometries that favor ring formation. For example, Constable and colleagues (2003) tailored the length of a polypeptide core unit to suit a thermodynamically favorable ring size. Newkome's group (Wang et al. 2005) synthesized bent ditopic monomers that predispose the system to ring formation upon addition of metal ions, also yielding macrocyclic structures (Fig. 7.6). Attaching reactive groups onto these bent ditopic ligands (**3**) yielded a method for accessing large macrocycles using a templating methodology (Wang et al. 2005).

Most MSPs are polyelectrolytes; as such, the nature of the counterions should be considered. For example, counterions can potentially act as competitive ligands for the metal ion and have a significant effect on the formation of an MSP. Schmatloch et al. (2004) demonstrated that the degree of complexation and subsequently the size of the MSP in aqueous solutions formed from Fe^{II} and terpy end-capped PEO depends on the counterion present: chloride > sulfate >> acetate. Lipophilic counterions offer the ability to access amphiphilic MSPs, which in turn potentially allow the polymeric materials themselves to assemble into hierarchical structures. Liu and coworkers (2004) used such a material to create higher ordered structures in a variety of different ways. For example, addition of long alkyl chain counterions, such as dihexadecyl phosphate (DHP), to a preformed MSP, based on a terpy/Fe^{II} system (**4** $\cdot$ Fe(OAc)$_2$), results in counterion exchange and organization of the DHP around the metal–ligand binding site. These amphiphilic MSPs can assemble into monolayers at the air–water interface or into straight rods (length ~ 200 nm) on a graphite surface in the presence of long chain alkanes. These amphiphilic MSPs also assemble in the solid state to form thermotropic

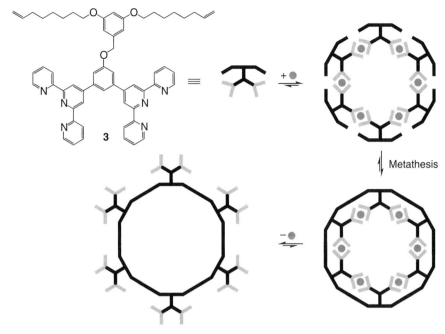

Figure 7.6 Structure of a bent ditopic ligand prepared by Wang et al. (2005) and a schematic representation of a metal-templated macrocycle.

phases that consist of alternating layers of the MSP and DHP. Bodenthin et al. (2005) went on to show that the phase transition in the amphiphilic mesophase (namely, melting of DHP's alkyl chains) can induce mechanical strain on the MSP backbone, resulting in a distortion of the metal ion's coordination geometry (Fig. 7.7). This results in a transition of the Fe^{II} from a low spin (diamagnetic) to a high spin (paramagnetic) state, as the crystal field splitting of the d-orbitals decreases. This interesting result nicely demonstrates how judicious choice of the counterions can be used to tailor the properties of polyelectrolyte MSPs.

One aspect of polyelectrolytes that has received a lot of attention in recent years is their utilization in the assembly of multilayered structures, which are formed via alternating adsorption of oppositely charged polymers. Using a terpy-end-capped ditopic monomer with, for example, Fe^{II} salts, Schütte et al. (1998) demonstrated that MSPs can be utilized in a layer-by-layer assembly with polyanions such as poly(styrene sulfonate).

A number of studies have begun to address the solid-state assembly and properties of MSPs. The nature of main chain MSPs is such that their backbone consists of alternating segments of a core unit and a metal–ligand complex, resulting in what can be considered a multiblock-like structure. If the core consists of a "soft" unit, such as poly(tetrahydrofuran) (Beck et al. 2005) or poly(ethylene glycol) (Schmatloch et al. 2003), and the metal–ligand complex is a "hard" ionic, aromatic species, for example, complexes of 2,6-bis(1′methylbenzimidazolyl) pyridine

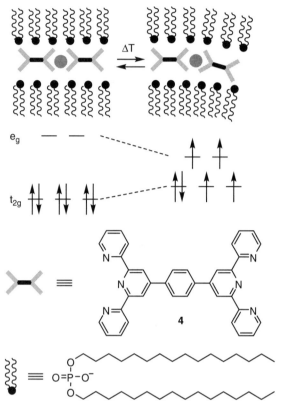

Figure 7.7 Schematic representation of the temperature-induced change in the MSP composed of terpy end-capped ditopic monomers and Fe^{II} dihexadecyl phosphate. The temperature-induced transition results in a distortion of the metal ion coordination geometry, which occurs because of the melting of the amphiphilic counterions, giving rise to a reversible transition from a diamagnetic low spin state to a paramagnetic high spin state (Bodenthin et al. 2005).

(Mebip) or terpy, then the different segments along the backbone of the MSP can phase segregate. This phase segregation can result in physically cross-linked films, which can exhibit thermoplastic elastomeric properties. Not all MSPs show this type of phase segregation. For example, MSPs based on polymeric aromatic species such as poly (*p*-phenylene ethynylene)s (**5**; Knapton, Rowan, et al. 2006) or poly (*p*-xylylene)s (**6**; Knapton, Iyer, et al. 2006) show very different behavior (Fig. 7.8). The self-assembly formation of an MSP offers an attractive approach to the assembly of high molecular weight conjugated macromolecules from well-defined, easy to process precursors (El-Ghayoury et al. 2002; Chen and Lin 2007). MSPs of **5** · Fe(ClO$_4$)$_2$ yielded soluble easy to process materials that formed mechanically stable films and fibers with metal ions [e.g., r.t. modulus: Fe(ClO$_4$)$_2$ ∼ 80 MPa]. Wide-angle X-ray scattering data of these films is not consistent with the phase segregation of the core and complex units, although a lamellae structure is observed

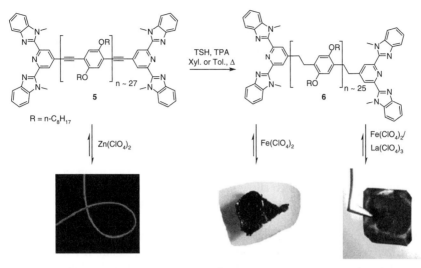

Figure 7.8 Chemical structures of 2,6-bis(1′methylbenzimidazolyl) pyridine (Mebip) end-capped poly(2,5-dioctyloxy-*p*-phenylene ethynylene) (PPE) (**5**) and poly(2,5-dioctyloxy-*p*-xylylene) (PPX; **6**) MSPs and pictures of the resulting fibers and films (Knapton, Iyer, et al. 2006; Knapton, Rowan, et al. 2006).

that appears to be mainly driven by the octyloxy side chains, commonly observed in other poly(*p*-phenylene ethylene) systems. In stark contrast to $5 \cdot Fe(ClO_4)_2$, $6 \cdot Fe(ClO_4)_2$ yields films with very poor mechanical properties, consistent with no physical cross-linking present in this system. As previously mentioned, Mebip ligands can bind to lanthanoid ions in a 3:1 ratio. Thus, replacing a percentage of the 2:1 complex forming an Fe^{II} ion with a lanthanide ion should result in a cross-linked system. Therefore, an MSP consisting of 9 mol% $La(ClO_4)_3$ and 91 mol% $Fe(ClO_4)_2$ (with respect to **6**) was able to be solution cast into purple films, which displayed appreciable mechanical properties (r.t. modulus = 220 MPa). This nicely demonstrated that changing the supramolecular behavior of the metal ions can result in the assembly of different polymeric architectures, which in turn is expressed in the mechanical properties of the MSP.

A different class of conjugated metal-containing polymers has recently been reported by Boydston et al. (2005). They investigated a range of ditopic *N*-heterocyclic carbenes (NHCs) as the ligand-containing monomer unit. Figure 7.9 depicts one example of a ditopic NHC that was investigated and shown to form organometallic polymers with a range of metal ions. For example, the reaction of **7** with Pd^{II} and Pt^{II} salts at 110 °C yielded kinetically inert metal-containing polymers. However, addition of Ni^{II} salts to **7** appears to yield dynamic polymeric materials (**8**) that depolymerize in polar protic solvents (H_2O and MeOH; Boydston et al. 2006).

Other more complex conjugated MSP architectures have also been investigated. One of the architectures that has received some attention is polyrotaxanes, in which a polymeric rodlike component is threaded through a macrocycle

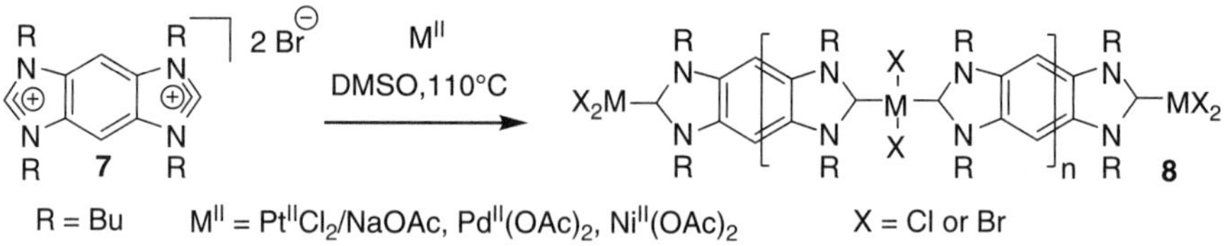

Figure 7.9   *N*-Heterocyclic carbine ligands polymerized by the addition of Pt^{II}, Pd^{II}, or Ni^{II} salts (Boydston et al. 2005, 2006).

(Divisia-Blohorn et al. 2003). Buey and Swager (2000) utilized a copper-containing side chain MSP as a template to allow access to a conjugated metal-containing three stranded polyrotaxane (Fig. 7.10). Metallorotaxane monomer **11** has thiophene moieties attached to an electron-poor bipyridine unit as the thread component **10** and electron-rich ethyelenedioxythiophene (EDOT) units, which can be oxidatively polymerized at low potentials, on macrocycle **9**. Thus, controlled oxidation of monomer **11** causes electropolymerization of the macrocycle unit while leaving the thread unreacted, yielding side chain pseudopolyrotaxane **12**. Applying higher

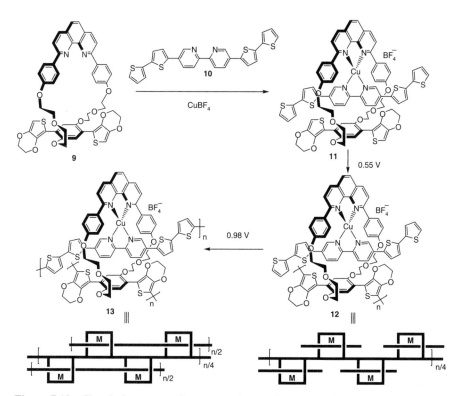

Figure 7.10   Chemical structure of a monomeric pseudorotaxane electropolymerized ladder polymer (Buey and Swager 2000).

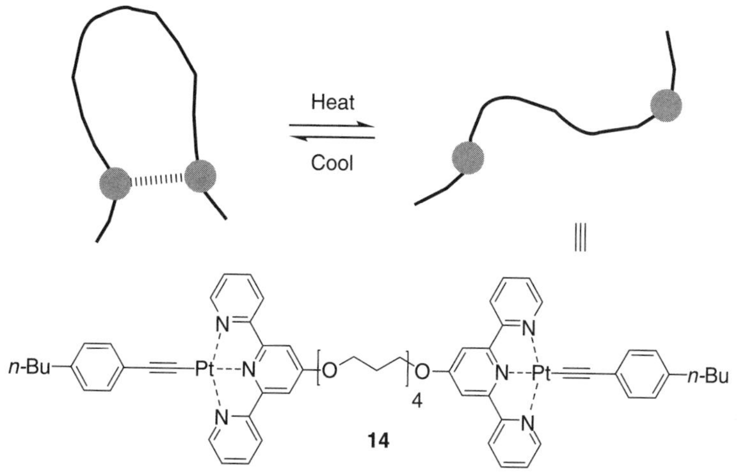

Figure 7.11 Schematic representation of the dynamic Pt–Pt interaction (Yam et al. 2006).

potentials to **12** results in the polymerization of threads, yielding conjugated three stranded metalloladder polymer **13**. Conductivity and cyclic voltammetry of the electropolymerized films suggest that the central polymer chain (the one derived from the EDOT units) is not only the main contributor to the conductivity of the ladder polymers but is also partially isolated when the outer two polymer strands are in their undoped (insulating) state. Furthermore, it is also thought that the metal centers play an important role in the conduction of this system by aiding interchain electron hopping. This is especially important at low potentials when the outer chains are insulating. This interesting example highlights the potential of the metal component of MSPs to aid functional processes if the system has been designed appropriately.

In some cases, MSPs can be accessed through the use of dynamic metal–metal bonds. For example, although terpy–Pt–acetylide complexes are themselves not dynamic, these complexes can interact with each other through a combination of Pt–Pt interactions and π–π stacking (Bailey et al. 1993). Using this interesting metal–metal binding motif, Yam and colleagues (2006) developed metal-containing monomers (e.g., **14**) that offer the potential to be thermoresponsive (Fig. 7.11).

7.4. METALLOSUPRAMOLECULAR NETWORKS AND GELS

Of course, appropriate design of the (polymeric) ligand can also result in nonlinear MSPs. Thus, a range of ligand-containing polymeric species can be imagined (e.g., Fig. 7.1), which upon addition of a metal ion can self-assemble into side chain, intra-/intercross-linked (e.g., Fig. 7.2), or other more complex MSPs architectures.

An interesting example of using metals to cross-link conjugated polymers was reported by Kokil et al. (2003). An organic-soluble PPE (**15**) was complexed with

Pt^0 via a ligand exchange process with Pt(styrene)$_3$ (**16**; Fig. 7.12), yielding a metal-cross-linked organogel (**17**) at low Pt loading. Gelation was not instantaneous, allowing the styrene solutions to be cast into mechanically stable films. These films were not soluble in common organic solvents, consistent with their cross-linked nature. The reversibility of the ligand exchange was demonstrated by redissolving the films in styrene, which presumably recomplexes the platinum cross-links. It is interesting that these films exhibit significantly enhanced charge transport characteristics relative to the "free" polymer, a functional advantage of using metallosupramolecular cross-linkers.

Yount et al. (2005) reported the formation of organogels using poly(vinylpyridine) (PVP) and ditopic metallopincer cross-linkers. This study provided particularly pertinent information on the dynamic elements of MSPs and elegantly demonstrated how they control the material's properties. PVP dissolved in dimethylsulfoxide (DMSO) is cross-linked with either bis-Pd^{II}- (**18a**) or Pt^{II}-pincer compounds (**18b**, Fig. 7.13; Yount et al. 2005). Addition of 5% **18a** · Pd^{II} to a PVP solution results in a viscous material ($\eta = 6.7$ Pa s), whereas the corresponding PVP **18b** · Pd^{II} is a gel ($\eta = 550$ Pa s). Changing R from methyl to ethyl does not affect the thermodynamics of the pyridine/Pd^{II} interaction; however, the rate of exchange decreases by approximately 2 orders of magnitude. Further studies on these materials and their

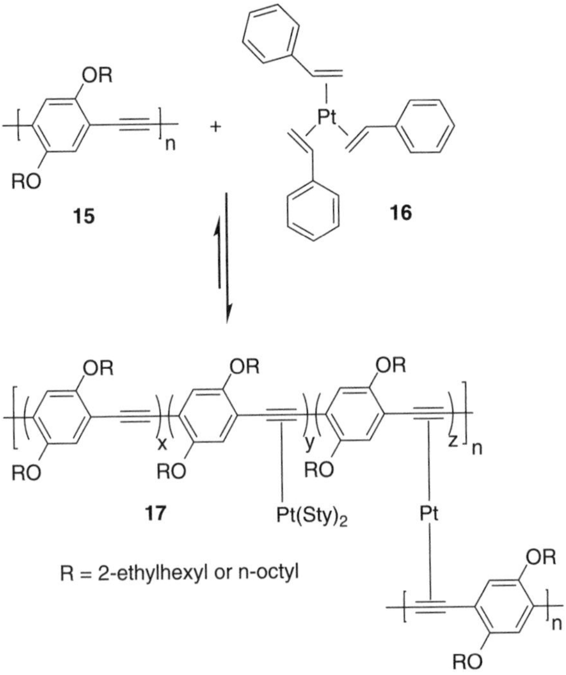

Figure 7.12 Chemical structure of a linear poly(*p*-phenylene ethynylene) (PPE) and the ligand exchange process utilized by Kokil et al. (2003) to yield a conjugated cross-linked MSP network and gel.

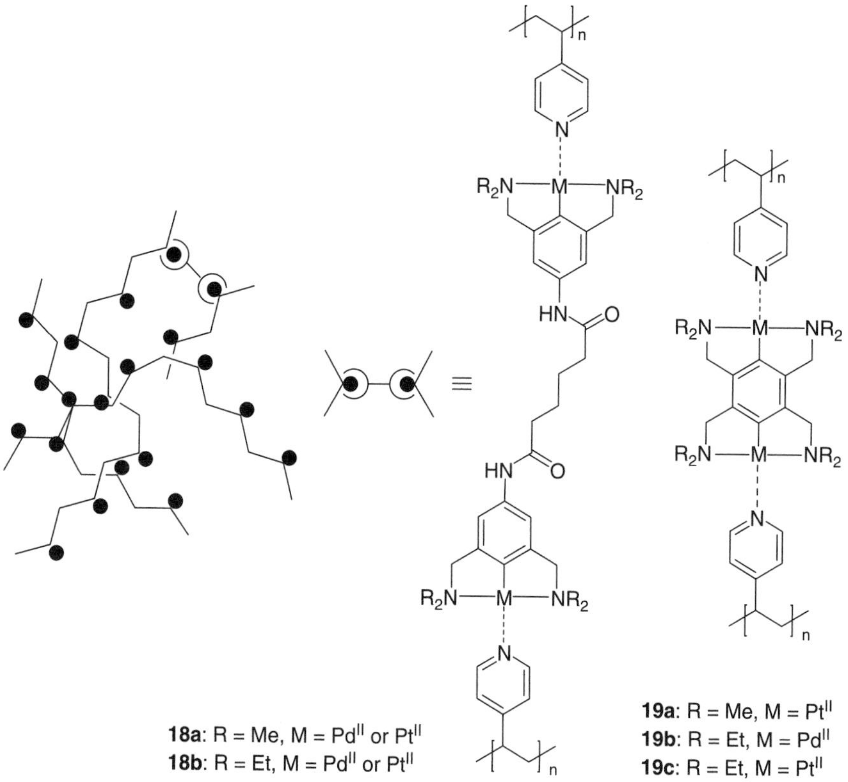

18a: R = Me, M = Pd^{II} or Pt^{II}
18b: R = Et, M = Pd^{II} or Pt^{II}

19a: R = Me, M = Pt^{II}
19b: R = Et, M = Pd^{II}
19c: R = Et, M = Pt^{II}

Figure 7.13 Schematic representation of the structure of a cross-linked MSP gel prepared by Loveless and coworkers (Yount et al. 2005; Loveless et al. 2007) from poly(vinyl pyridine) (PVP) cross-linked with bis-Pd^{II}– or Pt^{II}–pincer compounds (**18** and **19**).

more thermodynamically and kinetically stable Pt^{II} analogues demonstrate that it is the off-rate of the metal binding motif (and not the binding constant) that is responsible for the observed change in viscosity. This highlights the importance of kinetics in dictating the properties of MSPs and suggests that "slow" means "strong."

With the knowledge of how these systems behave, Loveless and coworkers (2007) set out to gain control of the sol–gel and gel–gel transitions of these materials. The reversible nature of both the pyridine/Pd^{II} and pyridine/Pt^{II} interactions was demonstrated by use of competitive interacting additives. For example, either acids (such as HCl and H$_2$SO$_4$), which can protonate the PVP, or competitive coordinating ligands [such as dimethylaminopyridine (DMAP) or Cl$^-$ ions], which can displace the pyridine in the complexes, reduce the cross-linking and weaken the gels. When acids were used, addition of a weak base reversed the effect and resulted in the reformation of the gel. The stimuli-responsive behavior of these gels was most profound close to the percolation threshold of the cross-linker. For example, for the 10 wt% PVP in DMSO system with the cross-linker **19a** at the percolation threshold (0.8 wt%), the addition of 0.004 wt% NaCl yielded a 200-fold drop in viscosity. With only slightly more cross-linker (1.0 wt%), the same amount of NaCl only halves the viscosity.

Mixing different cross-linkers (**19b** and **19c**) yielded systems with a strong gel–weak gel transition, rather than a distinct sol–gel shift. When the concentration of each of the cross-linkers was above the critical percolation threshold, the kinetically slower cross-linker (**19c**) dictated the gel properties. Upon addition of enough of a competitive binding additive, such as DMAP, to drop the concentration of the "active" cross-linking units below their individual percolation thresholds but still allowing the total amount of both active cross-linkers (**19b** and **19c**) to be above the percolation threshold results in a gel whose properties are now controlled by the kinetically faster cross-linker (**19c**).

If the polymeric macromonomer contains only one metal-binding center within its backbone, then addition of an appropriate metal ion will result in the self-assembly of a metallosupramolecular star polymer. Johnson and Fraser (2004) synthesized bipy-centered block copolymers that, upon addition of $Fe(BF_4)_2$, reversibly self-assemble into a hexa-arm star. Bender et al. (2002) also prepared star polymers utilizing Eu^{III} and the β-diketonate anion. A tri-arm star polymer could be self-assembled via complexation of Eu^{III} to three monotopic poly(lactic acid) macromonomers. The large coordination sphere of the Eu^{III} allows the further addition of a neutral bipy-containing poly(ε-caprolactone). The result is a penta-arm star, which exhibits block-copolymer-like properties, namely, phase segregation in the solid state. Atomic force microscopy studies show the formation of a lamellar structure comprising poly(ε-caprolactone) and poly(lactic acid) with the metal ions placed at the interface. Preliminary work on this system suggests that the temperature-sensitive nature of the film's morphology is partly a consequence of the dissociation of the kinetically labile bipy ligand at high temperature.

The above examples show one of the most appealing aspects of MSPs, namely that the dynamic nature of the metal–ligand bond potentially allows the development of stimuli-responsive materials, which exhibit dramatic changes in macroscopic properties upon exposure to environmental stresses. An elegant example of a linear stimuli-responsive MSP was reported by Paulusse et al. (2006). Mechanoresponsive polymers were synthesized from ditopic phosphine-terminated poly(tetrahydrofuran) monomers and $PdCl_2$ (Fig. 7.14a, **20**). In this system, they showed that ultrasound sonication can be used to induce chain scission of the MSP (**21** and **22**). The proposed mechanism of depolymerization involves shear-induced breakage of the coordination bonds caused by the collapse of the gas bubbles formed during sonication. This method provides a potentially facile means of processing, as chain scission only takes place at the reversible Pd^{II}–phosphine bonds. When the shear stress (or in this case sonication) is removed, the polymeric aggregates reform. In addition to the ease of processing afforded by this technique, potential catalytic sites are offered by the accumulation of unbound palladium chain ends, opening the door to mechanochemical reactions using similar systems. Replacing the Pd^{II} in these systems with metals ions that have up to four free coordination sites, for example, Ir^I or Rh^I, and using phosphinite rather than phosphine ligands (**23**), ultrasound-responsive gels have been prepared (Fig. 7.14b; Paulusse et al. 2007). These two

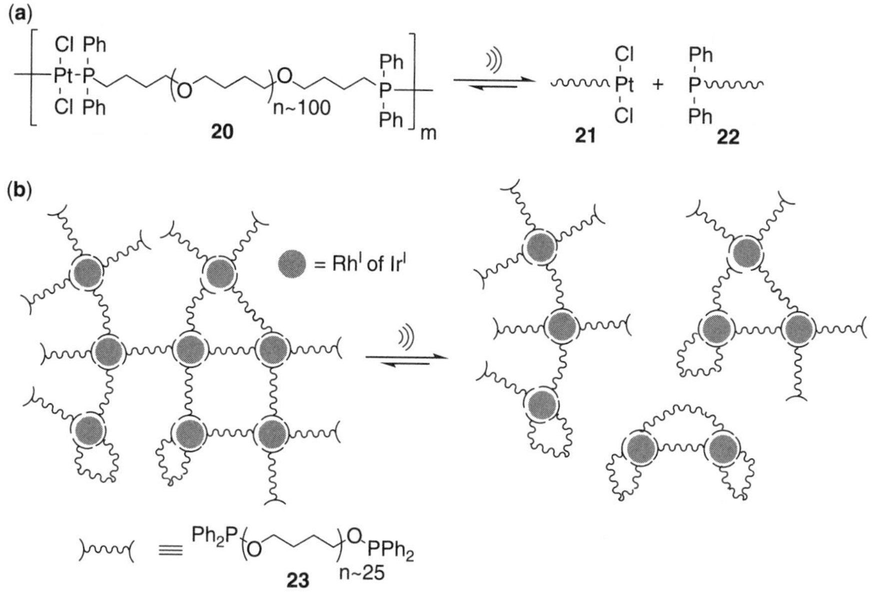

Figure 7.14 (a) Chemical structure of a phosphine–platinum mechanoresponsive MSP prepared by Paulusse et al. (2006, 2007) and (b) schematic representation of gels prepared by a similar polymer cross-linked with Ir^I or Rh^I salts.

different MSP gels show very different mechanical properties with the Ir^I gels exhibiting a higher modulus than the Rh^I gels. This is attributed to the different binding kinetics of these two metal ions. The exchange rate of Ir^I phosphinite complexes (~ 10 days) is much slower than that of the Rh^I phosphinite complexes ($= 26$ min), further illustrating the importance of kinetic considerations when designing dynamic systems and again highlighting the tunability inherent in MSPs.

Multistimuli/multiresponsive gels (Beck and Rowan 2003; Weng et al. 2006) were prepared by adding different metal ions to a ditopic monomer (Fig. 7.15, **24**). These materials form acetonitrile organogels that exhibit thermo-, chemo-, and mechanoresponses. Optical microscopy studies suggest that the material is composed of spherulitic particles, which aggregate to yield sample-spanning phases responsible for gelation. When subjected to increasing shear stress, the gels exhibit a yield point followed by a shear-thinning region. The study examined how different metal salts [e.g., $Zn(ClO_4)_2$ and $Zn(ClO_4)_2$ with 2% $La(ClO_4)_3$] could be used to tailor the nature of the self-assembly and macroscopic properties of the gel, probing the effect on the gel of metal salts with different coordination abilities. Although Zn^{II} ions bind two Mebip ligands, the larger lanthanide ion present in $La(ClO_4)_3$ can bind up to three. The particles in the Zn^{II} only gel are almost completely birefringent, suggesting that they are crystalline (Fig. 7.15a). However, in the Zn^{II}/La^{III} gel a birefringent core of the particle is surrounded by a large diffuse amorphous halo (Fig. 7.15b). It is speculated that the presence of the 3:1 Mebip/La^{III} complexes in

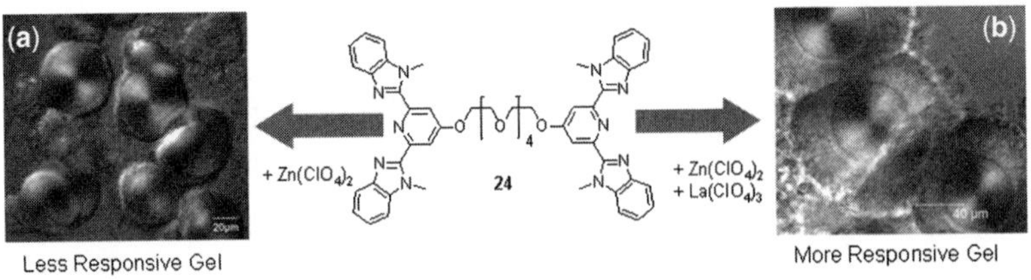

Figure 7.15 Chemical structure of the 2,6-bis(1′methylbenzimidazolyl) pyridine (Mebip) end-capped monomer (**24**) and optical microscopic images of its resulting MSP gels (8 wt% in acetonitrile) formed with (a) Zn(ClO$_4$)$_2$ and (b) Zn(ClO$_4$)$_2$/La(ClO$_4$)$_3$ (mole ratio = 97 : 2) obtained using a laser scanning confocal microscope operated in transmitted mode (Weng et al. 2006).

this gel results in the formation of nonlinear architectures that diminish the degree of crystallinity in the MSP, explaining the existence of the amorphous halo. Rheological studies show that the La^{III}-containing gel exhibits a much lower yield point than the gel with Zn^{II} only. This demonstrates that the responsive behavior of the gels can be tailored by the nature of the self-assembly, which in turn is controlled through judicious choice of the metal salts.

In a related system, Oh and Mirkin (2005) showed that carboxylate-functionalized binapthyl bis-metallotridentate Schiff base building blocks (Fig. 7.16, **25**), which form MSPs (**26a–c**) in the presence of appropriate metal salts [e.g., Zn(OAc)$_2$, Cu(OAc)$_2$, or Ni(OAc)$_2$], yield micron-sized particles upon the addition of a nonsolvent (e.g., diethyl ether or pentane). The growth of the particles could be monitored by microscopy. Scanning electron microscopy reveals that small particles (1–2 m) are initially formed that then aggregate into clusters before annealing into single larger particles (>5 m). Finally, the surfaces of these larger particles are smoothed, possibly illustrating the dynamic nature of the system. Additional studies with other carboxylate-functionalized Schiff base ligands have shown that more complex, solvent-switchable structures can be accessed, for example, particles that can be converted into crystalline rods upon the addition of methanol (Jeon et al. 2007) and molecular helices that are stable in the presence of methanol but can be converted into rings upon the addition of pyridine (Heo et al. 2007).

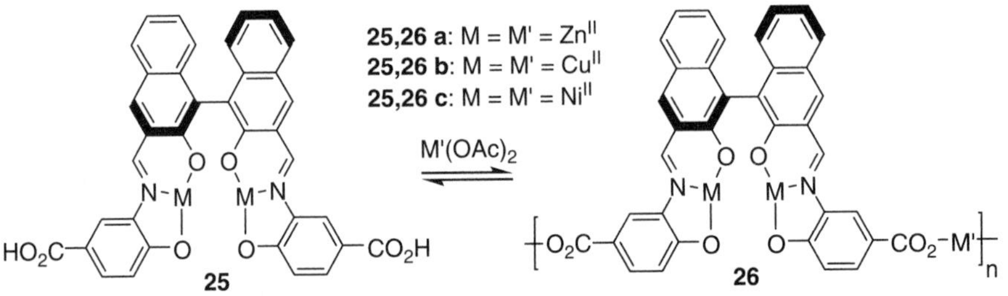

25,26 a: M = M' = Zn^{II}
25,26 b: M = M' = Cu^{II}
25,26 c: M = M' = Ni^{II}

Figure 7.16 Schiff base complexes utilized by Oh and Mirkin (2005) to prepare nanoparticles.

Kuroiwa et al. (2004) prepared metal-containing organogels by the addition of either $FeCl_2$ or $CoCl_2$ to chloroform solutions of a lipophilic triazole. The resulting linear "molecular wires" self-assemble into an interconnected, fibrous, gel-forming microstructure. The Co^{II}-based system shows an unusual thermoresponse by converting into a sol upon cooling, the opposite of normal supramolecular gel behavior. It is proposed that this occurs through a change in the coordination behavior of the metal ion. Alternatively, the stimuli-responsive component can be placed on the ligand (Kume et al. 2006). For example, photoresponsive gels have been prepared using $Fe(BF_4)_2$ and an azobenzene-containing triazole in chlorocyclohexane. Upon exposure to UV light the gels are converted in the sol state because of the trans–cis photoinduced conversion of the azobenzene in the ligand. The system reverts to the gel upon exposure to visible light and converts back to the *trans*-azobenzene conformation.

7.5. CONCLUSION AND OUTLOOK

The development of MSPs has come a long way in recent years, and an array of ligands and metal salts have been used to access a dramatic range of materials. In addition to the structures we described in this chapter, the versatility of this approach has allowed access to more complex architectures, such as helical polymers (Kimura et al. 1999; Ikeda et al. 2006) and block copolymers (Nair et al. 2006). The combination of metal ion functionality, ligand tuneability, and the potential to change the properties of both upon complexation should open the door to polymeric materials that exhibit a unique combination of properties or even to new material properties altogether. Thus, in addition to the metal ligand motifs that we described, there is still plenty of room for the development and utilization of new ligands that offer different properties (e.g., Chow et al. 2007). The variability in the dynamics of different coordination motifs can allow the development of materials that range from being under continuous equilibrium, and thus highly responsive to environmental stimuli, to being irreversible on a realistic timescale and behave more akin to traditional covalent polymers. MSPs, which exhibit rapid and controllable complexation/decomplexation, could permit access to a new generation of electronic, sensing, self-healing, and/or catalytic materials (Park et al. 2006). The investigation of these polymers is still in its infancy, but their potential as functional materials suggests an exciting future for MSPs.

REFERENCES

Abd-El-Aziz AS, Carraher Jr CE, Pittman Jr CU, Zeldin M. Macromolecules containing metal and metal-like elements. Hoboken (NJ): Wiley InterScience; 2006. Vols. 1–7.

Bailey JA, Miskowski VM, Gray HB. Spectroscopic and structural properties of binuclear platinum–terpyridine complexes. Inorg Chem 1993;32:369–370.

Beck JB, Rowan SJ. Multistimuli, multiresponsive metallo-supramolecular polymers. J Am Chem Soc 2003;125:13922–13923.

Beck JB, Ineman JM, Rowan SJ. Metal/ligand-induced formation of metallo-supramolecular polymers. Macromolecules 2005;38:5060–5068.

Bender JL, Corbin PS, Fraser CL, Metcalf DH, Richardson FS, Thomas EL, Urbas AM. Site-isolated luminescent europium complexes with polyester macroligands: metal-centered heteroarm stars and nanoscale assemblies with labile block junctions. J Am Chem Soc 2002;124:8526–8527.

Bodenthin Y, Pietsch U, Möhwald H, Kurth DG. Inducing spin crossover in metallo-supramolecular polyelectrolytes through an amphiphilic phase transition. J Am Chem Soc 2005;127:3110–3115.

Boydston AJ, Williams KA, Bielawski CW. A modular approach to main-chain organometallic polymers. J Am Chem Soc 2005;127:12496–12497.

Boydston AJ, Rice JD, Sanderson MD, Dykhno OL, Bielawski CW. Synthesis and study of bidentate benzimidazolylidene–group 10 metal complexes and related main-chain organometallic polymers. Organometallics 2006;25:6087–6098.

Buey J, Swager TM. Three-strand conducting ladder polymers: two-step electropolymerization of metallopolyrotaxanes. Angew Chem Int Ed 2000;39:608–612.

Chen C-C, Dormidontova EE. Supramolecular polymer formation by metal–ligand complexation: Monte Carlo simulations and analytical modeling J Am Chem Soc 2004;126: 14972–14978.

Chen Y-Y, Lin H-C. Synthesis and characterization of light-emitting main-chain metallopolymers containing bis-terpyridyl ligands with various lateral substituents. J Polym Sci Part A Polym Chem 2007;45:3243–3255.

Chiper M, Meier MAR, Kranenburg JM, Schubert US. New insights into nickel(II), iron(II), and cobalt(II) bis-complex-based metallo-supramolecular polymers. Macromol Chem Phys 2007;208:679–689.

Chow C-F, Fujii S, Lehn J-M. Metallodynamers: neutral dynamic metallosupramolecular polymers displaying transformation of mechanical and optical properties on constitutional exchange. Angew Chem Int Ed 2007;46:5007–5010.

Constable EC, Housecroft CE, Mundwiler S. Metal-directed assembly of cyclometallopeptides. Dalton Trans 2003;2112–2114.

Creus M, Ward TR. Designed evolution of artificial metalloenzymes: protein catalysts made to order. Org Biomol Chem 2007;5:1835–1844.

Divisia-Blohorn B, Genoud F, Borel C, Bidan G, Kern J-M, Sauvage J-P. Conjugated polymetallorotaxanes:in-situ ESR and conductivity investigations of metal–backbone interactions. J Phys Chem B 2003;107:5126–5132.

Dobrawa R, Würthner F. Metallosupramolecular approach toward functional coordination polymers. J Polym Sci Part A Polym Chem 2005;43:4981–4995.

El-Ghayoury A, Schenning APHJ, Meijer EW. Synthesis of π-conjugated oligomer that can form metallopolymers. J Polym Sci Part A Polym Chem 2002;40:4020–4023.

Ertesvåg H, Valla S. Biosynthesis and applications of alginates. Polym Degrad Stabil 1998;59:85–91.

Fustin C-A, Guillet P, Schubert US, Gohy J-F. Metallo-supramolecular block copolymers. Adv Mater 2007;19:1665–1673.

Heo J, Jeon Y-M, Mirkin CA. Reversible interconversion of homochiral triangular macrocycles and helical coordination polymers. J Am Chem Soc 2007;129:7712–7713.

Holyer RH, Hubbard CD, Kettle SFA, Wilkins RG. The kinetics of replacement reactions of complexes of the transition metals with 2,2′,2″-terpyridine. Inorg Chem 1966;5:622–625.

Ikeda M, Tanaka Y, Hasegawa T, Furusho Y, Yashima E. Construction of double-stranded metallosupramolecular polymers with a controlled helicity by combination of salt bridges and metal coordination. J Am Chem Soc 2006;128:6806–6807.

Jeon Y-M, Heo J, Mirkin CA. Dynamic interconversion of amorphous microparticles and crystalline rods in salen-based homochiral infinite coordination polymers. J Am Chem Soc 2007;129:7480–7481.

Johnson RM, Fraser CL. Iron tris(bipyridine)-centered star block copolymers: chelation of triblock macroligands generated by ROP and ATRP. Macromolecules 2004;37:2718–2727.

Kimura M, Sano M, Muto T, Hanabusa K, Shirai H. Self-assembly of twisted bridging ligands to helical coordination polymers. Macromolecules 1999;32:7951–7953.

Kitagawa S, Kitaura R, Noro S. Functional porous coordination polymers. Angew Chem Int Ed 2004;43:2334–2375.

Knapton D, Iyer P, Rowan SJ, Weder C. Synthesis and properties of metallo-supramolecular poly(p-xylylene)s. Macromolecules 2006;39:4069–4075.

Knapton D, Rowan SJ, Weder C. Synthesis and properties of metallo-supramolecular poly(p-phenylene ethynylene)s. Macromolecules 2006;39:651–657.

Kokil A, Huber C, Caseri WR, Weder C. Synthesis of π-conjugated organometallic polymer networks. Macromol Chem Phys 2003;204:40–45.

Kume S, Kuroiwa K, Kimizuka N. Photoresponsive molecular wires of Fe^{II} triazole complexes in organic media and light-induced morphological transformations. Chem Commun 2006;2442–2444.

Kuroiwa K, Shibata T, Takada A, Nemoto N, Kimizuka N. Heat-set gel-like networks of lipophilic Co(II) triazole complexes in organic media and their thermochromic structural transitions. J Am Chem Soc 2004;126:2016–2021.

Lahn B, Rehahn M. Coordination polymers from kinetically labile copper(I) and silver(I) complexes: true macromolecules or solution aggregates? Macromol Symp 2001;163:157–176.

Liu S, Volkmer D, Kurth DG. From molecular modules to modular materials. Pure Appl Chem 2004;76:1847–1867.

Lohmeijer BGG, Schubert US. Playing LEGO with macromolecules: design, synthesis, and self-organization with metal complexes. J Polym Sci Part A Polym Chem 2003;41:1413–1427.

Lohmeijer BGG, Schubert US. Water-soluble building blocks for terpyridine-containing supramolecular polymers: synthesis, complexation, and pH stability studies of poly(ethylene oxide) moieties. Macromol Chem Phys 2003;204:1072–1078.

Loveless DM, Jeon SL, Craig SL. Chemoresponsive viscosity switching of a metallo-supramolecular polymer network near the percolation threshold. J Mater Chem 2007;17:56–61.

Nair KP, Pollino JM, Weck M. Noncovalently functionalized block copolymers possessing both hydrogen bonding and metal coordination centers. Macromolecules 2006;39:931–940.

Oh M, Mirkin CA. Chemically tailorable colloidal particles from infinite coordination polymers. Nature 2005;438:651–654.

Park KH, Jang K, Son SU, Sweigart DA. Self-supported organometallic rhodium quinonoid nanocatalysts for stereoselective polymerization of phenylacetylene. J Am Chem Soc 2006;128:8740–8741.

Paulusse JMJ, Huijbers JPJ, Sijbesma RP. Quantification of ultrasound-induced chain scission in Pd^{II}–phosphine coordination polymers. Chem Eur J 2006;12:4928–4934.

Paulusse JMJ, van Beek DJM, Sijbesma RP. Reversible switching of the sol–gel transition with ultrasound in rhodium(I) and iridium(I) coordination networks. J Am Chem Soc 2007;129:2392–2397.

Rowan SJ, Cantrill SJ, Cousins GRL, Sanders JKM, Stoddart JF. Dynamic covalent chemistry. Angew Chem Int Ed 2002;41:899–952.

Schmatloch S, van den Berg AMJ, Alexeev AS, Hofmeier H, Schubert US. Soluble high-molecular-mass poly(ethylene oxide)s via self-organization. Macromolecules 2003;36: 9943–9949.

Schmatloch S, van den Berg AMJ, Fijten MWM, Schubert US. A high-throughput approach towards tailor-made water-soluble metallo-supramolecular polymers. Macromol Rapid Commun 2004;25:321–325.

Schubert US, Newkome GR, Manners I. Metal-containing and metallosupramolecular polymers and materials. Washington (DC): American Chemical Society; 2006.

Schütte M, Kurth DG, Linford MR, Cölfen H, Möhwald H. Metallosupramolecular thin polyelectrolyte films. Angew Chem Int Ed 1998;37:2891–2893.

Vermonden T, van der Gucht J, de Waard P, Marcelis ATM, Besseling NAM, Sudhölter EJR, Fleer GJ, Stuart MAC. Water-soluble reversible coordination polymers: chains and rings. Macromolecules 2003;36:7035–7044.

Wang P, Moorefield CN, Newkome GR. Nanofabrication:reversible self-assembly of an imbedded hexameric metallomacrocycle within a macromolecular superstructure. Angew Chem Int Ed 2005;44:1679–1683.

Weng W, Beck JB, Jamieson AM, Rowan SJ. Understanding the mechanism of gelation and stimuli-responsive nature of a class of metallo-supramolecular gels. J Am Chem Soc 2006;128:11663–11672.

Yaghi OM, O'Keeffe M, Ockwig NW, Chae HK, Eddaoudi M, Kim J. Reticular synthesis and the design of new materials. Nature 2003;423:705–714.

Yam VW-W, Chan KH-Y, Wong KM-C, Chu BW-K. Luminescent dinuclear platinum(II) ter-pyridine complexes with a flexible bridge and "sticky ends." Angew Chem Int Ed 2006;45:6169–6173.

Yount WC, Loveless DM, Craig SL. Strong means slow: dynamic contributions to the bulk mechanical properties of supramolecular networks. Angew Chem Int Ed 2005;44: 2746–2748.

CHAPTER 8

POLYMERIC CAPSULES: CATALYSIS AND DRUG DELIVERY

BRIAN P. MASON, JEREMY L. STEINBACHER, and D. TYLER MCQUADE

8.1. INTRODUCTION

Defining "inside" from "outside" is a fundamental trait of living organisms. The creation of nonnatural structures that can define in from out with nonpermeable or semipermeable barriers offers the potential of protecting the internal content from destruction, contamination, and unwanted dispersal until the content is delivered to a defined location. Small spherical structures that define in and out are well known and come in forms ranging from microcapsules to vesicles to micelles. We refer to these structures collectively as polymeric capsules.

Synthesis of polymeric capsules involves polymer self-assembly on many levels and length scales. Approaches to their synthesis include cases where monomers are polymerized around a liquid template, premade polymers are assembled around liquid and solid templates, premade particles are assembled around templates, and small molecules or polymers assemble spontaneously. Each method offers unique challenges and advantages, and each allows the production of capsules with different properties.

Complementing the many methods for capsule preparation, these materials have a wonderful array of applications including cosmetic use, drug delivery, passive displays, and healing materials. Because the field of synthetic polymer capsules is over 85 years old, we readily acknowledge that this chapter is in no way comprehensive. We will limit our discussion here to the two most relevant applications: catalytic capsules and drug delivery. Although existing capsules with these properties are well respected, we see a resurgence in these areas driven by the recent contributions of both new materials and ideas by synthetic chemists.

Molecular Recognition and Polymers: Control of Polymer Structure and Self-Assembly.
Edited by V. Rotello and S. Thayumanavan
Copyright © 2008 John Wiley & Sons, Inc.

Being recent research arrivals to the field ourselves, we wrote this chapter in a style that we hope will entice others to enter it as well. We begin with a brief overview of methods used to make capsules, making sure to reference other, more comprehensive sources in each case. We also survey the catalytic capsule field and capsules as drug-delivery vehicles, highlighting recent innovations in both areas. We hope this chapter will convince you that polymeric capsules are exciting materials with equally exciting applications.

8.2. METHODS OF ENCAPSULATION

Hollow polymeric capsules are usually produced using one of three methods (Fig. 8.1).

1. Emulsion-templated encapsulation: A polymeric shell is self-assembled or formed over a liquid emulsion droplet. This typically includes capsules made by coacervation and interfacial polymerization as well as many colloidosomes.
2. Solid-templated encapsulation: A solid or gel bead replaces the liquid droplet as the template. This typically includes colloidal templating, and it is not as widespread for hollow capsules because the solid template must be broken down and removed without harming the shell.
3. Vesicle-based capsules: The shell is self-assembled into a hydrophobic shell or bilayer surrounded inside and outside by solution. These materials are formed using either small amphiphiles or amphiphilic block copolymer vesicles.

We will discuss each of these general methods in turn, using examples that are historically significant or particularly representative. We typically mention only hollow capsules in which the core material (be it liquid or solid) is encapsulated by a polymeric shell that has a composition different from the core. "Microcapsule" diameters range from 1 to 1000 μm, where any capsules with diameters smaller than 1 μm are considered "nanocapsules." Because we intend for this discussion to be merely instructive for the general reader new to this field, it should not be regarded as comprehensive. In addition, although not all of the methods discussed here have been used for drug delivery and catalyst encapsulation (some are rather new); we offer them as possible candidates for such applications.

Many processes for forming polymeric capsules use emulsion droplets as templates for shell self-assembly. An emulsion is a mixture of two immiscible liquids where the minor (dispersed) phase forms small droplets in the major (continuous) phase. We generally think of emulsions as being oil-in-water (where the oil is the dispersed phase in the aqueous continuous phase) or water-in-oil (vice versa), although any two immiscible phases form emulsions under the appropriate conditions. Mayonnaise is a common example of an emulsion, in this case a water-in-oil emulsion of vinegar or lemon juice in up to 80 vol% oil. Emulsification is nonspontaneous, requiring shaking, stirring, or homogenization. Once formed, emulsions

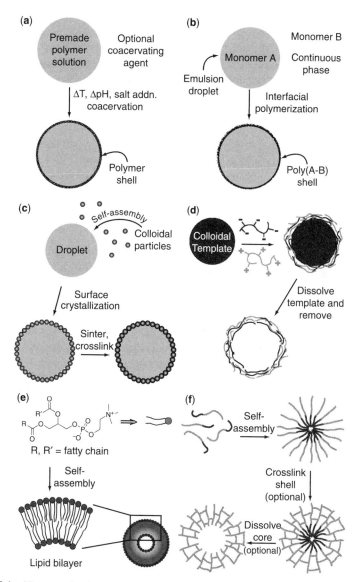

Figure 8.1 Nano- and microcapsule synthesis: (a) coacervation, (b) interfacial polymerization, (c) colloidosomes, (d) colloidal templating, and (e, f) vesicles.

often rapidly coalesce and must contain emulsifiers, usually surfactants, to stabilize them. In the case of mayonnaise, the emulsifiers are the diglyceride phospholipids in egg yolks, which form a protective and stabilizing layer around the dispersed phase droplets (McGee 2004). In the academic laboratory, emulsions are generally made by either rapid mechanical stirring or homogenization (McClements 1996; Hipp et al. 1999).

8.2.1. Coacervation

One of the first methods for making capsules involved polymer coacervation. In this method, macromolecules are dissolved in either the dispersed or continuous phase of an emulsion and are induced to precipitate as a shell around the dispersed phase. Coacervation can be brought about in several ways, such as changes in temperature or pH, addition of salts or a second macromolecular substance, or solvent evaporation (Bungenberg de Jong 1949).

Coacervates are either "simple" or "complex." Simple coacervations generally involve nonionized groups on the precipitated macromolecule. With complex coacervations, the precipitation is brought about by the formation of salt bridges. The distinction is nebulous, however, because both mechanisms can occur in the same system. An example is when an electrolyte and an alcohol are added to an emulsion containing gum arabic, which is made up of saccharides and glycoproteins. The alcohol serves to reduce the solubility of the nonionized groups on the macromolecule (simple), but it also strengthens the interaction between the cations of the electrolyte and the gum arabic carboxyl groups (complex; Bungenberg de Jong 1949).

Capsules made with this method must be optimized for whichever application they are intended. One example is the work of Barry Green at the National Cash Register Company. Often considered the father of microencapsulation, Green first developed microcapsules by coacervation for use in multiple-use carbonless paper (Lim 1984). The project began in 1939 with two goals: to produce better and cheaper NCR sales books (made of one-time use paper) and to manufacture paper for cash register receipts that contained all the chemicals needed to form images on impact (Thies 1999). Generally speaking, these materials were loosely based on Bungenberg de Jong's coacervation studies at Utrecht University in The Netherlands (Bungenberg de Jong 1949).

8.2.2. Interfacial Polymerization

A second classical method for making capsules from emulsions is to form the shell polymer in situ using interfacial polymerization (Morgan and Kwolek 1959; Wittbecker and Morgan 1959). This method is similar to the "nylon rope trick" often used as a demonstration, where a solution of diacid chloride in organic solvent (such as adipoyl chloride in hexanes) is layered in a beaker with a diamine aqueous phase (such as 1,6-hexadiamine in water; Friedli et al. 2005). Because the two monomers meet only at the interface of the two phases, the condensation polymerization to form the polyamide occurs only at the interface.

If the organic solution of diacid chloride is recast as a dispersed phase in an emulsion with the aqueous solution of diamine as the continuous phase, the polymer membrane forms around the dispersed phase droplets, effectively making polyamide shell capsules around the organic phase. Of course, the relative volumes could be reversed so that the aqueous phase was encapsulated if so desired.

The encapsulation of hemoglobin developed by Chang in the late 1950s is one excellent example of this process (Chang 1964, 1966, 1972). Chang's initial goal was to synthesize artificial red blood cells for subcutaneous implantation where the semipermeable membrane would deliver hemoglobin to the patient at a controlled rate. To generate the capsules, Chang combined hemoglobin with an alkaline aqueous dispersed phase of 1,6-hexamethylenediamine. Outside the emulsion droplets, the continuous phase was composed of cyclohexane and chloroform, with sebacoyl chloride acting as the second condensation monomer. The capsule size in this system was dependent on the stir rate and surfactant concentration. Chang found that increasing the stir rate and surfactant concentration caused the average microcapsule diameter to vary from about 80 to 20 μm.

It is also possible to generate microcapsules through interfacial polymerization using only one monomer to form the shell. In this class of encapsulations, polymerization must be performed with a surface-active catalyst, a temperature increase, or some other surface chemistry. Herbert Scher of Zeneca Ag Products (formerly Stauffer Chemical Company) developed an excellent example of the latter class of shell formation (Scher 1981; Scher et al. 1998). He used monomers featuring isocyanate groups, like poly(methylene)-poly(phenylisocyanate) (PMPPI), where the isocyanate reacts with water to reveal a free primary amine. Dissolved in the oil-dispersed phase of an oil-in-water emulsion, this monomer contacts water only at the phase boundary. The primary amine can then react with isocyanates to form a polyurea shell. Scher used this technique to encapsulate pesticides, which in their free state would be too volatile or toxic, and to control the rate of pesticide release.

Because the size of the emulsion droplets dictates the diameter of the resulting capsules, it is possible to use miniemulsions to make nanocapsules. To cite a recent example, Carlos Co and his group developed relatively monodisperse ~200-nm capsules by interfacial free-radical polymerization (Scott et al. 2005). Dibutyl maleate in hexadecane was dispersed in a miniemulsion of poly(ethylene glycol)-1000 (PEG-1000) divinyl ether in an aqueous phase. They generated the miniemulsion by sonication and used an interfacially active initiator, 2,2'-azobis(N-octyl-2-methyl-propionamidine) dihydrochloride, to initiate the reaction, coupled with UV irradiation.

A new variation of interfacial polymerization was developed by Russell and Emrick in which functionalized nanoparticles or premade oligomers self-assemble at the interface of droplets, stabilizing them against coalescence. The functional groups are then crosslinked, forming permanent capsule shells around the droplets to make water-in-oil (Lin et al. 2003; Skaff et al. 2005) and oil-in-water (Breitenkamp and Emrick 2003; Glogowski et al. 2007) microcapsules with elastic membranes.

8.2.3. Colloidosomes

Colloidosomes are a recent class of microcapsules and thus far have only been applied to catalysis and drug delivery in a few cases. The term colloidosomes was coined by Anthony Dinsmore and colleagues in 2002 to refer to capsules where the shells are composed of close-packed layers of monodisperse colloidal particles (usually micron-sized polymer beads) that have been linked together by sintering, cross-linking, or ionic interactions (Dinsmore et al. 2002). This class of capsules is notable because, unlike the previous two methods of capsule preparation, colloidosomes inherently have uniform pores, defined by the interstices between the colloid particles, a parameter tuned by varying the particle diameter.

Orlin Velev and colleagues are considered to have reported the first colloidosomes in the mid-1990s (Velev et al. 1996; Velev and Nagayama 1997). In this system, 1-μm diameter polystyrene (PS) latex beads were surface functionalized with sulfate groups to give them a negative charge and were dispersed in an oil-in-water emulsion of octanol in water. The authors found that if the surface charges on the PS beads were optimized, the beads could be made just hydrophobic enough to sit at the phase boundary of the oil and water. Addition of casein served to stabilize the capsules by mitigating adsorption of PS beads onto the surface of the templates or flocculation of the colloidosomes. Addition of HCl and $CaCl_2$, which act as coagulants for the latexes, provided more shell stabilization. As with previous methods, the size of the resulting microcapsules was dictated by the size of the original disperse-phase droplets, which in this case were generated by homogenization and were on the order of $10-100\,\mu m$.

Other groups have used alternate methods for stabilizing colloidosome shells, including chemical cross-linking (Cayre et al. 2004) and sintering (Dinsmore et al. 2002). Dinsmore and coworkers heated PS latex colloidosomes to 105 °C, just beyond the glass transition of PS, for 5 min and found by scanning electron microscopy that this caused 150-nm diameter bridges to form between the latex beads. Heating for 20 min caused the interstitial spaces to close completely.

Colloidosome formation can be generalized as long as the surface energies among the dispersed and continuous phases, and the colloid particles, are optimized experimentally. Thus far, colloidosome compositions have included unfunctionalized poly(methyl methacrylate) (Dinsmore et al. 2002), magnetite (Fe_3O_4) nanoparticles (Duan et al. 2005), polymeric microrods (Cayre et al. 2004), and temperature-responsive gels (Berkland et al. 2007).

A subcategory of colloidosomes and the colloidal templating of the next section is the use of "gel-trapping" methods for generating colloidosomes (Paunov 2003; Paunov and Cayre 2004). Gel trapping is similar to the general methods except that the disperse phase of the emulsion is an agarose (1.5%) aqueous gel, where the emulsion is stabilized by the PS latexes used to form the capsule shells (Binks and Lumsdon 2001). This gives the capsules more mechanical stability as they are washed and as the colloid shells are strengthened by cross-linking (Cayre et al. 2004). The encapsulated gel can be left in the microcapsules if so desired or heated out after shell stabilization.

8.2.4. Colloidal Templating

We can think of the previous methods for generating capsules as "liquid templating," where self-assembly occurs around a liquid. An alternative is to use a solid template, usually called colloidal templating. The shell is generated around a solid particle that is later digested, leaving a hollow sphere (Caruso 2000). Deposition of shell material takes place either by growing a polymer on the surface or precipitating a premade polymer on the surface. Capsule size is controlled by the templating colloid and the shell thickness is determined by polymer growth or deposition.

A popular method for colloidal templating is based on sequential electrostatic self-assembly of premade charged polymers onto an oppositely charged surface. This layer-by-layer (LBL) method stems from work done in the 1960s by Ralph Iler at Du Pont with two-dimensional surfaces (Iler 1966) and extended by Gero Decher and coworkers in the 1990s (Decher and Hong 1991a, 1991b; Decher et al. 1992). Decher et al. used single silicon crystals that had been given a positive charge with aminopropylsilyl groups as a thin film template. To this they added a solution of negatively charged amphiphiles, which adsorbed onto the surface because of the opposite charge. Alternating exposures to positively and negatively charged amphiphiles (interspersed with washings) yielded a 35-layer film (Decher and Hong 1991a).

To make capsules using LBL deposition, the solid sphere template must be degraded and washed away without harming the capsule shell (Voigt et al. 1999; Dahne et al. 2001; Sukhorukov et al. 2001). Frank Caruso and colleagues used weakly cross-linked melamine formaldehyde (MF) resin particles as a template (Caruso et al. 1998). The resin, a mainstay in commercial products, was functionalized with negatively charged poly(sodium styrenesulfonate) (PSS), which adsorbed onto the surface of the resin. The process was followed by deposition of oppositely charged poly(allylamine hydrochloride) and iterated until the desired thickness was achieved. Final treatment with acidic solution (below pH 1.6) digested the MF particles, leaving flexible, hollow capsules (Caruso et al. 1999).

8.2.5. Vesicles

Vesicles are capsules in which the shells are composed of amphiphilic small molecules or polymers. Generally, the shell is an amphiphilic bilayer with an aqueous interior. These differ fundamentally from capsules generated in a water-in-oil emulsion because the "oil phase" in the vesicle system is only in the shells, which are surrounded by an outer aqueous phase.

Vesicles for use as materials can be divided into two categories: naturally occurring vesicles, or liposomes, which are composed of natural amphiphiles, usually phospholipids; and polymer vesicles, which are generally composed of block copolymers.

Liposomes. The field of encapsulated drug delivery has been dominated by liposomic systems. Liposomes are naturally occurring vesicles composed of discrete phospholipid bilayers, particularly phosphatidylcholines, because they form lamella under wide salt and temperature ranges (New 1995). Liposomes were first observed by Alec Bangham and coworkers at the Institute of Animal Physiology in Babraham, Cambridge, in 1962. First called "lipid somes," these 50-nm vesicles were not firmly established until 1964 (Bangham and Horne 1964; Bangham 1983).

Along with phosphatidylcholines, cholesterol is often included in the vesicle bilayers in as high as a 1:1 ratio because it reduces permeability of the bilayer to aqueous solutes.

The study of liposomes as drug-delivery agents has been ongoing for decades. The lipids for liposome formation are typically harvested by extraction from egg yolks and soybeans, and a number of recipes exist for generating liposomes of various diameters. Because the shell material in liposomes is not polymeric, we will not discuss them in depth, limiting ourselves only to those aspects that are pertinent to synthetic analogues like the copolymer vesicles of the next section.

Copolymer Vesicles. Although lipids can self-assemble into useful thin-walled vesicles, their applications can be limited. Phospholipids form membranes in aqueous solutions, but they often fail to do so in organic solvents that synthetic chemists find convenient (Discher and Eisenberg 2002). If the vesicles do happen to form in these decidedly "unbiological" environments, they lack mechanical strength, or even basic stability (Antonietti and Forster 2003). To overcome this issue, researchers have synthesized various amphiphiles and have even encased classical vesicles in polymer shells (Regen et al. 1984; Fukuda et al. 1986). More recently researchers have turned to diblock copolymers to form tougher and more stable vesicles that can be more easily tuned for specific applications (Antonietti and Forster 2003).

Adi Eisenberg and coworkers have had great success synthesizing and characterizing simple block copolymer nanocapsule vesicles (Zhang and Eisenberg 1995; Zhang et al. 1996; Discher and Eisenberg 2002; Soo and Eisenberg 2004). For instance, Eisenberg and Zheng showed that diblocks of PS and poly(acrylic acid) (PAA) synthesized by anionic polymerization can form various morphologies, including spheres, rods, lamellae, and vesicles, depending on how much electrolyte (NaCl, HCl, or $CaCl_2$) is added (Zhang and Eisenberg 1995; Zhang et al. 1996). The general idea is that the PS block acts as a hydrophobic portion of the linear chain, analogous to the two greasy chain ends on a phosphatidylcholine in a liposome, and the PAA block acts as the hydrophilic ionic end (like the positively charged choline nitrogen). Thus, the relative lengths of the blocks and the overall molecular weight can be altered to tune the characteristics of the vesicle, such as size and overall hydrophilicity.

8.3. CATALYST ENCAPSULATION

8.3.1. General

Microcapsule properties make them attractive materials for a wide variety of practical applications. In the area of catalysts, microcapsules provide semipermeable membranes that are readily produced and dispersed. These properties, along with others, have inspired systems that include synthetic or man-made encapsulated catalysts, such as organocatalysts, metal particles, enzymes, and organometallic catalysts. This section will examine strategies and classes of catalysts encapsulated within microcapsules. Generally, there are two ways encapsulated catalysts may act in a reaction.

First, the catalyst is meant to leach out of the capsules into a reaction solution. In this case, the capsules are not meant to break open but are semipermeable to the catalyst, which diffuses into the reaction mixture over time. This method is typically used for metal catalysts or catalyst precursors where the metals leach out and perform the desired reaction. This method is useful because metal-catalyzed reactions typically require lower catalyst loading than organocatalysts (<1 mol%), and highly loaded capsules can be isolated and reused until exhausted. Such metal catalysts are often touted for their decreased pyrophoricity relative to such catalysts as palladium on carbon (Coleman and Royer 1980; Bremeyer et al. 2002). One could simply use resins, microspheres, or other solid supports as catalyst reservoirs, but capsules are well suited because of their inherently higher surface areas (Royer et al. 1985; Wang et al. 2006).

Second, the catalyst is meant to be site isolated. Site isolation implies that the catalyst is not meant to leave the interior of the capsule where all catalysis occurs. This requires that the catalyst be kept inside the capsule either by anchoring it to the inner shell wall or by anchoring it to something that cannot diffuse out through the shell under reaction conditions, like a polymer chain. Encapsulated enzymes fall into this category, because they are typically much larger than substrates and products, which can diffuse in and out of the capsules. The reasons for pursuing such materials are to facilitate catalyst recycling, to create a unique microenvironment inside the capsule, or to protect the catalyst from other fouling reagents throughout the reaction.

There are several ways to determine if a catalyst leaches or remains inside the capsule. It must first be determined by microscopy whether the capsule shell walls have ruptured. If the shell walls remain intact, strategies exist to determine if the catalysis occurs outside the capsule or within it. Of the many approaches (Hagen et al. 2005), we favor the "three-phase test" illustrated in Figure 8.2 (Rebek and Gavina 1974; Rebek et al. 1975; Rebek 1979; Hagen et al. 2005). One of the substrates in the reaction is bound to a resin and the reaction is carried out in the presence of the encapsulated catalyst. If the bound substrate undergoes reaction it is assumed that the catalyst has leached into solution, because the resin-bound substrate should not have been able to interact with the encapsulated catalyst (Davies et al. 2001; Steel and Teasdale 2004; Okamoto et al. 2005; Broadwater and McQuade 2006). To prevent false positives, the homogeneous reaction should be carried out in the same reaction vessel. This also serves as an internal control experiment to show if the reaction with encapsulated catalyst works at all.

Third and finally, one can use the catalytic capsules in a reaction or Soxhlet extract them in an appropriate solvent and then analyze the supernatant of the reaction for the presence of catalyst after the solvent and capsules have been removed. Some form of spectroscopy (NMR, IR, etc.) or elemental analysis is appropriate for this (Price et al. 2006). This type of evidence is, of course, circumstantial because the absence of catalyst does not necessarily prove that it did not return to the capsules before capsule removal.

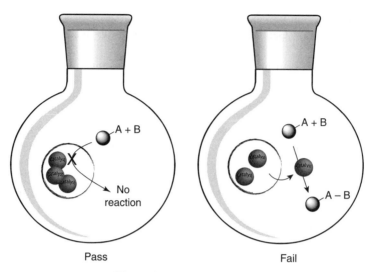

Figure 8.2 Three-phase test.

8.3.2. Encapsulated Catalysts

We chose to organize the discussion of catalyst encapsulation by separating examples into catalyst type as opposed to encapsulation method. As previously, we do not intend to present a comprehensive treatment of the literature and discuss only examples we consider particularly illustrative.

Metal Catalysts. An early example of encapsulated metal catalysts are the Pd-containing "ghosts" developed by Garfield Royer's group in the late 1970s to early 1980s. The process used colloidal templating (described in Section 8.2.4) to form capsules using porous alumina beads (Meyers and Royer 1977; Royer 1982) and later porous silica beads (Coleman and Royer 1980) to assemble a polyethyleneimine (PEI) shell. Specifically, PEI-600, which has the form $(—CH_2CH_2NH_x—)_n$, was dissolved in methanol and allowed to self-assemble onto porous silica beads (400-μm diameter). The coated beads were exposed to glutaraldehyde to cross-link the polymer, then more PEI, and finally sodium borohydride to reduce the imines formed by the cross-linking. The silica template was dissolved away by treatment with a concentrated sodium hydroxide solution. To add catalytic Pd(0) to the microcapsules (ghosts), the shells were soaked in a 1 M hydrochloric acid solution of $PdCl_2$, followed by neutralization and reduction of the palladium to Pd(0) with sodium borohydride. The catalysts were shown to be active in palladium-catalyzed transfer hydrogenolysis and were recycled 10 times without loss of activity (Coleman and Royer 1980). As of this writing, they are commercially available as the Royer® catalyst, with analogous platinum and rhodium versions.

Other well-known but more recent examples of encapsulated metal catalysts are the palladium-containing polyurea microcapsules of Steven Ley's group at Cambridge (commercialized as PdEnCat™; Bremeyer et al. 2002; Ley et al. 2002, 2003; Ramarao et al. 2002; Yu et al. 2003; Lee et al. 2005; Baxendale et al. 2006). The preparation was based on the interfacial polymerization of isocyanates by Herbert Scher discussed earlier (Scher 1981; Scher et al. 1998). To make the capsules, palladium acetate and PMPPI were dissolved in dichloroethane, which was then dispersed in an aqueous continuous-phase surfactant cocktail. Simple stirring at 800 rpm generated the emulsion, yielding capsules ranging from 50 to 250 μm (Ramarao et al. 2002). The resulting Pd(II)-embedded microcapsules were active in various palladium-catalyzed cross-coupling reactions, such as carbonylations and the Heck, Suzuki, and Stille reactions (Ley et al. 2002; Ramarao et al. 2002). With the capsules packed into a high-performance liquid chromatography column, the catalyst also carried out the Suzuki reaction in continuous flow, in organic solvent and supercritical CO_2 (Lee et al. 2005), and with microwave assistance (Baxendale et al. 2006).

The palladium(II) in the microcapsules can also be reduced directly to Pd(0). Typical of metal nanoparticle catalysis, the researchers found that the type of reducing agent affected the size of the nanoparticles and thus controlled the catalytic activity (Narayanan and El-Sayed 2003). Therefore, using hydrogen gas to reduce the palladium, the resulting nanoparticles were greater than 5 nm in size and inactive. Using formic acid as the reducing agent, however, gave particles of 2 nm or less that were quite active. The Pd(0)-containing microcapsules were then found to catalyze the H_2 hydrogenation of alkenes, alkynes, imines, and nitro groups (Bremeyer et al. 2002) and the transfer hydrogenation of aryl ketones and benzyl alcohols (Yu et al. 2003). Finally, the microcapsules catalyzed the ring-opening hydrogenolysis of benzylic epoxides (Ley et al. 2003).

Catalyst leaching from the capsules appears to occur, and it is quite solvent dependent. The capsules are catalyst reservoirs rather than truly site isolating the catalyst (Broadwater and McQuade 2006). The catalyst is recyclable (20 times for hydrogenation in cyclohexane), however, and capsules are easily isolated for reuse by simple filtration through a 20-μm frit (Bremeyer et al. 2002).

Organocatalysts. Sometimes it is desired that an encapsulated catalyst never leave the confines of the capsule because other reagents or catalysts in the reaction mixture would then interfere with it, or vice versa. In this case, the organocatalyst must be attached to a much larger molecule that cannot diffuse out of the shell or to the inside of the shell itself.

Our research group has developed a multicatalyst system around an encapsulated amine catalyst (Kobašlija and McQuade 2006) that promotes nitroalkene synthesis. Like the Royer catalysts, these microcapsules are based on a PEI shell. Unlike that system, however, templating is accomplished with a methanol in cyclohexane ("oil-in-oil") emulsion in an interfacial polymerization, with the PEI cross-linked

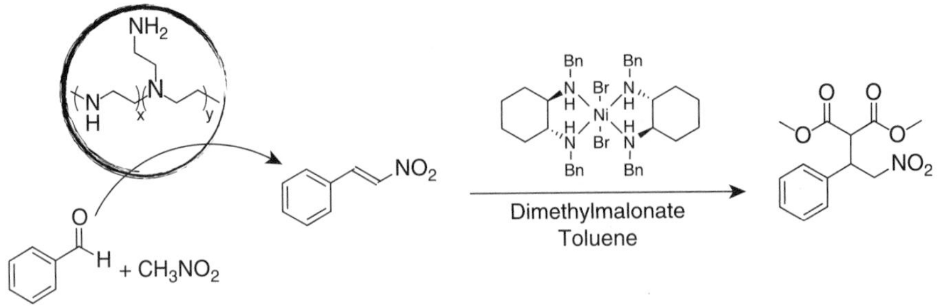

Scheme 8.1   Two-step reaction.

through some of its 1° and 2° amines by 2,4-toluene diisocyanate. In this system, the PEI is not only attached to the shell wall (comprising some of the wall itself) but also free and untethered inside the capsules.

Used in practice, the encapsulated amine catalyzes the coupling of an aldehyde with a nitroalkane to form the nitroalkene (Scheme 8.1) in methanol solvent, because methanol is optimal for the reaction (Poe et al. 2006, 2007). Coupled to this is a second, nickel-based catalyst to take the nitroalkene onto a Michael adduct in toluene, a good solvent for this reaction. Thus, the first reaction is carried out inside the microcapsules in a methanol environment, and the second reaction with the nickel catalyst takes place in the toluene phase outside the capsules. Interestingly, if the two catalysts are simply mixed without encapsulation, the PEI amines ligate the nickel, impairing both catalysts. In addition, inclusion of the second catalyst prevents the nitroalkene intermediate product from reacting again with the nitroalkane to form a dinitroalkane. Thus, encapsulation and coupling of two catalysts allows the formation of a product that otherwise would not have been possible.

Enzymes. Enzymes are particularly well suited for encapsulation because they often require a specialized microenvironment milder than the typically harsh conditions of an organic reaction. Although there are many examples of encapsulated enzymes (Watanabe and Royer 1983; Rao et al. 1993; Uludag et al. 1995; Hwang and Sefton 1997; Balabushevich et al. 2003; Shutava et al. 2004; Stein et al. 2006), there are few that have been used expressly for organic synthesis.

One example by Changyou Gao and colleagues illustrates the advantage of enzyme encapsulation for increased stability (Gao, Liu et al. 2002). Microcapsules were generated from polyelectrolytes using colloidal templating techniques (described in Section 8.2.4; Gao, Donath, et al. 2002). These polyelectrolyte multilayer (PEM) microcapsules are negatively charged, so that when horseradish peroxidase is introduced into solution it spontaneously deposits itself into the capsules and does not leach out even after multiple aqueous washes. Using the dehydrogenation of 2,2'-azino-bis-ethylbenzthiazoline-6-sulfonic acid with hydrogen peroxide as a test reaction (Holm 1995), the researchers found that the encapsulated enzyme remained more active than the free enzyme (50% activity vs. 90% activity after 3 h). They hypothesized that this was because the encapsulated enzyme was more resistant to denaturation in the 20% dioxane reaction solution.

8.4. DRUG DELIVERY WITH MICROCAPSULES

Delivering pharmaceutical agents to specific cells in the body is a difficult task involving complex interactions between many elements. Delivery systems have several fundamental requirements to achieve this task. The delivery vehicle must be ingestible, implantable, or injectable to introduce the drug into the body. The system must then protect the drug from the body's defense mechanisms in order to accumulate in selected cells. Once at the target, the delivery system should release the enclosed pharmaceutical agent with a controllable and predictable profile. Finally, the delivery vehicle should be biocompatible, nontoxic, and easily eliminated from the body.

Polymeric capsules can meet many of these requirements for drug-delivery vehicles. Most are small enough to be easily delivered into the body via ingestion or either intravenous or subcutaneous injection. Small microcapsules and nanocapsules circulate safely in the cardiovascular system, which can thoroughly mix the body's blood in minutes (Saltzman 2001). Generally, a microencapsulated pharmaceutical agent injected intravenously will reach every capillary in the body in roughly 5 min. Moreover, many microcapsule materials offer protection from the body's defense mechanisms, such as the liver's Kupffer cells (Hole 1989) and specialized immune cells called phagocytes (Becker et al. 1996) that break down foreign material into simpler components. Particulates are marked for elimination by antigens or specific complementary molecules that allow phagocytes to bind to the target material. Thus, a major theme in microcapsule drug-delivery applications is surface functionalization to avoid recognition by the immune system. Additional tuning of microcapsule characteristics can be used to change the biodistribution of pharmaceutical-bearing microcapsules, and tuning the shell composition and structure is an effective way to control drug-release profiles. Finally, many of the polymers used for drug microencapsulation are either derived from biological sources, and thus inherently biocompatible, or degrade easily into smaller molecules that are easily eliminated from the body. In the following sections, divided into specific encapsulation methods and materials, we will discuss how polymeric nano- and microcapsules have been successful in both laboratory and clinical settings as delivery vehicles for pharmaceutical agents.

8.4.1. Homogeneous Particles

Although homogeneous solid particles constitute a major class of microparticles used for the encapsulation of pharmaceutical agents, we will not discuss these materials in detail because of their noncapsular nature, mentioning them here to only provide perspective. Many solid particles are based on poly(lactic acid), poly(lactic-co-glycolic acid) (PLGA), and/or polyanhydrides and these have been developed especially by several research groups (Ogawa et al. 1988; Tabata and Ikada 1988; Pekarek et al.

1994; Langer 2000). Furthermore, homogeneous gel beads consisting of the naturally occurring polysaccharides alginate and chitosan have found use as drug-delivery vehicles and particularly in cell encapsulation. Seminal work by Franklin Lim and Anthony Son showed that cells could be encapsulated in polysaccharide particles and continue to function in vivo (Lim and Sun 1980), culminating in an implantable, artificial pancreas in at least one human subject (Soon-Shiong et al. 1994).

8.4.2. Interfacial Polymerization

As discussed in Section 8.2.2, hollow, liquid-core microcapsules were first used for a biomedical application by Chang in 1964 (Chang 1964). Chang used a stirred emulsion to create a dispersed phase of di- or polyamines in aqueous enzyme solution and a continuous phase of diacid chloride in an organic solvent. Interfacial polycondensation yielded polyamide microcapsules surrounding enzymes in an aqueous environment. Although additional encapsulated reagents were required in some cases to retain activity, in vitro tests showed that the enzymes usually retained significant activity relative to free enzymes. The microcapsules were also tested for in vivo compatibility. In some cases, such as encapsulated urease, enzyme activity was retained and the microcapsules showed little sign of toxicity when placed in vivo intravenously, subcutaneously, or intraperitoneally. Chang's early work (Chang 1964; Chang et al. 1966) still stands as a breakthrough in the field of encapsulated therapeutics. Chang continued to study these liquid-cored microcapsules for 5 years after his initial innovation. He showed how enzymes were stabilized (Chang 1971b) against degradation by the cell-like environment within microcapsules. In addition, microcapsules containing asparaginase (Chang 1971a) could stave off the onset of tumors in mice injected with lymphosarcoma cells for at least five times longer than free asparaginase.

Despite these early advances, the use of interfacial polymerization to make microcapsules for therapeutic agents slowed during the 1970s. Polyamide microcapsules were later used to create an implantable delivery system for insulin that responded rapidly to changes in external glucose levels (Makino et al. 1990), although this approach never reached clinical trials. In addition, poly(alkyl cyanoacrylate) (PACA), the major constituent of Super Glue, has been used to prepare water-cored nanocapsules via interfacial polymerization (Lambert, Fattal et al. 2000). These capsules have been used to encapsulate anticancer oligonucleotides (Lambert, Fattal et al. 2000), which had previously been difficult to use in therapeutic applications because of low stability and inadequate cellular penetration. PACA-encapsulated oligonucleotides have been shown to inhibit sarcoma-related tumors in mice when injected directly into tumors (Lambert, Bertrand et al. 2000; Lambert, Fattal et al. 2000).

8.4.3. Colloidal Templates

Colloidal templating techniques offer the potential of precise tuning for drug-release applications (see Section 8.2.4). For instance, the Feldheim group used gold nanoparticles as templates for the in situ polymerization of conducting polypyrrole and poly(*N*-methylpyrrole) capsules (Marinakos et al. 1999). The core sizes and shell thicknesses could be tuned independently by changing the size of the gold colloid templates and polymerization time of the shell polymer, respectively. The authors found that encapsulated dyes and proteins (Marinakos et al. 2001) were entrapped within the nanocapsules, and the inward diffusion of small molecules could be controlled by changes in the oxidation state of the shell.

Microcapsules composed of PEMs made by LBL deposition (Donath et al. 1998) also show promise for drug delivery. Small molecule drugs can be soaked into these microcapsules after removal of the templating colloid, providing control of drug-release rates (Mao et al. 2005). Alternatively, Caruso, Trau et al. (2000) used the LBL technique to prepare polyelectrolyte-encapsulated enzymes in which the templating colloid was actually a single crystal of the desired enzyme. Once encapsulated, the enzyme exhibited greatly improved resistance to external proteases. In addition to enzymes, they used LBL polyelectrolyte methods to encapsulate crystalline, uncharged small molecules such as dyes and model drugs (Caruso, Yang et al. 2000; Shi and Caruso 2001). The release rates could be tuned by varying the number of polyelectrolyte layers comprising the shells; as expected, thicker shells slowed the release from the microcapsules. In addition, the amphiphiles initially used to disperse the templating crystals affected the release rate by shielding the crystals from solvent to varying degrees, depending on the amphiphile structure (Shi and Caruso 2001). These reports demonstrate that PEM microcapsules are suitably tunable for drug-delivery applications, allowing control of microcapsule size, permeability, and release characteristics.

8.4.4. Vesicles

Liposomes. As discussed, liposomes are naturally occurring nanoscale vesicles comprising phospholipid bilayers surrounding an aqueous core. Liposomes have been successful as delivery vehicles for drugs, with several formulations on the market (Tollemar et al. 1995; Northfelt et al. 1996; Allen 1997). Because they self-assemble from relatively small molecules and are not cross-linked, they are not technically polymeric capsules. Nevertheless, we will discuss liposomes briefly because principles learned from their use can be applied to artificial copolymer vesicles discussed in the following section. Important work by Gregoriadis and Allison (1974) showed that liposomes could diminish primary immune response to encapsulated diptheria toxoid while simultaneously increasing immunogenicity, thus demonstrating the potential for liposomes as drug encapsulants.

A crucial advancement for the use of liposomes and other nanocapsules in drug-delivery applications was the finding that surface modification by hydrophilic polymers can drastically increase circulation time in the bloodstream and favorably alter biodistribution. This concept was first applied to liposomes by Huang's group in 1990, who used 5-kDa PEG conjugated to a dioleoyl phospholipid in their liposome formulations (Klibanov et al. 1990). The PEGylated phospholipids improved the blood elimination half-life by a factor of 10 without compromising the storage

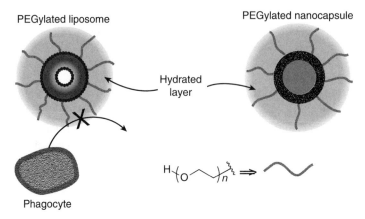

Figure 8.3 Schematic representation of the effects of liposome and nanoparticle PEGylation.

capacity of a dye, outperforming optimized formulations composed of naturally occurring phospholipids. In addition, the composition of the PEGylated chains can be tuned to control the effect of the polymer modification (Allen et al. 1991; Maruyama et al. 1991). The improvements in circulation and biodistribution due to PEGylation have been attributed to the hydrated polymer layer surrounding the modified liposomes (Fig. 8.3). The hydrated coating prevents the nonspecific phagocytosis of liposomes (Klibanov et al. 1990), prevents plasma proteins from marking the liposomes for elimination (Senior et al. 1991), and suppresses uptake of the particles by the liver and spleen (Illum and Davis 1984), both major pathways of particle elimination by the reticuloendothelial system.

The benefits of improved circulation time and altered biodistribution have important consequences for targeted drug delivery of nanocapsules. The enhanced permeability and retention (EPR) effect, for instance, causes nanoparticles and macromolecules to accumulate in malignant tumors at much higher concentrations than in normal cells (Saltzman 2001). The EPR effect occurs because the vasculature of tumors is generally very porous or "leaky," especially to nanoparticles and macromolecules, which then leak into the tumors rather than continuing to circulate in the bloodstream (Dvorak et al. 1988). PEGylated liposomes take advantage of this effect, as shown by Papahadjopoulos and coworkers (1991) in landmark work with liposomes encapsulating the anticancer drugs doxorubicin and epirubicin. The liposome-encapsulated drugs showed significant improvement in antitumor activity because the carrier liposomes preferentially accumulated in the tumors, whereas free drug or nonPEGylated liposomes were eliminated from the body before significant amounts could reach the cancerous cells.

In addition to the passive targeting of tumors due to the EPR effect, active targeting of PEGylated liposomes has also been successful. A study by Huwyler and coworkers (1996), for example, showed that coupling a monoclonal antibody to the surface of PEGylated liposomes resulted in significant transfer of the liposomes across the blood–brain barrier, which is difficult to achieve otherwise. The attached monoclonal antibodies were recognized by specific transfer receptors at the barrier, marking the drug-containing liposomes for transport into the brain. Other active targeting schemes rely on in situ stimulus to release liposome payloads. Irradiation with light (Pidgeon and Hunt 1983) and changes in temperature (Weinstein et al. 1979) or pH (Yatvin et al. 1980), for example, can cause modified liposomes to release pharmaceutical payloads. Similar targeting schemes have been adapted by others in creating synthetic copolymer micelles for drug delivery, as we will discuss in the next section.

Block Copolymer Micelles. Researchers have borrowed Nature's design of liposomes made from self-assembling amphiphilic molecules to create synthetic polymers that mimic this behavior. Although liposomes are immensely useful and researchers have learned many guiding principles from studying them, synthetic copolymer amphiphiles offer the potential of more control over the physical and chemical characteristics of nanocapsules, as well as more diverse functionality. Self-assembling polymers are usually diblock copolymers that contain components with significantly different hydrophilicity. Thus, the polymers self-assemble into hydrophobic-core and hydrophilic-shell nanoassemblies in aqueous environments. In this case, hydrophobic guest molecules, such as drugs, can be entrapped in the greasy core of the artificial vesicles.

One of the most widely used amphiphilic block copolymer architectures is the family of triblock copolymers with the trade name Pluronic®. These polymers, also called poloxamers, consist of two terminal PEG blocks with a middle poly(propylene glycol) (PPG) block. In aqueous solution, the chains aggregate so that the PPG blocks form a hydrophobic core and the PEG blocks point outward, remaining hydrated. Pharmaceutical agents trapped in the hydrophobic core and dispersed in the body enjoy the same benefits provided to PEGylated liposomes already discussed, such as increased circulation time and improved biodistribution. Research by Batrakova et al. (1996) showed that Pluronic copolymer micelles can deliver anthracycline drugs, which are noncovalently bound in the cores of the micelles, to tumors. Studies with mice showed complete disappearance of tumors in nearly 50% of subjects. Subsequent work showed that the composition of the triblock copolymers, such as the total molecular weight and relative block lengths between hydrophilic and hydrophobic units, strongly affects the cytotoxicity of encapsulated doxorubicin and the accumulation of micelles in cancerous cells (Batrakova et al. 1999). Hydrophobic chains of intermediate length and short hydrophilic chains are the most effective, and the use of Pluronic micelles for the delivery of doxorubicin has reached phase II clinical trials (Alakhov et al. 1999).

Another synthetic polymer that has shown promise in recent clinical trials for the micellar encapsulation of anticancer drugs is a block copolymer of PEG and poly (aspartic acid) [PEG-*b*-P(Asp)]. Doxorubicin can be covalently attached to PEG-*b*-P(Asp) through the free carboxylic acid groups on aspartic acid, and the block copolymer then forms micelles in solution with the hydrophobic aspartic acid and drug block forming the core (Yokoyama et al. 1991; Kataoka et al. 1993). As typically occurs, the hydrated PEG chains significantly increased blood circulation times and allowed the drug-encapsulating micelles to accumulate in tumors. Once in the tumors, the amide bond between the polymer and doxorubicin apparently hydrolyzes, releasing the drug. This treatment was very effective at reducing the volume of certain tumor types and notably increased the life span of treated mice (Yokoyama et al. 1991). Doxorubicin conjugated with PEG-*b*-P(Asp) micelles performed well in preclinical trials (Nakanishi et al. 2001) and has progressed past phase I trials (Matsumura et al. 2004).

In addition to Pluronic and PEG-*b*-P(Asp) micelles, a variety of other synthetic polymers has been used to create nanoassemblies with potential drug-delivery applications. Gref et al. (1994) used amphiphilic block copolymers of PEG and PLGA to form drug-carrying nanoparticles by solvent evaporation. The hydrophilic PEG chains enhanced the biodistribution characteristics of the micelles in mice, and they were also used to encapsulate prednisolone and lidocaine. The length of the PEG block could be used to tune the uptake of the particles by the liver as well as the release characteristics of the encapsulated drugs. Synthetic copolymer micelles have also been used in active and passive targeting nanoassemblies for the delivery of pharmaceutical agents. For instance, pH-responsive micelles were created from poly(2-ethyl acrylic acid)–phospholipid conjugates (Maeda et al. 1988) and PEGylated dendrimers (Gillies and Fréchet 2005). Both systems showed significant changes in permeability induced by only a slightly more acidic pH relative to normal physiological conditions. Finally, actively targeted synthetic micelles have been prepared by conjugating a tumor-specific peptide ligand to the end of a PEG-PEI diblock copolymer (Schiffelers et al. 2004). The peptide-derivatized polymers formed core–shell nanoassemblies in aqueous environments and were shown to effect selective tumor uptake and inhibition of tumor growth.

Shell Cross-Linked Artificial Micelles. A recent advance with great potential for drug-delivery applications is the development of shell cross-linked artificial micelles. Karen Wooley's group, who first created this class of nanoassemblies, named them shell cross-linked knedels (SCKs), after a Polish stuffed pastry (Thurmond et al. 1996). Briefly, amphiphilic block copolymers self-assemble into hydrophobic-core micelles, and the hydrophilic-shell blocks are crosslinked, forming permanent core–shell nanoparticles. The inner block may be subsequently removed to form hollow nanostructures. Wooley created SCK nanoparticles from several diblock copolymers (Wooley 2000; O'Reilly et al. 2006) and others have expanded the palette with further variations (Butun et al. 1998; Sanji et al. 2000). The shells have been crosslinked with a wide range of reactions including amidation, alkylation, radical polymerization, and Huisgen 1,3 dipolar cycloaddition (O'Reilly et al. 2006). Finally, methods such as ester hydrolysis, photochemical degradation, ozonolysis, and thermolysis have been used to remove the hydrophobic cores. Given the wide array of compositions and synthetic tools available for the preparation of SCKs, they offer considerable flexibility for the encapsulation and targeting of pharmaceutical agents.

Concentrating on targeted delivery applications, Wooley's group used SCKs as therapeutic tools. An early attempt showed that DNA can wrap around the outside of SCKs, protecting the biopolymer from enzymatic digestion and possibly allowing for gene therapy applications (Thurmond et al. 1999). In another study, the shells of SCKs were functionalized with an oligomeric peptide sequence known to bind with certain cells (Liu et al. 2001). In vitro experiments demonstrated that nanoparticles without the conjugated peptide did not associate with the cells. In contrast, functionalized particles aggregated at cell surfaces and some were taken into cells, suggesting potential for targeted delivery applications. Further work has shown that shell-immobilized biotin (Qi et al. 2004) and mannose (Joralemon et al. 2004) can be used by SCKs as targeting ligands for cells. Similar approaches have shown promise in delivering an antibacterial peptide to cells in vitro (Becker et al. 2005) as well as in creating artificial viruses with antigen-coated SCKs (Joralemon et al. 2005). Although high levels of nanoparticles can cause cell death, lower levels are well tolerated and do not cause death or large-scale morphological changes in mice (Becker et al. 2004). Wooley's groups has also conjugated SCKs with known cancer cell ligands in the hopes of targeted delivery of antitumor drugs (Pan et al. 2003), and "clickable" SCKs promise custom core and/or shell functionalization of premade nanoparticles (O'Reilly et al. 2005). In vivo results have yet to be reported for most of these formulations, however, making it difficult to reach conclusions regarding the ultimate usefulness of the SCK approach at this time.

8.5. CONCLUSION

Polymeric capsule preparation is achieved by self-assembly of small molecules, polymers, and particles into nanometer- and micron-sized objects. Despite the age of the field and number of materials produced thus far, researchers have only just begun to tap the full potential of polymeric capsules that are partially illustrated here with examples of catalytic capsules and drug-delivery vehicles.

We predict that polymeric capsules will be produced with more precision through the use of milli-, micro-, and nanofluidics. Next-generation capsules will be synthesized with greater chemical complexity, leading to more effective responses to stimuli, time, and their environments. Complex capsules will also act as microreactors, allowing chemists to synthesize small molecules with higher efficiency than current methods provide. To meet these expectations, more groups worldwide must dedicate themselves to creating, assembling, and testing new polymeric capsules.

REFERENCES

Alakhov V, Klinski E, Li SM, Pietrzynski G, Venne A, Batrakova E, Bronitch T, Kabanov A. Block copolymer-based formulation of doxorubicin. From cell screen to clinical trials. Colloid Surf B Biointerfaces 1999;16:113–134.

Allen TM. Liposomes—opportunities in drug delivery. Drugs 1997;54:8–14.

Allen TM, Hansen C, Martin F, Redemann C, Yau-Young A. Liposomes containing synthetic lipid derivatives of poly(ethylene glycol) show prolonged circulation half-lives in vivo. 1991;1066:29–36.

Antonietti M, Forster S. Vesicles and liposomes: a self-assembly principle beyond lipids. Adv Mater 2003;15:1323–1333.

Balabushevich NG, Tiourina OP, Volodkin DV, Larionova NI, Sukhorukov GB. Loading the multilayer dextran sulfate/protamine microsized capsules with peroxidase. Biomacromolecules 2003;4:1191–1197.

Bangham AD. Preface. In: Bangham AD, editor. Liposome letters. London: Academic Press; 1983. pp. xii–xiv.

Bangham AD, Horne RW. Negative staining of phospholipids and their structural modification by surface active agents as observed in electron microscope. J Mol Biol 1964;8:660–668.

Batrakova E, Lee S, Li S, Venne A, Alakhov V, Kabanov A. Fundamental relationships between the composition of Pluronic block copolymers and their hypersensitization effect in MDR cancer cells. Pharm Res 1999;16:1373–1379.

Batrakova EV, Dorodnych TY, Klinskii EY, Kliushnenkova EN, Shemchukova OB, Goncharova ON, Arjakov SA, Alakhov VY, Kabanov AV. Anthracycline antibiotics non-covalently incorporated into block copolymer micelles: in vivo evaluation of anti-cancer activity. Br J Cancer 1996;74:1545–1552.

Baxendale IR, Griffiths-Jones CM, Ley SV, Tranmer GK. Microwave-assisted Suzuki coupling reactions with an encapsulated palladium catalyst for batch and continuous-flow transformations. Chem Eur J 2006;12:4407–4416.

Becker ML, Bailey LO, Wooley KL. Peptide-derivatized shell-cross-linked nanoparticles. 2. Biocompatibility evaluation. Bioconjug Chem 2004;15:710–717.

Becker ML, Liu JQ, Wooley KL. Functionalized micellar assemblies prepared via block copolymers synthesized by living free radical polymerization upon peptide-loaded resins. Biomacromolecules 2005;6:220–228.

Becker WM, Reece JB, Poenie MF. The world of the cell. 3rd ed. Menlo Park (CA): Benjamin Cummings; 1996.

Berkland C, Pollauf E, Raman C, Silverman R, Kim K, Pack DW. Macromolecule release from monodisperse PLG microspheres: control of release rates and investigation of release mechanism. J Pharm Sci 2007;96:1176–1191.

Binks BP, Lumsdon SO. Pickering emulsions stabilized by monodisperse latex particles: effects of particle size. Langmuir 2001;17:4540–4547.

Breitenkamp K, Emrick T. Novel polymer capsules from amphiphilic graft copolymers and cross-metathesis. J Am Chem Soc 2003;125:12070–12071.

Bremeyer N, Ley SV, Ramarao C, Shirley IM, Smith SC. Palladium acetate in polyurea microcapsules: a recoverable and reusable catalyst for hydrogenations. Synlett 2002;1843–1844.

Broadwater SJ, McQuade DT. Investigating PdEnCat catalysis. J Org Chem 2006;71: 2131–2134.

Bungenberg de Jong HG. Crystallization–coacervation–flocculation. In: Kruyt HR, editor. Colloid science. Vol. II.New York: Elsevier; 1949. pp. 232–258.

Butun V, Billingham NC, Armes SP. Synthesis of shell cross-linked micelles with tunable hydrophilic/hydrophobic cores. J Am Chem Soc 1998;120:12135–12136.

Caruso F. Hollow capsule processing through colloidal templating and self-assembly. Chem Eur J 2000;6:413–419.

Caruso F, Caruso RA, Möhwald H. Nanoengineering of inorganic and hybrid hollow spheres by colloidal templating. Science 1998;282:1111–1114.

Caruso F, Schuler C, Kurth DG. Coreshell particles and hollow shells containing metallo-supramolecular components. Chem Mater 1999;11:3394–3399.

Caruso F, Trau D, Möhwald H, Renneberg R. Enzyme encapsulation in layer-by-layer engin-eered polymer multilayer capsules. Langmuir 2000;16:1485–1488.

Caruso F, Yang WJ, Trau D, Renneberg R. Microencapsulation of uncharged low molecular weight organic materials by polyelectrolyte multilayer self-assembly. Langmuir 2000;16:8932–8936.

Cayre OJ, Noble PF, Paunov VN. Fabrication of novel colloidosome microcapsules with gelled aqueous cores. J Mater Chem 2004;14:3351–3355.

Chang TMS. Semipermeable microcapsules. Science 1964;146:524–525.

Chang TMS. Semipermeable aqueous microcapsules (artificial cells)—with emphasis on experiments in an extracorporeal shunt system. Trans Am Soc Artif Internal Organs 1966;12:13–19.

Chang TMS. In-vivo effects of semipermeable microcapsules containing L-asparaginase on 6C3HED lymphosarcoma. Nature 1971a;229:117–118.

Chang TMS. Stablisation of enzymes by microencapsulation with a concentrated protein sol-ution or by microencapsulation followed by cross-linking with glutaraldehyde. Biochem Biophys Res Commun 1971b;44:1531–1536.

Chang TMS. 1972.Artificial cells. In: Kugelmass IN, editor. American lecture series. Springfield (IN): Charles C. Thomas.

Chang TMS, MacIntosh FC, Mason SG. Semipermeable aqueous microcapsules. I. Preparation and properties. Can J Physiol Pharmacol 1966;44:115–128.

Coleman DR, Royer GP. New hydrogenation catalyst–palladium poly(ethyleneimine) ghosts—applications in peptide synthesis. J Org Chem 1980;45:2268–2269.

Dahne L, Leporatti S, Donath E, Möhwald H. Fabrication of micro reaction cages with tailored properties. J Am Chem Soc 2001;123:5431–5436.

Davies IW, Matty L, Hughes DL, Reider PJ. Are heterogeneous catalysts precursors to homo-geneous catalysts? J Am Chem Soc 2001;123:10139–10140.

Decher G, Hong JD. Buildup of ultrathin multilayer films by a self-assembly process. 1. Consecutive adsorption of anionic and cationic bipolar amphiphiles on charged surfaces. Makromolekulare Chemie-Macromolecular Symposia 1991a;46:321–327.

Decher G, Hong JD. Buildup of ultrathin multilayer films by a self-assembly process. 2. Consecutive adsorption of anionic and cationic bipolar amphiphiles and polyelectrolytes on charge surfaces. Ber Bunsen Ges Phys Chem Chem Phys 1991b;95:1430–1434.

Decher G, Hong JD, Schmitt J. Buildup of ultrathin multilayer films by a self-assembly process. 3. Consecutively alternating adsorption of anionic and cationic polyelectrolytes on charged surfaces. Thin Solid Films 1992;210:831–835.

Dinsmore AD, Hsu MF, Nikolaides MG, Marquez M, Bausch AR, Weitz DA. Colloidosomes: selectively permeable capsules composed of colloidal particles. Science 2002;298:1006–1009.

Discher DE, Eisenberg A. Polymer vesicles. Science 2002;297:967–973.

Donath E, Sukhorukov GB, Caruso F, Davis SA, Möhwald H. Novel hollow polymer shells by colloid-templated assembly of polyelectrolytes. Angew Chem Int Ed 1998;37:2202–2205.

Duan HW, Wang DY, Sobal NS, Giersig M, Kurth DG, Möhwald H. Magnetic colloidosomes derived from nanoparticle interfacial self-assembly. Nano Lett 2005;5:949–952.

Dvorak HF, Nagy JA, Dvorak JT, Dvorak AM. Identification and characterization of the blood-vessels of solid tumors that are leaky to circulating macromolecules. Am J Pathol 1988;133:95–109.

Friedli AC, Schlager IR, Wright SW. Demonstrating encapsulation and release: a new take on alginate complexation and the nylon rope trick. J Chem Ed 2005;82:1017–1020.

Fukuda H, Diem T, Stefely J, Kezdy FJ, Regen SL. Polymer-encased vesicles derived from dioctadecyldimethylammonium methacrylate. J Am Chem Soc 1986;108:2321–2327.

Gao CY, Donath E, Möhwald H, Shen JC. Spontaneous deposition of water-soluble substances into microcapsules: phenomenon, mechanism, and application. Angew Chem Int Ed 2002;41:3789–3793.

Gao CY, Liu XY, Shen JC, Möhwald H. Spontaneous deposition of horseradish peroxidase into polyelectrolyte multilayer capsules to improve its activity and stability. Chem Commun 2002;1928–1929.

Gillies ER, Fréchet JMJ. pH-responsive copolymer assemblies for controlled release of doxo-rubicin. Bioconjug Chem 2005;16:361–368.

Glogowski E, Tangirala R, He JB, Russell TP, Emrick T. Microcapsules of PEGylated gold nanoparticles prepared by fluid–fluid interfacial assembly. Nano Lett 2007;7:389–393.

Gref R, Minamitake Y, Peracchia MT, Trubetskoy V, Torchilin V, Langer R. Biodegradable long-circulating polymeric nanospheres. Science 1994;263:1600–1603.

Gregoriadis G, Allison AC. Entrapment of proteins in liposomes prevents allergic reactions in pre-immunized mice. FEBS Lett 1974;45:71–74.

Hagen CM, Widegren JA, Maitlis PM, Finke RG. Is it homogeneous or heterogeneous cataly-sis? Compelling evidence for both types of catalysts derived from [Rh(η(5)-C5Me5) Cl-2](2)as a function of temperature and hydrogen pressure. J Am Chem Soc 2005;127: 4423–4432.

Hipp AK, Storti G, Morbidelli M. Particle sizing in colloidal dispersions by ultrasound. Model calibration and sensitivity analysis. Langmuir 1999;15:2338–2345.

Hole JW. Essentials of human anatomy and physiology. 3rd ed. Dubuque (IA): Wm. C. Brown; 1989.

Holm KA. Automated determination of microbial peroxidase activity in fermentation samples using hydrogen peroxide as the substrate and 2,2′-azino-bis(3-ethylbenzothiazoline-6-sulfo-nate) as the electron donor in a flow-injection system. Analyst 1995;120:2101–2105.

Huwyler J, Wu DF, Pardridge WM. Brain drug delivery of small molecules using immunoli-posomes. Proc Natl Acad Sci USA 1996;93:14164–14169.

Hwang JR, Sefton MV. Effect of capsule diameter on the permeability to horseradish peroxi-dase of individual HEMA–MMA microcapsules. J Controlled Release 1997;49:217–227.

Iler RK. Multilayers of colloidal particles. J Colloid Interface Sci 1966;21:569–594.

Illum L, Davis SS. The organ uptake of intravenously administered colloidal particles can be altered using a non-ionic surfactant (Poloxamer-338). FEBS Lett 1984;167:79–82.

Joralemon MJ, Murthy KS, Remsen EE, Becker ML, Wooley KL. Synthesis, characterization, and bioavailability of mannosylated shell cross-linked nanoparticles. Biomacromolecules 2004;5:903–913.

Joralemon MJ, Smith NL, Holowka D, Baird B, Wooley KL. Antigen-decorated shell cross-linked nanoparticles: synthesis, characterization, and antibody interactions. Bioconjug Chem 2005;16:1246–1256.

Kataoka K, Kwon GS, Yokoyama M, Okano T, Sakurai Y. Block-copolymer micelles as vehicles for drug delivery. J Controlled Release 1993;24:119–132.

Klibanov AL, Maruyama K, Torchilin VP, Huang L. Amphipathic polyethyleneglycols effectively prolong the circulation time of liposomes. FEBS Lett 1990;268:235–237.

Kobašlija M, McQuade DT. Polyurea microcapsules from oil-in-oil emulsions via interfacial polymerization. Macromolecules 2006;39:6371–6375.

Lambert G, Bertrand JR, Fattal E, Subra F, Pinto-Alphandary H, Malvy C, Auclair C, Couvreur P. EWS Fli-1 antisense nanocapsules inhibits Ewing sarcoma-related tumor in mice. Biochem Biophys Res Commun 2000;279:401–406.

Lambert G, Fattal E, Pinto-Alphandary H, Gulik A, Couvreur P. Polyisobutylcyanoacrylate nanocapsules containing an aqueous core as a novel colloidal carrier for the delivery of oligonucleotides. Pharm Res 2000;17:707–714.

Langer R. Biomaterials in drug delivery and tissue engineering: one laboratory's experience. Acc Chem Res 2000;33:94–101.

Lee CKY, Holmes AB, Ley SV, McConvey IF, Al-Duri B, Leeke GA, Santos RCD, Seville JPK. Efficient batch and continuous flow Suzuki cross-coupling reactions under mild conditions, catalysed by polyurea-encapsulated palladium(II) acetate and *tetra-n*-butylammonium salts. Chem Commun 2005;2175–2177.

Ley SV, Mitchell C, Pears D, Ramarao C, Yu JQ, Zhou WZ. Recyclable polyurea-microencapsulated Pd(0) nanoparticles: an efficient catalyst for hydrogenolysis of epoxides. Org Lett 2003;5:4665–4668.

Ley SV, Ramarao C, Gordon RS, Holmes AB, Morrison AJ, McConvey IF, Shirley IM, Smith SC, Smith MD. Polyurea-encapsulated palladium(II) acetate: a robust and recyclable catalyst for use in conventional and supercritical media. Chem Commun 2002;1134–1135.

Lim F. Preface. In: Lim F, editor. Biomedical applications of microencapsulation. Boca Raton (FL): CRC Press; 1984.

Lim F, Sun AM. Microencapsulated islets as bioartificial endocrine pancreas. Science 1980;210:908–910.

Lin Y, Skaff H, Boker A, Dinsmore AD, Emrick T, Russell TP. Ultrathin cross-linked nanoparticle membranes. J Am Chem Soc 2003;125:12690–12691.

Liu JQ, Zhang Q, Remsen EE, Wooley KL. Nanostructured materials designed for cell binding and transduction. Biomacromolecules 2001;2:362–368.

Maeda M, Kumano A, Tirrell DA. H + -Induced release of contents of phosphatidylcholine vesicles bearing surface-bound poly-electrolyte chains. J Am Chem Soc 1988;110:7455–7459.

Makino K, Mack EJ, Okano T, Kim SW. A microcapsule self-regulating delivery system for insulin. J Controlled Release 1990;12:235–239.

Mao ZW, Ma L, Gao CY, Shen JC. Preformed microcapsules for loading and sustained release of ciprofloxacin hydrochloride. J Controlled Release 2005;104:193–202.

Marinakos SM, Anderson MF, Ryan JA, Martin LD, Feldheim DL. Encapsulation, permeability, and cellular uptake characteristics of hollow nanometer-sized conductive polymer capsules. J Phys Chem B 2001;105:8872–8876.

Marinakos SM, Novak JP, Brousseau LC, House AB, Edeki EM, Feldhaus JC, Feldheim DL. Gold particles as templates for the synthesis of hollow polymer capsules. Control of capsule dimensions and guest encapsulation. J Am Chem Soc 1999;121:8518–8522.

Maruyama K, Yuda T, Okamoto A, Ishikura C, Kojima S, Iwatsuru M. Effect of molecular-weight in amphipathic polyethyleneglycol on prolonging the circulation time of large unilamellar liposomes. Chem Pharm Bull 1991;39:1620–1622.

Matsumura Y, Hamaguchi T, Ura T, Muro K, Yamada Y, Shimada Y, Shirao K, Okusaka T, Ueno H, Ikeda M, Watanabe N. Phase I clinical trial and pharmacokinetic evaluation of NK911, a micelle-encapsulated doxorubicin. Br J Cancer 2004;91:1775–1781.

McClements DJ. Principles of ultrasonic droplet size determination in emulsions. Langmuir 1996;12:3454–3461.

McGee H. On food and cooking: the science and lore of the kitchen. 2nd ed. New York: Scribner; 2004.

Meyers WE, Royer GP. Catalysis of *para*-nitrofluoroacetanilide hydrolysis by an imidazole derivative of polyethyleneimine ghosts. J Am Chem Soc 1977;99:6141–6142.

Morgan PW, Kwolek SL. Interfacial polycondensation. 2. Fundamentals of polymer formation at liqud interfaces. Journal of Polymer Science Part A—Chemistry 1959;40:299–327.

Nakanishi T, Fukushima S, Okamoto K, Suzuki M, Matsumura Y, Yokoyama M, Okano T, Sakurai Y, Kataoka K. Development of the polymer micelle carrier system for doxorubicin. J Controlled Release 2001;74:295–302.

Narayanan R, El-Sayed MA. Effect of catalysis on the stability of metallic nanoparticles: Suzuki reaction catalyzed by PVP–palladium nanoparticles. J Am Chem Soc 2003;125:8340–8347.

New RRC. Influence of liposome characteristics on their properties and fate. In: Philippot JR, Schuber F, editors. Liposomes as tools in basic research and industry. Boca Raton (FL): CRC Press; 1995. pp. 3–20.

Northfelt DW, Martin FJ, Working P, Volberding PA, Russell J, Newman M, Amantea MA, Kaplan LD. Doxorubicin encapsulated in liposomes containing surface-bound polyethylene glycol: pharmacokinetics, tumor localization, and safety in patients with AIDS-related Kaposi's sarcoma. J Clin Pharmacol 1996;36:55–63.

Ogawa Y, Yamamoto M, Okada H, Yashiki T, Shimamoto T. A new technique to efficiently entrap leuprolide acetate into microcapsules of polylactic acid or copoly(lactic glycolic) acid. Chem Pharm Bull 1988;36:1095–1103.

Okamoto K, Akiyama R, Yoshida H, Yoshida T, Kobayashi S. Formation of nanoarchitectures including subnanometer palladium clusters and their use as highly active catalysts. J Am Chem Soc 2005;127:2125–2135.

O'Reilly RK, Hawker CJ, Wooley KL. Cross-linked block copolymer micelles: functional nanostructures of great potential and versatility. Chem Soc Rev 2006;35:1068–1083.

O'Reilly RK, Joralemon MJ, Wooley KL, Hawker CJ. Functionalization of micelles and shell cross-linked nanoparticles using click chemistry. Chem Mater 2005;17:5976–5988.

Pan D, Turner JL, Wooley KL. Folic acid-conjugated nanostructured materials designed for cancer cell targeting. Chem Commun 2003;2400–2401.

Papahadjopoulos D, Allen TM, Gabizon A, Mayhew E, Matthay K, Huang SK, Lee KD, Woodle MC, Lasic DD, Redemann C, Martin FJ. Sterically stabilized liposomes—improvements in pharmacokinetics and antitumor therapeutic efficacy. Proc Natl Acad Sci USA 1991;88:11460–11464.

Paunov VN. Novel method for determining the three-phase contact angle of colloid particles adsorbed at air–water and oil–water interfaces. Langmuir 2003;19:7970–7976.

Paunov VN, Cayre OJ. Supraparticles and "Janus" particles fabricated by replication of particle monolayers at liquid surfaces using a gel trapping technique. Adv Mater 2004;16:788–791.

Pekarek KJ, Jacob JS, Mathiowitz E. Double-walled polymer microspheres for controlled drug-release. Nature 1994;367:258–260.

Pidgeon C, Hunt CA. Light sensitive liposomes. Photochem Photobiol 1983;37:491–494.

Poe SL, Kobašlija M, McQuade DT. Microcapsule enabled multicatalyst system. J Am Chem Soc 2006;128:15586–15587.

Poe SL, Kobašlija M, McQuade DT. Mechanism and application of a microcapsule enabled multicatalyst reaction. J Am Chem Soc 2007;129:9216–9221.

Price KE, Mason BP, Bogdan AR, Broadwater SJ, Steinbacher JL, McQuade DT. Microencapsulated linear polymers: "soluble" heterogeneous catalysts. J Am Chem Soc 2006;128:10376–10377.

Qi K, Ma QG, Remsen EE, Clark CG, Wooley KL. Determination of the bioavailability of biotin conjugated onto shell cross-linked (SCK) nanoparticles. J Am Chem Soc 2004;126:6599–6607.

Ramarao C, Ley SV, Smith SC, Shirley IM, DeAlmeida N. Encapsulation of palladium in poly-urea microcapsules. Chem Commun 2002;1132–1133.

Rao AM, John VT, Gonzalez RD, Akkara JA, Kaplan DL. Catalytic and interfacial aspects of enzymatic polymer synthesis in reversed micellar systems. Biotechnol Bioeng 1993;41: 531–540.

Rebek J. Mechanistic studies using solid supports—the 3-phase test. Tetrahedron 1979;35:723–731.

Rebek J, Brown D, Zimmerman S. 3-Phase test for reaction intermediates—nucleophilic cata-lysis and elimination reactions. J Am Chem Soc 1975;97:454–455.

Rebek J, Gavina F. 3-Phase test for reactive intermediates—cyclobutadiene. J Am Chem Soc 1974;96:7112–7114.

Regen SL, Shin JS, Yamaguchi K. Polymer-encased vesicles. J Am Chem Soc 1984;106: 2446–2447.

Royer GP. Polymeric particulate carrier. US Patent 4,326,009. 1982.

Royer GP, Chow WS, Hatton KS. Palladum polyethyleneimine catalysts. J Mol Catal 1985;31:1–13.

Saltzman WM. Drug delivery: engineering principles for drug therapy. In: Gubbins KE, editor. Topics in chemical engineering: a series of textbooks and monographs. Oxford: xford University Press; 2001.

Sanji T, Nakatsuka Y, Ohnishi S, Sakurai H. Preparation of nanometer-sized hollow particles by photochemical degradation of polysilane shell cross-linked micelles and reversible encapsulation of guest molecules. Macromolecules 2000;33:8524–8526.

Scher HB. Encapsulation process and capsules produced thereby. US Patent 4,285,720. 1981.

Scher HB, Rodson M, Lee KS. Microencapsulation of pesticides by interfacial polymerization utilizing isocyanate or aminoplast chemistry. Pest Sci 1998;54:394–400.

Schiffelers RM, Ansari A, Xu J, Zhou Q, Tang QQ, Storm G, Molema G, Lu PY, Scaria PV, Woodle MC. Cancer siRNA therapy by tumor selective delivery with ligand-targeted steri-cally stabilized nanoparticle. Nucleic Acids Res 2004;32:e149.

Scott C, Wu D, Ho CC, Co CC. Liquid-core capsules via interfacial polymerization: a free-radical analogy of the nylon rope trick. J Am Chem Soc 2005;127:4160–4161.

Senior J, Delgado C, Fisher D, Tilcock C, Gregoriadis G. Influence of surface hydrophilicity of liposomes on their interaction with plasma-protein and clearance from the circulation—studies with poly(ethylene glycol)-coated vesicles. Biochim Biophys Acta 1991;1062: 77–82.

Shi XY, Caruso F. Release behavior of thin-walled microcapsules composed of polyelectrolyte multilayers. Langmuir 2001;17:2036–2042.

Shutava T, Zheng ZG, John V, Lvov Y. Microcapsule modification with peroxidase-catalyzed phenol polymerization. Biomacromolecules 2004;5:914–921.

Skaff H, Lin Y, Tangirala R, Breitenkamp K, Boker A, Russell TP, Emrick T. Crosslinked capsules of quantum dots by interfacial assembly and ligand crosslinking. Adv Mater 2005;17:2082–2086.

Soo PL, Eisenberg A. Preparation of block copolymer vesicles in solution. J Polym Sci B Polym Phys 2004;42:923–938.

Soon-Shiong P, Heintz RE, Merideth N, Yao QX, Yao ZW, Zheng TL, Murphy M, Moloney MK, Schmehl M, Harris M, Mendez R, Mendez R, Sandford PA. Insulin independence in a type-1 diabetic patient after encapsulated islet transplantation. Lancet 1994;343:950–951.

Steel PG, Teasdale CWT. Polymer supported palladium N-heterocyclic carbene complexes: long lived recyclable catalysts for cross coupling reactions. Tetrahedron Lett 2004;45: 8977–8980.

Stein EW, Volodkin DV, McShane MJ, Sukhorukov GB. Real-time assessment of spatial and temporal coupled catalysis within polyelectrolyte microcapsules containing coimmobilized glucose oxidase and peroxidase. Biomacromolecules 2006;7:710–719.

Sukhorukov GB, Antipov AA, Voigt A, Donath E, Möhwald H. pH-Controlled macromolecule encapsulation in and release from polyelectrolyte multilayer nanocapsules. Macromol Rapid Commun 2001;22:44–46.

Tabata Y, Ikada Y. Effect of the size and surface-charge of polymer microspheres on their phagocytosis by macrophage. Biomaterials 1988;9:356–362.

Thies C. A short history of microencapsulation technology. In: Arshady R, editor. Preparation and chemical applications. Vol. 1. MML series.London: Citus Books; 1999. pp. 47–54.

Thurmond KB, Kowalewski T, Wooley KL. Water-soluble Knedel-like structures: the preparation of shell-cross-linked small particles. J Am Chem Soc 1996;118:7239–7240.

Thurmond KB, Remsen EE, Kowalewski T, Wooley KL. Packaging of DNA by shell cross-linked nanoparticles. Nucleic Acids Res 1999;27:2966–2971.

Tollemar J, Hockerstedt K, Ericzon BG, Jalanko H, Ringden O. Liposomal amphotericin-B prevents invasive fungal, infections in liver-transplant recipients—a randomized, placebo-controlled study. Transplantation 1995;59:45–50.

Uludag H, Hwang JR, Sefton MV. Microencapsulated human hepatoma (HEPG2) cells—capsule-to-capsule variations in protein secretion and permeability. J Controlled Release 1995;33:273–283.

Velev OD, Furusawa K, Nagayama K. Assembly of latex particles by using emulsion droplets as templates. 1. Microstructured hollow spheres. Langmuir 1996;12:2374–2384.

Velev OD, Nagayama K. Assembly of latex particles by using emulsion droplets. 3. Reverse (water in oil) system. Langmuir 1997;13:1856–1859.

Voigt A, Lichtenfeld H, Sukhorukov GB, Zastrow H, Donath E, Baumler H, Möhwald H. Membrane filtration for microencapsulation and microcapsules fabrication by layer-by-layer polyelectrolyte adsorption. Ind Eng Chem Res 1999;38:4037–4043.

Wang HG, Zheng XM, Ping C, Zheng XM. The fabrication of reactive hollow polysiloxane capsules and their application as a recyclable heterogeneous catalyst for the Heck reaction. J Mater Chem 2006;16:4701–4705.

Watanabe K, Royer GP. Polyethyleneimine silica-gel as an enzyme support. J Mol Catal 1983;22:145–152.

Weinstein JN, Magin RL, Yatvin MB, Zaharko DS. Liposomes and local hyperthermia—selective delivery of methotrexate to heated tumors. Science 1979;204:188–191.

Wittbecker EL, Morgan PW. Interfacial polycondensation. 1. J Polym Sci 1959;40:289–297.

Wooley KL. Shell crosslinked polymer assemblies: nanoscale constructs inspired from biological systems. J Polym Sci Polym Chem 2000;38:1397–1407.

Yatvin MB, Kreutz W, Horwitz BA, Shinitzky M. pH-Sensitive liposomes—possible clinical implications. Science 1980;210:1253–1254.

Yokoyama M, Okano T, Sakurai Y, Ekimoto H, Shibazaki C, Kataoka K. Toxicity and antitumor-activity against solid tumors of micelle-forming polymeric anticancer drug and its extremely long circulation in blood. Cancer Res 1991;51:3229–3236.

Yu JQ, Wu HC, Ramarao C, Spencer JB, Ley SV. Transfer hydrogenation using recyclable polyurea-encapsulated palladium: efficient and chemoselective reduction of aryl ketones. Chem Commun 2003;678–679.

Zhang LF, Eisenberg A. Multiple morphologies of crew-cut aggregates of polystyrene-*b*-poly (acrylic acid) block copolymers. Science 1995;268:1728–1731.

Zhang LF, Yu K, Eisenberg A. Ion-induced morphological changes in "crew-cut" aggregates of amphiphilic block copolymers. Science 1996;272:1777–1779.

CHAPTER 9

SEQUENCE-SPECIFIC HYDROGEN BONDED UNITS FOR DIRECTED ASSOCIATION, ASSEMBLY, AND LIGATION

BING GONG

9.1. INTRODUCTION

Specific association of various structures, either covalently or noncovalently, plays vital roles in both chemical and biological processes. In natural systems, the association of molecular components is realized through the cooperative action of multiple noncovalent interactions, leading to a highly specific formation of assemblies with high thermodynamic stability. In forming multicomponent assemblies, the unfavorable loss of entropy is usually offset by the presence of numerous cooperative enthalpic interactions within the assembled structures. Similar to examples found in Nature, many nanoscaled structures have been designed and constructed by forming multicomponent assemblies. Most multicomponent assemblies reported thus far were designed based on a case by case basis, because the information encoded in the corresponding components only defines the dimension and shape of the specific assembly.

Duplex DNA, consisting of two helical strands with complementary shapes and H bonding sequences, represents a system that best illustrates the action of multiple interactions. A unique feature of duplex DNA is its modular, digital nature in storing and retrieving information, which is realized by arranging the four different nucleotides along a linear backbone. Inspired by DNA molecules, a systematic approach to self-assembly has been proposed and tested, which involves associating units (modules) that store and retrieve information in a digital fashion (Zimmerman and Corbin, 2000). This field originated from investigation of the molecular association of nucleobases that later led to the development of highly specific H bonded

Molecular Recognition and Polymers: Control of Polymer Structure and Self-Assembly.
Edited by V. Rotello and S. Thayumanavan

pairs for specifying intermolecular interactions (Jorgensen and Pranata 1990; Pranata, et al. 1990; Murray and Zimmerman 1992). Subsequently, particularly stable hetero-cyclic complexes with arrays of H bond donors (D) and acceptors (A) were developed by the groups of Zimmerman (Fenlon et al. 1993; Corbin and Zimmerman 1998) and others (Beijer, Kooijman, et al. 1998; Beijer, Sijbesma, et al. 1998; Folmer et al. 1999). Many other groups also reported H bonded self-assembling aggregates consisting of multicomponents (Yang et al. 1993; Whitesides et al. 1995; Sessler and Wang 1996; Vreekamp et al. 1996; Conn and Rebek 1997; Hartgerink et al. 1998; Kolotuchin and Zimmerman 1998; Orr et al. 1999; Lehn 2000; Percec et al. 2000; Fenniri et al. 2002; Wu et al. 2002).

The development of DNA-like associating units with strength and specificity that can be systematically tuned is of great significance for the specific control of intermolecular association. Such units may serve as versatile "molecular glue" that allows the association or linking of various structural units, including those with no obvious complementarity. With such associating units, the formation of a self-assembling architecture can be readily carried out under thermodynamic control, that is, by simply mixing the components together. In addition to the development of noncovalent associating units, an even more exciting possibility is the combination of systematically tunable supramolecular linking units and reversible covalent bonds, leading to covalently linked structures under reversible conditions.

In recent years we have developed a series of information-storing molecular duplexes based on oligoamides consisting of simple building blocks (Gong et al. 1999; Zeng et al. 2000, 2001, 2002, 2003; Gong 2001; Yang et al. 2003, 2004; Sanford et al. 2004; Yang and Gong 2005). This system involves oligoamide strands carrying various arrays (sequences) of amide hydrogens and carbonyl oxygens as H bond acceptors and donors. These molecular duplexes are featured by tunable affinity, programmable sequence specificity, and convenient synthetic availability, offering a novel class of associating units for the instructed assembly of various discrete, oligomeric, or polymeric structures.

The discussion in this chapter is mostly focused on 1) the development of hydrogen bonded duplexes that form in a sequence-specific fashion, 2) application of the duplexes as information-storing molecules for specifying intermolecular association, and 3) converting the duplexes into sequence-specific dynamic covalent (Rowan et al. 2002) ligation units in highly competitive media. Many other important duplex systems (Bisson et al. 2000; Gabriel and Iverson, 2002; Zhao et al. 2003; Jiang et al. 2004; Gong and Krische, 2005) are noteworthy and will not be discussed here.

9.2. GENERAL DESIGN: INFORMATION-STORING MOLECULAR DUPLEXES BASED ON THE RECOMBINATION OF H BOND DONORS AND ACCEPTORS

The objective of this research is to develop easily modifiable associating units that can direct unnatural molecular systems to self-assemble in a highly predictable fashion. Specifically, hydrogen bonded duplexes with unnatural backbones are chosen to serve as information-storing molecules. These duplexes should show predictable and adjustable stabilities based on their number of H bonding sites and should simultaneously demonstrate high specificity in their formation. The specific design involves the combination of residues derived from substituted 3-aminobenzoic acid, isophthalic acid (1,3-benzenedicarboxylic acid), and *m*-phenylenediamine (1,3-diaminobenzene), leading to oligoamides having various arrays (i.e., H bonding sequences) of H bond donors and acceptors. The resulting molecular strands are expected to form duplexes via hydrogen bonding interactions between the backbone amide O and H atoms. A strand is expected to sequence a specific pair with another strand of its complementary sequence. By incorporating multiple H bonding sites into the design, the formation of the corresponding duplex should be highly specific and have tunable specificity. Depending on the hydrogen bonding sequence, either homo-dimers or heterodimers can be formed. The number of hydrogen bonding sites can be easily adjusted to afford duplexes of different stabilities. With n intermolecular H bonding sites, the number (N) of different duplexes can be calculated by the following equations:

$$N = 2^{n-2} + 2^{(n-3)/2} \qquad (9.1)$$

when n is an odd number and

$$N = 2^{n-2} + 2^{(n-2)/2} \qquad (9.2)$$

when n is an even number. When n is an odd number, only complementary (or hetero-) duplexes are possible. When n is an even number, both complementary and self-complementary duplexes become possible. The first term in the second equation is the number of complementary duplexes and the second term is the number of self-complementary (or homo-) duplexes. The number (N) of unique duplexes increases very rapidly as the number of intermolecular H bonding sites increases. For example, there should be 6 different quadruply H bonded duplexes. For duplexes having six intermolecular H bonds, the number of different duplexes becomes 20.

9.3. QUADRUPLY H BONDED DUPLEXES WITH SEQUENCE-INDEPENDENT STABILITY

Oligoamides **1** and **2** (Fig. 9.1) represent the first examples of the H bonded duplexes that we designed (Gong et al. 1999). The backbones of these oligoamides are preorganized by the favorable six-membered intramolecularly H bonded rings. The introduced intramolecular H bonds also serve to block undesired H bonding interactions that may otherwise lead to the formation of polymeric aggregates. As a result, oligoamides **1** and **2** have one H bonding edge and should dimerize via their self-complementary sequences of DADA and DDAA, respectively. In addition, the ether side chains allow convenient incorporation of various groups into the design, leading to duplexes that are compatible with different solvents.

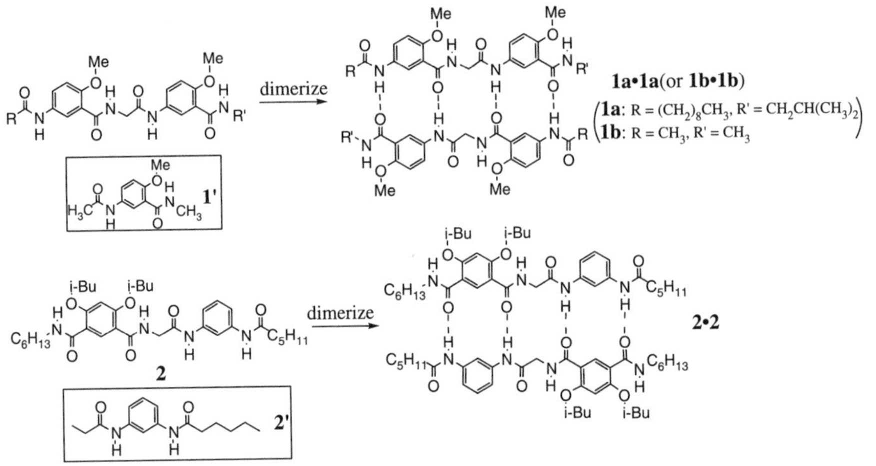

Figure 9.1   Quadruply H bonded duplexes **1•1** and **2•2**. In spite of their different donor (D)–acceptor (A) sequences, duplexes **1•1** and **2•2** have similar stabilities.

The structures and formation of homoduplexes **1•1** and **2•2** were confirmed by one-dimensional (1-D), two-dimensional (2-D), and variable temperature ¹H-NMR; vapor pressure osmometry (VPO) studies; and X-ray crystallography. The ¹H-NMR experiments in CDCl₃ revealed significant downfield (>2 ppm) shifts of the aniline NH signals of **1a** and **2** compared to those of the one-ring compounds **1'** and **2'**, consistent with the formation of duplexes **1•1** and **2•2**. The formation of the duplexes was also indicated by ¹H-NMR binding studies carried out in CDCl₃, which revealed dimerization constants of $\geq 4.4 \times 10^4$ M^{-1} for **1a•1a** and $\sim 6.5 \times 10^4$ M^{-1} for **2•2**. Considering the error of the NMR binding experiments ($\pm 10\%$), duplexes **1a•1a** and **2•2** can be regarded as having similar stabilities. In addition to the NMR binding experiments, the formation of dimers by **1a** and **2** were also indicated by VPO studies.

The involvement of the aniline NH groups in forming intermolecular H bonding was demonstrated by variable temperature ¹H-NMR, which showed that the amide H atoms of both of the two aniline NH signals of **1a** shifted upfield. In contrast, the two glycine NH signals, which formed the S(6) intramolecular hydrogen bonds, showed much smaller changes within the same concentration and temperature ranges.

The most conclusive evidence for the presence of duplexes **1a•1a** and **2•2** came from 2-D ¹H-NMR [nuclear Overhauser effect spectroscopy (NOESY)] and X-ray crystallography. The NOESY spectrum of **1b** in CDCl₃ contains interstrand NOEs (Fig. 9.2a) between protons *c* and *e*, *c* and *i*, and *c* and *j*, which are consistent with an H bonded dimer. The crystal structures of **1a** and **2** both revealed the expected dimeric structures held together by intermolecular H bonds (Fig. 9.2b).

It was found that α-amino acid linkers other than glycine, or different alkoxy side chains did not affect the stability of each of the duplexes, which lies in the range of 10⁴ M^{-1}. Because of its ready availability, the glycine residue has been adopted in most of the subsequently designed duplexes. The similar stability of these duplexes

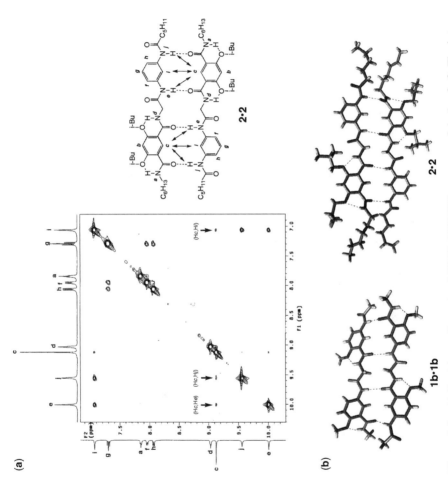

Figure 9.2 (a) Partial two-dimensional NMR (NOESY) spectrum of duplex **2•2**. The inter-strand NOEs are indicated by arrows. (b) The crystal structures of duplexes **1b•1b** and **2•2**.

211

suggests that the observed dimerization constants are independent of the sequence and are only proportional to the number of interstrand H bonds in each duplex. This is in contrast to many previously reported H bonded dimers based on rigid heterocycles, whose stability depends not only on the number of intermolecular H bonds but also on the arrangement of H bond donors and acceptors attributable to secondary electrostatic interactions (Jorgensen and Pranata 1990; Pranata et al. 1990; Murray and Zimmerman 1992).

9.4. TUNING BINDING STRENGTH BY VARYING THE NUMBER OF INTERSTRAND H BONDS

Because the stability of our duplexes depends only on the number of intermolecular H bonds, increasing the number of interstrand H bonds in a duplex, which can be readily realized by adding additional aromatic residues into the oligoamide strands, should lead to an increase in the stability of a duplex. Such an expectation was confirmed by the 6-H bonded, heteroduplex **3•4** (Fig. 9.3a), consisting of two different strands having complementary H bonding sequences DADDAD and ADAADA (Zeng et al. 2000). Probing the sequence-specific pairing of **3** and **4** into an H bonded duplex was investigated to address the following questions: 1) In addition to those with self-complementary arrays, could duplexes consisting of two different strands of complementary H bonding sequences still form? 2) Will the backbones of longer strands stay in register to allow pairing of the duplexes? 3) Compared to the homologous quadruply H bonded duplexes, what would be the expected enhancement of the binding strength (association constant) of the longer duplex?

The formation of duplex **3•4** was confirmed by 1-D and 2-D ^{1}H-NMR. In CDCl$_3$, the ^{1}H-NMR spectrum of the 1:1 mixture of **3** and **4** revealed very large (2–3 ppm) downfield shifts of the aniline NH signals compared to those of strands **3** or **4** alone. The association constant of **3•4** was so high that attempts to measure it by ^{1}H-NMR

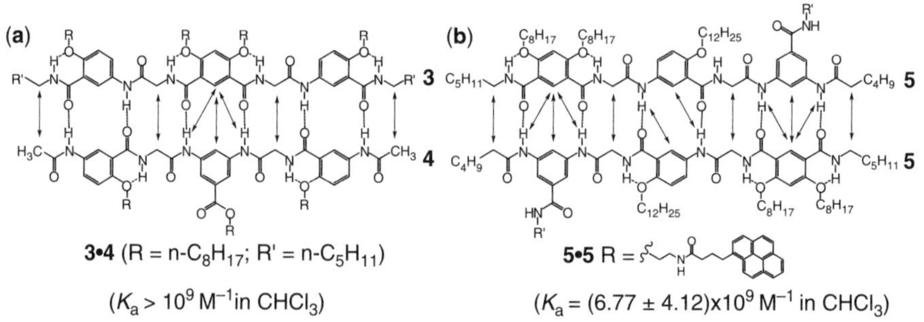

Figure 9.3 (a) Duplex **3•4** consisting of two different strands with complementary H bonding sequences. (b) Duplex **5•5** consisting of two identical single strands with a self-complementary H bonding sequence. Interstrand NOE contacts revealed by NOESY are indicated by arrows.

dilution experiments were unsuccessful. No significant upfield shifts of the aniline NH signals were observed across a broad concentration range (100 mM to 1 μM), suggesting a lower limit of $9 \times 10^7 \, M^{-1}$ for the association constant of duplex **3•4** and assuming a 10% dissociation at 1 μM. Isothermal titration calorimetry (ITC) could only give an estimated value of $\geq 10^9 \, M^{-1}$. The high stability of **3•4** even allowed its detection by straight-phase thin layer chromatography (TLC) under rather polar conditions [silica gel plate, 10% dimethylformamide (DMF) in chloroform]. The ^{1}H-NMR spectra of mixtures of **3** and **4** in stoichiometries other than 1:1 revealed separate sets of signals corresponding to both duplex **3•4** and the uncomplexed strands. This indicated that exchange between the assembled and the uncomplexed states was very slow on the timescale of NMR spectroscopy, which suggested high kinetic stability for the duplex. Although efforts to crystallize **3•4** have not succeeded, the presence of **3•4** in solution was unequivocally demonstrated by numerous interstrand NOEs in the NOESY spectrum of this compound (Fig. 9.3a).

Analysis of the pyrene-labeled homoduplex **5•5** (Fig. 9.3b) by NMR, mass spectrometry, and TLC suggested that **5•5** had a stability similar to that of **3•4**. NOESY spectra revealed cross-strand NOEs consistent with the formation of the self-dimer **5•5** (Zeng et al. 2003). Based on a fluorescence method described in the literature (Sontjens et al. 2000), the dimerization constant of the pyrene-labeled duplex **5•5** was found to be $(6.77 \pm 4.12) \times 10^9 \, M^{-1}$. The studies on duplexes **3•4** and **5•5** clearly demonstrated that the stabilities of our duplexes are indeed only determined by the number of intermolecular H bonds, and both hetero- and homoduplexes can be easily designed and constructed.

Thus, the stability is enhanced by more than 5 orders of magnitude with two additional H bonds (i.e., the quadruply vs. 6-H bonded duplexes), which implies the presence of positive cooperativity in this H bonded system. Comparing the stabilities of doubly ($K_a = 25 \, M^{-1}$, -0.9 kcal/mol for each H bond) and quadruply (dimers of **1** and **2**, $10^4 \, M^{-1} < K_a < 10^5 \, M^{-1}$, -1.4 to -1.7 kcal mol^{-1} for each H bond) H bonded duplexes with that of **3•4** ($1 \times 10^9 \, M^{-1}$, 2.0 kcal/mol for each H bond) indicates that the increase in stabilities is not just due to the additive effect from increased numbers of hydrogen bonds. Instead, these data clearly demonstrate that the self-assembly of **3•4** is highly cooperative: after the initial association of the two strands, which may involve one or, at most, two hydrogen bonds and is entropically unfavorable, the subsequent formation of H bonds during the growth of the duplex is enhanced by multiple, enthalpically favorable interactions. This type of cooperativity is one of the most prominent features of the self-assembly of DNA and lies at the heart of many self-assembling and self-organizing systems (Lindsey 1991).

9.5. PROBING SEQUENCE SPECIFICITY

Sequence-specific pairing of DNA and RNA strands is essential for the storage, transmission, and expression of genetic information, which forms the basis for techniques such as polymerase chain reaction, hybridization techniques, and DNA chip arrays. Having demonstrated the sequence-independent nature in the stabilities of

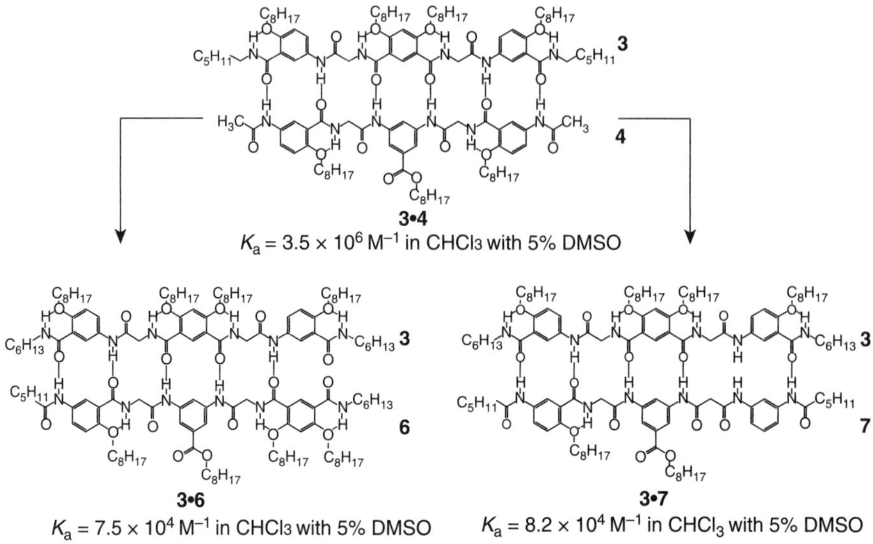

Figure 9.4   Duplex **3•4** and two duplexes **3•6** and **3•7** containing mismatched-binding sites. The repulsive mismatched binding sites in **3•6** (acceptor–acceptor) and in **3•7** (donor–donor) cause more than a 40-fold decrease in the stabilities of these duplexes compared to **3•4**.

our H bonded duplex, the next objective was to establish the sequence specificity of this class of information-storing molecules. This could be realized by incorporating mismatched binding sites into an otherwise fully matched duplex, leading to "mutant" duplexes that should provide an ideal platform for systematically probing the effect of sequence specificity on unnatural self-assembling processes, which had been difficult to investigate because of the lack of appropriate model systems.

Thus, to investigate the sequence specificity of our duplexes, mismatched binding sites were introduced into duplex **4•5** (Fig. 9.4; Zeng et al. 2001). Single strands **6** and **7**, with H bonding sequences of DADDAA and DADDDD, respectively, were designed and synthesized to pair with **3**. Each of the resultant pairs, **3•6** or **3•7**, should contain a mismatched binding site. With their mismatched sites, can duplexes **3•6** and **3•7** still form? Will the single strands in duplexes **3•6** and **3•7** be partially aligned to avoid the mismatched binding sites? What are the stabilities of **3•6** and **3•7** compared to that of the fully matched **3•4**?

^{1}H-NMR showed poorly defined resonances of individual strands **6** or **7**, indicating a slowly equilibrating mixture of many conformations. Upon mixing with 1 equiv of **3**, both **6** and **7** showed sharp sets of signals that can only be assigned to single conformers, suggesting the association of **6** or **7** with **3**. One interesting observation was that the aniline NH protons of **4**, which show one degenerate signal in the ^{1}H-NMR spectra of **3** or **3•4** due to the symmetry of the structures, became nonequivalent upon mixing **3** with **6** or **7**. At 25 mM in CDCl$_3$, ^{1}H-NMR showed that the aniline protons of strand **3** in **3•6** or **3•7** appeared as two signals that were separated

by >0.2 ppm. Obviously, the exchange of single strands **3**, **6**, and **7** with the corresponding duplex **3•6** or **3•7** was slow on the NMR timescale. This phenomenon is consistent with the formation of H bonded **3•6** and **3•7**, which have unsymmetrical structures and relatively high kinetic stabilities.

The NOESY spectrum of **3•6** showed that one end of this pair was locked by intermolecular H bonds, whereas the other end around the mismatched site consisting of the two amide carbonyl groups at that end was open. In contrast, the NOESY study indicated that both ends of **3•7** were locked, but no or very weak NOEs were detected for protons close to the internal mismatched site. These results indicated that, in spite of the presence of mismatched sites, these mutant duplexes still formed. More detailed analysis showed five interstrand H bonds in **3•6** or **3•7**, suggesting that these strands were still fully registered. However, the oligoamide backbones must be twisted near the mismatched sites to alleviate any unfavorable interactions.

The sequence specificity in the formation of duplex **3•4** was probed in the presence of strands **6** and **7** by ^{1}H-NMR. Thus, adding 1 equiv of **4** into the solution of either **3•6** or **3•7** in CDCl$_3$ containing 5% dimethylsulfoxide-d_6 (DMSO-d_6) led to the complete replacement of the "wrong" strands **6** or **7**, resulting in the appearance of new ^{1}H-NMR signals corresponding to those from duplex **3•4** and the free strand **6** or **7**.

The stabilities of **3•6** and **3•7** were compared to that of **3•4** by ITC (5% DMSO in CHCl$_3$). Results from ITC experiments showed that introducing one mismatched site led to a >40-fold drop in the association constants of the mismatched **4•7** and **4•8** relative to that of **4•5**. Like the binding of **4** to **3**, the binding of **6** or **7** to **3** was largely enthalpically driven. However, compared to

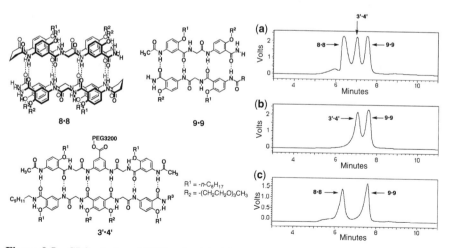

Figure 9.5 GPC studies (CHCl$_3$, 1 mL/min) on mixtures of (a) **8•8 + 9•9 + 3′•4′**, (b) **3′•4′ + 11•9•9**, and (c) **8•8 + 9•9**. The retention times of the individual duplexes are identical to those shown here.

3•4, the formation of **3•6** or **3•7** was entropically more costly, reflecting the fact that the presence of mismatched sites in these mutant duplexes led to more flexible backbones. The ITC results demonstrated that the decreased stabilities of **3•6** and **3•7** were largely attributable to the loss of one H bonding site in each pair, whereas increased entropic barriers also played a minor role. Note that the observed drop in stability was not due to partial alignment of the two strands, as indicated by NOESY studies. These results firmly established the high sequence fidelity of the fully H bonded **3•4**.

The high fidelity of the H bonded duplexes was again demonstrated by duplexes **8•8** (see Section 8.5), **9•9**, and **3′•4′**, with eight, six, and four intermolecular H bonds, by gel permeation chromatography (GPC). When mixed in solution, each of these three duplexes clearly existed independently (Fig. 9.5). When tethered to various structural units, these duplexes can be used in mixtures to sort out and assemble the corresponding units in a highly specific fashion.

9.6. UNEXPECTED DISCOVERY: DUPLEXES CONTAINING FOLDED STRANDS

Having demonstrated the adjustable strength and programmable sequence specificity of our H bonded duplexes, we decided to probe the application of these information-carrying associating units for constructing various supramolecular structures. Among various self-assembling structures, supramolecular polymers have attracted intense interest in recent years (Moore 1999; Brunsveld et al. 2001; Zimmerman and Lawless 2001). With the ready tunability of their sequences, the duplexes we developed should be able to direct not only the association of molecular components but also the regiospecificity of such an association.

As shown in Figure 9.6, oligoamide strand **10**, consisting of two 4-H bonding DDAD units linked in a head-to-head fashion, should associate with single strand **11**, formed by linking two AADA units in a head-to-head fashion (Yang et al. 2003). If **10** and **11** adopted extended conformations, they would only partially overlap each other because of their unsymmetrical 4-H bonding units, leading to

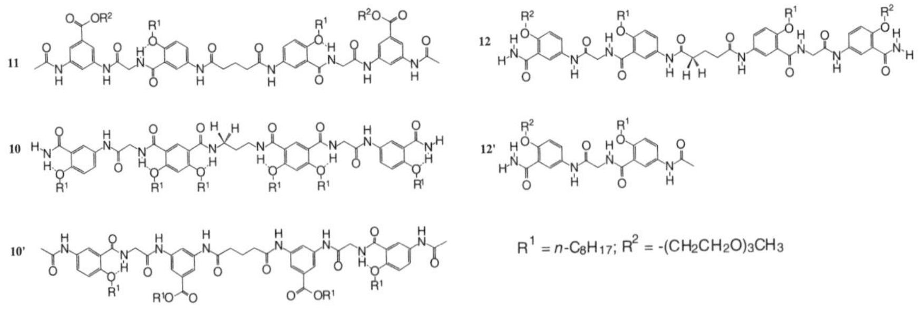

$R^1 = n\text{-}C_8H_{17}$; $R^2 = \text{-}(CH_2CH_2O)_3CH_3$

Figure 9.6 Oligoamide strands **10**, **10′**, **11**, and **12** containing two quadruply H bonding units linked in a head-to-head fashion. Control strand **12′** is essentially half of **12**.

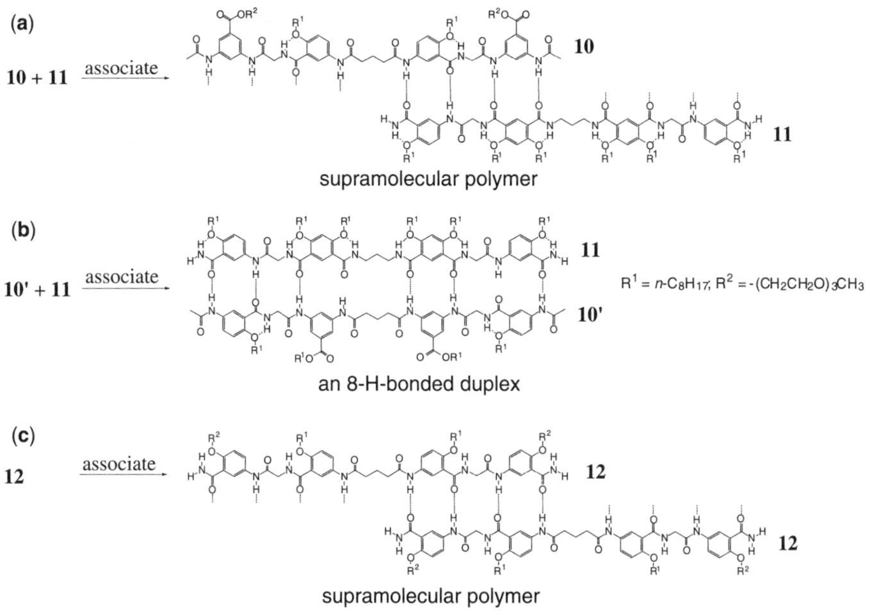

Figure 9.7 If adopting an extended conformation, (a) strands **10** and **11** should form H bonded polymer chains, (b) strands **10′** and **11** should form an 8-H bonded duplex, and (c) strand **12** should form H bonded polymer chains.

the formation of supramolecular AB copolymers (Fig. 9.7a). In contrast, if strand **10′** adopted an extended conformation, it would be completely complementary to an extended strand **11**, leading to the formation of an 8-H bonded, extended duplex (Fig. 9.7b). Similar to the design of **10** and **11**, linking two self-complementary ADAD units in a head-to-head fashion leads to oligoamide strand **12** (Fig. 9.6). Based on similar reasoning made with **10** and **11**, if strand **12** adopted an extended conformation, it would also self-associate into H bonded homopolymeric aggregates (Fig. 9.7c).

The signal of the aniline proton of **11** appeared at 9.66 ppm in the ^{1}H spectrum of **11** dissolved in CDCl$_3$ (1 mM, 23 °C). The signal of the same proton moved to 10.03 ppm in the spectrum of the 1:1 mixture of **1** and **2** in CDCl$_3$ (1 mM, 23 °C). Thus, strands **10** and **11** did associate via intermolecular H bonding. However, the ^{1}H-NMR spectrum of the mixture of **1** and **2** contained very well defined, sharp signals, which was inconsistent with the presence of polymeric aggregates. Similarly, the ^{1}H-NMR spectrum of strand **12**, which presumably would form H bonded polymers, also showed a set of sharp, well-resolved peaks, suggesting the presence of a single, discrete species in solution. Furthermore, comparing the ^{1}H-NMR spectrum of **12** with that of **12′** revealed a surprising result. Oligomer **12′**, with its self-complementary ADAD array, was known to dimerize into a self-complementary, 4-H bonded duplex (Gong et al. 1999). However, the signal of the (internal) aniline proton of **12** appeared at 10.59 ppm whereas the corresponding proton of **12′**

appeared at 9.89 ppm. These results suggest that the molecules of **12** associated through much stronger H bonding interactions than those of **12′**. This result was unexpected because if the molecules of **3** adopted an extended conformation and associated in a staggered fashion as required by its ADAD units, the strength of its intermolecular H bonds would be similar to those of the 4-H bonded self-dimer **12′ • 12′**.

Further ^{1}H-NMR experiments in CDCl$_3$ containing 10–20% DMSO-d_6 confirmed that the intermolecular H bonding interactions between **10** and **11** and among the molecules of **12** were very strong. For example, at 1 mM, the aniline NH signals of the 1:1 mixture of **10** and **11** and those of **12** showed insignificant shifts with increasing percentage (up to 20%) of DMSO-d_6 in CDCl$_3$. Diluting a sample of **10** and **11** (1:1) or that of **12** in a mixed solvent containing 10% DMSO-d_6 (down to 10 μM) in CDCl$_3$ did not lead to any apparent change in the chemical shifts of the aniline NH signals involved in intermolecular H bonding. The stabilities of the H bonds of **10** and **11** and those of **12** are in sharp contrast to those of the corresponding 4-H bonded (i.e., DDAD/AADA and DADA/ADAD) duplexes (Gong et al. 1999; Zeng et al. 2002), whose association constants were previously shown to be in the 10^4 M^{-1} range in chloroform and could be easily determined by detecting the concentration-dependent shifts of amide protons using NMR dilution experiments.

These observations suggested that strands **10** and **11** and strand **12** did not assemble into H bonded polymers as expected. Instead, they seemed to have formed highly stable, well-defined discrete species. Such a conclusion was confirmed by VPO and mass spectrometry measurements. VPO experiments showed that, in solution, **10** and

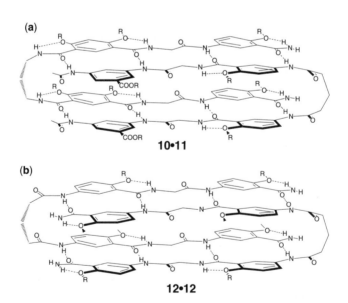

Figure 9.8   Hydrogen bonded duplex consisting of folded strands: (a) heteroduplex **10•11** and (b) homoduplex **12•12**.

11 together existed as a heterodimer and **12** as a homodimer. The ESI spectrum of the 1:1 mixture of **10** and **11** detected the doubly charged $[10 + 11 + 2Cl]^{2-}$ ($m/z =$ 1454.3) as the most abundant ion. For **12**, $[12 + 12 + Cl]^{-}$ ($m/z = 2636.3$), corresponding to the self-dimer **12•12**, and $[12 + Cl]^{-}$ ($m/z = 1335.6$), corresponding to single strand **12**, were the two major ions detected by electrospray ionization.

To account for the formation of dimers by **10** and **11**, and **12** itself, the best explanation is that, instead of being extended, strands **10** and **11** and strand **12** adopt folded conformations when associated into their corresponding H bonded assemblies. The specific (and most reasonable) model involves the folded (stacked) conformations as shown in Figure 9.8a adopted by strands **10** and **11**. Based on the same analysis, the molecules of **12** can associate into a homodimer only by adopting a similar folded conformation (Fig. 9.8b). This model of self-assembling foldamers is fully consistent with the above experimental results. Thus, these molecular duplexes have the novel feature of combining sequence-specific self-assembly and the folding of component strands. It is not yet clear whether the folded conformations are contingent upon the association of the two strands by forming, for example, the first four H bonds, or vice versa. Further study on this system should lead to new self-assembling and folding structures.

9.7. DIRECTED ASSEMBLY: FORMATION OF β-SHEETS AND SUPRAMOLECULAR BLOCK COPOLYMERS

9.7.1. Directed Assembly: Templated Formation of Two-Stranded β-Sheets

When tethered to structural units that may associate in more than one way, an H bonded duplex template will help specify the intermolecular interaction, leading to a single assembly. The possibility of directing the association of natural peptide strands was tested by using our H bonded duplexes as templates.

Because the interstrand distance of a duplex, based on the crystal structures of 4-H bonded duplexes (Gong et al. 1999), is the same as that ($\sim$5 Å) in a β-sheet, these hydrogen bonded duplexes may serve as specific, noncovalent templates for directing and nucleating β-sheet structures when attached to natural oligopeptide strands. Because the H bonding sequence of the duplex template can be designed to be unsymmetrical, two flexible peptide chains can be bought into proximity and thus forced to pair with each other, leading to a double-stranded, antiparallel β-sheet (Fig. 9.9).

Our results confirmed the feasibility of this strategy (Zeng et al. 2002). Peptide segments were attached to an unsymmetrical, 4-H bonded heteroduplex template with the complementary sequences of ADAA/DADD, leading to four hybrid strands **13a**, **13b**, **14a**, and **14b**. Combination of strands **13a** and **13b** with strands **14a** and **14b** resulted in four different pairs. The formation of well-defined β-sheets by different combinations of paired peptide sequences was confirmed by 1-D and 2-D NMR, ITC, and VPO studies. For example, 2-D NMR (NOESY) showed interstrand NOEs corresponding to protons of duplex templates and the

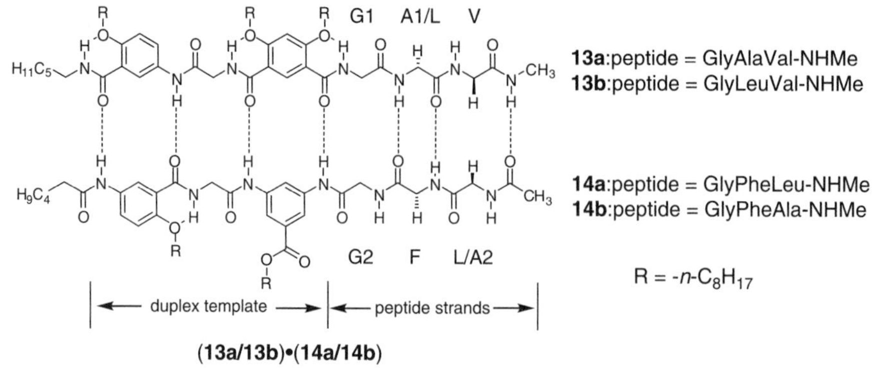

13a: peptide = GlyAlaVal-NHMe
13b: peptide = GlyLeuVal-NHMe

14a: peptide = GlyPheLeu-NHMe
14b: peptide = GlyPheAla-NHMe

R = -n-C$_8$H$_{17}$

(**13a/13b**)•(**14a/14b**)

Figure 9.9 When attached to a duplex template, two otherwise flexible peptide chains are directed to form a stably folded β-sheet.

peptide strands. No cross-strand NOEs were observed between the protons belonging to template and peptide segments, suggesting that the hybrid strands were registered as expected. Furthermore, ITC experiments showed that the hybrid strands showed enhanced stability compared to the duplex template or the peptides alone. VPO measurements showed that strands **13** and **14** formed heterodimers whereas the corresponding peptide strands form poorly defined, random oligomers.

Thus, by attaching different numbers of peptide strands to each of the two duplex templates, the supramolecular nature of this approach allows the rapid generation of a large number of combinations by simple mixing. In contrast to currently known model systems of templated β-sheets, most of which involve linear peptide segments linked by turns and are thus strongly sequence dependent, this system may simplify the study of factors stabilizing β-sheets by focusing on the interactions of linear peptide segments without having to rely on turn structures. Furthermore, the directed assembly of peptide chains may offer a powerful synthetic strategy for the site-specific cross-linking of sheet structures using disulfide and other covalent interactions.

9.7.2. Supramolecular Block Copolymers

We also examined the possibility of constructing supramolecular block copolymers using our duplexes as associating units (Fig. 9.10; Yang et al. 2004; Hua et al. 2005; Kim et al. 2005; Li et al. 2005). By attaching hydrophobic polystyrene (PS) and hydrophilic poly(ethylene glycol) (PEG) chains to the two strands of a duplex derived from the 6-H bonded **3•4**, noncovalent diblock copolymers were obtained by simply mixing the modified polymers in chloroform or benzene.

The successful noncovalent ligation of the PS and PEG chains was confirmed by GPC, which showed quantitative formation of the H bonded block copolymers (Fig. 9.11). Thus, using toluene containing 10% DMF as the eluent, the GPC results showed a single peak corresponding to the H bonded **15c•16c**, which

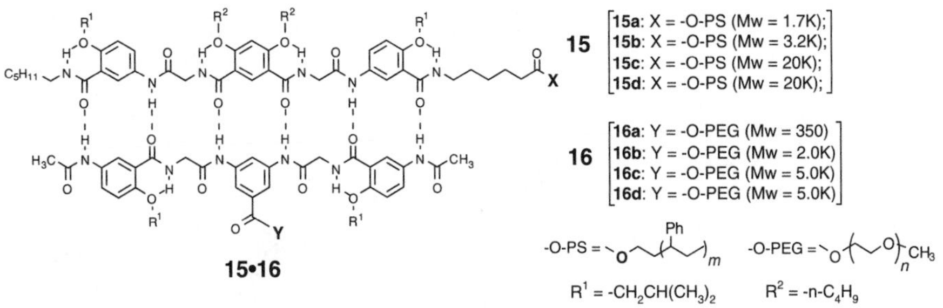

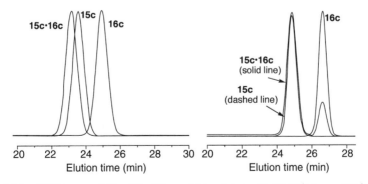

Figure 9.10 Design of supramolecular block copolymers based on the 6-H bonded hetero-duplex **15•16**. Mixing three templated PS chains with three complementarily templated PEG chains leads to nine block copolymers.

appeared earlier than **15c** and **16c**. When polar DMF was used as the eluent, **15c•16c** decomposed into two peaks that coincided with the peaks of **15c** and **16c**. This result demonstrates that the incorporation of polymer side chains DID NOT affect the sequence-specific formation of the H bonded duplex.

The NOESY study also confirmed that the duplex template was precisely registered as directed by the H bonding sequences. Differential scanning calorimetry analysis indicated the thermally reversible formation and dissociation of these supra-molecular block copolymers. The H bonded blocks only dissociate at temperatures above 150 °C. For pairs with sufficiently long PS and PEG chains (**15c•16c** and **15d•16d**), microphase separated nanodomains, typical of covalent diblock copoly-mers, were observed by atomic force microscopy (AFM). The hexagonally packed cylinders can be clearly identified in Figure 9.12a. In contrast, the AFM image of the 1:1 blend of the corresponding PS and PEG chains without the duplex template showed obvious macrophase separation.

For the pairs with relatively short PS and PEG chains (e.g., **15a•16a** and **15b•16b**), no microphase separation was observed in the corresponding spin-cast films (Li et al. 2005). The AFM image of **15b•16b** was drastically different, although

Figure 9.11 GPC traces of **15c•16c**, **15c**, and **16c** eluted with DMF/toluene (10/90, v/v, left) and DMF (right) at 60 °C.

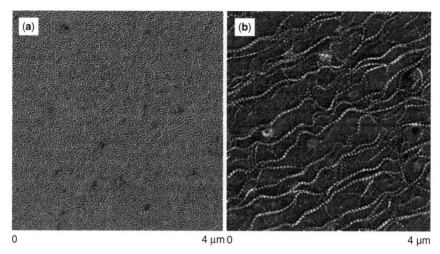

0   4 μm 0   4 μm

Figure 9.12 (a) AFM image of spin-cast **15d•16d** from benzene showing cylindrical nano-domains from the microphase separation of the supramolecular block copolymer and (b) AFM image of spin-cast **15b•16b** from benzene showing self-assembled fibers.

a well-defined morphology was observed in the thin film. As shown in Figure 9.12b, the self-assembly of **15b•16b** leads to nanostructures that have nothing in common with typical diblock copolymers. The morphology of **15b•16b** may be better understood by regarding it as a supramolecular coil–rod–coil triblock copolymer. This is reasonable because the 6-H bonded duplex is in fact a rigid duplex consisting of mostly aromatic residues. These results suggest that when the polymer blocks are sufficiently long, the effect of the H bonded duplex can be neglected and the behavior of the corresponding supramolecular block copolymer is only determined by the PS and PEG blocks. When the polymer blocks are short, the H bonded duplex unit plays a significant role in the self-assembly of the corresponding block copolymer. In this case, the property of the corresponding supramolecular block copolymer is determined by all of the components including the H bonded association unit.

This system has provided a diverse set of sites for creating a wide variety of block copolymers and for tuning their properties. Thus, by adjusting the lengths of the incorporated polymer chains and those of the duplex units, diblock, triblock, or multi-block copolymers can be easily generated. In addition to linear block copolymers, the multiple side chains of the duplex units offer sites for attaching different oligomer and polymer chains, which should lead to brushlike or starlike block copolymers that otherwise would be difficult to prepare with conventional methods. Oligomer and polymer chains that are difficult to link together based on traditional covalent methods can be modified separately with our duplex units and then linked via H bonds. Furthermore, the simplicity (i.e., mixing) of this method allows the construction of novel supramolecular combinatorial libraries, which should lead to rapid generation and screening of various materials.

9.8. INTEGRATING NONCOVALENT AND COVALENT INTERACTIONS: DIRECTED OLEFIN METATHESIS AND DISULFIDE BOND FORMATION

The previous examples demonstrate the power of H bonded duplexes as highly specific supramolecular assemblers for ligating various structural units. However, our H bonded duplexes, like most other H bonded complexes, are stable only in nonpolar solvents. Their intolerance to nonpolar media such as water hampers many applications, particularly those involving biological conditions. Although several systems with microenvironments that promote H bonding in polar media are known (Nowick and Chen 1992; Fan et al. 1993; Kato et al. 1995; Torneiro and Still 1995; Paleos and Tsiourvas 1997; Ariga and Kunitake 1998; Brunsveld et al. 2002), in competitive environments, including aqueous solutions, the association of artificial molecular components based on H bonding still represents a largely unsolved fundamental problem. In contrast to designed systems, all water-soluble natural self-assembling systems are stabilized by the cooperative interaction of multiple noncovalent forces. DNA duplexes provide one of the best known examples that are stabilized by H bonding, aromatic stacking, and other noncovalent interactions.

Using similar approaches adopted by Nature, one possible solution to constructing stable and specific assemblies in competitive media involves the design of molecular components capable of simultaneously specifying H bonding and other noncovalent forces. However, the successful design of such molecular components still represents a daunting challenge. Instead of contemplating the design of molecular structures that can assemble based on different types of noncovalent forces, we decided to integrate the superb specificity of H bond arrays and the strength of covalent interactions into the same duplex. Although most covalent interactions are irreversible and can only lead to structures that are formed under kinetically controlled conditions, a few reversible (dynamic) covalent bond formation reactions (Furlan et al. 2002) do exist. In fact, they have attracted intense interest in recent years in thermodynamically controlled covalent synthesis and in the creation of dynamic combinatorial libraries (Swann et al. 1996; Huc and Lehn 1997; Giger et al. 1998; Cousins et al. 1999; Lehn 1999; Polyakov et al. 1999; Rowan et al. 2002; Corbett et al. 2006; Saur et al. 2006). Therefore, the integration of H bonding and dynamic covalent interaction may lead to a simplified, easily manageable system.

9.8.1. Templated Olefin Cross-Metathesis

The two most used reversible covalent reactions are disulfide exchange and palladium-catalyzed olefin metathesis. We first probed the incorporation of olefin units into the H bonded duplexes by subjecting the modified duplexes to a Pd (Grubb's) catalyst. Based on a duplex template with the same unsymmetrical H bonding sequence used for directing the formation of the β-sheet structures, we prepared two groups (strands **17** and **18**) of five olefins covalently linked to the two template strands (Fig. 9.13). Mixing each one of components **17** with each one of components **18** in a 1:1 fashion results in a small library of 25 (5 × 5) members.

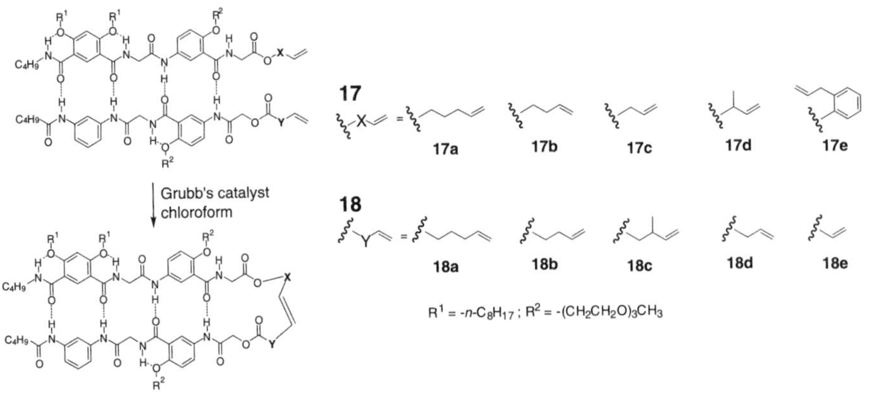

Figure 9.13 Templated olefin cross-metathesis of two groups of olefins tethered to the two duplex strands.

Formation of the ADAA/DADD duplex template should bring the olefin moieties into proximity, which lowers the $\Delta S^{\ddagger}$ of the reaction and thus facilitates the metathesis reactions. In the presence of Grubb's catalyst, most of the templated pairs were cross-linked in very high yields. In cases when the tethers were too short, such as the combination of **17e** and **18e**, no metathesis reactions were observed. With strands **17** or **18** alone, no homo-cross-linked product was observed, indicating that the H bonded duplex template promoted the metathesis reactions by increasing the effective molarity of the reacting olefins. It was also found that under the same conditions (r.t., 1–2 mM in CDCl$_3$) the same reactions did not happen with the corresponding free olefins. The fact that only hetero-cross-linked products were observed suggests that the otherwise symmetrical olefin metathesis reaction has been directed to proceed in an unsymmetrical fashion. In principle, this strategy should be applicable to a wide variety of bimolecular reactions.

9.8.2. Directed, Sequence-Specific Disulfide Cross-Linking in Water

Can the strategy of combining the sequence specificity of H bond arrays with reversible covalent interactions be extended into polar media? To answer this question, strands **3** and **4** were modified with *S*-trityl end groups capable of reversibly forming disulfide bonds in the presence of iodine (Kamber et al. 1980), leading to complementary strands **19** and **20** (Fig. 9.14a; Li et al. 2006).

The 1:1 mixtures of strands **19** and **20** (0.5 mM each) were prepared in CH$_2$Cl$_2$, methanol, and water. The samples were subjected to redox conditions (I$_2$) for various lengths of time and then examined by matrix assisted laser desorption ionization (MALDI). As expected, **19** and **20** sequence specifically associated with each other in CH$_2$Cl$_2$, leading to the disulfide cross-linked **19–20** as the major product. With an increasing ratio of methanol in CH$_2$Cl$_2$, the cross-linked **19–20** still appeared as the dominant product. The same phenomenon was observed in pure methanol and even in water (Fig. 9.14b). The sequence dependence of the cross-linking of **19** and

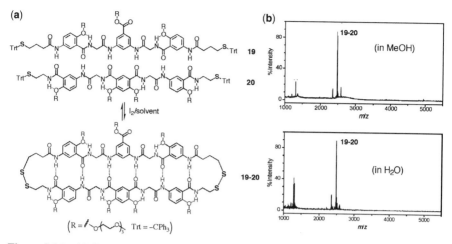

Figure 9.14 (a) Complementary strands **19** and **20** carrying trityl-protected thio groups can be cross-linked when subjected to reversible redox conditions. (b) MALDI spectra show that **19** and **20** were sequence specifically cross-linked into **19–20** in both methanol and water.

20 in water became obvious when control strand **21**, which had mismatched binding sites with both **19** and **20**, was mixed with **19** and **20** under the same redox condition in water. MALDI revealed that the control strand did not interfere with the formation of **19–20**, which was still the dominant product. The strict sequence dependence of the cross-linking process was further demonstrated by **21** and its complementary strand **22**. Although **21** failed to cross-link with either **19** or **20**, MALDI detected cross-linked **21–22** as the only product from the 1:1 mixture of **21** and **22** (0.5 mM each) in aqueous solution containing iodine (Fig. 9.15a), the same outcome as that observed for **19** and **20** (Li et al. 2006). Furthermore, when strands **19**, **20**, **21**, and **22** were mixed together in aqueous media in the presence of iodine, only the sequence specifically cross-linked **19–20** and **21–22** were detected by MALDI (Fig. 9.15b).

This result indicates that two different strands can be specifically cross-linked only when their H bonding sequences are complementary to each other. Strand **19** or **20** alone led to mostly disulfide-cyclized monomer and only a very small amount of cross-linked self-dimers, which provided further support for the H bond dependence of the disulfide cross-linking reaction. Under the same redox conditions, the cross-linking of **19** and **20** on a large (gram) scale in water led to the isolation of purified **19–20** in high (>85%) yield, confirming the high efficiency of the cross-linking reactions revealed by mass spectrometry and high-performance liquid chromatography.

The formation of the cross-linked products demonstrated an apparent correlation to H bonding sequences. What is the mechanism behind the observed sequence specificity? One of two possible mechanisms, one kinetic and the other thermodynamic, may be involved in aqueous media. The kinetic mechanism involves intermolecular

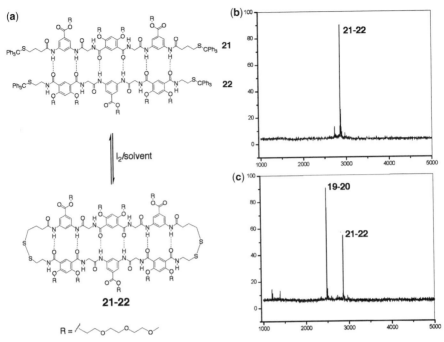

Figure 9.15 (a) Complementary strands **21** and **22** carrying trityl-protected thio groups were cross-linked in the presence of iodine. (b) MALDI spectra show that **21** and **22** were sequence specifically cross-linked into **21–22** in water containing 1% THF. (c) When placed in the same aqueous solution (water/THF $= 9/1$, v/v), strands **19, 20, 21,** and **22** sequence specifically cross-linked into **19–20** and **21–22**.

association of the two stands, which increases the rate of disulfide bond formation by increasing the effective molarity of the *S*-trityl groups. The thermodynamic mechanism is based on the selective stabilization of the cross-linked product consisting of the sequence-matched strands. To distinguish these two mechanisms, control strands **23, 24,** and **25** (Fig. 9.16) were designed and prepared (Li et al. 2008).

First, to investigate whether intermolecular association of complementary strands determines the final product distribution, the cross-linking of **19** and **20** was examined in the presence of control strand **23**. Sharing the same backbone and H bonding sequence, strands **23** and **19** should compete for H bonding or aromatic stacking interactions with **20**. The only difference between **19** and **23** is that the latter is incapable of forming a disulfide cross-linked product with **20**. If the sequence-specific formation of **19–20** relied on the intermolecular association of **19** and **20**, the presence of **23** should interfere with such an association, which would slow down the rate-determining step of the disulfide bond formation reaction and compromise the sequence selectivity. Examining the mixture containing **19, 20,** and **23** (0.5 mM each) in aqueous solution in the presence of iodine using MALDI revealed two peaks corresponding to cross-linked **19–20** and strand **23**

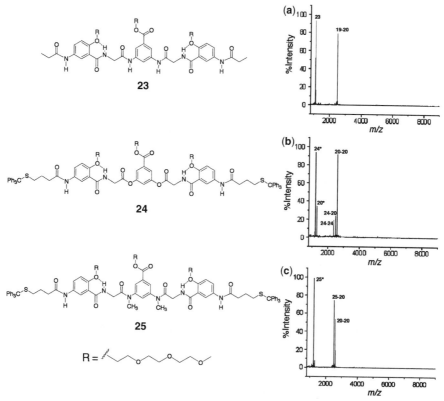

Figure 9.16 MALDI spectra of the 1:1 mixture of (a) **19**, **20**, and **23** (0.5 mM each), (c) **24** and **20** (0.5 mM each), and (d) **25** and **20** (0.5 mM each) in H_2O/THF (9/1, v/v) in the presence of iodine.

(Fig. 9.16a). Other possible products from the self-cyclization or homodimerization of **19** or **20** were not found. Therefore, the formation of cross-linked **19–20** was not affected to any detectable extent by **23**. Based on this result, it can be concluded that the sequence specificity shown in the formation of **19–20** was not due to the intermolecular association of **19** and **20**. This conclusion was further verified by investigating the cross-linking of **1** and **2** at different concentrations in water containing 10% THF, from 0.05 mM to 5 mM for each strand. It was found that the sequence specifically cross-linked **19–20** was formed as the only product at the concentrations examined, indicating that intermolecular association played no role in influencing the observed product distribution.

Based on these results, the possibility of a kinetic mechanism for the sequence-specific cross-linking of **19** and **20**, and similarly, **21** and **22**, was ruled out, which left the thermodynamic mechanism that requires the selective stabilization of the sequence-matched, cross-linked products. The observed sequence specificity is likely to be due to the stabilization of the products by intramolecular H bonds between two complementary strands. These intramolecular H bonds between two complementary strands come into existence upon the formation of the disulfide cross-links. However, given the strong competition from water molecules, whether H bonds, even intramolecular ones, could have any effect on the putative stabilization of the cross-linked product remains unclear. Thus, another possibility is that the observed sequence specificity was not attributable to H bonding but instead arose from π-stacking interactions between the complementary strands of the cross-linked products. One possibility is that, with their complementary H bonding sequences, strands **19** and **20**, and **21** and **22** also happen to have complementary surfaces that facilitate very specific intermolecular π-stacking interactions, leading to the observed selective formation of the cross-linked products.

To elucidate the role of H bonding or π-stacking in stabilizing the products, the cross-linking of oligomers **24** and **25** with **20** was examined. Strands **24** and **25** can be regarded as being derived from **19** by replacing the two central aniline NH groups of the latter with O atoms and N-Me groups. These two control strands have electronic structures similar to that of **19** but they carry fewer (four) H bonding sites. Therefore, in aqueous media, if aromatic stacking, rather than H bonding, played the major role in determining the relative stability of the cross-linked product, strand **24** or **25** should behave similarly to **19** by pairing with **20**, leading to the corresponding disulfide cross-linked **24–20** or **25–20** as the major product.

The 1:1 mixture of **24** and **20** (or **25** and **20**) in aqueous media (0.5 mM each) was subjected to the same conditions used for forming **19–20**. MALDI results showed that neither **24** nor **25** selectively cross-linked with **20**. In sharp contrast to the exclusive formation of **19–20** or **21–22**, multiple products were detected from the 1:1 mixture of **24** and **20**, among which **24***, formed from the self-cyclization of **24**, and **20–20** from the homodimerization of **20**, represented the major products (Fig. 9.16b). Product **24–20**, from the cross-linking of **24** and **20**, only appeared as a minor peak. Similarly, when mixed together, **25** and **20** could not be cross-linked exclusively into **25–20** (Fig. 9.16c). The observed distribution of products, that is, the self-cyclized **25***, heterodimer **25–20**, and homodimer **20–20**, can be regarded as being a statistical one.

Results from these experiments indicate that the aromatic stacking interaction between complementary strands is probably not the factor responsible for the stabilization of the cross-linked product in aqueous media. Instead, the critical role played by interstrand H bonding in the stabilization of products **19–20** (and similarly **21–22**) was clearly demonstrated by the high sequence specificity of the cross-linking reactions, as well as by the behavior of control strands **24** and **25** in the presence of **20**.

The above observations are consistent with a thermodynamically controlled process shown in Figure 9.17. Thus, when two strands, **A** and **B**, with complementary H bonding sequences and termini that can be reversible cross-linked, are present in the same solution under redox conditions, products **A–A** (or self-cyclized **A***), **B–B** (or self-cyclized **B***), and **A–B**, may be generated. Among these products, **A–B** gains the most stabilization from the newly generated, complementary intramolecular H bonds because of the formation of the two disulfide bonds. Thus,

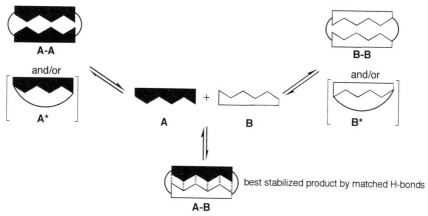

Figure 9.17 Two components bearing complementary H bonding sequences can reversibly form disulfide cross-linked products. **A–B** represents the most stable product because of the stabilization from its fully matched, interstrand (now intramolecular) H bonds.

sequence-matched product **A–B** represents the most stable combination. The reversibility of the dynamic covalent cross-linking allows the equilibria shown in Figure 9.17 to shift toward the formation of the most stable **A–B**. Therefore, the observed sequence specificity in the formation of the cross-linked pairs was due to the extra stabilization from the H bonding interactions that in turn were the result of the disulfide bond formation.

By examining pairs with four and two interstrand H bonds, it was found that as few as two H bonds were sufficient to shift the equilibria toward the formation of sequence-matched products in aqueous solutions. This discovery has opened a new avenue to the design of a variety of highly specific, dynamic covalent associating units.

The presence of intramolecular H bonds in the cross-linked products was probed by carrying out 2-D NMR (NOESY) experiments on cross-linked **26–27**, which contains two interstrand intramolecular H bonds. It was found that the same cross-strand NOEs were detected in CDCl$_3$ (Fig. 9.18a) and in H$_2$O/THF-d_8 (80/20, v/v; Fig. 9.18b). Furthermore, diluting **26–27** from 9 to 0.2 mM in the same aqueous solvent led to no shift in the two aninline NH signals, which remained at 9.75 ppm. These results suggest the persistence of an intramolecular H bonded conformation for **26–27** in aqueous media, which is consistent with the thermodynamic mechanism shown in Figure 9.17.

Thus, similar to the olefin cross-metathesis reactions, the symmetrical disulfide formation under reversible conditions has also been directed into an unsymmetrical process by a duplex consisting of two different strands. This result represents significant progress in specifying molecular association by H bonding in water, which thus far remains a major, largely unsolved problem. The availability of sequence-specific linking units in water should greatly facilitate the ligation of various structural units. For example, di-, tri-, and multiblock copolymers can be constructed based on the sequence-specific pairing and subsequent covalent cross-linking of the duplexes, which allows the incorporation of otherwise incompatible oligomer and polymer

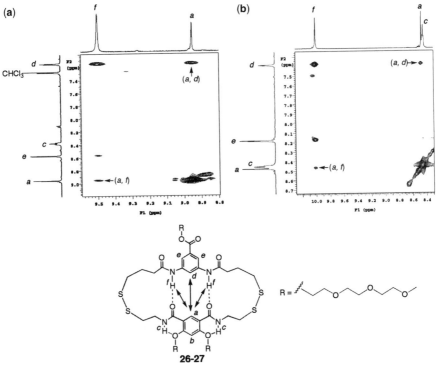

Figure 9.18 Partial 2-D ^{1}H-NMR spectra of disulfide cross-linked **26–27** in (a) CDCl$_3$ (rotational nuclear Overhauser effect spectroscopy, 278 000, 0.3-s mixing time) and (b) H$_2$O/THF-d_8 (80/20, v/v; NOESY, 278 000, 0.3-s mixing time). The same interstrand NOEs were detected in both CDCl$_3$ and H$_2$O/THF-d_8.

strands into the final block copolymers. Many structures that have not been utilized for designing block copolymers can now be adopted based on this strategy.

9.9. CONCLUSIONS AND FUTURE PERSPECTIVES

Research during the last several years has firmly established a class of information-storing molecular duplexes based on readily available oligoamide strands. The programmability of H bonding sequences and tunability of the binding strength offered by these duplexes provide a class of highly specific and highly stable linking units for systematically control of intermolecular association. This has already been demonstrated by the noncovalent conjugation of otherwise noncompatible polymer chains and by the directed association of peptide strands. The integration of these duplexes with reversible (dynamic) covalent bonding interactions represents exciting progress, which should lead to sequence specifically formed, covalently linked structures under thermodynamic conditions. The resulting dynamic covalent duplexes combine the advantages of a programmable system with the strength of covalent bonds, which provides a novel method of ligation under very mild conditions in both nonpolar and polar solvents. These self-assembling, dynamic covalent ligation units should open a new avenue to many previously unavailable architectures such as block copolymers that are difficult to prepare based on traditional methods.

ACKNOWLEDGMENTS

Financial support from the ACS-PRF, NSF, NIH, and ONR is greatly appreciated. The author thanks the students, postdoctoral fellows, and collaborators who have contributed to this research.

REFERENCES

Ariga K, Kunitake T. Molecular recognition at air–water and related interfaces: complementary hydrogen bonding and multisite interaction. Acc Chem Res 1998;31:371–378.

Beijer FH, Kooijman H, Spek AL, Sijbesma RP, Meijer EW. Self-complementarity achieved through quadruple hydrogen bonding. Angew Chem Int Ed Eng 1998;37:75–78.

Beijer FH, Sijbesma RP, Kooijman H, Spek AL, Meijer EW. Strong dimerization of ureidopyrimidones via quadruple hydrogen bonding. Self-assembly mediated by the donor–donor–acceptor•acceptor–acceptor–donor (DDA-AAD) hydrogen-bonding motif: formation of a robust hexameric aggregate. J Am Chem Soc 1998;120:6761–6769.

Bisson AP, Carver FJ, Eggleston DS, Haltiwanger RC, Hunter CA, Livingstone DL, McCabe JF, Rotger C, Rowan AE. Synthesis and recognition properties of aromatic amide oligomers: molecular zippers. J Am Chem Soc 2000;122:8856–8868.

Brunsveld L, Folmer BJB, Meijer EW, Sijbesma RP. Supramolecular polymers. Chem Rev 2001;101:4071–4097.

Brunsveld L, Vekemans JAJM, Hirschberg JHKK, Sijbesma RP, Meijer EW. Hierarchical formation of helical supramolecular polymers via stacking of hydrogen-bonded pairs in water. Proc Natl Acad Sci USA 2002;99:4977–4982.

Conn MM, Rebek J. Self-assembling capsules. Chem Rev 1997;97:1647–1668.

Corbett PT, Leclaire J, Vial L, West KR, Wietor JL, Sanders JKM, Otto S. Dynamic combinatorial chemistry. Chem Rev 2006;106:3652–3711.

Corbin PS, Zimmerman SC. Self-association without regard to prototropy. A heterocycle that forms extremely stable quadruply hydrogen-bonded dimers. J Am Chem Soc 1998;120:9710–9711.

Cousins GRL, Poulsen SA, Sanders JKM. Dynamic combinatorial libraries of pseudo-peptide hydrazone macrocycles. Chem Commun 1999;1575–1576.

Fan E, Vanarman SA, Kincaid S, Hamilton AD. Molecular recognition—hydrogen-bonding receptors that function in highly competitive solvents. J Am Chem Soc 1993;115:369–370.

Fenlon EE, Murray TJ, Baloga MH, Zimmerman SC. Convenient synthesis of 2-amino-1,8-naphthyridines, building-blocks for host–guest and self-assembling systems. J Org Chem 1993;58:6625–6628.

Fenniri H, Deng BL, Ribbe AE. Helical rosette nanotubes with tunable chiroptical properties. J Am Chem Soc 2002;124:11064–11072.

Folmer BJB, Sijbesma RP, Kooijman H, Spek AL, Meijer EW. Cooperative dynamics in duplexes of stacked hydrogen-bonded moieties. J Am Chem Soc 1999;121:9001–9007.

Furlan RLE, Otto S, Sanders JKM. Supramolecular templating in thermodynamically controlled synthesis. Proc Natl Acad Sci USA 2002;99:4801–4804.

Gabriel GJ, Iverson BL. Aromatic oligomers that form hetero duplexes in aqueous solution. J Am Chem Soc 2002;124:15174–15175.

Giger T, Wigger M, Audetat S, Benner SA. Libraries for receptor-assisted combinatorial synthesis (RACS). The olefin metathesis reaction. Synlett 1998;688–691.

Gong B. Specifying non-covalent interactions: sequence-specific assembly of hydrogen-bonded molecular duplexes. Synlett 2001;582–589.

Gong B, Yan YF, Zeng HQ, Skrzypczak-Jankunn E, Kim YW, Zhu J, Ickes H. A new approach for the design of supramolecular recognition units: hydrogen-bonded molecular duplexes. J Am Chem Soc 1999;121:5607–5608.

Gong HG, Krische MJ. Duplex molecular strands based on the 3,6-diaminopyridazine hydrogen bonding motif: amplifying small-molecule self-assembly preferences through preorganization and iterative arrangement of binding residues. J Am Chem Soc 2005;127:1719–1725.

Hartgerink JD, Clark TD, Ghadiri MR. Peptide nanotubes and beyond. Chem Eur J 1998;4:1367–1372.

Hua FJ, Yang XW, Gong B, Ruckenstein E. Preparation of oligoamide-ended poly(ethylene glycol) and hydrogen-bonding-assisted formation of aggregates and nanoscale fibers. J Polym Sci Part A Polym Chem 2005;43:1119–1128.

Huc I, Lehn JM. Virtual combinatorial libraries: dynamic generation of molecular and supramolecular diversity by self-assembly. Proc Natl Acad Sci USA 1997;94:2106–2110.

Jiang H, Maurizot V, Huc I. Double versus single helical structures of oligopyridine–dicarboxamide strands. Part 1: effect of oligomer length. Tetrahedron 2004;60:10029–10038.

Jorgensen WL, Pranata J. Importance of secondary interactions in triply hydrogen bonded complexes: guanine–cytosine vs uracil-2,6-diaminopyridine. J Am Chem Soc 1990;112: 2008–2010.

Kamber B, Hartmann A, Eisler K, Riniker B, Rink H, Sieber P, Rittel W. The synthesis of cystine peptides by iodine oxidation of S-trityl-cysteine and S-acetamidomethyl-cysteine peptides. Helv Chim Acta 1980;63:899–915.

Kato Y, Conn MM, Rebek J. Hydrogen-bonding in water using synthetic receptors. Proc Natl Acad Sci USA 1995;92:1208–1212.

Kim W, Yang HC, Ryu CY, Yang XW, Li MF, Gong B. Morphology study on supramolecular diblock copolymers associated via hydrogen-bonded molecular duplexes. Polym Prepr 2005;46:698–699.

Kolotuchin SV, Zimmerman SC. Self-assembly mediated by the donor–donor–acceptor• acceptor–acceptor–donor (DDA•AAD) hydrogen-bonding motif: formation of a robust hexameric aggregate. J Am Chem Soc 1998;120:9092–9093.

Lehn JM. Dynamic combinatorial chemistry and virtual combinatorial libraries. Chem Eur J 1999;5:2455–2463.

Lehn JM. Programmed chemical systems: multiple subprograms and multiple processing/expression of molecular information. Chem Eur J 2000;6:2097–2102.

Li MF, Yamato K, Ferguson JS, Gong B. Sequence-specific association in aqueous media by integrating hydrogen bonding and dynamic covalent interactions. J Am Chem Soc 2006;128:12628–12629.

Li MF, Yamato K, Ferguson JS, Singarapu KK, Szyperski T, Gong B. Sequence-specific, dynamic covalent crosslinking in aqueous media. J Am Chem Soc 2008;130:490–500.

Li MF, Yang XW, Ryu CR, Gong B. Molecular zippers for preparing supramolecular and dynamic covalent block copolymers. Polym Prepr 2005;46:1128–1129.

Lindsey JS. Self-assembly in synthetic routes to molecular devices—biological principles and chemical perspectives—a review. New J Chem 1991;15:153–180.

Moore JS. Supramolecular polymers. Curr Opin Colloid Interface Sci 1999;4:108–116.

Murray, TJ, Zimmerman, SC. New triply hydrogen-bonded complexes with highly variable stabilities. J Am Chem Soc 1992;114:4010–4011.

Nowick JS, Chen JS. Molecular recognition in aqueous micellar solution—adenine thymine base-pairing in SDS micelles. J Am Chem Soc 1992;114:1107–1108.

Orr GW, Barbour LJ, Atwood JL. Controlling molecular self-organization: formation of nanometer-scale spheres and tubules. Science 1999;285:1049–1052.

Paleos CM, Tsiourvas D. Molecular recognition of organized assemblies via hydrogen bonding in aqueous media. Adv Mater 1997;9:695–710.

Percec V, Cho WD, Ungar G. Increasing the diameter of cylindrical and spherical supramolecular dendrimers by decreasing the solid angle of their monodendrons via periphery functionalization. J Am Chem Soc 2000;122:10273–10281.

Polyakov VA, Nelen MI, Nazarpack-Kandlousy N, Ryabov AD, Eliseev AV. Imine exchange in *O*-aryl and *O*-alkyl oximes as a base reaction for aqueous "dynamic" combinatorial libraries. A kinetic and thermodynamic study. J Phys Org Chem 1999;12:357–363.

Pranata J, Wierschkem SG, Jorgensen WL. OPLS potential functions for nucleotide bases—relative association constants of hydrogen-bonded base-pairs in chloroform. J Am Chem Soc 1990;113:2810–2819.

Rowan SJ, Cantrill SJ, Cousins GRL, Sanders JKM, Stoddart JF. Dynamic covalent chemistry. Angew Chem Int Ed 2002;41:898–952.

Sanford AR, Yamato K, Yang XW, Yuan LH, Han YH, Gong B. Well-defined secondary structures: information-storing molecular duplexes and helical foldamers based on unnatural peptide backbones. Eur J Biochem 2004;271:1416–1425.

Saur B, Scopelliti R, Severin K. Utilization of self-sorting processes to generate dynamic combinatorial libraries with new network topologies. Chem Eur J 2006;12:1058–1066.

Sessler JL, Wang RZ. Self-assembly of an "artificial dinucleotide duplex." J Am Chem Soc 1996;118:9808–9809.

Sontjens SHM, Sijbesma RP, van Genderen MHP, Meijer EW. Stability and lifetime of quadruply hydrogen bonded 2-ureido-4[1H]-pyrimidinone dimers J Am Chem Soc 2000;122:7487–7493.

Swann PG, Casanova RA, Desai, A, Frauenhoff MM, Urbanic M, Slomczynska U, Hopfinger AJ, LeBreton GC, Venton DL. Nonspecific protease-catalyzed hydrolysis synthesis of a mixture of peptides: product diversity and ligand amplification by a molecular trap. Biopolymers 1996;40:617–625.

Torneiro M, Still WC. Sequence-selective binding of peptides in water by a synthetic receptor molecule. J Am Chem Soc 1995;117:5887–5888.

Vreekamp RH, van Duynhoven JPM, Hubert M, Verboom W, Reinhoudt DN. Molecular boxes based on calix[4]arene double rosettes. Angew Chem Int Ed Engl 1996;35:1215–1218.

Whitesides GM, Simanek EE, Mathias JP, Seto CT, Chin DN, Mammen M, Gordon DM. Noncovalent synthesis—using physical–organic chemistry to make aggregates. Acc Chem Res 1995;28:37–44.

Wu AX, Chakraborty A, Fettinger JC, Flowers RA, Isaacs L. Molecular clips that undergo heterochiral aggregation and self-sorting. Angew Chem Int Ed 2002;41:4028–4031.

Yang J, Fan EK, Geib SJ, Hamilton AD. Hydrogen-bonding control of molecular self-assembly—formation of a $2+2$ complex in solution and the solid-state. J Am Chem Soc 1993;115:5314–5315.

Yang XW, Gong B. Template-assisted cross olefin metathesis. Angew Chem Int Ed 2005;44:1352–1356.

Yang XW, Hua FJ, Yamato K, Ruckenstein E, Gong B, Kim W, Ryu CY. Supramolecular AB diblock copolymers. Angew Chem Int Ed 2004;43:6471–6474.

Yang XW, Martinovic S, Smith RD, Gong B. Duplex foldamers from assembly-induced folding. J Am Chem Soc 2003;125:9932–9933.

Zeng HQ, Ickes H, Flowers RA, Gong B. Sequence specificity of hydrogen-bonded molecular duplexes. J Org Chem 2001;66:3574–3583.

Zeng HQ, Miller RS, Flowers RA, Gong B. A highly stable, six-hydrogen-bonded molecular duplex. J Am Chem Soc 2000;122:2635–2644.

Zeng HQ, Yang XW, Brown AL, Martinovic S, Smith RD, Gong B. An extremely stable, self-complementary hydrogen-bonded duplex. Chem Commun 2003;1556–1557.

Zeng HQ, Yang XW, Flowers RA, Gong B. A noncovalent approach to antiparallel β-sheet formation. J Am Chem Soc 2002;124:2903–2910.

Zhao X, Wang XZ, Jiang XK, Chen YQ, Li ZT, Chen GJ. Hydrazide-based quadruply hydrogen-bonded heterodimers, structure, assembling selectivity, and supramolecular substitution. J Am Chem Soc 2003;125:15128–15139.

Zimmerman SC, Corbin PS. Heteroaromatic modules for self-assembly using multiple hydrogen bonds. Struct Bond 2000;96:63–94.

Zimmerman SC, Lawless LJ. Supramolecular chemistry of dendrimers. Topics Curr Chem 2001;217:95–120.

CHAPTER 10

BIOINSPIRED SUPRAMOLECULAR DESIGN IN POLYMERS FOR ADVANCED MECHANICAL PROPERTIES

ZHIBIN GUAN

10.1. INTRODUCTION

10.1.1. Design of Synthetic Polymers with High Order Structures

Polymeric materials remain the most important class of materials because they offer the advantages of structural flexibility and functional versatility. Mechanical properties are among the most fundamental properties of polymeric materials (Hearle 1982). For most applications polymers first have to meet the basic mechanical criteria such as strength (modulus), energy dissipating capacity (toughness), and elasticity. With the tremendous progress made in polymer science in the last century, a wide range of synthetic polymers with excellent mechanical properties have been developed for various applications including plastics, fibers, and elastomers (Krejchi et al. 1994; Rathore and Sogah 2001). Whereas man-made polymers can be prepared to meet particular mechanical parameters *one at a time*, to design advanced polymeric materials that can *combine* a number of mechanical properties is still challenging. For example, it remains a major challenge to design synthetic polymers that can *combine* high mechanical strength, high toughness values, and high elasticity because these properties are usually considered orthogonal to each other (Booth and Price 1989).

In contrast, through eons of evolution Nature has come up with many biopolymers that can *combine* important mechanical properties including strength, toughness, and elasticity. For example, silks (Oroudjev et al. 2002), cell adhesion proteins (Law et al. 2003), and connective proteins existing in both soft and hard tissues such as muscle (Kellermayer et al. 1997; Rief, Gautel, et al. 1997; Marszalek et al. 1999; Li et al. 2000), seashells (Smith et al. 1999), and bone (Thompson et al. 2001)

Molecular Recognition and Polymers: Control of Polymer Structure and Self-Assembly.
Edited by V. Rotello and S. Thayumanavan
Copyright © 2008 John Wiley & Sons, Inc.

exhibit a remarkable combination of high strength, toughness, and sometimes elasticity, three properties that are rarely found in one synthetic polymer (Booth and Price 1989). Recent advanced structural analysis and single molecule nano-mechanical studies revealed that the combination of these mechanical properties in natural materials arises from their unique molecular and nanoscopic structures. These mechanistic understandings at the molecular level provide inspiration to materials scientists for designing biomimetic polymers that have a balance of advanced mechanical properties.

10.1.2. Biomimetic Design of Oligomers with High Order Structures

Whereas it remains a major challenge for chemists to design polymers with high order structures comparable to their biological counterparts, significant progress has been made in the last decade in designing short oligomers having well-defined secondary and even tertiary structures (Cheng et al. 2001; Hill et al. 2001). The most basic rules of protein folding are now understood to a sufficient extent to allow for the de novo design and preparation of peptide structures containing α-helices, β-sheets, and β-turns (Regan and DeGrado 1988; DeGrado et al. 1989; Richardson and Richardson 1989; Richardson et al. 1992). By placing hydrophilic, hydrophobic, and turn-forming amino acid residues in a suitable sequence it is possible to generate structures that fold upon themselves or aggregate in helical or sheetlike conformations (Richardson and Richardson 1989; Richardson et al. 1992; Xu et al. 2001). Tremendous efforts have been devoted to the design and synthesis of peptidomimetics in which nonpeptide fragments are used as templates to induce structures in peptides or to mimic elements of protein secondary structures: most notably, α-helices and β-sheets. Many folded and potentially functional helical oligomers have been developed, including β-peptides (Seebach and Matthews 1997; Cheng et al. 2001), γ-peptides (Seebach et al. 2001), peptoid oligomers (Horwell et al. 1994), and other foldamers involving unnatural backbones (Gong 2001; Hill et al. 2001; Oh et al. 2001; Orner et al. 2001). A variety of artificial systems consisting of β-strands linked by unnatural templates have also been developed (Nowick 1999; Stigers et al. 1999; Phillips et al. 2002; Zeng et al. 2002).

The advancement in the understanding of protein structures and in peptidomimetics designs has inspired materials chemists to synthesize biomimetic biomaterials having well-defined secondary and tertiary structures. For example, dendrimers are designed to mimic the globular architecture of proteins (Tomalia 1996; Moore 1997; Fischer and Vogtle 1999; Newkome et al. 1999; Hawker and Piotti 2000; Hecht and Frechet 2001; Zimmerman et al. 2002). Supramolecular assembly was elegantly applied to construct polymers having two-dimensional and three-dimensional (3-D) molecular nano-objects (Ghadiri et al. 1994; Moore 1997; Stupp et al. 1997; Percec et al. 1998; Zubarev et al. 1999; Boal et al. 2000; Frankamp et al. 2002; Lu et al. 2003). Biosynthesis through genetic engineering has been utilized to synthesize artificial protein-based polymeric materials with unprecedented precision of molecular weight and primary sequence (Krejchi et al. 1994; McGrath and Kaplan 1997; Urry et al. 1997; McMillan and Conticello 2000; van Hest and Tirrell 2001).

Hybrid polymers containing peptide blocks were also reported to mimic the β-sheet structures in natural silk (Winningham and Sogah 1997; Qu et al. 2000; Rathore and Sogah 2001).

Despite this important progres, a huge gap still exists between synthetic and natural polymers (such as proteins) in terms of high order structures. Whereas advanced synthetic methodologies now allow us to make synthetic polymers with covalent bonds as strong as natural polymers, we are still at a very primitive stage in programming secondary molecular forces into polymers to control their organization into high order structures. Can we follow Nature's strategies by introducing secondary molecular forces into covalent polymers to achieve advanced properties that are so far beyond our reach? In addition, through molecular and supramolecular designs coupled with advanced property studies at various length scales (single molecule, nano- and microscopic, and bulk), can we connect the molecular structure of a polymer to its bulk properties so that we can start to rationally design new polymeric materials with desired properties?

10.2. BIOMIMETIC CONCEPT OF MODULAR POLYMER DESIGN

10.2.1. Titin as Model for Modular Polymer Designs

With these fundamental questions and goals in mind, my group uses exceptionally well-designed natural polymers as models for our biomimetic polymer designs. One specific model we have conducted extensive biomimetic studies on is titin, a giant protein (3000 kDa, 1 mm long) of the muscle sarcomere. Titin is composed of 300 modules in two motif types, immunoglobulin (Ig) and fibronectin type III domains (Fig. 10.1; Wang 1996; Maruyama 1997). Whereas actin and myosin are motor proteins responsible for muscle contraction, titin contributes to muscle's mechanical strength, toughness, and elasticity (Erickson 1994; Labeit and Kolmerer 1995; Granzier et al. 1996). Single molecule studies have shown that titin exhibits a remarkable combination of high mechanical strength, fracture toughness, and elasticity (Kellermayer et al. 1997; Rief et al. 1997a; Tskhovrebova et al. 1997; Marszalek et al. 1999; Li et al. 2000; Oberhauser et al. 2001). Further studies have revealed that titin's combination of these properties arises from its unique modular domain structures (Kellermayer et al. 1997; Rief, Gautel, et al. 1997; Tskhovrebova et al. 1997; Marszalek et al. 1999; Smith et al. 1999; Clausen-Schaumann et al. 2000; Li et al. 2000; Oesterhelt et al. 2000; Oberhauser et al. 2001; Li et al. 2002; Marszalek et al. 2002). Sequential unfolding of the domains results in the sawtooth

Figure 10.1 Modular multidomain structure of titin protein.

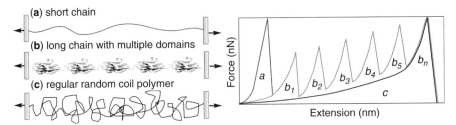

Figure 10.2   Single molecule force-extension curve of three different types of molecules: (a) a short chain or rigid rod molecule; (b) a long chain polymer with multiple domains, such as titin, where b_1-b_n represents the unfolding of individual domains; and (c) a regular random coil long chain. Adapted from Smith et al. (1999). Copyright 1999 Nature Publishing Group.

pattern in the force-extension curve, which provides the molecular basis for the combined high strength, high fracture toughness, and elasticity of these materials.

Figure 10.2 compares the single molecule force-extension curves for a short chain polymer or a rigid rod, a long chain with multiple domains, and a regular random coil polymer (Smith et al. 1999). For a short chain or a rigid rod, the force rises rapidly with relatively small extension. The energy required to break the chain, which is the integration of the area under the curve in Figure 10.2a, is small. Therefore, they are brittle even though they can be strong. In contrast, a regular long chain polymer shows that the entropic force increases gradually as the elastic material is stretched to its elastic limit (Fig. 10.2c). In the end the force increases rapidly for further extension until the molecule breaks. This material is tougher because it takes more energy to break; however, its mechanical strength at low extension is small and reaches high modulus only at very large extension.

Now let us consider a long chain built with many domains or modules that are folded by multiple weak interactions (Fig. 10.2b). For such a molecule, the force rises quickly with extension (Fig. 10.2b). However, when the force reaches a significant fraction of the force required to break the backbone of the chain, a domain unfolds, therefore avoiding the breaking of a covalent bond. The unfolding of one domain abruptly reduces the holding force. On further pulling, the force rises again until another domain unfolds. The process continues until all domains unfold (b_1-b_n) and eventually a covalent bond in the backbone breaks at maximum stretching. The result is a large force sustained over the whole extension, which makes the polymer *strong*, along with a large area under the force-extension curve, making it *tough* as well (Smith et al. 1999). In addition, when the external force is removed, the unfolded domains of modular proteins will refold automatically, making them *elastic*.

Titin demonstrates a fascinating strategy to combine high tensile strength, high toughness, and elasticity by using a modular domain design. By repetitive breaking of the reversible secondary interactions in each domain, the polymers can absorb a very large amount of energy without breaking the permanent covalent bonds. This tandem domain design appears to be a general mechanism used in nature to achieve a combination of mechanical properties, which have been observed in many other biological macromolecules that play mechanical roles (Fisher et al. 2000).

10.2.2. Biomimetic Modular Domain Strategy for Polymer Designs

My group has applied this elegant modular design to synthetic polymers for combined high strength, toughness, and elasticity. Although this modular domain design has been observed in many biopolymers (Fisher et al. 2000), to the best of our knowledge, it has not been applied to synthetic polymer design. Our work represents the first biomimetic design of modular polymers having covalent multidomain structures (Guan et al. 2003, 2004; Roland et al. 2003). By following Nature's strategy, we attempt to address a fundamental challenge in biomaterial design: to combine the three most fundamental mechanical properties—tensile strength, fracture toughness, and elasticity—into one structure (Booth and Price 1989).

The structure of the titin Ig domain exhibits a double β-sheet architecture. Based on molecular modeling and single molecule studies, the six hydrogen bonds between β-stands A′ and G play a critical role in the mechanical stability of the protein (Marszalek et al. 1999; Lu and Schulten 2000). In our design of synthetic mimics, multiple hydrogen bonds were used to direct the formation of loops or domains as modules. Specifically, the four designs shown in Chart 10.1 have been used to construct polymers having well-defined multiple loops or domains: strong quadruple H bonding motif to form loops, strong quadruple H bonding motif to form double-closed loop (DCL) modules, peptidomimetic β-sheet motifs to form DCL modules, and small mechanically stable proteins as domains.

The designs proceed hierarchically from relatively simple to more complex systems. We started with the quadruple H bonded loop system (Chart 10.1a) because it has strong dimerization force and is relatively straightforward to synthesize. To increase the fidelity for rebinding upon unfolding of modules, a DCL analog of the first system was also constructed in our laboratory (Chart 10.1b). One limitation of the quadruple H bond system is the fixed binding strength. The third system is designed to address this issue by using peptidomimetic β-sheet

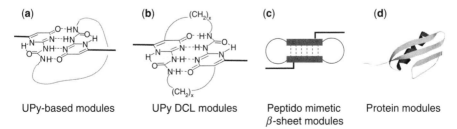

| (a) | (b) | (c) | (d) |
| UPy-based modules | UPy DCL modules | Peptido mimetic β-sheet modules | Protein modules |

Chart 10.1 Four biomimetic modules used in our design of modular polymers: (a) strong quadruple H bonding motif to form loops, (b) strong quadruple H bonding motif to form double-closed loop (DCL) modules, (c) peptidomimetic β-sheet motifs to form DCL modules, and (d) small mechanically stable proteins as domains.

motifs to form modules (Chart 10.1c). By choosing the sequence and length of the β-sheet strand, modules with tunable binding strength can be accessed. Finally, for biocompatibility and complete reversibility for retraction, mechanically stable small proteins have also been synthesized in our laboratory as modules to make protein-based polymers (Chart 10.1d). The thermodynamic reversibility of protein folding should lead to complete reversibility in retraction. In the next section we will selectively discuss the synthesis, single molecule nanomechanics, and bulk mechanics of some of our biomimetic systems.

10.3. RESULTS AND DISCUSSION

10.3.1. Synthesis and Studies of 2-Ureido-4-pyrimidone (UPy) Modular Polymers

We started this research program by making polymers having loops folded by strong hydrogen bonds (Guan et al. 2004). Based on molecular modeling and single molecule studies, the six hydrogen bonds between β-strands A$'$ and G in the Ig module of titin play a critical role in its mechanical stability (Marszalek et al. 1999; Lu and Schulten 2000). In our first biomimetic design, a strong quadruple hydrogen bonding motif, UPy, was employed to direct the formation of loops along a polymer chain. Sijbesma and coworkers (1997) have shown beautifully that UPy dimerizes strongly with a dissociation constant (K_d) of more than 10^{-8} M in toluene. Based on the magnitude of the dimerization constant, the free energy required to break the UPy dimer is more than 11 kcal/mol, which is comparable to protein unfolding energy and lower than typical covalent bond energies, and therefore suits our biomimetic study. We chose this system for several reasons: it dimerizes relatively strongly by precise quadruple H bonding, the breakage energy for the dimer is lower than that for breaking the macromolecule backbone, and it is relatively easy to synthesize and to functionalize at the R and R$'$ positions.

We chose urethane polymerization for constructing our modular polymers because polyurethanes (PUs) have excellent biocompatibility and are an important class of biomaterials (Pinchuk 1994). The synthesis of the UPy containing monomers began with the alkylation of ethyl acetoacetate with allyl bromide followed by condensation with guanidine carbonate to afford an isocytosine, which was further treated with allyl isocyanate to provide compound **4** (Scheme 10.1). The isocytosine ring was protected as benzyl ether for improved solubility and for further synthesis of a control polymer to be used in comparative studies. Hydroboration of the diolefin **3** afforded the protected UPy containing monomer **6**. The benzyl protection was removed by catalytic hydrogenation to give the free UPy monomer **7**. Both the protected UPy monomer **6** and the free UPy monomer **7** will be incorporated into polymer chains for comparative studies.

Polymerization reactions were carried out via a prepolymer formation followed by chain extension. One equivalent of poly(tetramethylene oxide) [number-average molecular weight (M_n) = 1400 g/mol] was injected slowly by a syringe pump into a

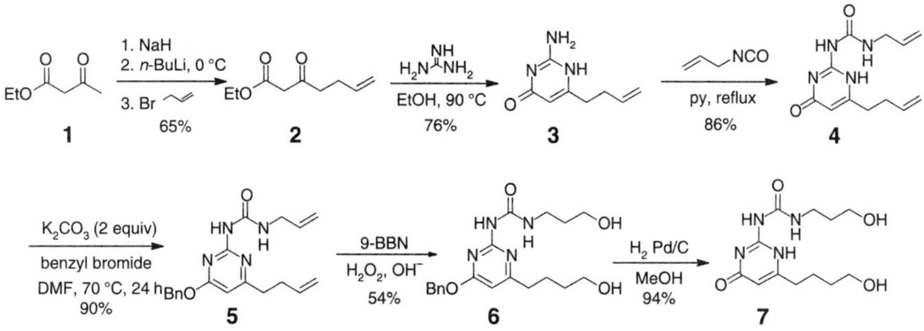

Scheme 10.1 UPy monomer synthesis.

solution of 2 equiv of 1,6-diisocyanatohexane (HDI) in *N,N*-dimethylacetamide to form a prepolymer. Benzyl protected UPy monomer **6** was added as a chain extender to complete the polymerization process, providing a fully protected control polymer that was unable to form UPy dimers, because the quadruple H bonding was prevented by the benzyl protection (**9**, Scheme 10.2). Polymerization reactions with the free UPy monomer **7** were carried out in the same manner, providing a deprotected poly-mers in which the free UPy units dimerize to form loops along polymer chains (**8**, Scheme 10.2). The formation of UPy dimers in polymer **8** was confirmed by ^{1}H-NMR. For polymer **9**, only one intramolecularly H bonded proton was observed at 9.2 ppm. For polymer **8**, three peaks were instead observed at 10.28, 11.95, and 13.20 ppm, which correspond to the three protons that form H bonds in UPy dimers. From comparing the NMR spectra with other UPy dimers reported by Ky Hirschberg et al. (1999), the UPy units all exists in keto tautomers in polymer **8**. This is consistent with their observation that alkyl-substituted UPys almost exclusively exist as keto tautomers.

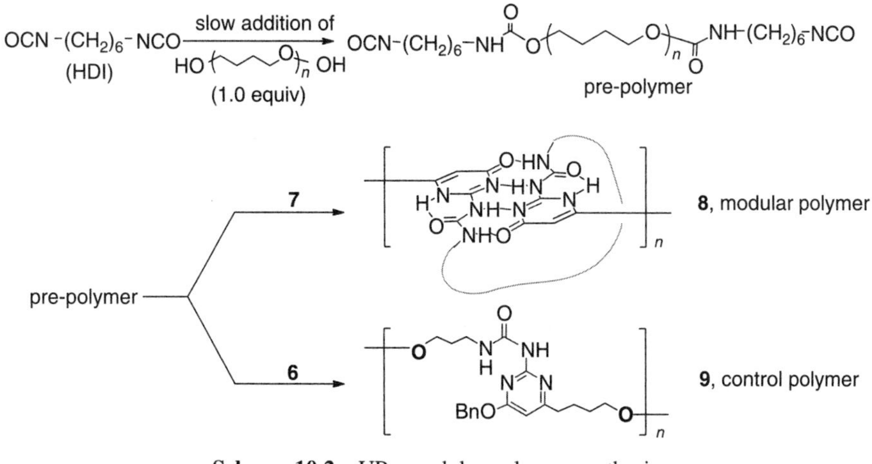

Scheme 10.2 UPy modular polymer synthesis.

Both the control polymer (**9**) and the modular polymer having multiple loops (**8**) were subjected to single molecule force-extension studies using atomic force microscopy (AFM). The single molecule nanomechanical properties for many biopolymers including proteins (Kellermayer et al. 1997; Rief, Gautel, et al. 1997; Marszalek et al. 1999; Li et al. 2000), polysaccharides (Rief, Oesterhelt, et al. 1997), DNAs, (Smith et al. 1999), and synthetic polymers (Ortiz and Hadziioannou 1999) have been studied using AFM. We followed similar protocols reported previously (Rief, Gautel, et al. 1997; Ortiz and Hadziioannou 1999; Smith et al. 1999; Li et al. 2002) for single chain studies of our polymers using AFM (Guan et al. 2003; Roland et al. 2003). Either the modular polymer **8** or the control polymer **9** was allowed to be adsorbed from a dilute solution (10^{-5}–10^{-7} M) onto a freshly coated gold surface. The sample was probed in our custom-built liquid cell using toluene as the solvent. We tried both physisorption and chemisorption with thiol ether terminally functionalized polymers, and both protocols gave comparable results. When the tip was retracted, extension curves like the ones shown in Figure 10.3 were recorded.

The single chain force-extension curves for both the control polymer (**9**) and the modular polymer (**8**) are shown in Figure 10.3 (Guan et al. 2004). For control

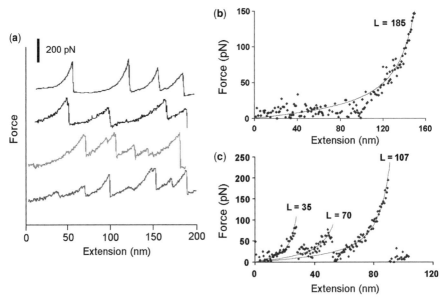

Figure 10.3   AFM single chain force-extension data. (a) An overlay of representative force-extension curves for modular polymer **8**. Sawtooth patterned curves were consistently obtained. (b) One representative single chain force-extension curve for control polymer **9**, in which only one peak was observed. (c) A single chain force-extension curve for modular polymer **8** shows the characteristic sawtooth pattern with three peaks. In Figure 10.2b and c, all scattered dots represent experimental data and the solid lines are results from WLC fitting. Adapted from Guan et al. (2004). Copyright 2004 American Chemical Society.

polymer **9** that does not have the modular structure, we only observed a single peak that is characteristic for the entropic extension of a random coil chain (Fig. 10.3b). In contrast, modular polymer **8** showed distinct sawtooth patterns in the force-extension curve (Fig. 10.3a, c), similar to those observed in titin and other modular biopolymers (Kellermayer et al. 1997; Rief, Gautel, et al. 1997; Marszalek et al. 1999; Li et al. 2000). The big contrast for the force-extension curves observed for polymers **8** and **9** strongly suggests the sequential unfolding of the UPy dimers along the modular polymer **8** chain. As the polymer chain was stretched, the force rose gradually until one UPy dimer could not sustain the force and broke instantaneously, adding additional length to the polymer chain. Therefore, the peak force corresponds to the force required to break a UPy dimer at a specific pulling rate and the spacing between the peaks represents the gain in length during the unfolding. The curves can be fit nicely with a wormlike chain (WLC) model, which predicts the relationship between the extension of a single polymer chain and the entropic restoring force generated (Rief, Gautel, et al. 1997; Ortiz and Hadziioannou 1999). This further confirms that the observed curves are from stretching individual polymer chains.

The mechanical properties of the polymers were tested in bulk as well to correlate with the single chain nanomechanical studies. Polymer samples were cast into thin films and subjected to stress–strain analysis. Figure 10.4 shows the comparison of

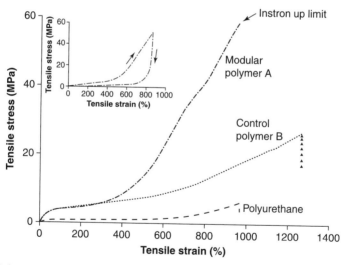

Figure 10.4 Stress–strain curves for control and modular polymers. The curve in the bottom (– – –) is for the polyurethane (PU) made from poly(tetramethylene glycol) and HDI without any UPy units. The curve in the middle (⋯⋯) is control polymer **9** containing the protected UPy units. The curve at the top (– ⋅ –) is for modular polymer **8** containing UPy dimerized loops. Modular polymer **8** could not be broken by the Mini-Instron at its maximum load. A stretching–retraction cycle for polymer **8** is shown in the inset, which shows a huge hysteresis, indicating great energy dissipation during the cycle. Adapted from Guan et al. (2004). Copyright 2004 American Chemical Society.

stress–strain data for modular polymer **8**, control polymer **9**, and the regular PU that does not contain any UPy units. On a stress–strain curve, the maximum stress at break gives the tensile strength of the material whereas the whole area covered under the curve reflects the total energy required to break the material, that is, the toughness. The introduction of protected UPy units into the PU enhanced its tensile strength and toughness. Even though the benzyl protection blocked the UPy from dimerization to form loops, π–π stacking, dipole–dipole interactions, and weak residual hydrogen bonding between the protected UPy units should increase the tensile strength and toughness of protected polymer **9**. For polymer **8** in which UPy units can dimerize to form loops along the chains, the stress–strain curve shows dramatic differences from the curve of the control polymer (**9**). It shows a sigmoid stress–strain curve characteristic for elastic polymers. After a pseudo yield region, the sample becomes much stiffer with a significant increase in modulus and strength. Because of its high strength and toughness, the ultimate tensile strength at break and the fracture toughness could not be obtained for polymer **8** because the sample could not be broken even at the maximum load of the Mini-Instron for the smallest sample specimen we could prepare. Nevertheless, quantitative comparisons can be made at 950% strain for the three samples. At 950% strain, the tensile stresses are 58.2, 17.4, and 6.0 MPa and the energy absorptions per unit volume are 191.5, 78.5, and 17.7 MPa for modular polymer **8**, control polymer **9**, and the PU samples, respectively. This comparison clearly shows that modular polymer **8** is significantly stronger and tougher than either control polymer **9** or the simple PU. Polymer **8** is also very elastomeric, as evidenced by the high strain up to 900% and the complete recovery to its original length in three consecutive extension–retraction experiments. The huge hysteresis accompanied in one extension–retraction cycle (inset, Fig. 10.4) further reveals the great energy dissipation capability of the system, an important feature for high toughness. The bulk mechanical data correlate well with our single chain force-extension observation, which successfully demonstrates our biomimetic concept: the introduction of modular structures held by sacrificial weak bonds into a polymer chain can successfully combine the three most fundamental mechanical properties (high tensile strength, toughness, and elasticity) into one polymer.

10.3.2. Synthesis and Single Molecule Nanomechanical Studies of Peptidomimetic β-Sheet Modular Polymers

Although demonstrating our biomimetic concept, the first-generation UPy system had a few limitations. The structure of the polymer had nonuniformity that arose from the polydispersed poly(tetramethylene oxide) loop and the different enchainment of the UPy units (head–head, head–tail, and tail–tail, Chart 10.2). The UPy units could also randomly bind to each other within a chain or between different chains. Finally, the binding strength of UPy is not tunable. For further exploration of modular biomimetic materials with high strength and toughness, more uniform and higher ordered polymer systems would be desirable. To address these issues, we

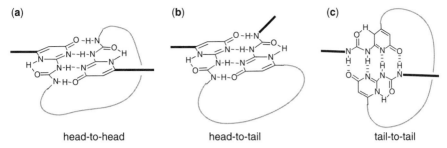

head-to-head   head-to-tail   tail-to-tail

Chart 10.2 Different folding topologies in the first-generation UPy modular polymer.

used a peptidomimetic β-sheet motif to construct a modular polymer that has a DCL structure.

The module in this system is composed of a β-sheet like duplex that is connected at both ends with hydrocarbon loops (Fig. 10.5; Roland and Guan 2004). We envision that this system should overcome the limitations of our previously reported UPy system. First, the DCL topology will enhance the possibility for each hydrogen bonding unit to bind to its original counterpart, therefore minimizing dimerization of nonadjacent units on the same chain or from different chains. Second, by choosing a monodispersed alkyl linker in the monomer synthesis, the loop size is monodispersed. Third, by choosing the sequence and adjusting the length, the binding strength between β-sheets can be tuned.

Among the many peptidomimetic β-sheets reported in the literature (Hill et al. 2001), we began with the quadruple H bonding duplex developed by Gong and coworkers (1999) because of its simplicity of synthesis. Construction of the DCL monomer began with the synthesis of the self-complementary hydrogen bonding module **10**. Monomer **10** was synthesized by iterative amide coupling using modified literature procedures (Scheme 10.3; Zeng et al. 2000). Alkyl ethers next to the amides serve the purpose to form six-membered ring intramolecular hydrogen bonds. This will confine the conformation of the monomer and promote duplex formation between two monomers. An *n*-octyl group was introduced to the first aromatic ring to enhance the solubility of the monomer. We chose ring-closing metathesis (RCM) for cyclization because it has been shown to be a very powerful methodology to close medium to giant rings (Grubbs et al. 1995; Schuster and Blechert 1997; Maier 2000). Covalent capture of the hydrogen bonded dimer via RCM provided

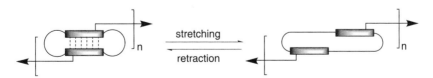

stretching

retraction

Figure 10.5 Schematic representation of β-sheet peptidomimetic modular polymer composed of double-closed loop monomers. Adapted from Roland and Guan (2004). Copyright 2004 American Chemical Society.

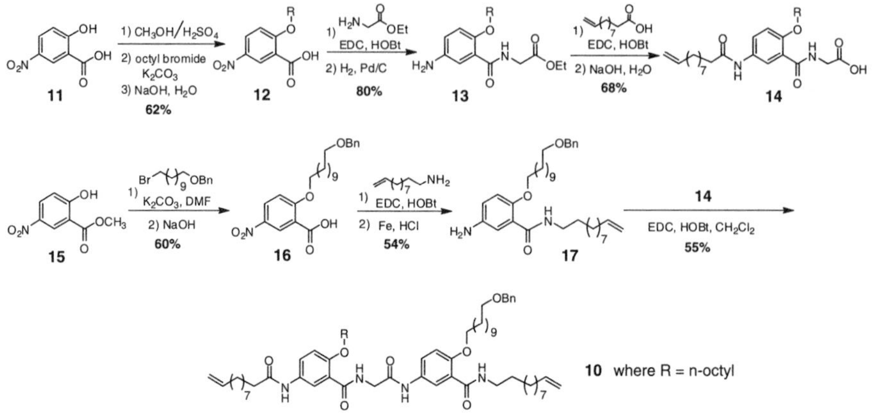

Scheme 10.3 Synthesis of quadruple H bonded β-sheet peptidomimetic monomer.

the protected DCL monomer as a mixture of E/Z double bond isomers (Clark and Ghadiri 1995). Hydrogenation reduced the double bonds and removed the benzyl protecting groups simultaneously, affording DCL monomer **11** (Scheme 10.4). Polymerization reactions were performed by mixing equimolar amounts of the DCL monomer and 4,4′-methylenebis(phenyl-isocyanate) in chloroform at 45 °C. Gel permeation chromatography analysis showed the number-average molecular weights for the DCL polymers were as high as 89 000 g/mol (Scheme 10.4).

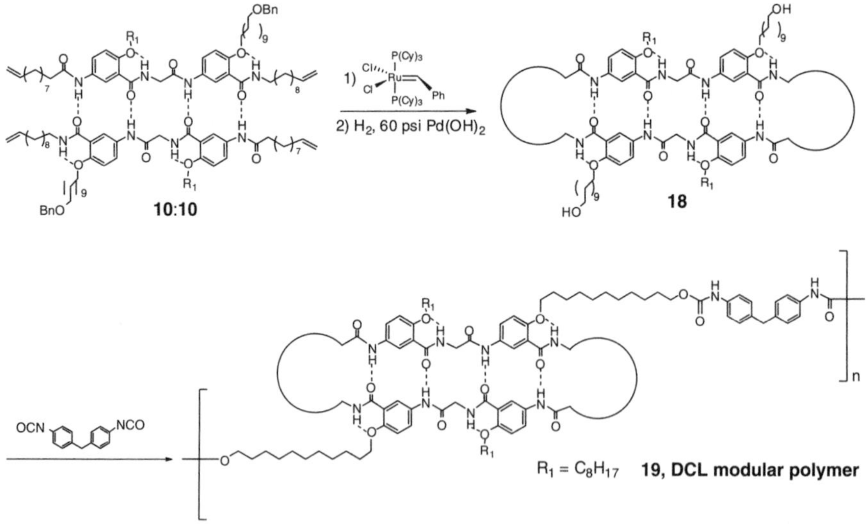

Scheme 10.4 Synthesis of peptidomimetic double-closed loop (DCL) monomer and modular polymer.

(a)

(b)

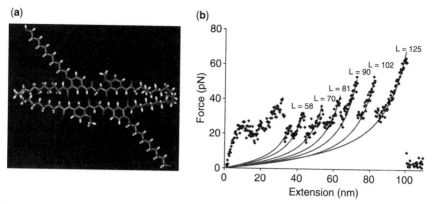

Figure 10.6   (a) A molecular model for one double-closed loop (DCL) module used in the study. (b) An AFM single molecule force-extension curve for the DCL modular polymer. The solid line is the fitting with the WLC model at a 0.55-nm persistence length (L is the contour length during stretching). Adapted from Roland and Guan (2004). Copyright 2004 American Chemical Society.

   Following the synthesis and characterizations, the DCL polymers (**19**) were subjected to single molecule force-extension experiments using contact mode AFM (Roland and Guan 2004). The force-extension profile observed for our DCL modular polymers was much more uniform than those observed for our first-generation UPy polymers (Fig. 10.6). This higher level of regularity is attributed to the more uniform structure of this polymer system. The consistent sawtooth pattern observed in the force-extension curves suggests that the DCL modules unfold sequentially as the polymer chain is stretched. The force-extension curves obtained for the DCL polymers were fit using the WLC model. This further confirms that the sawtooth patterned force spectra represent sequential unfolding events of single polymer chains.

   One nice feature of this peptidomimetic modular system is that the binding strength can be fine-tuned by changing the number of hydrogen bonds between a duplex pair. For structure–property correlation, we recently prepared a series of β-sheet peptidomimetic modules with increasing number of hydrogen bonds: 4H (**20**, DADA), 6H (**21**, AADADD), and 8H (**22**, AADADADD; Chart 10.3). We used both steered molecular dynamic (SMD) simulation and experimental single molecule force spectroscopy (SMFS) to understand the relationship between weak molecular interactions and mechanical stability (Guzman et al. 2007). Their unfolding forces were analyzed by SMD and compared to results obtained by SMFS. The computational model for the 8H dimer offered insight into a possible dimer intermediate after an unexpected decrease in mechanical stability that was observed by AFM force studies. In addition, atomic level analysis of the rupture mechanism verified the dependence of the mechanical strength on the pulling trajectory because of the directional nature of chemical bonding upon an applied force. The knowledge gained from this basic study will be used to guide our further design of modular

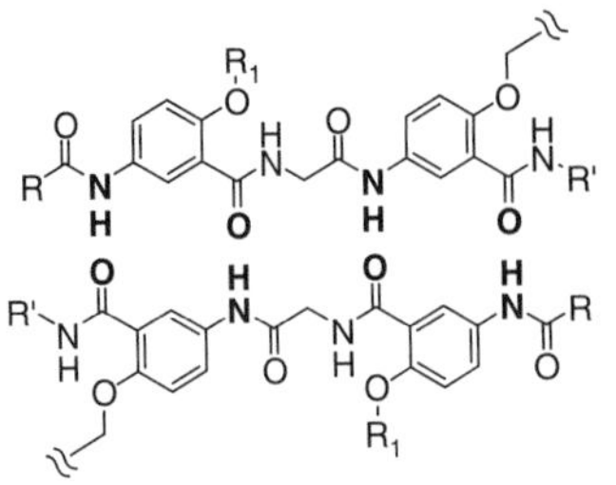

20, DADA

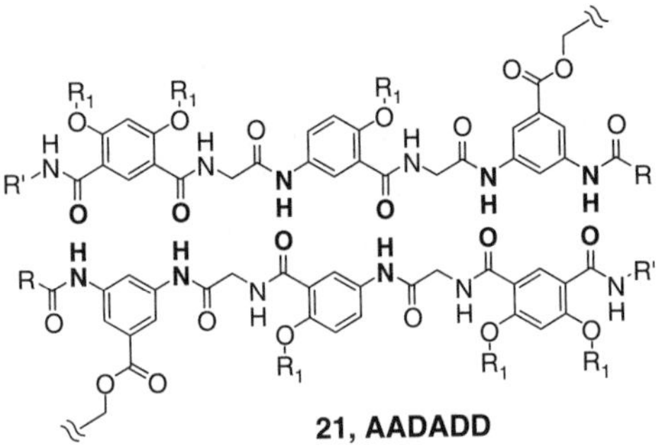

21, AADADD

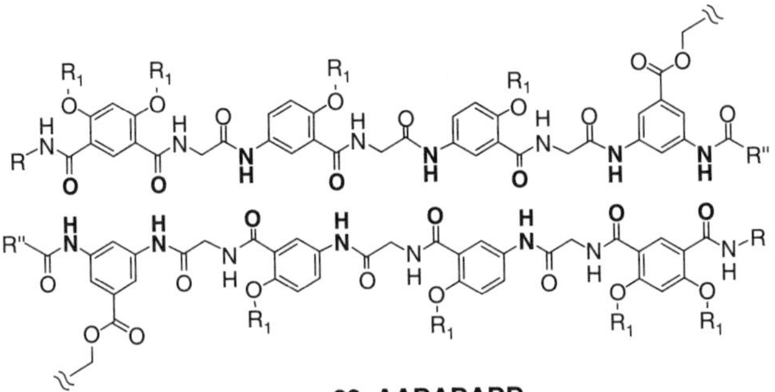

$R_1 = C_8H_{17}$

22, AADADADD

Chart 10.3 Peptidomimetic β-sheet modules with varying numbers of H bonding sites.

polymers having folded nanostructures through programming strong and sacrificial molecular forces.

10.3.3. 3-D Network Polymers Containing Biomimetic Reversibly Unfolding Crosslinkers for Advanced Mechanical Properties

Following these basic studies, we applied the modular design concept to bulk polymer design to enhance its mechanical properties. In our previous studies we synthesized linear PUs and polyesters containing these modules (Guan et al. 2004; Roland and Guan 2004). Bulk mechanical tests demonstrated that the modular polymers have significantly enhanced mechanical properties. Whereas these systems demonstrate our initial biomimetic concept, the linear architecture and the lack of chain orientation may prevent them from manifesting the maximum biomimetic effects. One important design feature of natural modular polymers is their high order organization and chain orientation. For example, in collagen (Buehler 2006) and muscle fibers (McNally et al. 2006), the proteins are oriented uniaxially along the fiber axis, which ensures cooperative actions of the modules to achieve high mechanical strength. For practical reasons, this is very difficult to achieve in linear synthetic polymers. In addition, for the linear polymer system every repeat unit contains one biomimetic module. For practical applications, it would be advantageous to use a relatively small amount of biomimetic modules to achieve maximum property enhancement. With these considerations in mind, we recently developed 3-D network polymers having biomimetic modules as cross-linkers (Kushner et al. 2007). As explained in more detail in the following paragraph, the network structure makes it less critical for chains to orient in any specific direction because the stress from any direction should be transferred to the cross-linking junction points. In addition, only a relatively small amount of modular cross-linker is needed to enhance the network properties.

Although many engineering approaches and chemical modifications (Mark et al. 1994) have been developed to improve the mechanical properties of elastomers, it remains a challenge to design ideal elastomers that simultaneously possess high modulus, high tensile strength, and high extensibility. Typically rigid elastomers tend to fail after only a short extension. Although flexible elastomers are more extensible, they usually have low moduli and exhibit shallow stress response. We envisioned that the incorporation of reversibly unfolding modular cross-linkers into a network should result in elastomers with combined high moduli, toughness, and elasticity. As illustrated in Figure 10.7, a stress applied from any direction will be ultimately transferred across the individual network junctions, where biomimetic modules can be reversibly unfolded (Kushner et al. 2007). Because it requires significant forces to unfold the modules held by strong multiple hydrogen bonds, further extension can be gained without sacrificing the strength. Interesting work has been reported on using interchain hydrogen bonding to improve polymer physical properties (Stadler and Burgert 1986; de Lucca Freitas and Stadler 1987; Mueller et al. 1995; Yamauchi et al. 2003; Elkins et al. 2005; Park et al. 2005). However, most of these studies are focused on the investigation of solution and melt rheological properties. To the best of our knowledge, there is no report in the literature on using molecularly engineered cross-linkers to enhance elastomer mechanical properties.

Based on our and other's single molecule nanomechanical studies (Guan et al. 2004; Roland and Guan 2004; Zou et al. 2005), the UPy module still has significantly higher mechanical stability than other peptidomimetic β-sheets that we investigated. This could be partially attributed to the compact structure of UPy that favors high

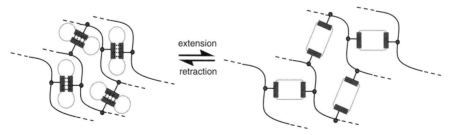

Figure 10.7 Introducing biomimetic modular cross-linkers to enhance elastomer mechanical properties. Adapted from Kushner et al. (2007). Copyright 2007 American Chemical Society.

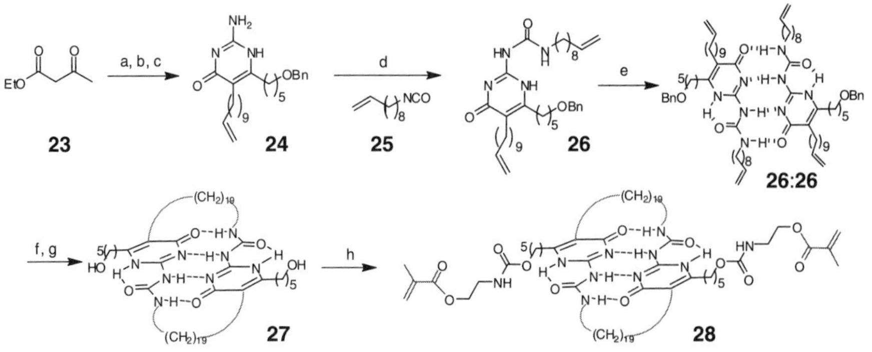

a) NaH, n-BuLi, THF, Br(CH$_2$)$_4$OBn, 52%. b) Br(CH$_2$)$_9$CH=CH$_2$, K$_2$CO$_3$, DMF, rt, 12h, 47%. c) Guanidine carbonate, EtOH, reflux, 12h, 72%. d) 4, Py, reflux, 6 h, 83%. e) DCM. f) (Cy$_3$P)$_2$RuCl$_2$=CHPh, DCM, reflux, 1-2 h, 65%. g) Pd(OH), 1 atm H$_2$, THF, 68%. h) CHCl$_3$, 2-isocyanatoethyl methacrylate, DBTDL.

Scheme 10.5   Synthesis of the UPy reversibly unfolding modular cross-linker.

folding cooperativity. Therefore, we designed UPy-based reversibly unfolding modules for our bulk polymer designs (Kushner et al. 2007). To ensure good reversibility for the refolding, we followed the design of our peptidomimetic DCL system and designed UPy-based DCL cross-linkers.

The synthesis of the module is provided in Scheme 10.5 (Kushner et al. 2007). Double alkylation of ethyl acetoacetate followed by guanidine condensation afforded alkenyl–pyrimidone intermediate **24** (Kushner et al. 2007). Isocyanate **25** was coupled to pyrimidone **24** to yield **26**. Upon dimerization in DCM, RCM effectively cyclized the two UPy units (Mohr et al. 1997; Weck et al. 1999). A one-pot reduction and deprotection through hydrogenation using Pearlman's catalyst gave diol module **27**. Finally, capping **27** with 2-isocyanatoethyl methacrylate at both ends provided the UPy sacrificial cross-linker **28**, which was thoroughly characterized by ^{1}H- and ^{13}C-NMR, Fourier transform IR (FTIR), and mass spectrometry.

In our initial study we chose poly(n-butyl acrylate) as the backbone because this polymer exhibits elastomeric properties upon cross-linking (glass-transition temperature = −54 °C; Brandrup and Immergut 1998) and can be easily synthesized by free-radical polymerization. A series of transparent rubbery films was prepared by radical copolymerization of n-butyl acrylate with desired amounts of cross-linker **28** using AIBN as the initiator. In control experiments, a poly(ethylene glycol) (PEG) dimethacrylate ($M_n = 750$) was used as the cross-linker instead. When fully stretched, this molecule has approximately the same length as the fully unfolded **28**. The thermosets were characterized by FTIR and differential scanning calorimetry. The films were then cut with a "dogbone" die and subjected to tensile testing on an MTS tester.

The comparison of the mechanical properties of the UPy samples and the PEG controls demonstrates that the introduction of our biomimetic module into the network dramatically enhanced the polymer mechanical properties. As shown in the stress–strain curves (Fig. 10.8), the network containing the biomimetic cross-linker has significantly higher modulus, tensile strength, and toughness than the

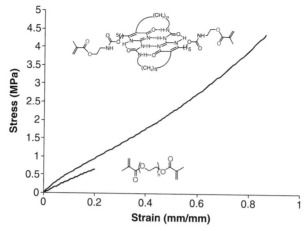

Figure 10.8 Stress–strain curves for 6% crosslinked poly(*n*-butyl acrylate) elastomer for the sample and control specimen (strain rate = 100 mm/min, room temperature). Adapted from Kushner et al. (2007). Copyright 2007 American Chemical Society.

control polymer (Kushner et al. 2007). Furthermore, for the PEG control samples, the increase of the modulus at higher cross-linker levels trades off with the maximum elongation. This is typical for regular thermoset elastomers: higher cross-linking density results in more rigid and less elastic rubbers (Mark et al. 1994; Treloar 1975). In contrast, in our UPy samples we observed a consistent increase in the modulus and tensile strength without sacrificing the extensibility. We attribute this to the unique structure of our modular cross-linker. Although the modulus is increased at higher cross-linking density, the reversibly folding cross-linkers can act as energy dissipating units to prevent fracture formation. To extend the control polymer (an entropic elastomer) by 100%, the free energy change (ΔG) is roughly $-T\Delta S$, which is estimated to be ~ 0.6 kcal/mol (Heimenz 1984). In contrast, it takes ~ 11 kcal/mol to fully unfold the UPy folded module (based on $K_d = \sim 10^{-8}$ M in toluene). The enhancement of the energy dissipation by the modular cross-linker was also supported by the much larger loss moduli measured by dynamic mechanical analysis for the real samples as compared to the controls. This work introduces a novel biomimetic concept to enhance rubber properties through design of molecularly engineered cross-linkers. Further studies are currently ongoing in our laboratory to extend this concept to other network systems and to probe the mechanisms for property enhancement.

10.4. CONCLUSION AND PERSPECTIVE

Nature has evolved many excellent materials, including both organic and polymeric materials such as silks (Oroudjev et al. 2002), adhesion proteins (Law et al. 2003), bioadhesives (Lin et al. 2007; Waite et al. 2005; Zhao et al. 2006), and connective proteins existing in both soft and hard tissues (Kellermayer et al. 1997; Rief, Gautel, et al. 1997; Marszalek et al. 1999; Li et al. 2000) and composite materials such as seashells (Smith et al. 1999) and bone (Thompson et al. 2001) that can combine important mechanical properties including strength, toughness, and elasticity. Recent single molecule and nanomechanical studies of these natural materials began to reveal the molecular origins for the combination of these mechanical properties. These mechanistic understandings at the molecular level provide inspiration to materials scientists for designing biomimetic polymers that have a balance of advanced mechanical properties (Winningham and Sogah 1997; Rathore and Sogah 2001; Lee, Dellatore, et al. 2007; Lee, Lee, et al. 2007; Westwood et al. 2007).

One common strategy employed in natural material design is the programming of secondary molecular forces into strong covalent polymers to further enhance the physical performance. A key mechanism used in natural materials is to program weak molecular forces, intermolecularly or intramolecularly, that serve as reversible sacrificial bonds to dissipate energy while gaining further extension. This sacrificial bonding concept has generated significant interest in the biophysics and materials science communities (Smith et al. 1999; Fantner et al. 2006). My group has been using titin as the model for developing biomimetic polymers consisting of a linear array of modules folded by sacrificial weak hydrogen bonds (Guan et al. 2003, 2004; Roland and Guan 2004; Guan 2007; Guzman et al. 2007; Kushner et al. 2007). Titin uses intramolecular weak forces to fold a long chain polymer into a bead-on-string architecture for combining mechanical strength, toughness, and elasticity into one system. In the first generation of biomimetic design, we synthesized polymers having loops held by UPy quadruple H bonding units and successfully conducted both single molecule force-extension and bulk stress–strain studies for this modular polymer (Guan et al. 2004). AFM single chain stretching results show that the mechanical properties of individual chains can be precisely measured at a single molecule level. The bulk tensile testing tests show that the modular polymer indeed possesses a combination of high tensile strength and high toughness.

In our second-generation biomimetic design, we successfully synthesized a series of peptidomimetic β-sheet modules (Roland and Guan 2004). Both computational SMD simulation and experimental single molecule force microscopy were used to understand the relationship between weak molecular interactions and mechanical stability. The DCL modules exhibit significantly improved regularity in sequential unfolding events as revealed by AFM single molecule force microscopy. It was further demonstrated that other factors in addition to thermodynamic stability, such as the trajectory of the pulling force relative to the duplex axis and possible kinetic folding intermediates, play an important role in the mechanical stability of the supramolecular assemblies (Guzman et al. 2007).

Finally, we successfully demonstrated one particular application of our biomimetic modular concept in improving bulk polymer properties. Through introduction of a small amount of reversibly unfolding modular cross-linker into 3-D network polymers, the mechanical properties of the network were drastically enhanced (Kushner et al. 2007). Most strikingly, at increasing cross-linking density both the modulus and tensile strength were significantly improved without sacrificing the extensibility.

These results demonstrate our biomimetic concept: the introduction of modular structures held by sacrificial weak bonds into a polymer chain can successfully combine the three most fundamental mechanical properties (high tensile strength, toughness, and elasticity) into one polymer.

Despite the significant progress that has been made in bioinspired or biomimetic synthesis of polymeric materials, because of the complexity and subtlety of biopolymer structures (Guan 2007), it remains a major challenge to develop truly effective biomimetic polymers that can rival the performance of their natural analogs. Thanks to advanced analytical techniques including single molecule methods, our understanding of biopolymers is increasing exponentially. Facilitated by the deeper insight from studying biopolymer systems and our sharpened synthetic tools, the future of biomimetic polymer design is bright. If strong, covalent bonding was the major concern for polymer chemists in the last century, we believe the programming of weak, noncovalent interactions into covalent polymers will be the central topic for polymer chemists in this century. Supramolecular chemistry, which is mainly focused on investigation of systems involving weak molecular interactions, should play a critical role in designing the next generation of advanced polymeric materials. We will continue our journey on programming weak forces, both inter- and intramolecularly, into covalent polymers for achieving advanced properties.

ACKNOWLEDGMENTS

Thanks to the National Institute of Health (R01EB04936) and the Department of Energy (DE-FG02-04ER46162) for financial support. I am also indebted to our collaborators and many dedicated coworkers who have contributed to this research project. I gratefully acknowledge a Beckman Young Investigator Award, a Camille Dreyfus Teacher–Scholar Award, a DuPont Young Investigator Award, a 3M New Faculty Award, a Friedrich Wilhem Bessel Research Award from the Alexander von Humboldt Foundation, and a CAREER Award from the National Science Foundation.

REFERENCES

Boal AK, Ilhan F, DeRouchey JE, Thurn-Albrecht T, Russell TP, Rotello VM. Self-assembly of nanoparticles into structured spherical and network aggregates. Nature (Lond) 2000;404:746–748.

Booth C, Price C, editors. Comprehensive polymer science: the synthesis, characterization, reactions, and applications of polymers. Volume 2, polymer properties. Oxford, UK: Pergamon Press; 1989.

Brandrup J, Immergut EH, editors. Polymer handbook. 4th ed. New York: Wiley; 1998.

Buehler MJ. Nature designs tough collagen: explaining the nanostructure of collagen fibrils. Proc Natl Acad Sci USA 2006;103:12285–12290.

Cheng RP, Gellman SH, DeGrado WF. β-Peptides: from structure to function. Chem Rev 2001;101:3219–3232.

Clark TD, Ghadiri MR. Supramolecular design by covalent capture. Design of a peptide cylinder via hydrogen-bond-promoted intermolecular olefin metathesis. J Am Chem Soc 1995;117:12364–12365.

Clausen-Schaumann H, Rief M, Tolksdorf C, Gaub HE. Mechanical stability of single DNA molecules. Biophys J 2000;78:1997–2007.

de Lucca Freitas LL, Stadler R. Thermoplastic elastomers by hydrogen bonding. 3. Interrelations between molecular parameters and rheological properties. Macromolecules 1987;20:2478–2485.

DeGrado WF, Wasserman ZR, Lear JD. Protein design, a minimalist approach. Science 1989;243:622–628.

Elkins CL, Park T, McKee MG, Long TE. Synthesis and characterization of poly(2-ethylhexyl methacrylate) copolymers containing pendant, self-complementary multiple-hydrogen-bonding sites. J Polym Sci Part A Polym Chem 2005;43:4618–4631.

Erickson HP. Reversible unfolding of fibronectin type III and immunoglobulin domains provides the structural basis for stretch and elasticity of titin and fibronectin. Proc Natl Acad Sci USA 1994;91:10114–10118.

Fantner GE, Oroudjev E, Schitter G, Golde LS, Thurner P, Finch MM, Turner P, Gutsmann T, Morse DE, Hansma H, Hansma PK. Sacrificial bonds and hidden length: unraveling molecular mesostructures in tough materials. Biophys J 2006;90:1411–1418.

Fischer M, Vogtle F. Daendrimers: from design to application—a progress report. Angew Chem Int Ed 1999;38:885–905.

Fisher TE, Carrion-Vazquez M, Oberhauser AF, Li H, Marszalek PE, Fernandez JM. Single molecule force spectroscopy of modular proteins in the nervous system. Neuron 2000;27:435–446.

Frankamp BL, Boal AK, Rotello VM. Controlled interparticle spacing through self-assembly of Au nanoparticles and poly(amidoamine) dendrimers. J Am Chem Soc 2002;124:15146–15147.

Ghadiri MR, Granja JR, Buehler LK. Artificial transmembrane ion channels from self-assembling peptide nanotubes. Nature (Lond) 1994;369:301–304.

Gong B. Crescent oligoamides: from acyclic "macrocycles" to folding nanotubes. Chem Eur J 2001;7:4336–4342.

Gong B, Yan Y, Zeng H, Skrzypczak-Jankunn E, Kim YW, Zhu J, Ickes H. A new approach for the design of supramolecular recognition units: hydrogen-bonded molecular duplexes. J Am Chem Soc 1999;121:5607–5608.

Granzier H, Helmes M, Trombitas K. Nonuniform elasticity of titin in cardiac myocytes: a study using immunoelectron microscopy and cellular mechanics. Biophys J 1996;70:430–442.

Grubbs RH, Miller SJ, Fu GC. Ring-closing metathesis and related processes in organic synthesis. Acc Chem Res 1995;28:446–452.

Guan Z. Supramolecular design in biopolymers and biomimetic polymers for advanced mechanical properties. Polym Int 2007;56:467–473.

Guan Z, Roland J, Ma SX, Kong X, McIntire TM, Brant DA. Design and single molecule studies of titin-mimicking modular macromolecules having precise secondary structures formed by hydrogen bonding. Polym Prepr 2003;44:592–593.

Guan Z, Roland JT, Bai JZ, Ma SX, McIntire TM, Nguyen M. Modular domain structure: a biomimetic strategy for advanced polymeric materials. J Am Chem Soc 2004;126: 2058–2065.

Guzman DL, Roland JT, Keer H, Kong YP, Ritz T, Yee AF, Guan Z. Using computational single molecule force spectroscopy in rational design of strong, modular nanostructures for use in materials synthesis. Submitted 2007.

Hawker CJ, Piotti M. Dendritic macromolecules: hype or unique specialty materials. ACS Symp Ser 2000;755:107–118.

Hearle JWS. Polymers and their properties. Volume 1. Fundamentals of structure and mechanics. Chichester, UK: Ellis Horwood; 1982.

Hecht S, Frechet JMJ. Dendritic encapsulation of function: applying nature's site isolation principle from biomimetics to materials science. Angew Chem Int Ed 2001;40:74–91.

Heimenz PC. Polymer chemistry: the basic concepts. New York: Marcel Dekker; 1984.

Hill DJ, Mio MJ, Prince RB, Hughes TS, Moore JS. A field guide to foldamers. Chem Rev 2001;101:3893–4011.

Horwell DC, Howson W, Rees DC. "Peptoid" design. Drug Design Discov 1994;12:63–75.

Kellermayer MSZ, Smith SB, Granzier HL, Bustamante C. Folding–unfolding transitions in single titin molecules characterized with laser tweezers. Science 1997;276:1112–1126.

Krejchi MT, Atkins EDT, Waddon AJ, Fournier MJ, Mason TL, Tirrell DA. Chemical sequence control of β-sheet assembly in macromolecular crystals of periodic polypeptides. Science 1994;265:1427–1432.

Kushner AM, Gabuchian V, Johnson EG, Guan Z. Biomimetic design of reversibly unfolding modular cross-linker to enhance mechanical properties of 3D network polymers. J Am Chem Soc 2007;129:14110.

Ky Hirschberg JHK, Beijer FH, van Aert HA, Magusin PCMM, Sijbesma RP, Meijer EW. Supramolecular polymers from linear telechelic siloxanes with quadruple-hydrogen-bonded units. Macromolecules 1999;32:2696–2705.

Labeit S, Kolmerer B. Titins: giant proteins in charge of muscle ultrastructure and elasticity. Science 1995;270:293–296.

Law R, Carl P, Harper S, Dalhaimer P, Speicher DW, Discher DE. Cooperativity in forced unfolding of tandem spectrin repeats. Biophys J 2003;84:533–544.

Lee H, Dellatore SM, Miller WM, Messersmith PB. Mussel-inspired surface chemistry for multifunctional coatings. Science 2007;318:426–430.

Lee H, Lee BP, Messersmith PB. A reversible wet/dry adhesive inspired by mussels and geckos. Nature (Lond) 2007;448:338–341.

Li H, Linke WA, Oberhauser AF, Carrion-Vazquez M, Kerkvliet JG, Lu H, Marszalek PE, Fernandez JM. Reverse engineering of the giant muscle protein titin. Nature (Lond) 2002;418:998–1002.

Li H, Oberhauser AF, Fowler SB, Clarke J, Fernandez JM. Atomic force microscopy reveals the mechanical design of a modular protein. Proc Natl Acad Sci USA 2000;97:6527–6531.

Lin Q, Gourdon D, Sun C, Holten-Andersen N, Anderson TH, Waite JH, Israelachvili JN. Adhesion mechanisms of the mussel foot proteins mfp-1 and mfp-3. Proc Natl Acad Sci USA 2007;104:3782–3786.

Lu H, Schulten K. The key event in force-induced unfolding of titin's immunoglobulin domains. Biophys J 2000;79:51–65.

Lu K, Jacob J, Thiyagarajan P, Conticello VP, Lynn DG. Exploiting amyloid fibril lamination for nanotube self-assembly. J Am Chem Soc 2003;125:6391–6393.

Maier ME. Synthesis of medium-sized rings by the ring-closing metathesis reaction. Angew Chem Int Ed 2000;39:2073–2077.

Mark JE, Erman B, Eirlich FR. Science and technology of rubber. 2nd ed. San Diego (CA): Academic Press; 1994. p 751.

Marszalek PE, Lu H, Li H, Carrion-Vazquez M, Oberhauser AF, Schulten K, Fernandez JM. Mechanical unfolding intermediates in titin modules. Nature (Lond) 1999;402:100–103.

Marszalek PE, Li H, Oberhauser AF, Fernandez JM. Chair–boat transitions in single polysaccharide molecules observed with force-ramp atomic force microscopy. Proc Natl Acad Sci USA 2002;99:4278–4283.

Maruyama K. Connectin/titin, giant elastic protein of muscle. FASEB J 1997;11:341–345.

McGrath K, Kaplan D, editors. Protein-based materials. Cambridge, MA: Birkhauser; 1997.

McMillan RA, Conticello VP. Synthesis and characterization of elastin-mimetic protein gels derived from a well-defined polypeptide precursor. Macromolecules 2000;33:4809–4821.

McNally EM, Lapidos KA, Wheeler MT. Skeletal muscle structure and function. Principles of molecular medicine. 2nd ed. Totowa: Humana Press Inc.; 2006. pp. 674–681.

Mohr B, Weck M, Sauvage J-P, Grubbs RH. High-yield synthesis of [2]catenanes by intramolecular ring-closing metathesis. Angew Chem Int Ed Engl 1997;36:1308–1310.

Moore JS. Shape-persistent molecular architectures of nanoscale dimension. Acc Chem Res 1997;30:402–413.

Mueller M, Seidel U, Stadler R. Influence of hydrogen bonding on the viscoelastic properties of thermoreversible networks: analysis of the local complex dynamics. Polymer 1995;36:3143–3150.

Newkome GR, He E, Moorefield CN. Suprasupermolecules with novel properties: metallodendrimers. Chem Rev 1999;99:1689–1746.

Nowick JS. Chemical models of protein β-sheets. Acc Chem Res 1999;32:287–296.

Oberhauser AF, Hansma PK, Carrion-Vazquez M, Fernandez JM. Stepwise unfolding of titin under force-clamp atomic force microscopy. Proc Natl Acad Sci USA 2001;98:468–472.

Oesterhelt F, Oesterhelt D, Pfeiffer M, Engel A, Gaub HE, Muller DJ. Unfolding pathways of individual bacteriorhodopsins. Science 2000;288:143–146.

Oh K, Jeong K-S, Moore JS. Folding-driven synthesis of oligomers. Nature (Lond) 2001;414:889–893.

Orner BP, Ernst JT, Hamilton AD. Toward proteomimetics: terphenyl derivatives as structural and functional mimics of extended regions of an α-helix. J Am Chem Soc 2001;123:5382–5383.

Oroudjev E, Soares J, Arcdiacono S, Thompson JB, Fossey SA, Hansma HG. Segmented nanofibers of spider dragline silk: atomic force microscopy and single-molecule force spectroscopy. Proc Natl Acad Sci USA 2002;99:6460–6465.

Ortiz C, Hadziioannou G. Entropic elasticity of single polymer chains of poly(methacrylic acid) measured by atomic force microscopy. Macromolecules 1999;32:780–787.

Park T, Zimmerman SC, Nakashima S. A highly stable quadruply hydrogen-bonded heterocomplex useful for supramolecular polymer blends. J Am Chem Soc 2005;127:6520–6521.

Percec V, Ahn CH, Ungar G, Yeardley DJP, Moller M, Sheiko SS. Controlling polymer shape through the self-assembly of dendritic side-groups. Nature (Lond) 1998;391:161–164.

Phillips ST, Rezac M, Abel U, Kossenjans M, Bartlett PA. "@-Tides": the 1,2-dihydro-3(6H)-pyridinone unit as a β-strand mimic. J Am Chem Soc 2002;124:58–66.

Pinchuk L. A review of the biostability and carcinogenicity of polyurethanes in medicine and the new generation of "biostable" polyurethanes. J Biomater Sci Polym Ed 1994;6: 225–267.

Qu Y, Payne SC, Apkarian RP, Conticello VP. Self-assembly of a polypeptide multi-block copolymer modeled on dragline silk proteins. J Am Chem Soc 2000;122:5014–5015.

Rathore O, Sogah DY. Self-assembly of β-sheets into nanostructures by poly(alanine) segments incorporated in multiblock copolymers inspired by spider silk. J Am Chem Soc 2001;123:5231–5239.

Regan L, DeGrado WF. Characterization of a helical protein designed from first principles. Science 1988;241:976–978.

Richardson JS, Richardson DC. The de novo design of protein structures. Trends Biochem Sci 1989;14:304–309.

Richardson JS, Richardson DC, Tweedy NB, Gernert KM, Quinn TP, Hecht MH, Erickson BW, Yan Y, McClain RD, Donlan ME. Looking at proteins: representations, folding, packing, and design. Biophys J 1992;63:1185–1209.

Rief M, Gautel M, Oesterhelt F, Fernandez JM, Gaub HE. Reversible unfolding of individual titin immunoglobulin domains by AFM. Science 1997;276:1109–1112.

Rief M, Oesterhelt F, Heymann B, Gaub HE. Single molecule force spectroscopy on polysaccharides by atomic force microscopy. Science 1997;275:1295–1297.

Roland JT, Guan Z. Synthesis and single-molecule studies of a well-defined biomimetic modular multidomain polymer using a peptidomimetic β-sheet module. J Am Chem Soc 2004;126:14328–14329.

Roland JT, Ma SX, Nguyen M, Guan Z. Synthesis of titin-mimicking polymers having modular structures by using noncovalent interactions. Polym Prepr 2003;44:726–727.

Schuster M, Blechert S. Olefin metathesis in organic chemistry. Angew Chem Int Ed Engl 1997;36:2037–2056.

Seebach D, Beck AK, Brenner M, Gaul C, Heckel A. From synthetic methods to γ-peptides— from chemistry to biology. Chimia 2001;55:831–838.

Seebach D, Matthews JL. β-Peptides: a surprise at every turn. Chem Commun 1997;2015–2022.

Sijbesma RP, Beijer FH, Brunsveld L, Folmer BJB, Hirschberg JHKK, Lange RFM, Lowe JKL, Meijer EW. Reversible polymers formed from self-complementary monomers using quadruple hydrogen bonding. Science 1997;278:1601–1604.

Smith BL, Schaffer TE, Viani M, Thompson JB, Frederick NA, Kind J, Belcher A, Stucky GD, Morse DE, Hansma PK. Molecular mechanistic origin of the toughness of natural adhesives, fibers and composites. Nature (Lond) 1999;399:761–763.

Stadler R, Burgert J. Influence of hydrogen bonding on the properties of elastomers and elastomeric blends. Makromol Chem 1986;187:1681–1690.

Stigers KD, Soth MJ, Nowick JS. Designed molecules that fold to mimic protein secondary structures. Curr Opin Chem Biol 1999;3:714–723.

Stupp SI, LeBonheur V, Walker K, Li LS, Huggins KE, Keser M, Amstutz A. Supramolecular materials: self-organized nanostructures. Science 1997;276:384–389.

Thompson JB, Kindt JH, Drake B, Hansma HG, Morse DE, Hansma PK. Bone indentation recovery time correlates with bond reforming time. Nature (Lond) 2001;414:773–776.

Tomalia DA. Starburst dendrimers—nanoscopic supermolecules according to dendritic rules and principles. Macromol Symp 1996;101:243–255.

Treloar LRG. The physics of rubber elasticity. 3rd ed. Oxford, UK: Clarendon; 1975.

Tskhovrebova L, Trinick J, Sleep JA, Simmons RM. Elasticity and unfolding of single molecules of the giant muscle protein titin. Nature (Lond) 1997;387:308–312.

Urry DW, Luan CH, Harris CM, Parker TM. Protein-based materials with a profound range of properties and applications: the elastin DTt hydrophobic paradigm. Prot Based Mater 1997;133–177.

van Hest JCM, Tirrell DA. Protein-based materials, toward a new level of structural control. Chem Commun 2001;1897–1904.

Waite J, Andersen N, Jewhurst S, Sun C. Mussel adhesion: finding the tricks worth mimicking. J Adhes 2005;81:297–317.

Wang K. Titin/connectin and nebulin: giant protein rulers of muscle structure and function. Adv Biophys 1996;33:123–134.

Weck M, Mohr B, Sauvage J-P, Grubbs RH. Synthesis of catenane structures via ring-closing metathesis. J Org Chem 1999;64:5463–5471.

Westwood G, Horton TN, Wilker JJ. Simplified polymer mimics of cross-linking adhesive proteins. Macromolecules 2007;40:3960–3964.

Winningham MJ, Sogah DY. A modular approach to polymer architecture control via catenation of prefabricated biomolecular segments: polymers containing parallel β-sheets templated by a phenoxathiin-based reverse turn mimic. Macromolecules 1997;30:862–876.

Xu G, Wang W, Groves JT, Hecht MH. Self-assembled monolayers from a designed combinatorial library of de novo beta-sheet proteins. Proc Natl Acad Sci USA 2001;98:3652–3657.

Yamauchi K, Lizotte JR, Long TE. Thermoreversible poly(alkyl acrylates) consisting of self-complementary multiple hydrogen bonding. Macromolecules 2003;36:1083–1088.

Zeng H, Miller RS, Flowers II RA, Gong B. A highly stable, six-hydrogen-bonded molecular duplex. J Am Chem Soc 2000;122:2635–2644.

Zeng H, Yang X, Flowers II RA, Gong B. A noncovalent approach to antiparallel β-sheet formation. J Am Chem Soc 2002;124:2903–2910.

Zhao H, Robertson NB, Jewhurst SA, Waite JH. Probing the adhesive footprints of *Mytilus californianus byssus*. J Biol Chem 2006;281:11090–11096.

Zimmerman SC, Wendland MS, Rakow NA, Zharov I, Suslick KS. Synthetic hosts by monomolecular imprinting inside dendrimers. Nature (Lond) 2002;418:399–403.

Zou S, Schoenherr H, Vancso GJ. Force spectroscopy of quadruple H-bonded dimers by AFM: dynamic bond rupture and molecular time–temperature superposition. J Am Chem Soc 2005;127:11230–11231.

Zubarev ER, Pralle MU, Li L, Stupp SI. Conversion of supramolecular clusters to macromolecular objects. Science 1999;283:523–526.

CHAPTER 11

STRUCTURE AND SELF-ASSEMBLY OF AMPHIPHILIC DENDRIMERS IN WATER

HUI SHAO and JON R. PARQUETTE

11.1. INTRODUCTION

The spontaneous segregation of polar and apolar segments of an amphiphilic molecule in water is an extremely powerful driving force for the self-assembly and folding of biological systems (Lins and Brasseur 1995; Tsai et al. 2002). The self-assembly of lipids into micelles and bilayers represents a particularly important example of amphiphilic self-assembly that is mediated by the phase separation of polar and apolar regions in water. In this case, the self-assembly process sequesters the hydrophobic lipid chains together at the interior of the micelle/bilayer, away from both the polar head groups and the aqueous phase (Fig. 11.1).

Many biological events such as membrane translocation (Deshayes et al. 2006) and antibiotic activity (Yount et al. 2006) are mediated by peptides that attain their functional properties via a self-assembled state that is attained by amphiphilic phase segregation. Molecular amphiphilicity also plays a critical role in mediating the assembly of nonnatural amphiphiles. Many of these synthetic systems are capable of assembling into ordered structures that mimic the functional compartmentalization of biological systems, and they have emerged as a new class of supramolecular materials displaying biomedical functions (Fuhrhop and Wang 2004; Loewik and van Hest 2004; Jun et al. 2006; McNally et al. 2007). The structural impact of creating amphiphilicity in dendritic structures has been of intense interest for many years because their branched connectivity induces a globular morphology with distinct internal and external regions, apparently similar to the structures of micelles and globular proteins (Tomalia et al. 1990). Amphiphilic dendrimer systems are beginning to play important functional roles as artificial enzymes (Habicher et al.

Molecular Recognition and Polymers: Control of Polymer Structure and Self-Assembly.
Edited by V. Rotello and S. Thayumanavan
Copyright © 2008 John Wiley & Sons, Inc.

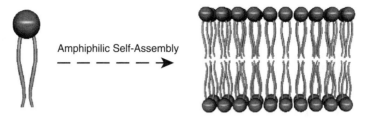

Amphiphilic Self-Assembly
— — — — — →

Figure 11.1 The amphiphilic self-assembly of lipids into bilayers.

1999; Hecht and Frechet 2001a; Kofoed and Reymond 2005; Darbre and Reymond 2006) and catalysts (O'Connor 1987; Lee et al. 1994; Piotti et al. 1999; Schmitzer et al. 1999; Liu et al. 2000; Hecht and Frechet 2001b; Kono 2002; Murugan et al. 2004; Ambade et al. 2005; Yin et al. 2007), drug delivery vehicles (Liu et al. 2000; Kono 2002; Boas and Heegaard 2004; Al-Jamal et al. 2005; Ambade et al. 2005; Gillies and Frechet 2005; Lee et al. 2005), mediators of DNA transfection (Choi and Baker 2005; Dufes et al. 2005; Luo et al. 2006; Pietersz et al. 2006; Gajbhiye et al. 2007; Paleos et al. 2007), and building blocks for biomaterials for tissue repair (Grinstaff 2002). Many of these areas have been the subject of detailed reviews in the past and will not be addressed in detail here. In this chapter, we specifically focus on the impact of amphiphilicity on the structure and self-assembly of dendrimers, dendrons, and their conjugates in water.

11.2. STRUCTURE

11.2.1. Unimolecular Micelle Analogy

Arborol Dendrimers. The spherical topology of dendrimers resembles the size and shape of the Hartley model (1936) of a micellar aggregate formed by surfactant molecules (Fig. 11.2; Tomalia et al. 1990). Micellar structures have a dynamic, spherical structure possessing a close-packed, solvent-incompatible core surrounded by an open, solvent-compatible layer.

This structural model segregates the polar and nonpolar regions by sequestering the apolar regions within a spherical aggregate displaying polar groups on the surface to interact with solvent. This resemblance was noted in some of the early reports on dendrimer chemistry in the mid-1980s. For example, Newkome's group noted that a hydrophobic dendrimer decorated with polar terminal groups could be considered as a covalently stabilized, nonequilibrating "unimolecular micelle" (Kim and Webster 1990) having a critical micelle concentration (CMC) of zero (Newkome et al. 1985). In 1986 Newkome and coworkers reported an "arborol" dendrimer based on tris(hydroxymethyl)aminomethane displaying polar hydroxyl groups at the periphery (Fig. 11.3a; Newkome et al. 1986).

Spherical aggregates ($\sim$200-Å diameter) of these dendrimers were formed in aqueous media, and the dendrimers exhibited a nonzero CMC of 2.02 mM. This

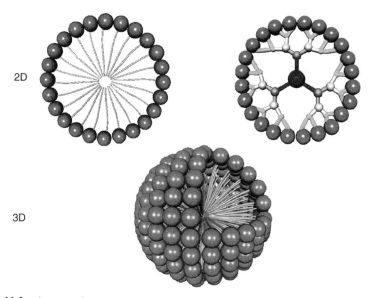

2D

3D

Figure 11.2 A comparison of the Hartley micellar model with the structure of an amphiphilic dendrimer.

behavior was likely a consequence of the small size of the dendrimer, which precluded the formation of a truly spherical structure capable of intramolecularly segregating polar and apolar regions. In 1991 Newkome's group observed that an all-hydrocarbon dendrimer containing 36 tetramethylammonium carboxylate termini

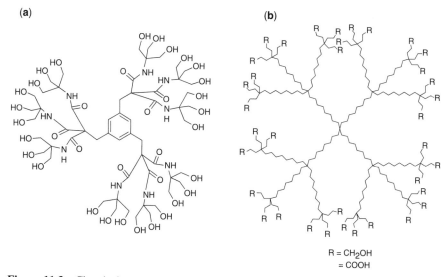

(a)

(b)

R = CH$_2$OH
= COOH

Figure 11.3 Chemical structures of the (a) arborol dendrimer and (b) water-soluble, all-hydrocarbon dendrimer.

formed uniformly monomeric particles (30 ± 10 Å; Newkome et al. 1991). Indeed, their fluorescence studies indicated that these dendrimers maintained a lipophilic internal environment capable of binding hydrophobic probes with an apparent CMC of $<3.9 \times 10^{-7}$ M (Fig. 11.3b; Newkome, Moorefield, Baker, Saunders, et al. 1991).

Poly(amido amine) (PAMAM) Dendrimers. These dendrimers have a hydrophilic polyamide interior that is relatively open to solvent, in contrast to a micellar assembly. Yet, the picture of a unimolecular micelle also emerged for a generation 4.5 PAMAM dendrimers having sodium carboxylate terminal groups (Fig. 11.4).

Individual molecules of this dendrimer could actually be observed by electron microscopy as spheroid particles having a diameter of 88 ± 10 Å, comparing favorably with a predicted diameter of 78 Å for a single, fully extended dendrimer structure (Tomalia et al. 1987). The similarity of the surface of these dendrimers to anionic

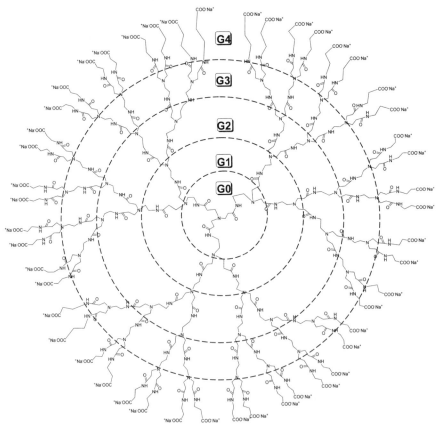

Figure 11.4   Generation 4.5 poly(amido amine) dendrimer having sodium carboxylate terminal groups.

micelles was elucidated by monitoring the Stern–Volmer quenching rates (K_{SV}) of *tris*(2,2′-bipyridyl)ruthenium(II) probes by methyl viologen in water (Gopidas et al. 1991). A plot of the K_{SV} versus dendrimer generation revealed an abrupt increase in K_{SV} going from generation 2.5 to 3.5, reminiscent of the CMC plot of a micelle. At lower generations, the quenching kinetics were "bimolecular" in nature, similar to that of small C-7/C-8 micelles. However, at generation 3.5 and higher, the kinetic behavior was an "intramicellar" type similar to larger micelles (C-9 and above). This behavior of the dendrimers was attributed to a cooperative association of the terminal groups that occurs concomitant with a change from a disklike conformation to a spherical topology. The conformational transition to a dense-packed sphere was confirmed by electron paramagnetic resonance studies on the corresponding copper complexes (Ottaviani et al. 1994, 1997). The copper complexes exhibited a decrease in mobility with generation, consistent with a change in dendrimer morphology.

11.2.2. Location of Terminal Groups

Theoretical Studies. A majority of experimental (Mansfield and Klushin 1992; Meltzer et al. 1992; Mourey et al. 1992; Wooley et al. 1997; Gorman et al. 1998) and theoretical studies (De Gennes and Hervet 1983; Lescanec and Muthukumar 1990; Mansfield and Klushin 1993; Mansfield 1994; Boris and Rubinstein 1996; Chen and Cui 1996; Murat and Grest 1996; Lue and Prausnitz 1997) indicate that the terminal groups of nonamphiphilic dendrimers are distributed throughout the molecular volume, resulting in a segmental density maximum near the core. In contrast, molecular simulation studies of dendrimers constructed with terminal groups that interact with solvent differently compared with the internal monomers have revealed a very different situation. These studies suggest that amphiphilic dendrimers will adopt phase-segregated conformations placing the structural segments that interact poorly with solvent near the core, regardless of whether the segment is located at the periphery or at the interior of the dendrimer. Connolly et al. (2004) used Metropolis Monte Carlo to vary the nature of the monomer–monomer interactions in amphiphilic dendrimers. Two types of codendrimers were considered: "outer H," which consisted of hydrophobic (H) groups positioned at the periphery of an otherwise polar (P) dendrimer; and an "inner H" system with the opposite monomer polarity. Generally, they found that the inner H system formed well-defined micellar structures at a lower generation, which projected the polar monomers at the surface (Fig. 11.5a, c). In contrast, the outer H dendrimers adopted a "loopy micelle" structure that backfolded the hydrophobic terminal groups toward the core to simultaneously sequester the H monomers from solvent at the core while presenting the internal P monomers at the periphery (Fig. 11.5b, d). At higher generations, both types of dendrimers formed elongated cigarlike cylindrical structures, which presumably formed to balance the developing steric congestion at the periphery with the need to segregate the P and H monomers (Fig. 11.5e, f).

Giupponi and Buzza (2005) explored how the magnitude of the monomer–solvent interactions impacted the dendrimer structure using a lattice-based

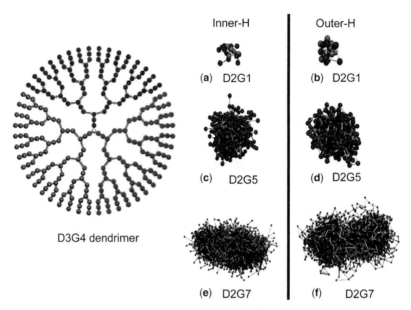

Figure 11.5   Two-dimensional representation of a D3G4 [D = spacers (3), G = dendrimer generation (4)] dendrimer. Typical snapshots of various generations of codendrimers, from Monte Carlo simulations, with the same number of spacers (D = 2) between each branch point. Inner H and outer H topologies are compared for three different degrees of branching (G = 1, 5, or 7). Reprinted from Connelly et al. (2004). Copyright 2004 American Chemical Society.

configurational-bias Monte Carlo method. They found that at later generations when the internal monomers were in a poor solvent and the terminal monomers were in an athermal solvent, a spherical micellar structure was adopted. Under opposite solvation conditions (i.e., poor solvent for termini, athermal solvent for internal monomers), a spherically symmetric loopy micelle structure emerged whereby the end groups folded back to minimize the unfavorable interactions with solvent (Fig. 11.6).

Lower generation dendrimers exhibited highly asymmetric conformations unless both the internal and terminal groups interact unfavorably with the solvent, under which conditions a compact globule conformation surrounded by a corona of termini was formed. More recently, Suek and Lamm (2006) studied dendrimers composed of a solvophobic interior surrounded by terminal monomers ranging in polarity from all solvophobic to all solvophilic. This study also found a tendency for the solvophilic groups to congregate at the surface. However, at high generations, steric crowding at the periphery forced some solvophilic termini to backfold into the dendrimer interior, consistent with the Connolly study. Increasing the solvophilic monomer–monomer interaction by lowering the simulation temperature reduces the tendency of the termini to backfold.

(a) 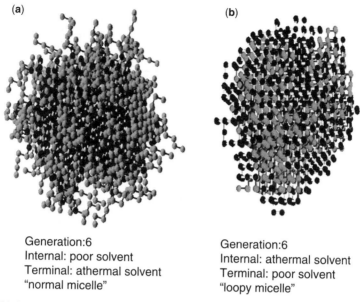 (b)

Generation:6
Internal: poor solvent
Terminal: athermal solvent
"normal micelle"

Generation:6
Internal: athermal solvent
Terminal: poor solvent
"loopy micelle"

Figure 11.6 Snapshots of calculated structures of dendrimers. For clarity, internal and terminal monomers are shaded darker and lighter, respectively, and the size of the core monomer has been made larger than the other monomers. Reprinted from Giupponi and Buzza (2005). Copyright 2005 American Institute of Physics.

Experimental Systems. Experimentally, amphiphilic Fréchet-type poly(aryl ether) dendrimers displaying poly(ethylene glycol) (PEG) peripheral arms were shown to form monomolecular micellar structures exhibiting solvent-sensitive conformational changes (Fig. 11.7; Gitsov and Frechet 1996).

Similarly, a poly(propylene imine) (PPI) dendrimer fitted with triethyleneoxy methyl ether and octyl groups at every terminal position was soluble in both organic and aqueous solvents, indicating sufficient structural flexibility to present either the hydrophilic or hydrophobic termini toward the solvent (Fig. 11.8; Pan and Ford 1999, 2000).

Thayumanavan's group reported dendrimers containing polar and apolar groups at every repeat unit that inverts from a normal to inverted micellar-type conformation going from aqueous to organic media (Fig. 11.9; Basu et al. 2004; Aathimanikandan et al. 2005; Klaikherd et al. 2006). This solvent-sensitive conformational behavior affords a unique capability to solubilize hydrophobic molecules in water and hydrophillic molecules in lipophilic media.

11.2.3. Polyelectrolyte Dendrimers

The addition of charge to a dendrimer structure could be expected to cause a structural expansion that minimizes the ensuing coulombic repulsions, similar to the

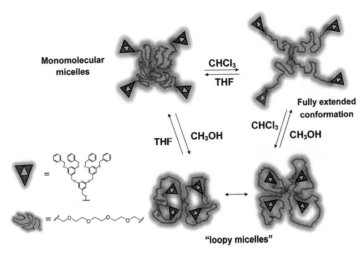

Figure 11.7 Solvent-sensitive conformational changes of Fréchet-type poly(aryl ether) dendrimers with poly(ethylene glycol) termini.

behavior of polyelectrolyte star polymers (Jusufi et al. 2002a, 2002b). Indeed, the hydrodynamic radii of carboxylic acid terminated arborol dendrimers, determined by ^{1}H diffusion ordered spectroscopy NMR, expand as the acids are ionized with increasing pH (Newkome, Young, et al. 1993; Young et al. 1994). Similarly, amine-terminated dendrimers expand as the terminal amines are protonated with decreasing pH (Young et al. 1994). The pH-dependent behavior of these dendrimers has been attributed to the development of coulombic repulsions between charged

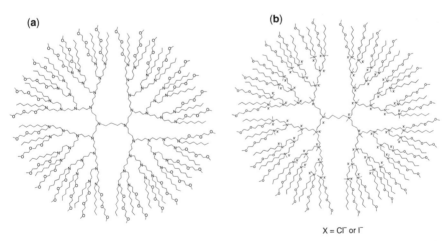

Figure 11.8 Structures of (a) parent poly(propylene imine) dendrimer and (b) fully quaternized form, which presents both hydrophilic triethyleneoxy methyl ether (TEO) and hydrophobic octyl chains at every terminal position.

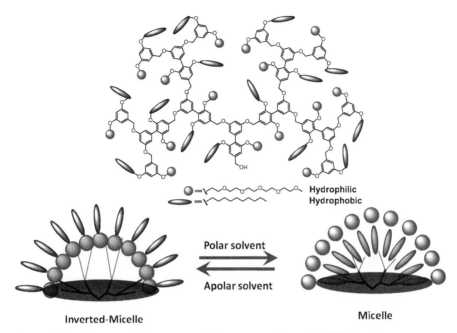

Figure 11.9   (Top) A typical structure of Thayumanavan's amphiphilic dendrimers. (Bottom) Schematic representation of micelle-type and inverse micelle-type structural organization.

terminal groups that were generated at low or high pH for the amine- or acid-terminated dendrimers, respectively (Fig. 11.10).

In contrast, neutral hydroxy-terminated dendrimers generally exhibit hydrodynamic volumes that are invariant with pH (Young et al. 1994). The structure of carboxyl-terminated arborol dendrimers has been extensively characterized by

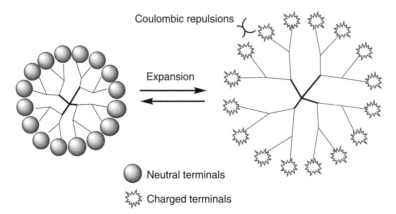

Figure 11.10   Schematic representations of the structural impact of charge repulsion on dendrimer size.

dynamic light scattering (Zhang et al. 1997; Miura et al. 1999; Zhang et al. 1999), capillary electrophoresis (Seyrek et al. 2004), and potentiometric titration (Zhang et al. 1997). These studies indicate that the dendrimers adopt a spherical structure having a uniform charge density that is modulated by both pH and salt conditions. Small-angle neutron scattering (SANS) studies confirm the localization of counterions at the surface of these dendrimers at high pH (Huang et al. 2005).

The conformational behavior of polyelectrolyte dendrimers having charges distributed throughout the volume of the dendrimers has been somewhat controversial. Welch and Muthukumar (1998) predicted a reversible conformational transition from a compact, dense core to a hollow, dense shell structure (up to a 180% increase in volume) as the ionic strength of the medium was changed from high to low. Their predictions were based on a series of Monte Carlo simulations on dendritic polyelectrolytes, using a mean-field approach to *implicitly* treat water and counterion molecules. The structural variation with the ionic strength was attributed to the ability of ions to screen the coulombic interactions between charged groups, thereby allowing for the usual dense core structure of neutral dendrimers. At low ionic strength, higher effective charge produces a structural expansion that minimizes charge repulsion (Fig. 11.11).

Similar conclusions were reported by Lyulin et al. (2004) using Brownian dynamics simulations, which treated the solvent as an effective viscous continuum. In contrast to these predictions, Nisato et al. (2000) observed experimentally using SANS that the size of a generation 8 PAMAM dendrimer was invariant with the pH and ionic strength. The discordance between experiment and theory arose from the implicit treatment of water and ions by a mean-field approach. Several molecular dynamics (MD) attempts to *explicitly* treat the water and counterions have been reported that qualitatively predict a small conformational change with pH (Lee, Athey, et al. 2002; Terao and Nakayama 2004; Maiti et al. 2005; Maiti and Goddard 2006; Opitz and Wagner 2006; Terao 2006; Giupponi et al. 2007; Lin et al. 2007) Moreover, the explicit treatment of water molecules in these MD

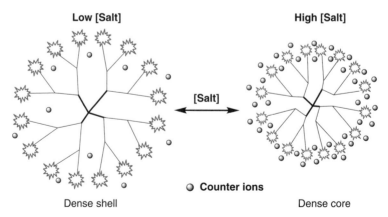

Figure 11.11 Schematic representation of the structural impact of counterion screening on the size of the dendrimer.

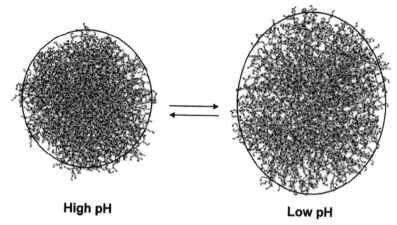

High pH   **Low pH**

Figure 11.12   Snapshots of the calculated structure of a poly(amido amine) dendrimer at high and low pH. The size was predicted to increase by 13% upon progression from high to low pH.

studies revealed that the dendrimer swelling is more accurately attributed to the incorporation of water molecules within the dendrimer interior along with a localization of counterions at the periphery, rather than merely a pH effect (Fig. 11.12; Maiti and Goddard 2006).

A more recent SANS study revealed a 4% increase in dendrimer volume with decreasing pH (Chen et al. 2007). Although these structural variations were significantly less than predicted by the MD studies, this study confirmed the occurrence of both counterion condensation and water penetration of the dendrimer interior.

11.3. SELF-ASSEMBLY AND AGGREGATION

11.3.1. Dendrimers and Dendrons

The phase segregation of polar and nonpolar segments of amphiphilic dendrimers that produces structural features resembling unimolecular micelles actually creates a strong propensity for self-association. High-generation dendrimers with high spherical symmetry achieve phase separation via an intramolecular folding process. However, when intramolecular segregation is structurally impeded, intermolecular aggregation produces assemblies consisting of a core of insoluble segments surrounded by a solvated shell of solvophilic segments. The tendency for intermolecular phase segregation is most pronounced for lower generation dendrimers exhibiting highly dynamic conformations ranging from globular to disklike (Fig. 11.13).

For example, PPI dendrimers functionalized with azobenzene chromophores appended with aliphatic side chains assemble into large spherical vesicles in water below pH 8 (Fig. 11.14; Tsuda et al. 2000).

Cryo-transmission electron spectroscopy (TEM), scanning electron spectroscopy, and confocal laser scanning microscopy studies indicated the presence of large,

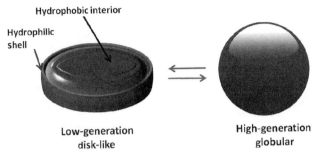

Figure 11.13 Schematic representation of dynamic conformational change from disklike to spherical morphologies in dendrimers.

spherical vesicles with an onionlike architecture containing water molecules. Small angle X-ray measurements confirmed the 4.9-nm thickness of the bilayer structures. In order to form the bilayer assembly, the individual dendrimers adopt a flattened, disklike structure consisting of a core of protonated nitrogens with the aliphatic chains protruding from one face. The flattened dendrimers assemble further into multilaminar vesicles that arrange the disks with the polar faces packed together and separated by an aqueous phase. The aliphatic end groups interdigitate with the end groups of an adjacent bilayer and exhibit restricted mobility due to $\pi-\pi$ stacking of the azobenzene chromophores. The combined effects of hydrogen bonding

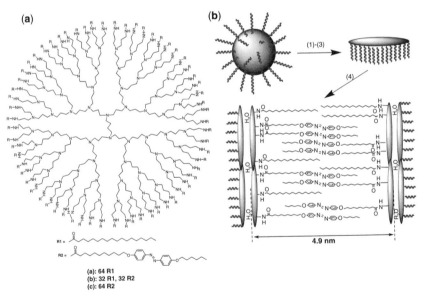

Figure 11.14 (a) Chemical structures of poly(propylene imine) dendrimers. (b) Illustration of the formation of the bilayer structure: (1) injection of the dendrimer in an aqueous solution of pH < 8; (2) protonation of the nitrogen atoms; (3) structural inversion of the dendrimer; (4) self-aggregation of the protonated dendrimers to form multilaminar vesicles containing interdigitated bilayers.

between adjacent protonated dendrimer cores and stacking of the azobenzene end groups stabilize the bilayer assemblies. At the air–water interface, these systems also adopt flattened disklike shapes and assemble into monolayers with the polar dendrimer core oriented toward the aqueous phase and the alkyl chains pointing toward the air (Fig. 11.15; Schenning et al. 1998).

Low generations of hydrophobically modified PAMAM dendrimers also exhibit a tendency to deform at air–water interfaces in order to present the hydrophilic dendrimer interior toward water and the aliphatic chains toward the air (Sayed-Sweet et al. 1997). Similarly, the tendency of poly(aryl ether) dendrimers with peripheral carboxylate end groups to aggregate in aqueous media decreased with increasing generation (Fig. 11.16; Laufersweiler et al. 2001).

The generational dependence of the aggregation state emerged from the greater conformational flexibility of lower generation dendrimers. At low dendrimer generation, the accessibility of a disklike morphology having a large exposed hydrophobic surface produces aggregates to minimize the surface area exposed to the aqueous phase. As the generational level increases, the conformational equilibria shift from a disklike structure toward a spherical morphology that intramolecularly segregates the hydrophobic interior segments from the aqueous phase, thereby circumventing the need to aggregate.

Similar generation-dependent aggregation behavior was reported for a series of dendrons having a phosphate focal group and functionalized at the periphery with azobenzene chromophores in water (Zhang et al. 2007). In this system, the stability and nature of the aggregate changed as the generation was increased. The G0 dendron forms stable H-type aggregates featuring a head–head alignment of the azobenzene chromophores. However, at higher generations, the stability of the H aggregate decreases and a "looser" J-type aggregate (head–tail alignment) emerges, resulting in a mixed aggregation state (Fig. 11.17).

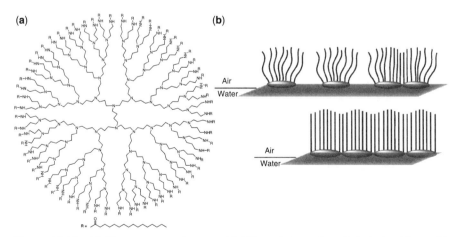

Figure 11.15 (a) The structure of an amphiphilic poly(propylene imine) dendrimer. (b) Schematic representation of the monolayer organization of amphiphilic dendrimers on the water surface.

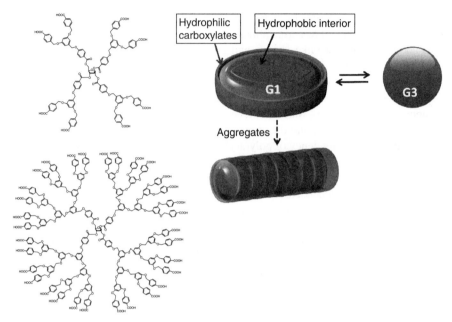

Figure 11.16 (Left) Chemical structures of G1 and G3 poly(aryl ether) dendrimers. (Right) Notional depiction of generation-dependent dynamic conformational equilibrium of amphiphilic dendrimers in water interconverting between disklike and spherical morphologies.

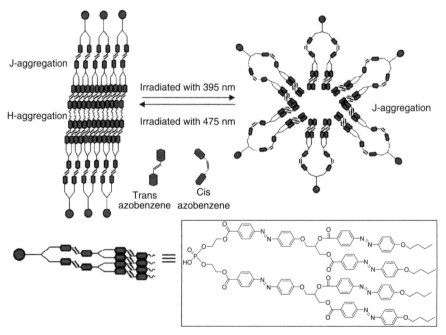

Figure 11.17 Aggregation behavior and photoresponsive properties of an amphiphilic azobenzene dendrimer.

The decreased stability of the H aggregate that occurs at higher generation is attributable to the attendant increase in peripheral congestion, which impedes the ability of the azobenzene chromophores to attain a proper orientation for efficient H-type packing. The $E \rightarrow Z$ photoisomerization disrupts the H aggregate and drives the assembly toward the looser J aggregate. The vesicles formed by the G1 dendron were capable of binding calcein in the E form and releasing calcein upon photoisomerization to the Z form. A transition toward a globular morphology at higher generations has been reported to lower the crystallinity of docsyl-terminated polyether dendrons in the melt (Cho et al. 2004). The reduction in the melting-transition temperature is also a consequence of a change from a flat shape to a globular structure with the generation, which increases the interfacial spacing between the lamellar hydrocarbon region and the amorphous dendritic core.

11.3.2. Linear–Dendritic Multiblock Copolymers

The addition of large linear blocks to dendrons with opposite polarity creates a desymmetrized structure predisposed to sequester insoluble components by aggregation rather than intramolecular hydrogen-bonding. Amphiphilic, linear–dendritic diblock (AB) and triblock (ABA) copolymers self-assemble into multimolecular micelles with CMC values that are well below those of low molecular weight surfactants. Typically, a hydrophilic linear block such as PEG is attached to the focal point

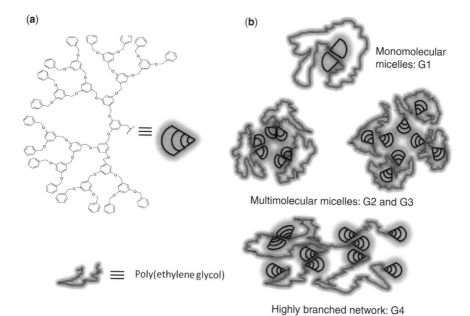

(a)

(b)

Monomolecular micelles: G1

Multimolecular micelles: G2 and G3

Highly branched network: G4

≡ Poly(ethylene glycol)

Figure 11.18 Dendron chain copolymers. (a) A polyether dendron displaying a focal PEG chain and (b) notional representation of the self-assembly of the dendron chain copolymers.

of one or two hydrophobic dendron blocks. For example, Gitsov and colleagues were the first to construct a series of AB and ABA block copolymers from apolar polyether dendrons (A) featuring PEG chains (B) at the focal point (Gitsov et al. 1992; Gitsov and Frechet 1993). They found that the AB hybrids were only soluble in water–methanol when the hydrophilic component was sufficiently larger than the dendritic component. Generally, the ABA copolymers were soluble when the dendritic block constituted approximately 30 wt % of the hybrid (Fig. 11.18).

Above certain concentrations, the hybrids assembled into multimolecule micellar aggregates, in which the dendritic block was protected from the polar media. Similar micellar behavior was reported for carbosilane dendrons with a focal PEG chain (Chang et al. 2000). Reversed polarity copolymers having a hydrophobic linear block such as polystyrene (PS) appended to a hydrophilic PPI dendron also assemble into multimolecular micelles in water and reversed micelles in CDCl$_3$ with low CMC values (Fig. 11.19a; van Hest, Baars, et al. 1995). The tendency to form aggregates in which the soluble block wraps the insoluble block was observed by NMR in triblock PEG and triazine dendron copolymers (Namazi and Adeli 2005). Similarly, Chapman et al. (1994) observed a CMC value of 8×10^{-5} M for a generation 4 t-butyloxycarbonyl (Boc)-protected lysine dendron having a focal PEG chain (number-average molecular weight = 4717; Fig. 11.19b).

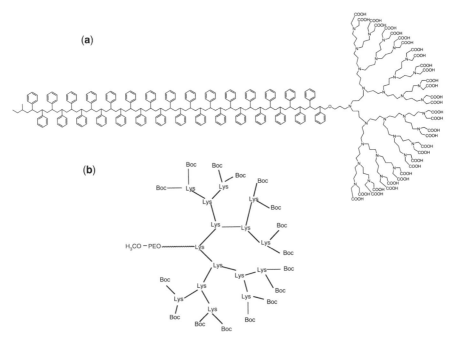

Figure 11.19 (a) Structure of a hydrophilic fifth-generation poly(propylene imine) dendron containing a focal hydrophobic polystyrene block. (b) Structure of a fourth-generation t-butyloxycarbonyl (Boc)-protected polylysine dendron having a focal poly(ethylene glycol) chain.

11.3.3. Hydrophilic–Hydrophobic Balance

The hydrophilic–hydrophobic size balance largely determines the size, shape, and physical properties of the aggregate. For example, whereas natural phospholipids have extremely low CMC values ($<10^{-10}$ M), synthetic PEG-phospholipids exhibit much higher CMCs because of the larger hydrophilic block. Tsuchida's group (Takeoka et al. 2000) explored the impact of the hydrophilic–hydrophobic balance by varying the dendron generation of related lipid-terminated lysine dendrons with a focal PEG-OMe chain. The CMCs of the resulting conjugates generally decreased as the number of terminal acyl groups increased at higher generations, but they were still in the micromolar range (Takeoka et al. 2000). Recently, Nguyen and Hammond (2006) observed that triblock copolymers of a poly(ethylene oxide) (PEO) linear block terminated on both ends by PAMAM dendrons form micellar aggregates, capable of encapsulating the antibiotic triclosan, with CMC values ranging from 10^{-5} to 10^{-6} M (Fig. 11.20).

Amphiphilic peptide–dendritic diblock hybrids, composed of a dodecyl polyglutamic acid hydrophobic linear segment appended to a hydrophilic PEO-terminated polyester dendron, were recently reported by this group to form micellar aggregates with remarkably low CMC values in the range of 10^{-8} M (Tian and Hammond 2006). The efficient self-assembly of these diblock systems was attributed to the more efficient packing of a conelike morphology of the amphiphile, which emerges from the globular shape of the dendron and the helical conformation of the polyglutamate segment (Fig. 11.21).

11.3.4. Steric Impact of Dendritic Block

Israelachvilli (1991) predicted that small, high curvature aggregates are preferentially formed by cone-shaped amphiphiles with polar, sterically demanding head groups and relatively small hydrophobic blocks. Several studies reported variations in aggregate morphology with dendron generation that generally confirm these predictions. For example, Stupp's group constructed a series of surfactants composed of a poly(L-lactic acid) chain functionalized on one end with an ammonium-terminated lysine dendron and on the other end with cholesterol (Fig. 11.22; Klok et al. 2002).

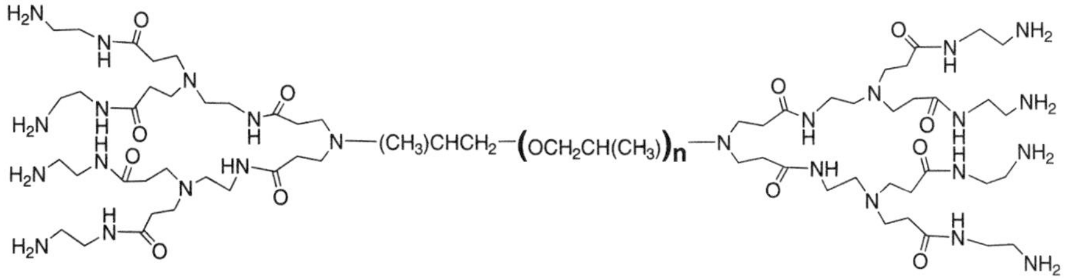

Figure 11.20 Structure of poly(amido amine) (PAMAM) (G2)-poly(propylene oxide)-PAMAM (G2) triblock copolymers.

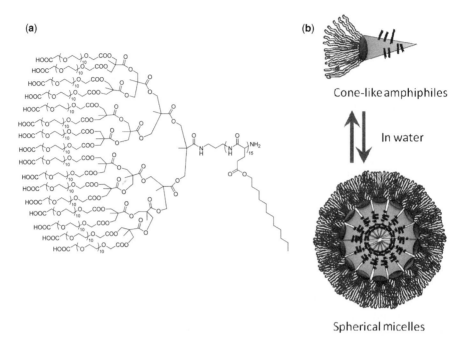

Figure 11.21 (a) Structure and (b) self-assembly of conelike amphiphilic glutamic acid–dendritic diblock hybrids.

These "rodcoil" dendrons assemble in the solid state, under hydrating conditions, into aggregates composed of a lamellar bilayer of cholesterol molecules surrounded by a shell of L-lysine dendrons. The L-lysine dendrons become more sterically demanding as the level of hydration is increased, which destabilizes the lamellar organization. This increased steric bulk is not sufficiently destabilizing at the first and second generations; however, the third-generation dendrons preclude formation of the bilayer in favor of nanosized aggregates. In a series of PS–PPI copolymers, Meijer found that the aggregate morphology similarly followed the theory of Israelachvilli and coworkers (1977, 1980) as the dendritic head group increased in size (van Hest, Baars, et al. 1995; van Hest, Delnoye, et al. 1995; van hest, Delnoye, et al. 1996). As the head group increased in size going from PS-dendr-(NH$_2$)$_8$ and PS-dendr-(NH$_2$)$_{16}$ to PS-dendr-(NH$_2$)$_{32}$, the aggregate morphology transitioned from vesicular structures and micellar rods to spherical micelles (Fig. 11.23).

Based on Israelachvilli's predictions, Kellermann et al. (2004) designed a cone-shaped dendro-calixarene molecule displaying two polar carboxylate-terminated G2 dendrons on one face and four C$_{12}$ chains extending from the opposite face (Fig. 11.24).

Remarkably, a highly uniform and highly stable micellar aggregate of exactly seven molecules was formed in water. Although these aggregates resisted

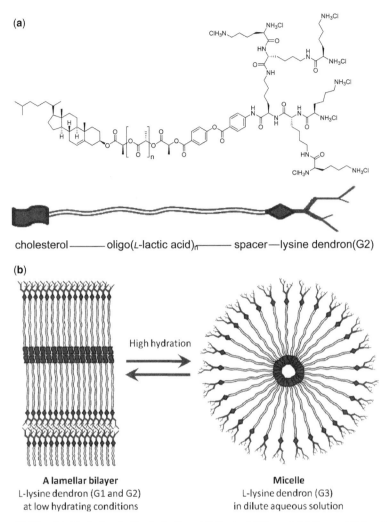

Figure 11.22   (a) Cholesteryl-oligo(L-lactic acid)-(lysine) G2 amphiphiles and (b) variations in aggregate morphology with dendron generation.

deformation even upon drying, they were capable of reversibly encapsulating porphyrin. More recently, dendritic–helical diblock copolypeptides composed of a hydrophobic poly(γ-benzyl-L-glutamate) block terminated with a hydrophilic poly(L-lysine) dendron were similarly observed to aggregate into micelles whose morphologies changed with the generation following Israelachvilli's predictions (Fig. 11.25; Kim et al. 2006).

Accordingly, the morphology of the micellar assemblies transitioned from large objects with low curvature at a low dendron generation to smaller vesicular/spherical structures with high curvature at higher generations.

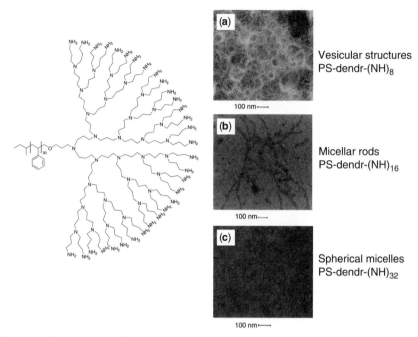

Figure 11.23 Structure of polystyrene (PS)-dendr-$(NH)_{32}$ and aggregate morphology as a function of dendron generation. Reprinted from van Hest et al. (1995). Reprinted with permission from the American Association for the Advancement of Science.

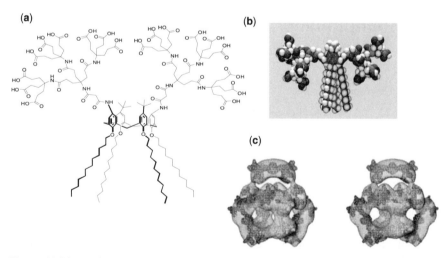

Figure 11.24 (a) Structure and (b) space-filling model of amphiphilic dendro-calixarene. (c) The arrangement of seven calixarene head groups on a spherical shell. (d) Stereo view of variant (c). Reprinted from Kellermann et al. (2004). Copyright 2004 Wiley–VCH Verlag GmbH & Co. KGaA.

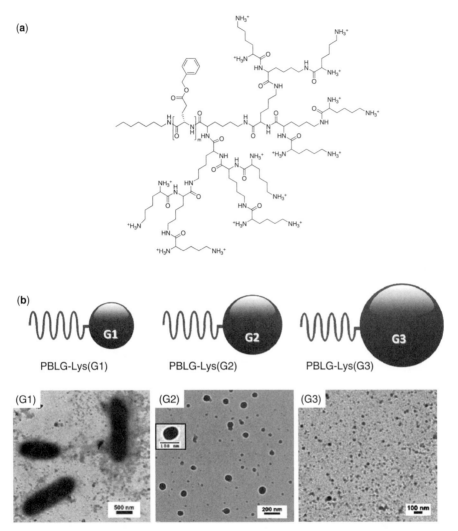

Figure 11.25   (a) Structure of poly(γ-benzyl-L-glutamate) (PBLG)-Lys (G3). (b) A graphic representation of amphiphilic copolypeptides PBLG-Lys (G1)–(G3) and TEM images of corresponding aggregates in water. Reprinted from Kim et al. (2006). Copyright 2006 The Royal Society of Chemistry.

11.3.5. Rigid Linear Block Copolymers

Amphiphilic systems containing an extended rigid-rod block, rather than a more flexible linear chain, have also been shown to self-assemble into well-defined structures. Stupp's group explored a series of rod–dendron and dendron–rod–dendron hybrids containing two to three biphenyl ester moieties as a rigid rod that was capped by one or two 3,4,5-*tris*-alkoxy benzoate dendrons (Lecommandoux et al. 2003). The molecules assembled into cubic, columnar, or smectic phases, depending on the rod length and dendrimer generation (Fig. 11.26).

Jang et al. (2004) observed that ABA triblock copolymers composed of a docosyl chain, a rigid aromatic segment, and a flexible PEO dendrimer assemble into a hexagonal columnar or body-centered cubic structure in the solid state for the first- and second-generation dendrons, respectively (Fig. 11.27).

In aqueous media, they assemble into spherical capsules capable of encapsulating Nile Red with remarkably low CMC values (2×10^{-7} M for G1 and 3×10^{-8} M for G2). Similarly, the size of the aggregate derived from a related dendron–rod–dendron copolymer, composed of a phenylene vinylene rod capped on each end by amphiphilic polyether dendrons, could be tuned in the melt state by modifying the generation level of the dendron (Fig. 11.28; Lee et al. 2002).

Presumably, at higher generations the increased bulk of the dendrons sterically impedes aggregation, resulting in smaller aggregate structures.

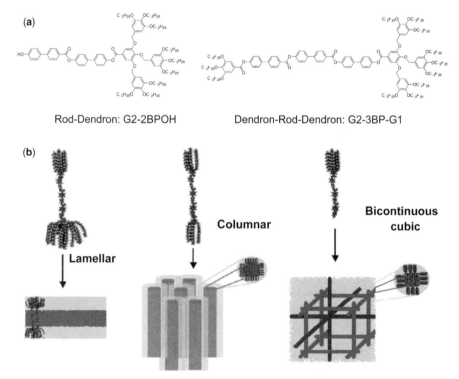

Figure 11.26 (a) Chemical structures of rod–dendron and dendron–rod–dendron hybrids. (b) Representations of the supramolecular organizations obtained from rod–dendron and dendron–rod–dendron molecules. Reprinted from Lecommandoux et al. (2003). Copyright 2003 Wiley-VCH Verlag GmbH & Co. KGaA.

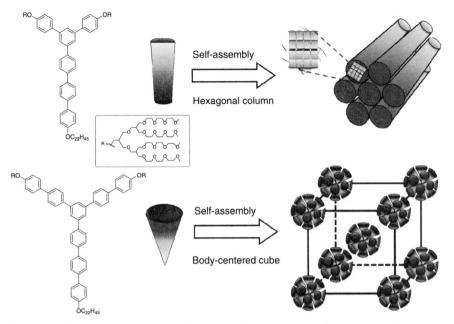

Figure 11.27 Schematic representation of the hexagonal columnar structure of an ABA triblock copolymer with a G1 dendron and the body-centered cubic structure of an ABA triblock with a G2 dendron.

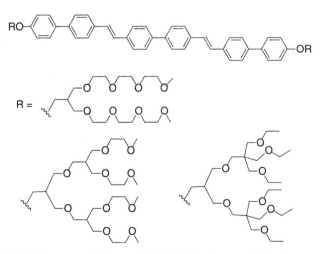

Figure 11.28 Structures of Lee et al.'s (2002) dendron–rod–dendron copolymers.

11.3.6. Supramolecular Gels

Supramolecular gels are solidlike materials composed of an interpenetrating network of self-assembled molecules and large quantities of solvent than can be up to 99% of the mass (Yang and Xu 2007). The tendency of amphiphilic dendrimers to phase segregate via self-assembly ideally suits them as precursors for supramolecular gels in aqueous and organic media (Hirst and Smith 2005; Smith 2006a, 2006b; Yang and Xu 2007). Furthermore, the biodegradability and reversibility of supramolecularly formed hydrogels makes these assemblies particularly attractive for a number of biomedical applications (Terech and Weiss 1997; Gronwald et al. 2002; Zhu et al. 2005; Jun et al. 2006; Yang and Xu 2007). The formation of a gel in water requires the proper hydrophobic/hydrophilic balance to ensure that neither precipitation nor dissolution of the monomers can occur. The formation of dendritic gels in organic solvents was reviewed by Smith (2006a, 2006b). Newkome and colleagues reported the first example of a series of two-directional arboral molecules that gelled aqueous and MeOH–H₂O mixtures at concentrations ranging from 2 to 10 wt% (Fig. 11.29; Newkome et al. 1990; Newkome, Lin, et al. 1993).

The monomers were composed of a hydrocarbon linkage that was functionalized on both ends by a hydrophilic, hydroxyl-terminated dendron. TEM revealed the presence of rodlike aggregates with diameters of 34–36 Å and variable lengths in the range of 2000 Å. Such rodlike assemblies are considered to be ideally suited for the immobilization of large quantities of solvent (Terech and Weiss 1997).

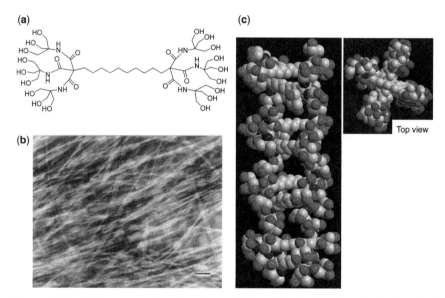

Figure 11.29   (a) Structure of Newkome's two-directional arboral molecule. (b) Negatively stained transmission electron micrograph of the arborol gel. (c) Schematic representation of the formation of a long fibrous rod structure via orthogonal stacking of the arboral molecules. Reprinted from Newkome et al. (1990). Copyright 1990 American Chemical Society.

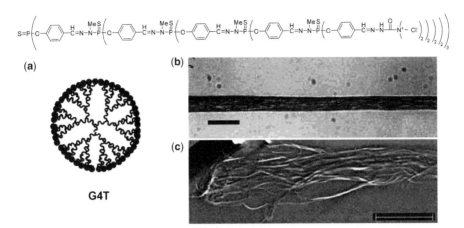

Figure 11.30 (a) Structure of a 4th generation alkylammonium-terminated organophosphorus dendrimer (G4T). (b) Optical micrograph of G4T (scale bar = 100 μm). (c) Scanning electron micrograph of one tip of a single macroscopic dendrimer fiber after dilution with 1 M NaOH to show the fibrillar substructure (scale bar = 100 μm). Reprinted from El Ghzaoui et al. (2004). Copyright 2004 American Chemical Society.

Similarly, Majoral's group reported a series of cationic organophosphorous dendrimers that formed noncovalent gels capable of immobilizing aqueous solutions at concentrations of 1.2–1.5 wt% after 11–13 days at 65 °C (Marmillon et al. 2001). Gelation time was significantly reduced when various buffer and metal salts were added. The impact of salt on the gelation process was attributed to a colloidal flocculation process whereby added salts screened repulsive electrostatic interactions, resulting in aggregation (El Ghzaoui et al. 2004). Freeze-fracture TEM and optical microscopy revealed dendrimer aggregates and fiberlike networks within the gels (Fig. 11.30).

11.4. FOLDED AMPHIPHILIC DENDRIMERS

The Parquette group developed a series of dendrimers that fold into a well-defined, helical structure capable of amplifying chiral perturbations (Recker et al. 2000; Gandhi et al. 2001; Huang and Parquette 2001; Parquette 2003; Lockman et al. 2005). The structural preorganization of the dendrimers arises from the preference of the branched pyridine-2,6-dicarboxamide repeat unit to exist predominantly in the syn–syn conformation rather than either the higher energy syn–anti or anti–anti forms (Fig. 11.31; Gabriel et al. 2006).

This conformational preference in conjunction with the preference of the terminal amides to exist in an s-trans conformation constrains the system such that the terminal anthranilamide substituents are positioned above and below the plane of the molecule, resulting in a local helical structure. The helical antipodes experience a

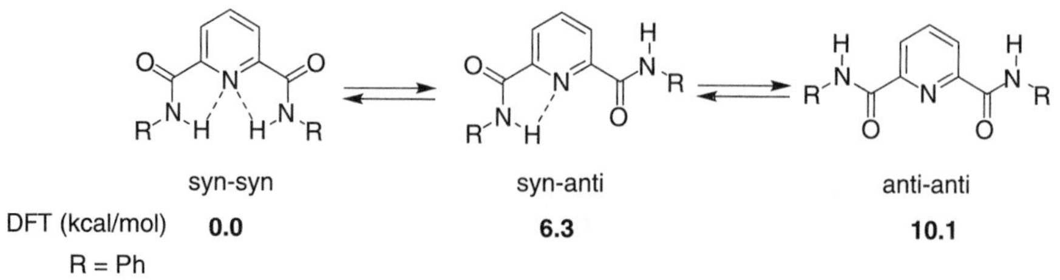

	syn-syn	syn-anti	anti-anti
DFT (kcal/mol)	0.0	6.3	10.1

R = Ph

Figure 11.31   Conformational equilibria of pyridine-2,6-dicarboxamides, as calculated using density functional theory (B3LYP/6-311 + G**//B3LYP/6-31G*), favors a syn–syn conformation.

highly dynamic equilibrium that interconverts the *M* and *P* conformations quickly relative to the NMR timescale (Preston et al. 2003). In solution, circular dichroism (CD) studies indicated that the helical secondary structure at the periphery of the dendrons becomes increasingly biased toward a single helix sense as the dendron generation was increased and in poor solvents (Recker et al. 2000; Huang and Parquette 2001; Fig. 11.32).

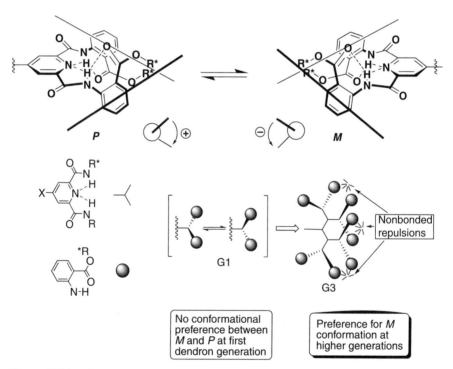

Figure 11.32   (Top) Helical conformational equilibria of dendron branching units and (bottom) chiral amplification in dendrons.

11.4.1. Conformational Switching in Water

The overall impact of water on the folding propensity of the dendrons emerges from the interplay of several competing noncovalent interactions. The surface area exposed to the aqueous media is minimal in the syn–syn conformational state, compared with other unfolded conformations. However, water could also be expected to disrupt the hydrogen bonding interactions that stabilize the syn–syn conformation and reduce the energetic cost of the higher dipole moments of the syn–anti and anti–anti conformations of the pyridine-2,6-dicarboxamide repeat unit. Recent computational studies suggested that dipole minimization effects and nonspecific solvophobic compression were more important factors in stabilizing the syn–syn conformation of these dendrons than intramolecular hydrogenizing interactions (Gabriel et al. 2006). A series of water-soluble dendrons with chiral pentaethylene glycol terminal groups were constructed to explore the potential for these systems to fold in water (Fig. 11.33; Hofacker and Parquette 2005).

Evidence for folding was obtained by CD spectroscopy, which displayed a negative excitonic couplet centered at 316 nm, demonstrating that the first-generation dendron adopted an *M*-type helical conformation relating the anthranilate chromophores in both tetrahydrofuran (THF) and water. The G2 dendron exhibited a negative couplet in THF of comparable magnitude to that of the G1 dendron, also indicating an *M* helical bias. The negative couplet transitioned to an equally intense positive couplet in aqueous media, indicating that an *M* → *P* helical transition occurred in water. The third-generation dendron exhibited identical behavior. IR spectroscopy

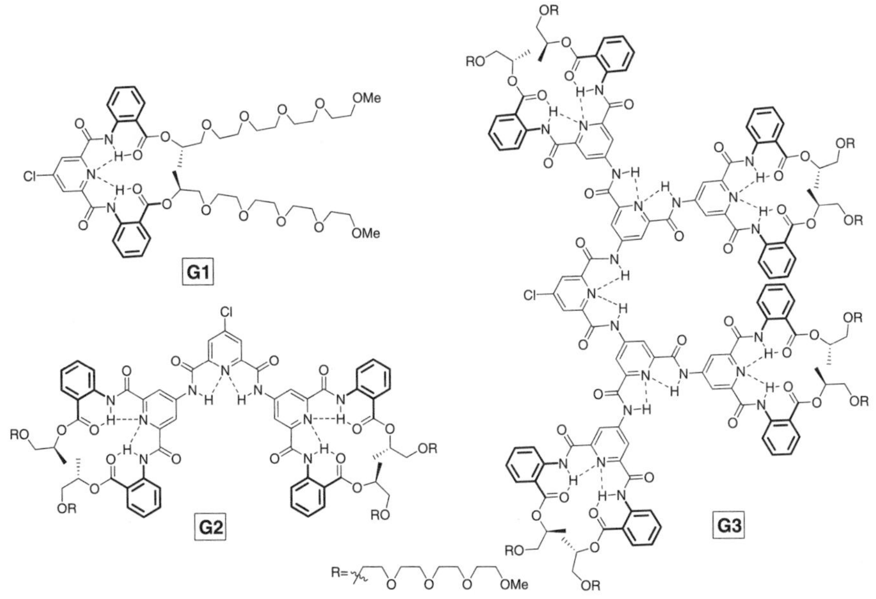

Figure 11.33 Structures of water-soluble dendrons with chiral pentaethylene glycols.

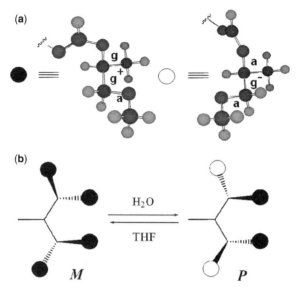

Figure 11.34   (a) Conformation of terminal glycol chains in (left) $CHCl_3$ and (right) H_2O as predicted by conformational searching (MM3). (b) Schematic depiction of the lowest energy conformers of the water-soluble G3 dendron. Reprinted from Hofacker and Parquette (2005). Copyright 2005 Wiley–VCH Verlag GmbH & Co. KGaA.

in conjunction with Monte Carlo conformational searching studies indicated that the $M \rightarrow P$ helical inversion occurs as a consequence of only two of the four glycol chains shifting from gauche-gauche$^+$–anti to anti–gauche$^-$–anti conformations around the respective O-C-C-OMe bonds (Fig. 11.34).

These observations indicated that 1) the dendrons adopt a stable folded state in aqueous media, 2) the terminal chains experience a shift in equilibrium toward the lower energy gauche conformational state (as compared to the pentaethylene glycol side chain), and 3) the dendron secondary structure and the terminal chain conformations are strongly correlated.

11.4.2. Amphiphilic Self-Assembly of Peptide–Dendrons

The design of synthetic systems displaying the hierarchical structural order present in natural systems remains a significant challenge. Achieving a level of structural organization similar to natural systems requires the ability to design conjugate structures that couple the equilibria of multiple, unique secondary structural elements over long distances. The allostery observed in the isolated dendrons in water represents a short-range example of a hierarchical organization wherein coupled motions relating the terminal pentaethylene glycol chain conformations with dendron helicity induced a solvent-mediated $M \rightarrow P$ helical inversion. To further explore this potential, a series of peptide–dendron conjugates having two locally achiral, PEG-terminated, water-soluble dendrons appended to a polyalanine peptide backbone

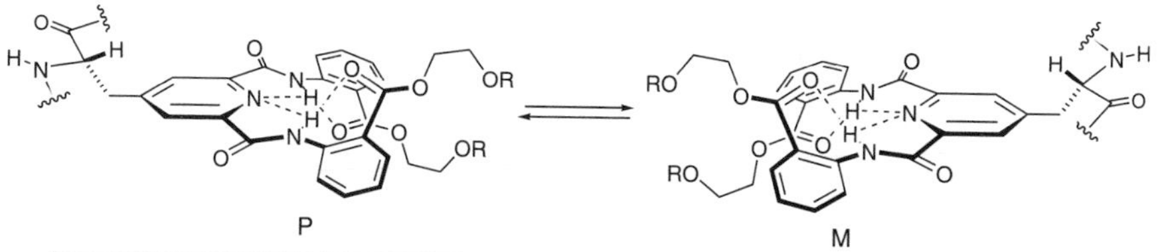

Peptide	Dendrons	Sequence
1	Control	Ac-AAAAKAAAAKAAAAYA-NH$_2$
2	i, i+4	Ac-AAA^{D}AKAA^{D}AAKAAAAYA- NH$_2$
3	i, i+5	Ac-AAA^{D}AKAAA^{D}AKAAAAYA- NH$_2$
4	i, i+6	Ac-AAA^{D}AKAAAA^{D}KAAAAYA- NH$_2$
5	i, i+7	Ac-AAA^{D}AAKAAAA^{D}KAAAYA- NH$_2$
6	i, i+8	Ac-AAA^{D}AKAAAAKA^{D}AAAYA- NH$_2$
7	i, i+9	Ac-AAA^{D}AKAAAAKAA^{D}AAYA- NH$_2$
8	i, i+10	Ac-AAA^{D}AKAAAAKAAA^{D}AYA- NH$_2$
9	i, i+11	Ac-AAA^{D}AKAAAAKAAAA^{D}YA-NH$_2$

Figure 11.35 Helical conformational equilibria of dendron-modified alanine residue (A^D) within a peptide–dendron conjugate. Peptide–dendron sequences are shown in the table form.

were studied (Shao et al. 2007). Interdendron spacing along the peptide backbone was progressively increased from $i, i + 4$ to $i, i + 11$ (Fig. 11.35).

Such polyalanine sequences generally display a high propensity to adopt an α-helical conformation, even in aqueous media. The majority of the peptide–dendrons exhibited α-helical/random coil conformations in water and tetrafluoroethylene (TFE) with slightly lower helicity compared with a control peptide lacking dendron side chains. These peptides did not exhibit any evidence for peptide–dendron coupled equilibria in the form of a helical bias in the dendron. An α-helix to β-sheet conformational transition occurred in water for two of the peptide–dendron conjugates: $(i, i + 6)$ and $(i, i + 10)$. Strong chiral transfer from the peptide backbone to the dendron emerged exclusively in the β-sheet forms of $(i, i + 6)$ and $(i, i + 10)$, revealing a synergistic coupling of the conformations of both structural elements. Atomic force microscopy (AFM) of both peptides revealed amyloid-like nanofibrils in PBS buffer at 10 μM, whereas in 60% TFE/PBS globular aggregates were observed (Fig. 11.36). In addition, $(i, i + 6)$ forms a hydrogel in water at concentrations as low as 1.7 mM.

These studies support the notion that the occurrence of coupled conformational equilibria between the peptide and dendron structures requires peptide folding to be strongly coupled with dendron packing. Satisfying this criterion ensures that the peptide secondary structure is stabilized by dendron packing; further, this packing concomitantly promotes interdendron motions that amplify helical biases. It is noteworthy that the peptide–dendrons exhibiting a lower α-helicity, relative to the control, lack evidence for peptide to dendron chiral communication. In contrast, the two sequences that exhibit β-sheet conformations exhibit strongly coupled

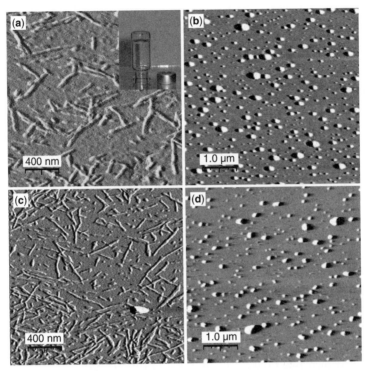

Figure 11.36   Tapping-mode AFM images of $(i, i + 6)$ and $(i, i + 10)$ peptide conjugates: (a) $(i, i + 6)$ in PBS $(10 \, \mu M)$ with the insert showing the photograph of the hydrogel formed by $(i, i + 6)$ in PBS at concentration of 1.7 mM; (b) $(i, i + 6)$ in 60% (v/v) TFE/PBS $(5 \, \mu M)$; (c) $(i, i + 10)$ in PBS $(10 \, \mu M)$; (c) $(i, i + 10)$ in 60% (v/v) TFE/PBS $(5 \, \mu M)$. Reprinted from Shao et al. (2007). Copyright 2007 American Chemical Society.

peptide and dendron equilibria. This coupling is a consequence of the intermolecular hydrophobic association of the dendritic side chains, which drives the α-helix to β-sheet interconversion going from TFE to water. Such intra- and interstrand interactions among laterally and diagonally paired hydrophobic side chains contribute

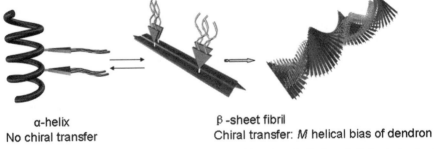

α-helix	β-sheet fibril
No chiral transfer	Chiral transfer: *M* helical bias of dendron

Figure 11.37   Notional depiction of dendron packing in α-helical and β-sheet structural forms. Reprinted from Shao et al. (2007). Copyright 2007 American Chemical Society.

significantly to the stability of β-sheet structures (Griffiths-Jones and Searle 2000; Syud et al. 2001; Phillips et al. 2005). Overall, these studies point to the importance of controlled self-assembly (of β-strands into β-sheet fibrils) for chiral communication; conversely, structures lacking peptide → dendron chirality transfer do not form stable β-sheet fibrils (Fig. 11.37). Therefore, the occurrence of strong peptide → dendron chirality transfer correlates strongly with β-sheet fibril stability.

11.5. LANGMUIR–BLODGETT MONOLAYERS

The formation of Langmuir–Blodgett monolayers from amphiphilic dendrimers at the air–water interface generally confirms the highly flexible structure of most dendrimers (Tully and Frechet 2001). In order to form monolayers on the surface of water, a conformation that extends the hydrophilic regions into the water phase and the hydrophobic groups upward from the interface must be accessible. For example, early work demonstrated that G2–G4 Fréchet-type poly(benzyl ether) dendrons readily formed monolayers (Saville et al. 1993, 1995). Pressure–area (Π-A) measurements suggest that the dendrons adopt a prolate shape oriented on the surface to point the focal benzyl alcohol group into the water phase and to project the hydrophobic poly(benzyl ether) sequences upward from the surface (Fig. 11.38; Kampf et al. 1999).

The G5 and G6 dendrons did not assemble into monolayers because of the tendency of the focal point to be encapsulated by the hydrophobic dendritic branches at higher generations. Similarly, Liskamp's group reported a series of amino acid based dendrons capped with Boc end groups and a focal fatty acid chain also formed stable monolayers; however, the fatty acid group was not a significant improvement over a shorter linkage (Mulders et al. 1998). In the condensed monolayer phase, the dendrons covered a smaller area than predicted for a fully spread

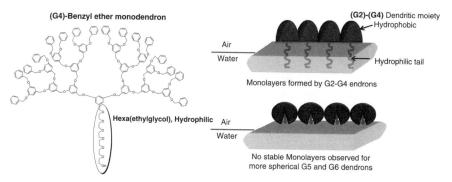

Figure 11.38 (Left) Fourth generation monodendron with a hexaethylene glycol tail. (Right) Proposed conformations of poly(benzyl ethyl ether) monodendrons oriented at the air–water interface.

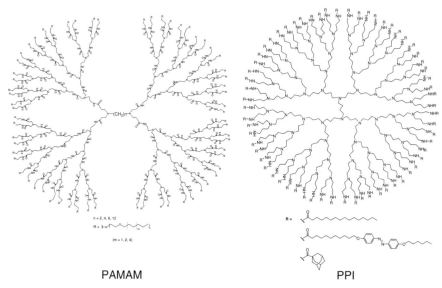

PAMAM   PPI

Figure 11.39 Structures of poly(amido amine) (PAMAM) and poly(propylene imine) (PPI) dendrimers with peripheral aliphatic chains.

dendron structure. In contrast, both PAMAM (Sayed-Sweet et al. 1997) and PPI dendrimers (Schenning et al. 1998; Tsuda et al. 2000; Weener and Meijer 2000), functionalized with peripheral aliphatic chains, adopt a flattened, disklike conformation to permit the polar dendritic interior to interact with the aqueous phase and to project the hydrophobic chains away from the surface (Figs 11.14, 11.15, 11.39).

When a more globular conformation is enforced by functionalizing the PPI dendrimers with bulky adamantyl terminal groups, or at higher generations (G4 and G5) of the PAMAM system, the dendrimers behave like nonamphiphilic, hydrophobic spheroids.

11.5.1. Effect of Hydrophobic/Hydrophilic Balance

The stability of dendritic Langmuir monolayers can be significantly enhanced by providing a hydrophilic anchor capable of extending into the water phase in a manner that secures the monolayer. However, optimal stability requires a balanced ratio of hydrophobic and hydrophilic regions within the dendritic amphiphile. For example, Frank and Hawker's groups found that poly(benzyl ether) dendrons having hydrophilic ethylenoxy (EO) chains installed at the focal point form more stable monolayers compared with dendrons having a only a focal alcohol (Fig. 11.38; Saville et al. 1995; Kampf et al. 1999). The EO chains anchor the monolayer by extending into the aqueous phase, incrementally increasing the collapse pressure by 3.1 and 4.1 mN/ m per additional EO unit for the G3 and G4 dendrons, respectively. However, the stability ultimately depended on a subtle balance between the relative sizes of the hydrophobic and hydrophilic regions. Hence, the G5 dendron displaying a hexaethylene glycol focal chain did not form a stable monolayer, similar to the G5-OH dendron. Pao et al. (2001) studied third-generation poly(aryl ether) dendrons displaying hydrophobic $C_{12}H_{25}$ alkyl terminal chains and a hydrophilic crown moiety at the focal point (Fig. 11.40).

X-ray reflectivity studies and Π-A isotherms revealed a similar model in which the dendrons project the hydrophilic crowns into the water phase and the hydrophobic chains perpendicular to the interface. Amphiphilic dendronized polymers that adopt a cylindrical, rodlike morphology, projecting hydrophilic and hydrophobic dendrons on opposite faces of the structure, form stable monolayers exhibiting collapse pressures of approximately 20 mN/m (Fig. 11.41).

The Π-A isotherms suggest an orientation that aligns the long axis of the polymer rod in the plane of the interface with the EO chains of the hydrophilic dendrons extended into the water layer and the hydrophobic dendrons pointed upward from the surface. The stability of the dendronized polymers likely arises from the balance between the polar and nonpolar regions exposed at the surface of the polymeric cylinder (Bo et al. 1999; Schluter and Rabe 2000). Ariga and colleagues (2004) constructed a series of "spider web" dendritic amphiphiles that project hydrophilic Lys-Lys-Glu tripeptides and hydrophobic chains at each generational level (Fig. 11.42).

This arrangement projects hydrophobic and hydrophilic surfaces above and below the plane of the dendron in a display that mimics the two-dimensional structure of lipid clusters. Accordingly, the amphiphile can efficiently spread out along the surface like a spider web. The Π-A isotherms revealed that these molecules form stable monolayers with collapse pressures in the range of 40–60 mN/m. The addition

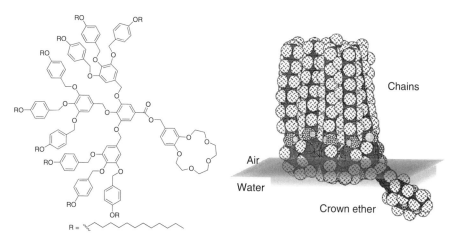

Figure 11.40 (Left) Structure of the Pao et al.'s (2001) amphiphilic monodendron and the space-filling model showing a possible conformation at the water–air interface. (Right) The chains are assumed to extend vertically away from the water surface, and the crown ether group is assumed to extend into the water. Reprinted from Pao et al. (2001). Copyright 2001 American Chemical Society.

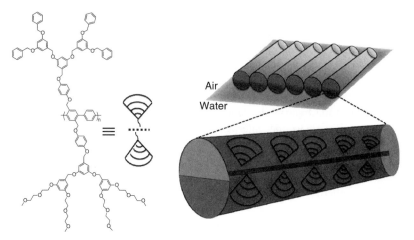

Figure 11.41 (Left) Structure of amphiphilic dendronized polymer and (right) schematic representations of the Langmuir monolayer formed by the amphiphilic polymeric cylinders.

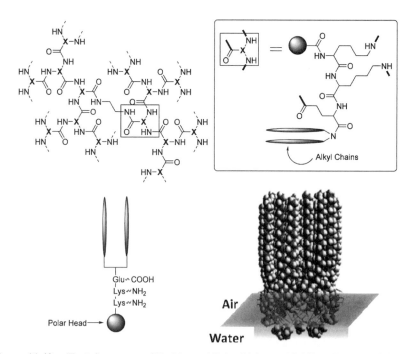

Figure 11.42 (Top) Structures of "spider web" dendritic amphiphiles. (Bottom) Schematic representation of the molecular orientation at the water–air interface. Reprinted from Ariga et al. (2004). Copyright 2004 American Chemical Society.

of octadecanoic acid as a component of the G2 and G3 amphiphiles appears to integrate into the hydrophobic region of the dendrons and increase the monolayer stability without significantly increasing the molecular areas.

11.5.2. Packing Effect

A recent report by Bury et al. (2007) highlights the importance of efficient packing of hydrophobic chains in addition to the need for a correct hydrophobic/hydrophilic balance in amphiphilic block codendrimers. Three series of dendrimers were constructed that varied the amount and connectivity of the hydrophobic aklyl groups (**Ia–c to VIa–c**); each series was also studied as a function of generation (Fig. 11.43).

Going from **Ia** to **IIIa**, the collapse pressure ($p\pi_c$) decreases because the increase in dendron generation contributes hydrophobicity to the amphiphile, resulting in a poorer hydrophobic balance. The series having all three alkyl tails emanating from a single carbon atom (**Ib–VIb**) exhibits significantly different behavior compared with the other two series in which the alkyl chains emanate from different carbon atoms (**Ia/b–VIa/c**). Whereas the **a** and **c** series exhibit linear increases in molecular area with more alkyl chains, the **b** series reveals not only nonlinear increases in area

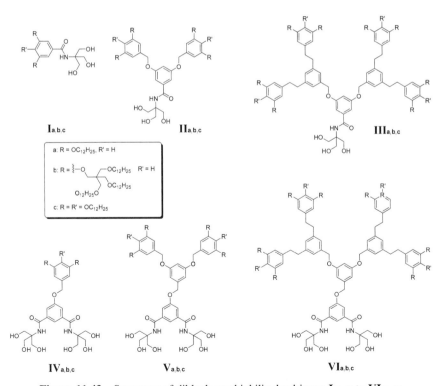

Figure 11.43   Structures of diblock amphiphilic dendrimers **Ia–c** to **VIa–c**.

but also larger areas than predicted from close-packed hexagonally ordered chains. The nonlinear behavior of series **b** is a consequence of the connectivity of the alkyl chains via a singe carbon atom precluding normal paraffin-like packing. This inefficient packing increases the molecular area taken up by the amphiphiles and greatly reduces the stability of the monolayers.

11.5.3. Photoresponsive Monolayers

Amphiphilic dendrimers functionalized with azobenzene groups have also been exploited to produce photoresponsive films. Generally, $E \rightarrow Z$ photoisomerization is often suppressed in Langmuir–Blodgett films because of aggregation of the azobenzene chromophores, as evidenced by peak broadening in the UV visible (UV–vis) spectrum (Nishiyama and Fujihira 1988; Yabe et al. 1988). Meijer demonstrated that this self-association can be hampered by incorporating the azobenzene chromophore at the periphery of the PPI dendrimer in the presence of varying ratios of palmitolyl groups to reduce the tendency of the azobenzenes to pack. Although dendrimers having either 1:1 or 1:0 palmitoyl/azobenzene ratios exhibited UV–vis spectra indicating monomeric azobenzene groups, their behavior at the air–water interface was quite different. The dendrimers orient themselves with the polar dendrimer core flattened to optimize the surface area anchored in the water phase and the alkyl-azobenzene groups extended perpendicular to the surface (Fig. 11.44; Weener and Meijer 2000; Su et al. 2007).

Because of some aggregation (H aggregation) of the azobenzene chromophores, the collapse pressure increased with higher azobenzene/palmitoyl ratios. However, only the palmitoyl functionalized dendrimer exhibited a change in surface area upon irradiation with 365-nm light. The lack of photoinduced surface area changes

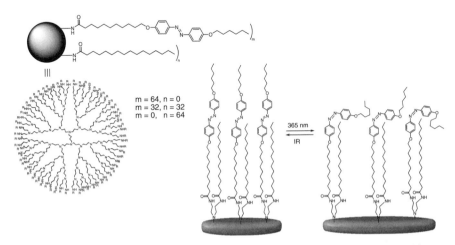

Figure 11.44 (Left) Structures of photoresponsive poly(propylene imine) dendrimers. (Right) The isomerization behavior of palmitoyl-functionalized amphiphilic dendrimers at the water surface.

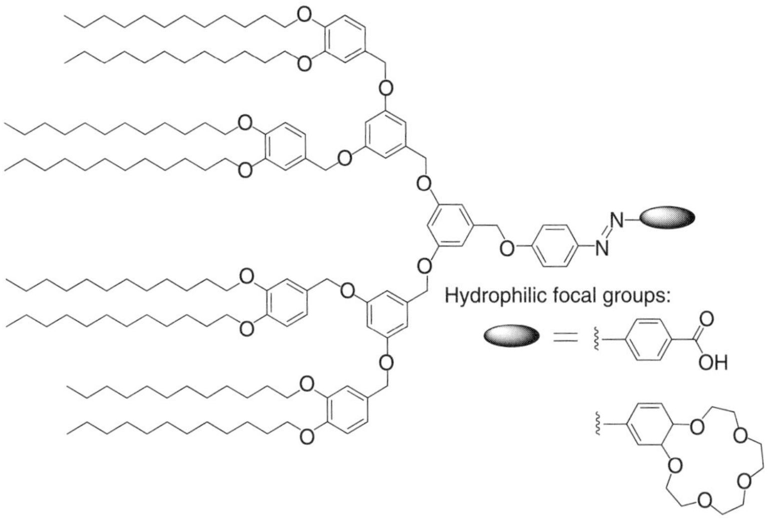

Figure 11.45 Structures of amphiphilic monodendrons containing a hydrophilic azobenzene focal group.

in the all-azobenzene dendrimer was attributed to greater aggregation in the Langmuir–Blodgett films, which inhibit $E \rightarrow Z$ isomerization. Another approach to enhance the extent and reversibility of photoisomerization focused on creating amphiphiles with a cross-sectional mismatch between a hydrophilic azobenzene focal group and a large dendritic hydrophobic region (Fig. 11.45; Peleshanko et al. 2002; Genson et al. 2005).

This cross-sectional mismatch between the small anchor and the large dendritic component modified the intermolecular packing in the Langmuir–Blodgett monolayer and film so that sufficient mobility was present near the azobenzene group. This mobility greatly facilitated efficient molecular reorganization in response to $E \rightarrow Z$ photoisomerization.

11.5.4. Fullerene Monolayers

The preparation of thin films of Buckminsterfullerene (C_{60}) is severely hampered by their tendency to aggregate at the air–water surface and form ill-defined films of three-dimensional aggregates at the air–water surface (Mirkin and Caldwell 1996). The attachment of a hydrophilic head group to the fullerene significantly improves the spreading ability by providing an attractive interaction with the water phase. However, successive compression–expansion cycles exhibit hysteresis due to the occurrence of strong fullerene–fullerene interactions that induce irreversible aggregation upon compression of the monolayer. Many of these problems were circumvented by encapsulation of the fullerene within amphiphilic glycodendrons to suppress self-association by sterically isolating the C_{60} molecules (Fig. 11.46a;

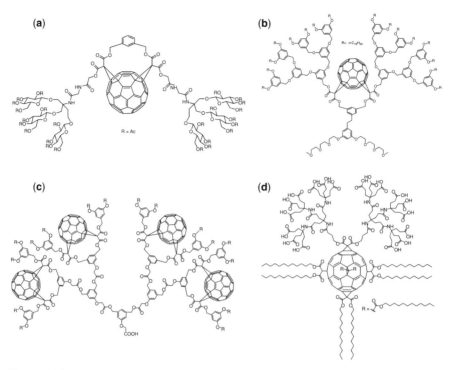

Figure 11.46 Typical structures of amphiphilic fullerene dendritic derivatives that form Langmuir–Blodgett films.

Cardullo et al. 1998). These conjugates formed stable monomolecular films exhibiting reversibility under repeated compression–expansion cycles (Maierhofer et al. 2000). Although these monolayers could be transferred onto quartz slides, the transfer ratios were low (0.7), suggesting that the hydrophobic/hydrophilic balance was not optimal. Nierengarten's group explored a series of dendritic C_{60} derivatives displaying both a polar head group to anchor the molecule in the aqueous phase and hydrophobic poly(benzyl ether) dendrons with terminal alkyl chains (Felder et al. 2002; Gallani et al. 2002; Hahn et al. 2007) to reduce fullerene–fullerene interactions (Fig. 11.46b; Nierengarten 2004). Varying the hydrophilic/hydrophobic balance by increasing the size of the polar head profoundly improved the stability and transfer properties of the monolayers. The molecular areas of the amphiphiles decreased with increasing size of the polar head. This behavior was attributed to a deeper anchoring of the molecules in the aqueous phase, which forces the hydrophobic dendron branches to collapse upon the fullerene. This conformational change reduces the measured molecular area and inhibits the fullerene–fullerene interactions that complicate the transfer of the monolayers to solid substrates (Zhang et al. 2003). Alternatively, the same group demonstrated that dendrons displaying multiple fullerenes at the periphery and a small, focal carboxylic acid were able to form Langmuir films (Fig. 11.46c; Felder et al. 2000; Hahn et al. 2005).

However, transfer of the monolayers to solid substrates was inefficient because of the small size of the hydrophilic anchor, which imparted a hydrophobic/hydrophilic imbalance. Accordingly, increasing the size of the hydrophilic anchor facilitated the formation of stable Langmuir films that could be transferred to solid substrates with a transfer ratio of 1 (Nierengarten et al. 2001). Bayer's research group reported a related approach in which the fullerene was modified with carboxylate-terminated dendrons and alkyl chains (Maierhofer et al. 2000). This amphiphilic fullerene derivative formed stable monolayers that could be compressed and expanded without hysteresis (Fig. 11.46d; Maierhofer et al. 2000).

11.6. CONCLUSION

The localization of molecular components having similar polarity in water is a powerful, nondirectional driving force for the self-assembly and folding of amphiphilic systems. The structure and self-assembly of the amphiphilic dendrimers discussed here emerge from a complex interplay of topological factors with the need for phase segregation. This interplay imparts amphiphilic dendrimers with properties that differ significantly from those of traditional surfactants. Further, the ability to readily adjust the dendrimer topology allows these physical properties to be adjusted by minor structural changes. Many of the functional superstructures produced in natural systems are constructed by a hierarchical self-assembly process that is mediated by a synergistic array of directional and nondirectional intermolecular forces. Although our understanding of how a particular noncovalent force contributes to the assembly process is reasonably mature, the design of superstructures that assemble via the action of multiple, interdependent forces is usually serendipitous. Nevertheless, our capability to design molecular systems that assemble via multiple interactions is rapidly increasing. Further progress toward the design of superstructures with a sophistication that rivals biological systems will require a better understanding of how directional interactions, such as hydrogen bonding, can be synergized with amphiphilic phase separation.

REFERENCES

Aathimanikandan SV, Savariar EN, Thayumanavan S. Temperature-sensitive dendritic micelles. J Am Chem Soc 2005;127:14922–14929.

Al-Jamal KT, Ramaswamy C, Florence AT. Supramolecular structures from dendrons and dendrimers. Adv Drug Deliv Rev 2005;57:2238–2270.

Ambade AV, Savariar EN, Thayumanavan S. Dendrimeric micelles for controlled drug release and targeted delivery. Mol Pharm 2005;2:264–272.

Ariga K, Urakawa T, Michiue A, Kikuchi J-I. Spider-net amphiphiles as artificial lipid clusters: design, synthesis, and accommodation of lipid components at the air–water interface. Langmuir 2004;20:6762–6769.

Basu S, Vutukuri DR, Shyamroy S, Sandanaraj BS, Thayumanavan S. Invertible amphiphilic homopolymers. J Am Chem Soc 2004;126:9890–9891.

Bo Z, Rabe JP, Schluter AD. A poly(*para*-phenylene) with hydrophobic and hydrophilic dendrons: prototype of an amphiphilic cylinder with the potential to segregate lengthwise. Angew Chem Int Ed 1999;38:2370–2372.

Boas U, Heegaard PMH. Dendrimers in drug research. Chem Soc Rev 2004;33:43–63.

Boris D, Rubinstein M. A self-consistent mean field model of a starburst dendrimer: dense core vs dense shell. Macromolecules 1996;29:7251–7260.

Bury I, Donnio B, Gallani J-L, Guillon D. Interfacial behavior of a series of amphiphilic block co-dendrimers. Langmuir 2007;23:619–625.

Cardullo F, Diederich F, Echegoyen L, Habicher T, Jayaraman N, Leblanc RM, Stoddart JF, Wang S. Stable Langmuir and Langmuir–Blodgett films of fullerene–glycodendron conjugates. Langmuir 1998;14:1955–1959.

Chang Y, Kwon YC, Lee SC, Kim C. Amphiphilic linear PEO–dendritic carbosilane block copolymers. Macromolecules 2000;33:4496–4500.

Chapman TM, Hillyer GL, Mahan EJ, Shaffer KA. Hydraamphiphiles: novel linear dendritic block copolymer surfactants. J Am Chem Soc 1994;116:11195–11196.

Chen W-R, Porcar L, Liu Y, Butler PD, Magid LJ. Small angle neutron scattering studies of the counterion effects on the molecular conformation and structure of charged G4 PAMAM dendrimers in aqueous solutions. Macromolecules 2007;40:5887–5898.

Chen ZY, Cui S-M. Monte Carlo simulations of star-burst dendrimers. Macromolecules 1996;29:7943–7952.

Cho B-K, Jain A, Nieberle J, Mahajan S, Wiesner U, Gruner SM, Tuerk S, Raeder HJ. Synthesis and self-assembly of amphiphilic dendrimers based on aliphatic polyether-type dendritic cores. Macromolecules 2004;37:4227–4234.

Choi Y, Baker JR Jr. Targeting cancer cells with DNA-assembled dendrimers. A mix and match strategy for cancer. Cell Cycle 2005;4:669–671.

Connolly R, Timoshenko EG, Kuznetsov YA. Monte Carlo simulations of amphiphilic co-dendrimers in dilute solution. Macromolecules 2004;37:7381–7392.

Darbre T, Reymond J-L. Peptide dendrimers as artificial enzymes, receptors, and drug-delivery agents. Acc Chem Res 2006;39:925–934.

De Gennes PG, Hervet H. Statistics of "starburst" polymers. J Phys Lett 1983;44:351–360.

Deshayes S, Morris MC, Divita G, Heitz F. Interactions of amphipathic peptides with membrane components and consequences on the ability to deliver therapeutics. J Peptide Sci 2006;12:758–765.

Dufes C, Uchegbu IF, Schaetzlein AG. Dendrimers in gene delivery. Adv Drug Deliv Rev 2005;57:2177–2202.

El Ghzaoui A, Gauffre F, Caminade A-M, Majoral J-P, Lannibois-Drean H. Self-assembly of water-soluble dendrimers into thermoreversible hydrogels and macroscopic fibers. Langmuir 2004;20:9348–9353.

Felder D, Gallani J-L, Guillon D, Heinrich B, Nicoud J-F, Nierengarten J-F. Investigations of thin films with amphiphilic dendrimers bearing peripheral fullerene subunits. Angew Chem Int Ed 2000;39:201–204.

Felder D, Nava MG, Del Pilar Carreon M, Eckert J-F, Luccisano M, Schall C, Masson P, Gallani J-L, Heinrich B, Guillon D, Nierengarten J-F. Synthesis of amphiphilic fullerene derivatives and their incorporation in Langmuir and Langmuir–Blodgett films. Helv Chim Acta 2002;85:288–319.

Fuhrhop J-H, Wang T. Bolaamphiphiles. Chem Rev 2004;104:2901–2937.

Gabriel CJ, DeMatteo MP, Paul NM, Takaya T, Gustafson TL, Hadad CM, Parquette JR. A new class of intramolecularly hydrogen-bonded dendrons based on a 2-methoxyisoph-thalamide repeat unit. J Org Chem 2006;71:9035–9044.

Gajbhiye V, Kumar PV, Tekade RK, Jain NK. Pharmaceutical and biomedical potential of PEGylated dendrimers. Curr Pharm Des 2007;13:415–429.

Gallani J-L, Felder D, Guillon D, Heinrich B, Nierengarten J-F. Micelle formation in Langmuir films of C60 derivatives. Langmuir 2002;18:2908–2913.

Gandhi P, Huang B, Gallucci JC, Parquette JR. Effect of terminal group sterics and dendron packing on chirality transfer from the central core of a dendrimer. Org Lett 2001;3: 3129–3132.

Genson KL, Holzmuller J, Villacencio OF, McGrath DV, Vaknin D, Tsukruk VV. Langmuir and grafted monolayers of photochromic amphiphilic monodendrons of low generations. J Phys Chem B 2005;109:20393–20402.

Gillies ER, Frechet JMJ. Dendrimers and dendritic polymers in drug delivery. Drug Discov Today 2005;10:35–43.

Gitsov I, Frechet JMJ. Solution and solid-state properties of hybrid linear–dendritic block copolymers. Macromolecules 1993;26:6536–6546.

Gitsov I, Frechet JMJ. Stimuli-responsive hybrid macromolecules: novel amphiphilic star copolymers with dendritic groups at the periphery. J Am Chem Soc 1996;118:3785–3786.

Gitsov I, Wooley KL, Frechet JMJ. Novel polyether copolymers with a linear central unit and dendritic end groups. Angew Chem 1992;104:1282–1285.

Giupponi G, Buzza DMA. A Monte Carlo study of amphiphilic dendrimers: spontaneous asymmetry and dendron separation. J Chem Phys 2005;122:194903/1–194903/13.

Giupponi G, Buzza DMA, Adolf DB. Are polyelectrolyte dendrimers stimuli responsive? Macromolecules 2007;40:5959–5965.

Gopidas KR, Leheny AR, Caminati G, Turro NJ, Tomalia DA. Photophysical investigation of similarities between starburst dendrimers and anionic micelles. J Am Chem Soc 1991;113: 7335–7342.

Gorman CB, Hager MW, Parkhurst BL, Smith JC. Use of a paramagnetic core to affect longi-tudinal nuclear relaxation in dendrimers—a tool for probing dendrimer conformation. Macromolecules 1998;31:815–822.

Griffiths-Jones SR, Searle MS. Structure, folding, and energetics of cooperative interactions between the beta-strands of a de novo designed three-stranded antiparallel beta-sheet peptide. J Am Chem Soc 2000;122:8350–8356.

Grinstaff MW. Biodendrimers: new polymeric biomaterials for tissue engineering. Chem Eur J 2002;8:2838–2846.

Gronwald O, Snip E, Shinkai S. Gelators for organic liquids based on self-assembly: a new facet of supramolecular and combinatorial chemistry. Curr Opin Colloid Interface Sci 2002;7:148–156.

Habicher T, Diederich F, Gramlich V. Catalytic dendrophanes as enzyme mimics. Synthesis, binding properties, micropolarity effect, and catalytic activity of dendritic thiazolio-cyclophanes. Helv Chim Acta 1999;82:1066–1095.

Hahn U, Cardinali F, Nierengarten J-F. Supramolecular chemistry for the self-assembly of fullerene-rich dendrimers. New J Chem 2007;31:1128–1138.

Hahn U, Hosomizu K, Imahori H, Nierengarten J-F. Synthesis of dendritic branches with peripheral fullerene subunits. Eur J Org Chem 2005;1:85–91.

Hartley GS. Aqueous solutions of paraffin-chain salts: a study in micelle formation. Paris: Hermann & Cie; 1936.

Hecht S, Frechet JMJ. Dendritic encapsulation of function: applying nature's site isolation principle from biomimetics to materials science. Angew Chem Int Ed 2001a;40:74–91.

Hecht S, Frechet JMJ. Light-driven catalysis within dendrimers: designing amphiphilic singlet oxygen sensitizers. J Am Chem Soc 2001b;123:6959–6960.

Hirst AR, Smith DK. Two-component gel-phase materials—highly tunable self-assembling systems. Chem Eur J 2005;11:5496–5508.

Hofacker AL, Parquette JR. Dendrimer folding in aqueous media: an example of solvent-mediated chirality switching. Angew Chem Int Ed 2005;44:1053–1057, S1053/1–S1053/17.

Huang B, Parquette JR. Effect of an internal anthranilamide turn unit on the structure and conformational stability of helically biased intramolecularly hydrogen-bonded dendrons. J Am Chem Soc 2001;123:2689–2690.

Huang QR, Dubin PL, Lal J, Moorefield CN, Newkome GR. Small-angle neutron scattering studies of charged carboxyl-terminated dendrimers in solutions. Langmuir 2005;21:2737–2742.

Israelachvilli JN, editor. Intermolecular and surface forces. San Diego (CA): Academic Press; 1991.

Israelachvilli JN, Mitchell DJ, Ninham BW. Theory of self-assembly of lipid bilayers and vesicles. Biochim Biophys Acta Biomembr 1977;470:185–201.

Israelachvilli JN, Marcelja S, Horn RG. Physical principles of membrane organization. Q Rev Biophys 1980;13:121–200.

Jang C-J, Ryu J-H, Lee J-D, Sohn D, Lee M. Synthesis and supramolecular nanostructure of amphiphilic rigid aromatic-flexible dendritic block molecules. Chem Mater 2004;16:4226–4231.

Jun H-W, Paramonov SE, Hartgerink JD. Biomimetic self-assembled nanofibers. Soft Matter 2006;2:177–181.

Jusufi A, Likos CN, Lowen H. Conformations and interactions of star-branched polyelectrolytes. Phys Rev Lett 2002a;88:018301.

Jusufi A, Likos CN, Lowen H. Counterion-induced entropic interactions in solutions of strongly stretched, osmotic polyelectrolyte stars. J Chem Phys 2002b;116:11011–11027.

Kampf JP, Frank CW, Malmstroem EE, Hawker CJ. Stability and molecular conformation of poly(benzyl ether) monodendrons with oligo(ethylene glycol) tails at the air–water interface. Langmuir 1999;15:227–233.

Kellermann M, Bauer W, Hirsch A, Schade B, Ludwig K, Boettcher C. The first account of a structurally persistent micelle. Angew Chem Int Ed 2004;43:2959–2962.

Kim KT, Winnik MA, Manners I. Synthesis and self-assembly of dendritic-helical block polypeptide copolymers. Soft Matter 2006;2:957–965.

Kim YH, Webster OW. Water soluble hyperbranched polyphenylene: "a unimolecular micelle?" J Am Chem Soc 1990;112:4592–4593.

Klaikherd A, Sandanaraj BS, Vutukuri DR, Thayumanavan S. Comparison of facially amphiphilic biaryl dendrimers with classical amphiphilic ones using protein surface recognition as the tool. J Am Chem Soc 2006;128:9231–9237.

Klok H-A, Hwang JJ, Hartgerink JD, Stupp SI. Self-assembling biomaterials: L-lysine-dendron-substituted cholesteryl-(L-lactic acid). Macromolecules 2002;35:6101–6111.

Kofoed J, Reymond J-L. Dendrimers as artificial enzymes. Curr Opin Chem Biol 2005;9: 656–664.

Kono K. Application of dendrimers to drug delivery systems—from the view point of carrier design based on nanotechnology. Drug Deliv Syst 2002;17:462–470.

Laufersweiler MJ, Rohde JM, Chaumette J-L, Sarazin D, Parquette JR. Synthesis, aggregation, and chiroptical properties of chiral, amphiphilic dendrimers. J Org Chem 2001;66: 6440–6452.

Lecommandoux S, Klok H-A, Sayar M, Stupp SI. Synthesis and self-organization of rod–dendron and dendron–rod–dendron molecules. J Polym Sci Part A Polym Chem 2003;41:3501–3518.

Lee CC, MacKay JA, Frechet JMJ, Szoka FC. Designing dendrimers for biological applications. Nat Biotechnol 2005;23:1517–1526.

Lee I, Athey BD, Wetzel AW, Meixner W, Baker JR Jr. Structural molecular dynamics studies on polyamidoamine dendrimers for a therapeutic application: effects of pH and generation. Macromolecules 2002;35:4510–4520.

Lee J-J, Ford WT, Moore JA, Li Y. Reactivity of organic anions promoted by a quaternary ammonium ion dendrimer. Macromolecules 1994;27:4632–4634.

Lee M, Jeong Y-S, Cho B-K, Oh N-K, Zin W-C. Self-assembly of molecular dumbbells into organized bundles with tunable size. Chem Eur J 2002;8:876–883.

Lescanec RL, Muthukumar M. Configurational characteristics and scaling behavior of starburst molecules: a computational study. Macromolecules 1990;23:2280–2288.

Lin Y, Liao Q, Jin X. Molecular dynamics simulations of dendritic polyelectrolytes with flexible spacers in salt free solution. J Phys Chem B 2007;111:5819–5828.

Lins L, Brasseur R. The hydrophobic effect in protein folding. FASEB J 1995;9:535–540.

Liu M, Kono K, Frechet JMJ. Water-soluble dendritic unimolecular micelles: their potential as drug delivery agents. J Controlled Release 2000;65:121–131.

Lockman JW, Paul NM, Parquette JR. The role of dynamically correlated conformational equilibria in the folding of macromolecular structures. A model for the design of folded dendrimers. Prog Polym Sci 2005;30:423–452.

Loewik DWPM, van Hest JCM. Peptide based amphiphiles. Chem Soc Rev 2004;33:234–245.

Lue L, Prausnitz JM. Structure and thermodynamics of homogeneous-dendritic-polymer solutions: computer simulation, integral-equation, and lattice-cluster theory. Macromolecules 1997;30:6650–6657.

Luo D, Li Y, Um SH, Cu Y. A dendrimer-like DNA-based vector for DNA delivery. A viral and nonviral hybrid approach. Methods Mol Med 2006;127:115–125.

Lyulin SV, Darinskii AA, Lyulin AV, Michels MAJ. Computer simulation of the dynamics of neutral and charged dendrimers. Macromolecules 2004;37:4676–4685.

Maierhofer AP, Brettreich M, Burghardt S, Vostrowsky O, Hirsch A, Langridge S, Bayer TM. Structure and electrostatic interaction properties of monolayers of amphiphilic molecules derived from C60-fullerenes: a film balance, neutron-, and infrared reflection study. Langmuir 2000;16:8884–8891.

Maiti PK, Cagin T, Lin S-T, Goddard III WA. Effect of solvent and pH on the structure of PAMAM dendrimers. Macromolecules 2005;38:979–991.

Maiti PK, Goddard III WA. Solvent quality changes the structure of G8 PAMAM dendrimer, a disagreement with some experimental interpretations. J Phys Chem B 2006;110: 25628–25632.

Mansfield ML. Dendron segregation in model dendrimers. Polymer 1994;35:1827–1830.

Mansfield ML, Klushin LI. Intrinsic viscosity of model starburst dendrimers. J Phys Chem 1992;96:3994–3998.

Mansfield ML, Klushin LI. Monte Carlo studies of dendrimer macromolecules. Macromolecules 1993;26:4262–4268.

Marmillon C, Gauffre F, Gulik-Krzywicki T, Loup C, Caminade A-M, Majoral J-P, Vors J-P, Rump E. Organophosphorus dendrimers as new gelating materials for hydrogels. Angew Chem Int Ed 2001;40:2626–2629.

McNally BA, Leevy WM, Smith BD. Recent advances in synthetic membrane transporters. Supramol Chem 2007;19:29–37.

Meltzer AD, Tirrell DA, Jones AA, Inglefield PT, Hedstrand DM, Tomalia DA. Chain dynamics in poly(amidoamine) dendrimers: a study of carbon-13 NMR relaxation parameters. Macromolecules 1992;25:4541–4548.

Mirkin CA, Caldwell WB. Thin film, fullerene-based materials. Tetrahedron 1996;52: 5113–5130.

Miura N, Dubin PL, Moorefield CN, Newkome GR. Complex formation by electrostatic interaction between carboxyl-terminated dendrimers and oppositely charged polyelectrolytes. Langmuir 1999;15:4245–4250.

Mourey TH, Turner SR, Rubinstein M, Frechet JMJ, Hawker CJ, Wooley KL. Unique behavior of dendritic macromolecules: intrinsic viscosity of polyether dendrimers. Macromolecules 1992;25:2401–2406.

Mulders SJE, Brouwer AJ, Kimkes P, Sudholter EJR, Liskamp RMJ. Sizing of amino acid based dendrimers in Langmuir monolayers. J Chem Soc Perkin Trans 1998;2: 1535–1538.

Murat M, Grest GS. Molecular dynamics study of dendrimer molecules in solvents of varying quality. Macromolecules 1996;29:1278–1285.

Murugan E, Sherman RL Jr, Spivey HO, Ford WT. Catalysis by hydrophobically modified poly(propylenimine) dendrimers having quaternary ammonium and tertiary amine functionality. Langmuir 2004;20:8307–8312.

Namazi H, Adeli M. Solution properties of dendritic triazine/poly(ethylene glycol)/dendritic triazine block copolymers. J Polym Sci Part A Polym Chem 2005;43:28–41.

Newkome GR, Baker GR, Arai S, Saunders MJ, Russo PS, Theriot KJ, Moorefield CN, Rogers LE, Miller JE, Lieux TR, Murray ME, Phillips B, Pascal L. Cascade molecules. Part 6. Synthesis and characterization of two-directional cascade molecules and formation of aqueous gels. J Am Chem Soc 1990;112:8458–8465.

Newkome GR, Lin X, Yaxiong C, Escamilla GH. Cascade polymer series. 27. Two-directional cascade polymer synthesis: effects of core variation. J Org Chem 1993;58:3123–3129.

Newkome GR, Moorefield CN, Baker GR, Johnson AL, Behera RK. Chemistry of micelles. 11. Alkane cascade polymers with a micellar topology: micelle acid derivatives. Angew Chem 1991;103:1205–1207.

Newkome GR, Moorefield CN, Baker GR, Saunders MJ, Grossman SH. Chemistry of micelles. 13. Monomolecular micelles. Angew Chem 1991;103:1207–1209.

Newkome GR, Yao Z, Baker GR, Gupta VK. Micelles. Part 1. Cascade molecules: a new approach to micelles. A [27]-arborol. J Org Chem 1985;50:2003–2004.

Newkome GR, Yao Z, Baker GR, Gupta VK, Russo PS, Saunders MJ. Chemistry of micelles series. Part 2. Cascade molecules. Synthesis and characterization of a benzene[9]3-arborol. J Am Chem Soc 1986;108:849–850.

Newkome GR, Young JK, Baker GR, Potter RL, Audoly L, Cooper D, Weis CD, Morris K, Johnson CS Jr. Cascade polymers. 35. pH dependence of hydrodynamic radii of acid-terminated dendrimers. Macromolecules 1993;26:2394–2396.

Nguyen PM, Hammond PT. Amphiphilic linear–dendritic triblock copolymers composed of poly(amidoamine) and poly(propylene oxide) and their micellar-phase and encapsulation properties. Langmuir 2006;22:7825–7832.

Nierengarten J-F. Chemical modification of C60 for materials science applications. New J Chem 2004;28:1177–1191.

Nierengarten J-F, Eckert J-F, Rio Y, del Pilar Carreon M, Gallani J-L, Guillon D. Amphiphilic diblock dendrimers: synthesis and incorporation in Langmuir and Langmuir–Blodgett films. J Am Chem Soc 2001;123:9743–9748.

Nisato G, Ivkov R, Amis EJ. Size invariance of polyelectrolyte dendrimers. Macromolecules 2000;33:4172–4176.

Nishiyama K, Fujihira M. The cis–trans reversible photoisomerization of an amphiphilic azo-benzene derivative in its pure LB film prepared as polyion complexes with polyallylamine. Chem Lett 1988;1257–1260.

O'Connor CJ. Interfacial catalysis by microphases in apolar media. Surfactant Sci Ser 1987;21: 187–255.

Opitz AW, Wagner NJ. Structural investigations of poly(amido amine) dendrimers in methanol using molecular dynamics. J Polym Sci Part B Polym Phys 2006;44:3062–3077.

Ottaviani MF, Bossmann S, Turro NJ, Tomalia DA. Characterization of starburst dendrimers by the EPR technique. 1. Copper complexes in water solution. J Am Chem Soc 1994;116:661–671.

Ottaviani MF, Montalti F, Turro NJ, Tomalia DA. Characterization of starburst dendrimers by the EPR technique. Copper(II) ions binding full generation dendrimers. J Phys Chem B 1997;101:158–166.

Paleos CM, Tsiourvas D, Sideratou Z. Molecular engineering of dendritic polymers and their application as drug and gene delivery systems. Mol Pharm 2007;4:169–188.

Pan Y, Ford WT. Dendrimers with both hydrophilic and hydrophobic chains at every end. Macromolecules 1999;32:5468–5470.

Pan Y, Ford WT. Amphiphilic dendrimers with both octyl and triethylenoxy methyl ether chain ends. Macromolecules 2000;33:3731–3738.

Pao W-J, Stetzer MR, Heiney PA, Cho W-D, Percec V. X-ray reflectivity study of Langmuir films of amphiphilic monodendrons. J Phys Chem B 2001;105:2170–2176.

Parquette JR. The intramolecular self-organization of dendrimers. C R Chim 2003;6:779–789.

Peleshanko S, Sidorenko A, Larson K, Villavicencio O, Ornatska M, McGrath DV, Tsukruk VV. Langmuir–Blodgett monolayers from lower generation amphiphilic monodendrons. Thin Solid Films 2002;406:233–240.

Phillips ST, Piersanti G, Bartlett PA. Quantifying amino acid conformational preferences and side-chain–side-chain interactions in beta-hairpins. Proc Natl Acad Sci USA 2005;102: 13737–13742.

Pietersz GA, Tang C-K, Apostolopoulos V. Structure and design of polycationic carriers for gene delivery. Mini-Rev Med Chem 2006;6:1285–1298.

Piotti ME, Rivera F Jr, Bond R, Hawker CJ, Frechet JMJ. Synthesis and catalytic activity of unimolecular dendritic reverse micelles with "internal" functional groups. J Am Chem Soc 1999;121:9471–9472.

Preston AJ, Fraenkel G, Chow A, Gallucci JC, Parquette JR. Dynamic helical chirality of an intramolecularly hydrogen-bonded bisoxazoline. J Org Chem 2003;68:22–26.

Recker J, Tomcik DJ, Parquette JR. Folding dendrons: the development of solvent-, temperature-, and generation-dependent chiral conformational order in intramolecularly hydrogen-bonded dendrons. J Am Chem Soc 2000;122:10298–10307.

Saville PM, Reynolds PA, White JW, Hawker CJ, Frechet JMJ, Wooley KL, Penfold J, Webster JRP. Neutron reflectivity and structure of polyether dendrimers as Langmuir films. J Phys Chem 1995;99:8283–8239.

Saville PM, White JW, Hawker CJ, Wooley KL, Frechet JMJ. Dendrimer and polystyrene surfactant structure at the air–water interface. J Phys Chem 1993;97:293–294.

Sayed-Sweet Y, Hedstrand DM, Spinder R, Tomalia DA. Hydrophobically modified poly(amidoamine) (PAMAM) dendrimers: their properties at the air–water interface and use as nanoscopic container molecules. J Mater Chem 1997;7:1199–1205.

Schenning APHJ, Elissen-Roman C, Weener J-W, Baars MWPL, Van der Gaast SJ, Meijer EW. Amphiphilic dendrimers as building blocks in supramolecular assemblies. J Am Chem Soc 1998;120:8199–8208.

Schluter AD, Rabe JP. Dendronized polymers: synthesis, characterization, assembly at interfaces, and manipulation. Angew Chem Int Ed 2000;39:864–883.

Schmitzer A, Perez E, Rico-Lattes I, Lattes A. First example of high asymmetric induction at the "pseudo-micellar" interface of a chiral amphiphilic dendrimer. Tetrahedron Lett 1999;40:2947–2950.

Seyrek E, Dubin PL, Newkome GR. Effect of electric field on the mobility of carboxyl-terminated dendrimers. J Phys Chem B 2004;108:10168–10171.

Shao H, Lockman JW, Parquette JR. Coupled conformational equilibria in beta-sheet peptide–dendron conjugates. J Am Chem Soc 2007;129:1884–1885.

Smith DK. Dendritic gels—many arms make light work. Adv Mater 2006a;18:2773–2778.

Smith DK. Dendritic supermolecules—towards controllable nanomaterials. Chem Commun 2006b;34–44.

Su A, Tan S, Thapa P, Flanders BN, Ford WT. Highly ordered Langmuir–Blodgett films of amphiphilic poly(propylene imine) dendrimers. J Phys Chem C 2007;111: 4695–4701.

Suek NW, Lamm MH. Effect of terminal group modification on the solution properties of dendrimers: a molecular dynamics simulation study. Macromolecules 2006;39: 4247–4255.

Syud FA, Stanger HE, Gellman SH. Interstrand side chain–side chain interactions in a designed beta-hairpin: significance of both lateral and diagonal pairings. J Am Chem Soc 2001;123:8667–8677.

Takeoka S, Mori K, Ohkawa H, Sou K, Tsuchida E. Synthesis and assembly of poly(ethylene glycol)-lipids with mono-, di-, and tetraacyl chains and a poly(ethylene glycol) chain of various molecular weights. J Am Chem Soc 2000;122:7927–7935.

Terao T. Counterion distribution and many-body interaction in charged dendrimer solutions. Mol Phys 2006;104:2507–2513.

Terao T, Nakayama T. Molecular dynamics study of dendrimers: structure and effective interaction. Macromolecules 2004;37:4686–4694.

Terech P, Weiss RG. Low-molecular mass gelators of organic liquids and the properties of their gels. Chem Rev 1997;97:3133–3159.

Tian L, Hammond PT. Comb–dendritic block copolymers as tree-shaped macromolecular amphiphiles for nanoparticle self-assembly. Chem Mater 2006;18:3976–3984.

Tomalia DA, Berry V, Hall M, Hedstrand DM. Starburst dendrimers. 4. Covalently fixed unimolecular assemblages reiminiscent of spheroidal micelles. Macromolecules 1987;20:1164–1167.

Tomalia DA, Naylor AM, Goddard III WA. Starburst dendrimers: control of size, shape, surface chemistry, topology and flexibility in the conversion of atoms to macroscopic materials. Angew Chem 1990;102:119–157.

Tsai C-J, Maizel JV Jr, Nussinov R. The hydrophobic effect: a new insight from cold denaturation and a two-state water structure. Crit Rev Biochem Mol Biol 2002;37:55–69.

Tsuda K, Dol GC, Gensch T, Hofkens J, Latterini L, Weener JW, Meijer EW, De Schryver FC. Fluorescence from azobenzene functionalized poly(propylene imine) dendrimers in self-assembled supramolecular structures. J Am Chem Soc 2000;122:3445–3452.

Tully DC, Frechet JMJ. Dendrimers at surfaces and interfaces: chemistry and applications. Chem Commun 2001;1229–1239.

van Hest JCM, Baars MWPL, Elissen-Roman C, van Genderen MHP, Meijer EW. Acid-functionalized amphiphiles, derived from polystyrene–poly(propylene imine) dendrimers, with a pH-dependent aggregation. Macromolecules 1995;28:6689–6691.

van hest JCM, Delnoye DAP, Baars MWPL, Elissen-Roman C, van Genderen MHP, Meijer EW. Polystyrene–poly(propylene imine) dendrimers: synthesis, characterization, and association behavior of a new class of amphiphiles. Chem Eur J 1996;2:1616–1626.

van Hest JCM, Delnoye DAP, Baars MWPL, van Genderen MHP, Meijer EW. Polystyrene–dendrimer amphiphilic block copolymers with a generation-dependent aggregation. Science 1995;268:1592–1595.

Weener J-W, Meijer EW. Photoresponsive dendritic monolayers. Adv Mater 2000;12:741–746.

Welch P, Muthukumar M. Tuning the density profile of dendritic polyelectrolytes. Macromolecules 1998;31:5892–5897.

Wooley KL, Klug CA, Tasaki K, Schaefer J. Shapes of dendrimers from rotational-echo double-resonance NMR. J Am Chem Soc 1997;119:53–58.

Yabe A, Kawabata Y, Niino H, Tanaka M, Ouchi A, Takahashi H, Tamura S, Tagaki W, Nakahara H, Fukuda K. Cis–trans isomerization of azobenzenes included as guests in Langmuir–Blodgett films of amphiphilic beta-cyclodextrin. Chem Lett 1988;1–4.

Yang Z, Xu B. Supramolecular hydrogels based on biofunctional nanofibers of self-assembled small molecules. J Mater Chem 2007;17:2385–2393.

Yin M, Bauer R, Klapper M, Muellen K. Amphiphilic multicore-shell particles based on polyphenylene dendrimers. Macromol Chem Phys 2007;208:1646–1656.

Young JK, Baker GR, Newkome GR, Morris KF, Johnson CS Jr. "Smart" cascade polymers. Modular syntheses of four-directional dendritic macromolecules with acidic, neutral, or basic terminal groups and the effect of pH changes on their hydrodynamic radii. Macromolecules 1994;27:3464–3471.

Yount NY, Bayer AS, Xiong YQ, Yeaman MR. Advances in antimicrobial peptide immuno-biology. Biopolymers 2006;84:435–458.

Zhang H, Dubin PL, Kaplan J, Moorefield CN, Newkome GR. Dissociation of carboxyl-terminated cascade polymers: comparison with theory. J Phys Chem B 1997;101:3494–3497.

Zhang H, Dubin PL, Ray J, Manning GS, Moorefield CN, Newkome GR. Interaction of a poly-cation with small oppositely charged dendrimers. J Phys Chem B 1999;103:2347–2354.

Zhang S, Rio Y, Cardinali F, Bourgogne C, Gallani J-L, Nierengarten J-F. Amphiphilic diblock dendrimers with a fullerene core. J Org Chem 2003;68:9787–9797.

Zhang W, Xie J, Yang Z, Shi W. Aggregation behaviors and photoresponsive properties of azobenzene constructed phosphate dendrimers. Polymer 2007;48:4466–4481.

Zhu C, Hard C, Lin C, Gitsov I. Novel materials for bioanalytical and biomedical applications: environmental response and binding/release capabilities of amphiphilic hydrogels with shape-persistent dendritic junctions. J Polym Sci Part A Polym Chem 2005;43:4017–4029.

BIOMOLECULAR RECOGNITION USING POLYMERS

CHAPTER 12

COLORIMETRIC SENSING AND BIOSENSING USING FUNCTIONALIZED CONJUGATED POLYMERS

AMIT BASU

12.1. INTRODUCTION

A sensor is a device or material that generates a measurable output when an external stimulus is applied. Living systems routinely use molecular recognition, the association of two or more molecules, as a stimulus for sensing in the biological context. In these cases the output typically consists of subsequent molecular recognition events, such as modulation of enzymatic activity, activation of signaling pathways, or communication with neighboring cells. Abiotic sensors, the goal of which is analyte detection, have been successfully developed by exploiting many natural molecular recognition systems. Synthetic recognition systems based on biomimicry and host–guest systems have also been used. The outputs of these synthetic sensors include changes in electrochemical properties, refractive indices, or spectroscopic signatures.

Conjugated polymers have been extensively studied for their applications in sensing (Mcquade et al. 2000). They can be readily modified with molecular recognition elements before or after polymerization, affording a modular platform on which sensors may be built or constructed. In addition, analyte binding often perturbs the optical and/or conductive properties of conjugated polymers, making them useful for the development of optical or electrochemical sensors. Although numerous examples of conjugated polymers that change their fluorescence properties upon analyte binding have been reported, several types of conjugated polymers also undergo striking visible color changes. Colorimetric sensing has the advantage that it frequently requires nothing more than the human eye for detection, although further quantification can be carried out using absorption spectroscopy.

Molecular Recognition and Polymers: Control of Polymer Structure and Self-Assembly.
Edited by V. Rotello and S. Thayumanavan
Copyright © 2008 John Wiley & Sons, Inc.

In this chapter we discuss applications of conjugated polymers in which the readout is observed colorimetrically. We provide an overview of the platforms that have been employed and the diversity of screened analytes. The chapter focuses primarily on polydiacetylenes (PDAs), reflecting the large body of literature on PDA-based molecular recognition and sensor development (Okada et al. 1998; Jelinek and Kolusheva 2001). We also discuss polythiophene, a type of conjugated polymer that is finding increasing use in sensor applications (McCullough 1998; Leclerc 1999).

12.2. PDA

Diacetylenes undergo photochemical polymerization to provide a conjugated enyne polymer, as shown in Figure 12.1. The polymerization has a topochemical requirement: the α-carbon of one diyne must be positioned in proximity to the δ-carbon of an adjacent diyne. The resulting PDAs are frequently colored a deep blue or purple. The blue/purple PDAs can be converted to a red material by a variety of external stimuli, including pH, irradiation time and intensity, solvent (solvatochromism), and temperature (thermochromism). The specific mechanisms of the blue to red chromatic transition are context dependent and have not been established unambiguously (Carpick et al. 2004; Schott 2006). These chromatic transitions have been studied most thoroughly in the context of thermochromism, and they generally involve a perturbation of one or both of the following two features: 1) head group packing and orientation and 2) conformation of main chain rotamers along the enyne backbone. These perturbations change the core chromophore of the PDA by affecting the planarity of the enyne, which is manifested by the color change. The chromatic transition can also be followed using fluorescence spectroscopy because the red form of PDA is fluorescent whereas the blue form is not.

Early work in diyne polymerizations focused on the formation of PDA from diynes that adopted the proper geometry in the crystalline state (Baughman 1972; Wegner 1972; Bässler 1984). If the diacetylene is amphiphilic, a variety of self-assembled structures, such as monolayers, vesicles, and fibrillar superstructures, can form. Many of these self-assemblies orient the diyne in a geometry that allows photochemical polymerization. The synthesis of phospholipid analogs in which the naturally occurring fatty acids on the glycerol were replaced with diacetylenic acids was reported in 1980 (Hub et al. 1980). These materials were assembled into liposomes that were subsequently photochemically polymerized to yield blue or red solutions, depending on the extent of irradiation. The efficiency of this polymerization is correlated with vesicle size (Peek et al. 1994).

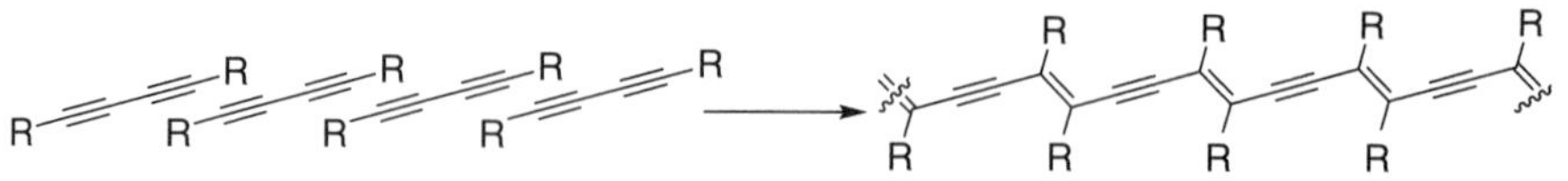

Figure 12.1 Topochemical polymerization of 1,3-diacetylenes.

Studies with liposomes prepared using amino acid terminated diacetylenes indicate that head-group rearrangements are capable of triggering lipid conformational changes, which in turn perturb the conformation of the PDA backbone (Cheng and Stevens 1998). These amino acid terminated diacetylenes can also self-assemble into fibrous structures under different conditions (Cheng et al. 2000). These fibers can be polymerized to provide a blue colored material that undergoes chromatic transitions upon changes in pH or temperature. The IR spectra of these materials indicate that different mechanisms of chromatic transition may be operative in the thermochromatic and pH-dependent responses.

12.2.1. Glycolipids

Despite the rich chemistry of the diacetylenes, their application in chemical and biological sensing was not extensively explored before a seminal report in 1993 describing the use of solid-supported PDA monolayers for colorimetric detection of the influenza virus (Charych et al. 1993; Fig. 12.2). A Langmuir–Blodgett monolayer containing diyne acid **1** and sialolipid **2** was prepared on a glass slide and subsequently irradiated to provide a blue film. The viral surface was coated with a lectin (a carbohydrate-binding protein) that recognizes sialic acid residues. Consequently, incubation of the coated slides in a solution of the X31 influenza A virus resulted in a color change of the film from blue to red, as shown in Figure 12.3a. The magnitude of this color change could be quantified by determining the colorimetric response (CR value). The CR value is calculated from the ratio of the

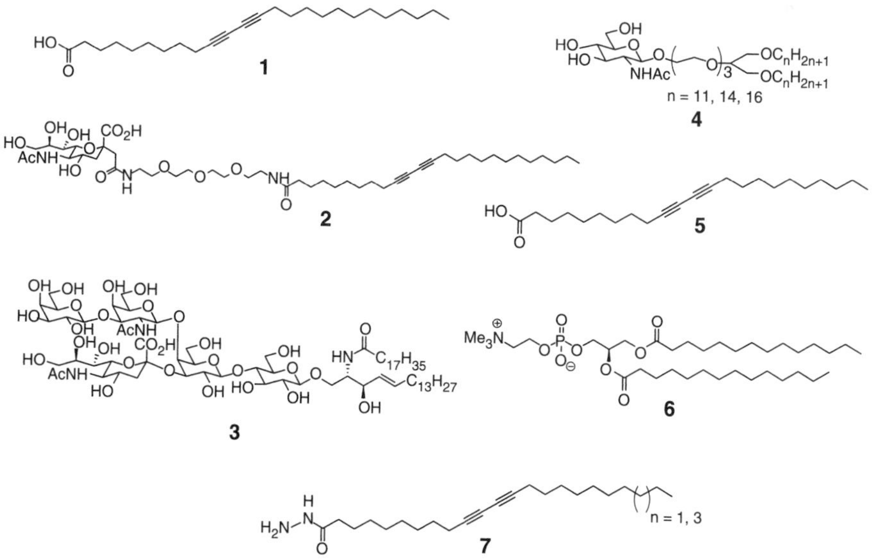

Figure 12.2 Structures of glycolipids and other lipids used in polydiacetylene sensors.

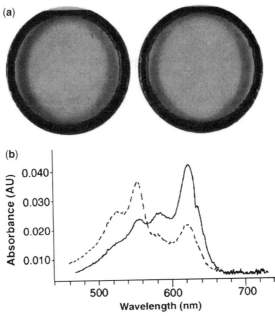

Figure 12.3 Colorimetric detection of influenza by sialoside bilayer assembly (2% sialoside lipid **2** and 98% matrix lipid **1**). (a) The colorimetric response of the film, supported on a glass microscope slide, is readily visible to the naked eye for qualitative evaluation of the presence of virus. The film on the left (blue) has been exposed to a blank solution of PBS. The film on the right (red) has been exposed to 100 hemagglutinin units (HAU) of virus (CR = 77%). (b) The visible absorption spectrum of a bilayer assembly (—) prior to and (- - -) after viral incubation. Reprinted from Charych et al. (1993). Copyright 1993 American Association for the Advancement of Science. (See color insert.)

absorbances of the "blue" and "red" forms of the thin film before and after the chromatic transition (Fig. 12.3b), as given by Eq. (1):

$$CR = \frac{B_o - B_{virus}}{B_o} \times 100\%$$

$$B_o = \frac{A_{620}}{A_{620} + A_{550}} \text{ and } B_{virus} = \frac{A_{620}}{A_{620} + A_{550}} \tag{1}$$

The chromatic transition was completely blocked by α-methyl sialoside, a soluble competitive inhibitor of the influenza virus, demonstrating that binding of the virus to the film was due to a specific carbohydrate–protein recognition event. This experiment demonstrated that the chromatic transition of the PDA backbone could be triggered by molecular recognition of a membrane surface ligand that was covalently attached to the polymerized lipid.

A subsequent experiment extended this approach to the use of liposomes instead of monolayers (Reichert et al. 1995; Charych et al. 1996). PDA vesicles containing

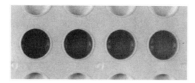

Figure 12.4   Colorimetric detection of influenza virus using polymerized liposomes containing sialic acid. (a) Photograph of liposomes to which have been added increasing amounts (from left to right) of influenza virus. Liposomes were **1** and 95% **5**. To each well was added the following amounts of influenza virus (left to right): 0, 8, 16, and 32 hemagglutinin units (HAU). Reprinted from Charych et al. (1996). Copyright 1996 Elsevier Science. (See color insert.)

2 were prepared using probe sonication at 80 °C and subsequently polymerized to provide a blue liposome solution. Addition of increasing amounts of the influenza virus afforded a dose-dependent color change in the solution, easily discernable by the naked eye (Fig. 12.4).

As seen with the thin films, addition of a competitive small molecule ligand for the virus reduced the CR. Longer UV irradiation resulted in the formation of purple liposomes that were more sensitive to the influenza virus. The increased sensitivity was suggested to arise as a consequence of the greater degree of polymerization induced by the longer irradiation time.

In both of these cases, the ligand (sialic acid) for the analyte of interest (influenza virus) was covalently linked to the PDA backbone generated upon photopolymerization. Functional sensors based on ligands that are noncovalently incorporated into liposomes have also been reported (Charych et al. 1996; Pan and Charych 1997). Mixed liposomes as well as mixed thin films on glass containing a combination of the ganglioside GM1 and diacetylene lipids detect the presence of cholera toxin, a protein that binds to GM1.

Glycerol derived lipids terminated with *N*-acetyl glucosamine have been noncovalently incorporated into PDA liposomes and used to study the binding of the lectin concanavalin A (ConA; Guo et al. 2005). The CR values in this study were dependent on 1) the length of the lipid chain and 2) the size of the liposomes. It was suggested that shorter glycolipids were not as effective at transducing a signal to the PDA chromophore upon molecular recognition. In contrast, longer lipids were in the gel phase at the temperature used for binding, which attenuated signal transduction. When the polymerized vesicles were extruded through membranes with various pore sizes to prepare smaller vesicles, the CR was inversely correlated to pore size. The higher CR values are believed to arise from the higher surface curvature of the smaller vesicles, which sensitizes the PDA crosslinking to the surface molecular recognition events by enabling the binding of more ConA.

Glycolipid incorporated liposomes have found extensive use as sensors for the detection of *Escherichia coli* bacteria. Liposomes prepared using a diacetylene and a glucosyl lipid underwent a chromatic transition upon the addition of *E. coli* (Ma et al. 1998). The chromatic transition is sensitive to the diyne and glycolipid structure (Ma et al. 2000). An optimized vesicle assembly, consisting of a maltotriosyl lipid, phospholipid, and diyne, detected *E. coli* at a concentration of 2×10^7 cells/mL

(Su et al. 2005). Mannosyl lipid containing liposomes have also been used for this purpose (Zhang et al. 2005). The glycolipids are presumed to interact with lectins present on the surface of the bacterial cells.

12.2.2. Other Ligands

PDA liposomes have been functionalized with a variety of ligands besides carbohydrates. Liposomes containing a mixture of biotin-terminated diyne lipids and carboxy-terminated diyne lipids were prepared (Jung et al. 2006). The irradiation time required for cross-linking decreased as the mole fraction of the biotinylated lipid was increased. Addition of streptavidin to these liposomes resulted in a chromatic transition from blue to red. Because of the high concentrations of streptavidin used in this experiment, the resulting red liposomes precipitated as a result of liposome cross-linking induced by streptavidin, which contains four binding sites for biotin. A 24-mer target DNA sequence was detected colorimetrically using two liposomes, each functionalized with a 12-mer probe sequence complementary to the $5'$ or $3'$ end of the target sequence (Wang and Ma 2005; Wang et al. 2006). The probe sequences were anchored to the liposome using a cholesterol conjugate of the oligonucleotide. A diyne lipid terminated in a hexapeptide sequence containing the Arg-Gly-Asp tripeptide sequence was used to detect a soluble form of the $\alpha_5\beta_1$ integrin receptor (Biesalski et al. 2005).

Several approaches for using PDA liposomes to detect antibody–antigen interactions were reported. One approach fused peptide epitopes to membrane spanning α-helical peptides as a method for incorporating antigens into mixed diyne/phospholipid liposomes (Kolusheva et al. 2001). A synthetic peptide anchor as well as one derived from a phage coat protein were successfully used as membrane anchors. A variety of peptide epitopes were detected, and the specificities obtained from the colorimetric assay compared favorably to the results from a more conventional ELISA assay. A second strategy conjugated the antibodies directly to PDA liposomes using succinimide ester bioconjugation chemistry (Su et al. 2004). Anti-human immunoglobulin (Ig) antibody from goat was conjugated to PDA liposomes that were subsequently photopolymerized. Addition of human Ig to a solution of these liposomes induced a chromatic transition in a concentration-dependent fashion.

Mixed PDA/phospholipid vesicles, most frequently prepared with a 3:2 ratio of tricosadiynoic acid (**5**) and dimyristoyl phosphatidylcholine (**6**, DMPC), have been used extensively for sensing applications. These vesicles have been characterized in detail using a variety of techniques, including small-angle X-ray scattering (SAXS), differential scanning calorimetry (DSC), and electron spin resonance (Kolusheva et al. 2003). These studies suggest that the phospholipids and diacetylenes assemble into distinct domains. Formation of ternary liposomes using cholesterol as the third component reduces the thermochromic response of the vesicles, consistent with decreased phospholipid fluidity upon cholesterol addition.

12.2.3. Small Molecules

PDA vesicles have been used to screen small molecules that are known to interact with membranes. Screening of almost 40 compounds against PDA/DMPC liposomes afforded three sets of distinctive CR profiles (Katz et al. 2006). One set of compounds provided uniformly low CR values, independent of the small molecule concentration. A second class of compounds afforded high CR values but only at high micromolar or millimolar concentrations of the small molecule. The last set of molecules also gave rise to high CR values but at low micromolar concentrations. Fluorescence, DSC, and SAXS measurements of compound binding to the PDA vesicles was carried out to determine the mechanism of action for these three compound classes. The low CR values with the first set of compounds was attributed to the weak membrane binding of these molecules. Compounds that resulted in high CR values at high concentrations were molecules that insert into the phospholipid bilayer but do not induce large structural lipid rearrangement (the trigger for a chromatic transition) until higher concentrations. Compounds that bind and aggregate at the surface of the membrane bilayer induce perturbations in lipid packing at very low concentrations, triggering the color change of the vesicle. As a result, we can rapidly categorize a given molecule into one of the three compound classes by measuring the CR at three different compound concentrations (Fig. 12.5).

The lipid-binding and perturbing activities of small molecules that facilitate drug delivery through skin and cellular membranes, known as membrane permeation enhancers, have been studied using PDA liposomes (Evrard et al. 2001; Valenta et al. 2004). Some enhancers such as oleic acid exert their effects directly and induce chromatic transitions when added to PDA/phospholipid vesicles. Other enhancers such as the oligoethylene glycol Transcutol only give rise to a chromatic transition when added to vesicles in the presence of a drug, suggesting a synergistic mode of action. Addition of α-cyclodextrin, a cyclic oligosaccharide that forms inclusion complexes with hydrophobic molecules, to PDA vesicles triggers chromatic transitions (Kim, Lee, Lee, et al. 2005).

12.2.4. Cation and pH Sensors

Liposomes prepared from diacetylene hydrazide **7** undergo chromatic transitions in response to pH changes (Jonas et al. 1999). Liposomes of **7**, which form only in

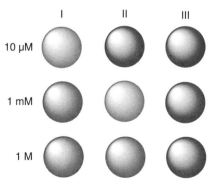

Figure 12.5 Patterns generated upon addition of a membrane surface-binding molecule (I), membrane penetrating compound (II), and a compound that does not interact with membranes (III). Adapted from Katz et al. (2006). Copyright 2006 Springer. (See color insert.)

acidic solution, undergo a color change from blue to red upon the addition of ammonia. Notably, addition of HCl to the red solution, which reprotonates the terminal nitrogen on the hydrazide, regenerates a blue solution. This is one of the few examples reported of a reversible chromatic transition of PDA sensors (Ahn et al. 2003; Kim, Lee, Choi, et al. 2005). Furthermore, the hydrazides form gels in organic solution at low temperatures, and even dilute organic solutions can be polymerized, indicative of ordered self-assembly even under dilute conditions. IR and computational studies suggest that this self-assembly is mediated by hydrogen bonding within the hydrazide head groups.

The chromatic transition can also be triggered by changes in pH or the addition of cations. Titration of vesicles composed of **5** with sodium hydroxide induces a chromatic transition (Kew and Hall 2006). The color changes can also be induced using calcium, cesium, and potassium as counterions, with higher CR values obtained with the larger cations cesium and potassium. Smaller cations such as lithium and magnesium afford lower CR values. The chromatic transition was accompanied by a transformation in vesicle morphology from globular to sheetlike structures, as determined by transmission electron miscroscopy (TEM). Similar results were observed for the binding of various metal cations to PDA/phospholipid vesicles enriched in anionic phospholipids (Rozner et al. 2003). Incorporation of metal-binding ligands and ionophores into the vesicles sensitizes them to different cations, depending on the cation-binding selectivities of the ligands. PDA/phospholipid vesicles containing the ionophore valinomycin undergo chromatic transitions upon the addition of alkali metal cations (Kolusheva, Shahal, et al. 2000). The magnitude of the resulting CR is cation dependent and parallels the known relative affinities of the cations for valinomycin itself. Similar results were also obtained with the ionophores monensin and A23187. Fluorescence anisotropy and ^{13}C-NMR studies suggest that the chromatic transition is induced by increasing the fluidity of the phospholipids upon cation binding to the ionophore.

12.2.5. Membrane-Binding Peptides

A variety of antibacterial peptides and related toxins function by binding to phospholipid membranes and weakening them by forming transmembrane pores or other structural perturbations. Vesicles prepared from a mixture of PDA lipids and phospholipids have been used extensively to detect the binding of these peptides and toxins. α-Helical membrane-binding peptides such as melittin, magainin, and alamethicin induce a chromatic transition when added to PDA/DMPC liposomes (Kolusheva, Boyer, et al. 2000). The assay is sensitive enough to distinguish point mutants of the peptides that do not have antibacterial activity or do not adopt the proper fold upon membrane binding. Nonspecific binding to the liposomes can be a potential problem, but it can be determined and corrected for by using vesicles lacking phospholipids. Other peptides and toxins that have been examined include indolicidin, an antibiotic derived from bovine neutrophils (Halevy et al. 2003), defensins, found in mammalian immune cells (Satchell et al. 2003), and synthetic lysine-rich amphipathic α-helical polypeptides (Oren et al. 2002; Sheynis et al. 2003).

When alamethicin is added to a ternary vesicle system comprising PDA, phospholipid, and lipopolysaccharide (LPS), the addition of polymyxin, an LPS-binding antibiotic, sensitizes the vesicles to alamethicin (Katz et al. 2003). Cholesterol-containing PDA liposomes have been used to colorimetrically detect streptolysin O, a cholesterol-dependent pore-forming toxin (Ma and Cheng 2005).

12.2.6. Membrane Active Enzymes

Enzymes that act upon membrane-bound lipids can be conveniently assayed using PDA liposomes. Vesicles of PDA and DMPC were prepared and incubated with phospholipase A2, an enzyme that hydrolyzes the myristoyl ester in DMPC (Jelinek et al. 1998). The CR of the solution increases in a time-dependent manner after addition of the enzyme. Other enzymes that used phospholipids as substrates, such as bungaratoxin and phospholipases C and D, all exhibit similar time-dependent increases in the CR values. These chromatic transitions arise as a result of membrane perturbations caused by the various hydrolyzed products of the enzymatic reactions. This assay can be used to distinguish between lipids from bacteria and archaea (Rozner et al. 2003). In contrast to bacterial and eukaryotic membranes, archaeal membranes contain ether linked lipids that are resistant to phospholipase mediated hydrolysis. Thus, when liposomes containing membrane extracts from the prokaryote *E. coli* and the archaea *Haloferax volcanii* are incubated with phospholipase A2, the CR from the former liposomes is three times higher than that from the vesicles containing lipids from *H. volcanii*. Other enzymes examined included sphingomyelinase and galactosidase, using sphingomyelin and galactosyl ceramide containing vesicles, respectively (Rozner et al. 2003).

12.2.7. Pattern Recognition

The use of pattern recognition principles in molecular recognition has had a tremendous influence on sensor development (Albert et al. 2000; Jurs et al. 2000; Collins and Anslyn 2007). Several reports describe the use of these principles with both PDA sensors that detect analytes using these principles. Vesicles prepared from tyrosine (Tyr) or tryptophan (Trp) terminated diyne lipids were used to prepare sensors for bacterial LPSs (Rangin and Basu 2004). Exposure of the Tyr or Trp containing liposomes to LPS from a given bacterial species under a variety of experimental conditions (temperature, additives) afforded CR values for each of the eight assay conditions. This panel of CR values served as a diagnostic fingerprint for each species of LPS examined (Fig. 12.6).

Pattern recognition using calixarene receptors that bind to proteins via surface electrostatic interaction has been used to identify a variety of proteins (Kolusheva et al. 2006). Amphiphilic calixarenes terminated with either amino or phosphate groups were incorporated into mixed PDA/phospholipid vesicles, which were incubated with various proteins that differed in their isoelectric points (pI). As expected, proteins with low pI values resulted in a large CR with liposomes containing cationic calixarene receptors. Each protein was characterized by a unique ΔCR value, where

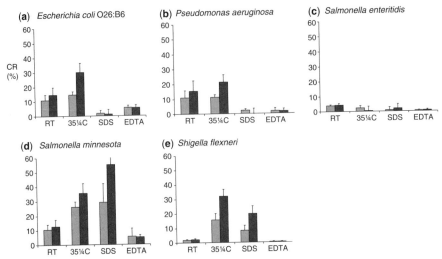

Figure 12.6 Colorimetric response (CR) values obtained upon exposure of the liposome array to lipopolysaccharides from different Gram negative bacteria for (▨) Trp and (■) Tyr. All values are the average of at least four experiments. RT, room temperature; [liposome] ≈ 0.6 mM, [LPS] ≈ 2.2 mg/mL, [SDS] = 2 mM, [EDTA] = 1 mM. See supporting information for details.

ΔCR corresponds to the difference between the CR in the presence of the calixarene in the vesicles and the CR in the absence of the calixarene receptor (to correct for nonspecific interactions). A plot of ΔCR for the cationic calixarene versus ΔCR for the anionic calixarene gave rise to a unique diagnostic point for each protein that was examined.

12.2.8. Whole Cells

The propensity of diyne lipids to phase separate from phospholipids has been successfully exploited to incorporate PDA patches into whole cells as functioning sensors. Addition of unpolymerized PDA/phospholipid vesicles to erythrocyte ghosts (red blood cells that have had their hemoglobin removed) resulted in incorporation of the vesicle lipids into the erythrocyte cell membrane (Shtelman et al. 2006). Subsequent UV irradiation of the solution resulted in the formation of the characteristic blue color of PDA. Addition of membrane binding or disrupting agents resulted in a chromatic transition to the red form. Moreover, because the red form of PDA is fluorescent, these chromatic transitions could also be followed by fluorescence microscopy. A similar set of experiments was carried out using U937 cells, a human leukemic monocyte lymphoma cell line (Orynbayeva et al. 2005). The viability of cells containing these PDA patches remained unchanged soon after treatment but decreased after 3 h.

12.2.9. Supported PDAs

A considerable amount of effort has been directed toward increasing the portability of PDA-based sensors by immobilizing them in various matrices. The original 1993 report described formation of sialic acid containing PDA monolayers on glass slides for sensing influenza virus (vide supra; Charych et al. 1993). Subsequently, several groups reported the preparation of supported PDA monolayers containing glycolipids or a biotinylated diyne (Geiger et al. 2002; Guo et al. 2007). The addition of a lectin or *E. coli* to the glycolipid monolayers or the addition of streptavidin to the biotin-containing monolayer induced chromatic transitions that were detected optically. Spotting of polymyxin B on mixed PDA/phospholipid films deposited on glass slides via Langmuir–Schaefer deposition resulted in the formation of red spots on a blue background (Volinsky et al. 2007). These slides were scanned and the color change was quantitated by digital image analysis. PDA/phospholipid vesicles fused onto silica beads was recently reported (Nie et al. 2006). The chromatic transition on the beads was verified by addition of phospholipase A2, a known analyte for these vesicles (vide supra). The color change on individual beads was readily discernable via standard optical microscopy.

A PDA-based glucose sensor was created by treating a Langmuir–Blodgett film containing a diyne acid activated ester with the enzyme hexokinase, immobilizing the enzyme via amide bond formation (Cheng and Stevens 1997). Hexokinase, which catalyzes the phosphorylation of glucose to provide glucose-6-phosphate, undergoes a large conformational change upon glucose binding. Addition of glucose to the immobilized enzyme induces a chromatic transition in the film, indicating that the conformational change can trigger the color change. The enzyme enolase has also been immobilized using this approach, and substrate binding was detected colorimetrically (Sadagopan et al. 2006).

Liposomes containing the sialic acid containing lipid **2** were immobilized in a silica sol–gel matrix (Yamanaka et al. 1997). A chromatic transition was observed upon addition of influenza virus to the matrix, but the response time was much slower than that of liposome solutions or monolayer coated slides. Sol–gel entrapment of PDA liposomes has since been studied extensively and optimized for high-throughput analysis (Gill and Ballesteros 2003). Six diacetylene lipids, seven phospholipids, and three sol–gel precursors were screened in a combinatorial manner to identify a composition that permitted efficient and reproducible incorporation of PDA–IgG conjugates within the sol–gel. Large CRs using picograms per milliliter concentrations of antigens were obtained within 10 min. PDA vesicles were incorporated into PEG-based hydrogels, and chromatic transitions mediated by cyclodextrin–diacetylene binding were demonstrated (Lee et al. 2006). These hydrogels can also be prepared inside microfluidic channels, facilitating high-throughput analyses. Diffusion of the cyclodextrin through the hydrogel was found to be a limiting factor.

PDA lipids embedded in silica fibers have been used for the detection of volatile organic compounds (Yoon et al. 2007). The fibers are prepared by subjecting a syringe tip containing a mixture of a diacetylene lipid, poly(ethylene oxide), and tetraethyl orthosilicate to a high voltage, resulting in the formation of a mat of thin silica fibers containing PDA assemblies within them. Subsequent irradiation produces the blue form of the mat. Addition of various organic solvents to a panel of four mats, each containing a different diacetylene, results in the formation of a slightly different color in each mat, generating a diagnostic fingerprint for the identification of these volatile organic materials (Fig. 12.7).

Polymerized PDA/phospholipid vesicles embedded in agar were used to detect bacterial growth in the agar plate (Silbert et al. 2006). Zones of the liposome-doped agar that were closer to the site of bacterial growth turned red, while the rest of the agar remained blue. Control experiments using solutions of liposomes indicated that the trigger for the chromatic transition was not the bacteria themselves. Instead, it was suggested that the color change was attributable to interactions of the liposomes with molecules shed from bacterial cell surfaces, such as secreted membrane-binding peptides and LPSs (Fig. 12.8).

Although the agar assay permits the detection of bacteria, it does not allow one to distinguish between different bacterial species. A subsequent version of this assay reported the use of Langmuir–Blodgett monolayers of mixed phospholipid/PDA films for bacterial detection and identification (Scindia et al. 2007). Four different slides, each coated with a monolayer of PDA and different phospholipids, were prepared. As a result of the different surface charges and lipid compositions, each monolayer afforded a slightly different CR value when exposed to different species of bacteria. Consequently, each species generated a distinct pattern of CRs, which could be used diagnostically (Fig. 12.9).

A variety of methods for immobilizing polymerized liposomes have been reported. Multiple layers of vesicles were deposited on quartz slides by layer-by-layer assembly (Su 2005). The slides were first coated with a layer of poly(ethylene imine), a

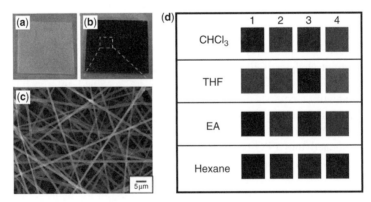

Figure 12.7 Photographs of electrospun fiber mats embedded with **1** (a) before and (b) after 254-nm UV irradiation (1 mW/cm^2) for 3 min. (c) Scanning electron microscopy image of the microfibers containing polymerized **1**. (c) Photographs of the polydiacetylene-embedded electrospun fiber mats prepared with various diacetylene monomers after exposure to organic solvent. Reprinted from Yoon et al. (2007). Copyright 2007 American Chemical Society. (See color insert.)

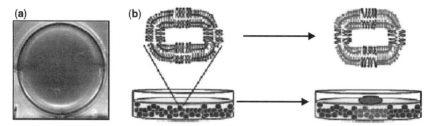

Figure 12.8 (a) Image of diacetylene liposome-embedded agar plate with three *E. coli* BL colonies after 18 h of growth. (b) Chromatic sensor concept. (Left) Agar scaffold containing vesicular nanoparticles composed of phospholipids (green) and PDA (blue). (Right) Bacterial proliferation (green oval) on the agar surface causes a blue to red transformation of embedded vesicles due to bacterially secreted compounds that diffuse through the agar. Reprinted from Silbert et al. (2006). Copyright 2006 American Society for Microbiology. (See color insert.)

polycation, followed by a polyanion layer of unpolymerized vesicles composed of **1**. Additional rounds of polycation and polyanion deposition were carried out. Photopolymerization of the slides gave rise to the characteristic blue color of PDA, which underwent a chromatic transition when the slides were heated to 60°C. Layer-by-layer assembly of already polymerized vesicles has also been reported (Potisatityuenyong et al. 2006). These slides undergo a chromatic transition in response to the addition of increasing amounts of ethanol and to changes in pH (Fig. 12.10).

Arrayed PDA liposomes have been prepared via microcontact printing of NHS ester-containing liposomes onto aminopropyl silane terminated glass surfaces

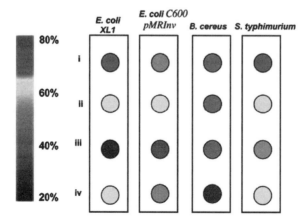

Figure 12.9 Colorimetric bacterial fingerprinting. The color combination for each bacterium reflects the chromatic transitions (RCS) recorded 7 h after the start of growth at a bacterial concentration of 1×10^9/mL. The RCS color key is shown on the left: (i) 1-α-dioleoylphosphatidylethanolamine (DOPE)/PDA (1:9 mole ratio), (ii) sphingomyelin/cholesterol/PDA (7:3:90), (iii) DMPC/PDA (1:9), and (iv) 1-palmitoyl-2-oleoyl-*sn*-glycero-3-[phospho-*rac*-(1-glycerol)] (POPG)/PDA (1:9). Reprinted from Scindia et al. (2007). Copyright 2007 American Chemical Society. (See color insert.)

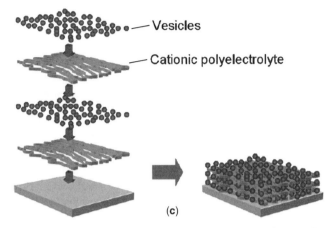

Figure 12.10   Illustration of layer-by-layer assembly on glass substrate. Reprinted from Potisatityuenyong et al. (2006). Copyright 2006 Elsevier Science.

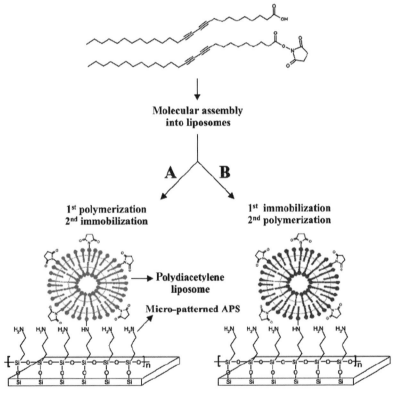

Figure 12.11   Strategies for immobilization of diacetylene liposomes on micropatterned glass substrates: (a) immobilization after polymerization of monomeric liposomes and (b) polymerization after immobilization of monomeric liposomes. Reprinted from Shim et al. (2004). Copyright 2004 Elsevier Science. (See color insert.)

(Shim et al. 2004). When the NHS ester-containing liposomes are polymerized prior to immobilization on the glass, the blue liposomes turn red upon contact with the slide. However, if the liposomes are immobilized prior to polymerization and subsequently irradiated, the immobilized vesicles are blue and can undergo a subsequent chromatic transition in response to heat (Fig. 12.11).

Although the bulk of PDA sensors involve vesicles and Langmuir monolayers, a few examples of responsive PDA assemblies based on bolaamphiphiles and diyne silica nanocomposites have been reported (Lu et al. 2001; Song et al. 2001, 2004; Yang et al. 2003; Peng et al. 2006). Although these materials have not been broadly utilized for analyte sensing, they do exhibit the thermochromic, solvatochromic, and pH responsive behavior seen with monolayers and liposomes and hold promise for future development.

12.3. POLYTHIOPHENES

2,5-Unsubstituted thiophenes undergo polymerization to provide polythiophenes, a class of conjugated polymers that have absorbances in the visible region of the spectrum. Polymerization is carried out using iron(III) salts or initiated electrochemically. The absorbance maxima of the polythiophenes are dependent on the aggregation state of the polymer, mediated by π-stacking interactions, as well as the degree of coplanarity between consecutive thiophene monomers. Numerous examples of ionochromic polythiophenes have been reported where addition of various cations to functionalized polythiophenes induces a color change. An early report described the synthesis of the bridged crown ether shown in Figure 12.12, which exhibits a 91-nm shift in λ_{max} upon binding of sodium (Marsella and Swager 1993). The structurally related oligoethylene polymer **8** was reported to be potassium selective (Levesque and Leclerc 1995).

Polymers of thiophene carboxylic acids (**9**) undergo different chromatic transitions upon the addition of a variety of cations (McCullough et al. 1997). Although the acid form of the polymer was not water soluble, addition of various ammonium or metal

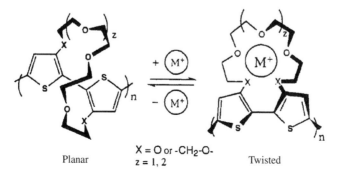

Figure 12.12 Modulation of polythiophene coplanarity using metal binding. Reprinted from Marsella and Swager (1993). Copyright 1993 American Chemical Society.

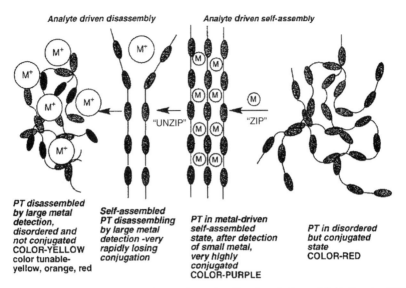

Analyte driven disassembly   *Analyte driven self-assembly*

"UNZIP"   "ZIP"

**PT disassembled
by large metal
detection,
disordered and
not conjugated
COLOR-YELLOW
color tunable-
yellow, orange, red**

**Self-assembled
PT disassembling
by large metal
detection -very
rapidly losing
conjugation**

**PT in metal-driven
self-assembled
state, after detection
of small metal,
very highly
conjugated
COLOR-PURPLE**

**PT in disordered
but conjugated
state
COLOR-RED**

Figure 12.13 Polythiophene sensing schematic. Reprinted from McCullough (1998). Copyright 1998 Wiley–VCH Verlag GmbH and Co. KGaA.

hydroxides resulted in the formation of colored solutions. Larger cations resulted in the formation of red, orange, or yellow solutions, whereas small cations afforded solutions that were purple, the color of the polymer in the solid state. The smaller cations are believed to favor formation of organized stacked assemblies of polymers, whereas larger cations form more disordered aggregates (Fig. 12.13).

Polythiophenes functionalized with monosaccharides have been evaluated for their ability to detect the influenza virus and *E. coli* (Baek et al. 2000). Copolymers of thiophene acetic acid **10** and carbohydrate-modified thiophenes **11** have been prepared via iron(III) chloride mediated polymerization. Addition of influenza virus to a sialic acid containing copolymer resulted in a blue shift of the polymer absorption maximum, resulting in an orange to red chromatic transition. Mannose-containing polythiophenes underwent color changes upon the addition of the lectin ConA or *E. coli* cells that contain cell surface mannose-binding receptors. A similar biotinylated polythiophene afforded a streptavidin responsive material (Faid and Leclerc 1996).

A modular and flexible approach to polythiophene sensors based on the polymerization of a thiophene-activated ester has been reported (Bernier et al. 2002). Subsequent reaction of the polymerized NHS ester with a variety of diamines permits the synthesis of sensors for different analytes from a common platform. For example, reaction of the NHS polymer with an aminomethyl-modified 15-crown-5 derivative yielded a polymer that underwent color changes in the presence of alkali cations (Fig. 12.14).

The carboxylic acid functionalized polythiophene **9** has been used to detect and identify a variety of different diamines (Nelson et al. 2006). Addition of various diamines to a solution of the carboxy polythiophene induces crosslinking of the

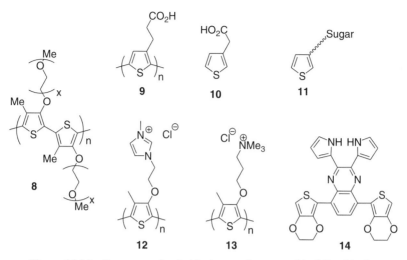

Figure 12.14 Structures of polythiophene polymers and building blocks.

polymer strands, resulting in a chromatic transition. Each diamine induces a slightly different aggregate with very subtle differences in the corresponding absorption spectra. Although the chromatic transitions are not distinct enough for the diamines to be differentiated upon visual inspection, application of pattern recognition algorithms enables one to distinguish between similar diamines such as ethlyene diamine, propane diamine, butane diamine, pentane diamine, and hexane diamine with greater than 99% accuracy. This approach has been used to develop a food freshness sensor based on the detection of histamine, a marker of tuna spoilage (Maynor et al. 2007; Fig. 12.15).

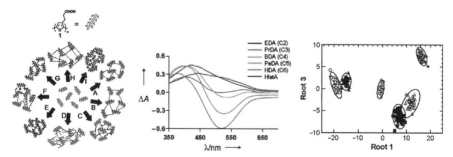

Figure 12.15 (a) Schematic representation of the multidimensional response from polymer **9** (yellow rods) upon analyte-induced aggregation with different amines (A–I). (b) Changes in the absorption spectra of polymer **9** (1 mM) upon addition of different diamines (1 mM each) in aqueous HEPES buffer (40 mM, pH 7.4). (c) Two-dimensional linear discriminant analysis plot of polymer **9** response to various diamines. Reprinted from Maynor et al. (2007). Copyright 2007 American Chemical Society. (See color insert.)

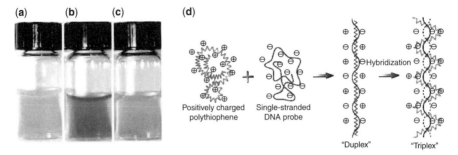

Figure 12.16 Photographs of solutions of (a) polymer **12**, (b) **12** + single-stranded DNA, and (c) **12** + double-stranded DNA. (d) Schematic description of the formation of polythiophene/single-stranded nucleic acid duplex and polythiophene/hybridized nucleic acid triplex forms. Reprinted from Ho et al. (2002). Copyright 2002 Wiley–VCH Verlag GmbH and Co. KGaA.

Conversely, cationic polythiophenes have found applications as chromogenic sensors for a variety of anions, including nucleic acids. The binding of the imidazolium salt functionalized polymer **12** to single-stranded DNA induces a yellow to red chromatic transition. Addition of the complementary DNA strand to this solution regenerates the yellow color. The initial color change is believed to arise from increased linearization of the polythiophene upon binding to DNA. Addition of the complementary strand generates duplex DNA, which the polythiophene can also bind to, but only in a nonlinear fashion (Ho et al. 2002; Fig. 12.16).

Polythiophene binding to a sequence of DNA capable of forming a G-quartet was demonstrated (Ho and Leclerc 2004). Addition of polythiophene **12** to the single-stranded DNA resulted in a chromatic transition from yellow to pink/red.

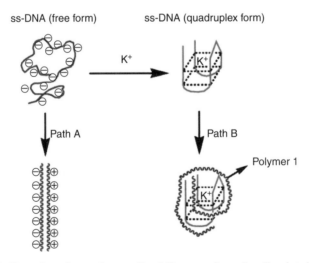

Figure 12.17 Detection of potassium mediated G-quartet formation. Reprinted from Ho and Leclerc (2004). Copyright 2004 American Chemical Society.

The color change is attributed to the formation of a complex between the cationic polythiophene and the polyanionic DNA. Complexation arises through increased linearization and unfolding of both the polythiophene and the nucleic acid, followed by aggregation-induced color changes. The color change is attenuated in the presence of potassium that promotes folding of the nucleic acid into a G-quartet structure, preventing linearization. A related strategy, employing a G-quartet that binds to the enzyme thrombin, has been used for thrombin detection. When a complementary DNA sequence is added to the solution, the color reverts to yellow (Ho et al. 2005). Cationic polythiophenes have been used to detect cleavage of single-stranded DNA by nucleases or hydroxyl radicals. The cleaved strands of DNA are not as effective in aggregating polythiophene as the uncleaved DNA, resulting in the restoration of the yellow color of uncomplexed polythiophene upon cleavage (Tang et al. 2006; Fig. 12.17).

The imidazolium salt functionalized polymer **12** also functions as an iodide sensor, rapidly changing color in the presence of sodium iodide (Ho and Leclerc 2003). Iodide-mediated aggregation was highly sensitive to the length of the side chain connecting the imidazolium salt to the polythiophene backbone. A variety of

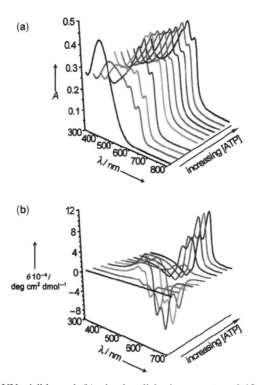

Figure 12.18 (a) UV visible and (b) circular dichroism spectra of **13** (0.10 mM) in the absence and presence of various amounts of adenosine triphosphate (ATP) in water at 20 °C. ATP concentrations (from front to back): 0, 0.001, 0.0125, 0.025, 0.10, 0.15, 0.175, 0.20, 0.25, 0.375, and 0.50 mM. Reprinted from Li et al. (2006). Copyright 2006 Wiley–VCH Verlag GmbH and Co. KGaA.

phosphate responsive polythiophenes have been reported. Tetra-alkylammonium-functionalized polythiophene **13** underwent chromatic transitions in the presence of a variety of phosphate anions, including triphosphate and adenosine mono-, di-, and triphosphates (ATP; Li et al. 2005). The polymers were most selective for ATP, generating a bright pink color when the nucleoside was added to the yellow polymer solution. Addition of ATP to the polythiophene gave rise to new absorbances in the circular dichroism (CD) spectrum of the solution, indicating formation of a chiral aggregate (Li et al. 2006). It is interesting that the CD spectrum of the solution underwent an inversion at increasing concentrations of ATP. Additional TEM and atomic force miscroscopy studies of the polymers suggested that the different CD spectra arose from the formation of structurally different aggregates at different ATP concentrations, possibly modulated by adenine–adenine interactions (Fig. 12.18).

Quinaxoline-containing monomer **14** was electrochemically polymerized to yield a polythiophene that changed from a yellow to orange color upon the addition of fluoride or pyrophosphate anions (Aldakov and Anzenbacher 2004). Analyte binding could be detected spectroscopically and electrochemically.

12.4. OTHER MATERIALS

Polyphenylene and polyfluorene have been extensively used as fluorescence-based sensors, and several chromogenic forms of these polymers have been reported. Incorporation of monomers with additional coordination sites into these polymers has led to the development of a variety of different anion sensors, mostly for halide ions (Lee et al. 2004; Zhou et al. 2005; Vetrichelvan et al. 2006; Kim et al. 2007). Extension of these materials toward recognition of more complex analytes should be possible.

12.5. CONCLUSION

Colorimetric sensors based on conjugated polymers detect a wide range of analytes. Because the chromatic transition is an intrinsic property of the polymer, a unique signal transduction mechanism for each analyte is not required. The ability to combine known ligand receptor interactions with preexisting polymer platforms has enabled the rapid development of this field. The work described in this chapter highlights the advances in colorimetric sensing over the past few decades. PDAs and polythiophenes have emerged as conjugate polymers that are well suited for a wide range of sensing applications. Increasing the sensitivity of detection remains a challenge for many colorimetric sensing systems. Sensors made of these polymers must be robust and durable so that they can be deployed in a variety of field conditions. Portability is also an important issue in sensor design, and here colorimetric sensing has the advantage in that sophisticated instrumentation is not necessarily required. Having a reversible sensor response is important for some sensor applications, and this is currently a limitation for most PDA sensors. Much work remains to be done in this area, and the next few decades should see the continued development of colorimetric sensing based on existing as well as new conjugated polymer platforms.

REFERENCES

Ahn DJ, Chae EH, Lee GS, Shim HY, Chang TE, Ahn KD, Kim JM. Colorimetric reversibility of polydiacetylene supramolecules having enhanced hydrogen-bonding under thermal and pH stimuli. J Am Chem Soc 2003;125:8976–8977.

Albert KJ, Lewis NS, Schauer CL, Sotzing GA, Stitzel SE, Vaid TP, Walt DR. Cross-reactive chemical sensor arrays. Chem Rev 2000;100:2595–2626.

Aldakov D, Anzenbacher P Jr. Sensing of aqueous phosphates by polymers with dual modes of signal transduction. J Am Chem Soc 2004;126:4752–4753.

Baek MG, Stevens RC, Charych DH. Design and synthesis of novel glycopolythiophene assemblies for colorimetric detection of influenza virus and *E. coli*. Bioconjug Chem 2000;11:777–788.

Bässler H. Photopolymerization of diacetylenes. Adv Polym Sci 1984;63:1–48.

Baughman RH. Solid-state polymerization of diacetylenes. J Appl Phys 1972;43:4362–4370.

Bernier S, Garreau S, Bera-Aberem M, Gravel C, Leclerc M. A versatile approach to affinity chromic polythiophenes. J Am Chem Soc 2002;124:12463–12468.

Biesalski M, Tu R, Tirrell MV. Polymerized vesicles containing molecular recognition sites. Langmuir 2005;21:5663–5666.

Carpick RW, Sasaki DY, Marcus MS, Eriksson MA, Burns AR. Polydiacetylene films: a review of recent investigations into chromogenic transitions and nanomechanical properties. J Phys Condens Matter 2004;16:R679–R697.

Charych D, Cheng Q, Reichert A, Kuziemko G, Stroh M, Nagy JO, Spevak W, Stevens RC. A "litmus test" for molecular recognition using artificial membranes. Chem Biol 1996;3:113–120.

Charych DH, Nagy JO, Spevak W, Bednarski MD. Direct colorimetric detection of a receptor–ligand interaction by a polymerized bilayer assembly. Science 1993;261:585–588.

Cheng Q, Stevens RC. Coupling of an induced fit enzyme to polydiacetylene thin films: colorimetric detection of glucose. Adv Mater 1997;9:481–483.

Cheng Q, Stevens RC. Charge-induced chromatic transition of amino acid-derivatized polydiacetylene liposomes. Langmuir 1998;14:1974–1976.

Cheng Q, Yamamoto M, Stevens RC. Amino acid terminated polydiacetylene lipid microstructures: morphology and chromatic transition. Langmuir 2000;16:5333–5342.

Collins BE, Anslyn EV. Pattern-based peptide recognition. Chem Eur J 2007;13:4700–4708.

Evrard D, Touitou E, Kolusheva S, Fishov Y, Jelinek R. A new colorimetric assay for studying and rapid screening of membrane penetration enhancers. Pharm Res 2001;18:943–949.

Faid K, Leclerc M. Functionalized regioregular polythiophenes: towards the development of biochromic sensors. Chem Commun 1996;2761–2762.

Geiger E, Hug P, Keller BA. Chromatic transitions in polydiacetylene Langmuir–Blodgett films due to molecular recognition at the film surface studied by spectroscopic methods and surface analysis. Macromol Chem Phys 2002;203:2422–2431.

Gill I, Ballesteros A. Immunoglobulin–polydiacetylene sol–gel nanocomposites as solid-state chromatic biosensors. Angew Chem Int Ed 2003;42:3264–3267.

Guo CX, Boullanger P, Jiang L, Liu T. Colorimetric detection of WGA in carbohydrate-functionalized polydiacetylene Langmuir–Schaefer films. Colloids Surf Physicochem Eng Asp 2007;293:152–156.

Guo CX, Boullanger P, Liu T, Jiang L. Size effect of polydiacetylene vesicles functionalized with glycolipids on their colorimetric detection ability. J Phys Chem B Condens Matter Mater Surf Interfaces Biophys 2005;109:18765–18771.

Halevy R, Rozek A, Kolusheva S, Hancock RE, Jelinek R. Membrane binding and permeation by indolicidin analogs studied by a biomimetic lipid/polydiacetylene vesicle assay. Peptides 2003;24:1753–1761.

Ho HA, Bera-Aberem M, Leclerc M. Optical sensors based on hybrid DNA/conjugated polymer complexes. Chemistry 2005;11:1718–1724.

Ho HA, Boissinot M, Bergeron MG, Corbeil G, Dore K, Boudreau D, Leclerc M. Colorimetric and fluorometric detection of nucleic acids using cationic polythiophene derivatives. Angew Chem Int Ed 2002;41:1548–1551.

Ho HA, Leclerc M. New colorimetric and fluorometric chemosensor based on a cationic polythiophene derivative for iodide-specific detection. J Am Chem Soc 2003;125:4412–4413.

Ho HA, Leclerc M. Optical sensors based on hybrid aptamer/conjugated polymer complexes. J Am Chem Soc 2004;126:1384–1387.

Hub HH, Hupfer B, Koch H, Ringsdorf H. Polyreactions in ordered systems. 20. Polymerizable phospholipid analogs—new stable biomembrane and cell models. Angew Chem Int Ed 1980;19:938–940.

Jelinek R, Kolusheva S. Polymerized lipid vesicles as colorimetric biosensors for biotechnological applications. Biotechnol Adv 2001;19:109–118.

Jelinek R, Okada S, Norvez S, Charych D. Interfacial catalysis by phospholipases at conjugated lipid vesicles: colorimetric detection and NMR spectroscopy. Chem Biol 1998;5:619–629.

Jonas U, Shah K, Norvez S, Charych DH. Reversible color switching and unusual solution polymerization of hydrazide-modified diacetylene lipids. J Am Chem Soc 1999;121:4580–4588.

Jung YK, Park HG, Kim JM. Polydiacetylene (PDA)-based colorimetric detection of biotin–streptavidin interactions. Biosens Bioelectron 2006;21:1536–1544.

Jurs PC, Bakken GA, McClelland HE. Computational methods for the analysis of chemical sensor array data from volatile analytes. Chem Rev 2000;100:2649–2678.

Katz M, Ben-Shlush I, Kolusheva S, Jelinek R. Rapid colorimetric screening of drug interaction and penetration through lipid barriers. Pharm Res 2006;23:580–588.

Katz M, Tsubery H, Kolusheva S, Shames A, Fridkin M, Jelinek R. Lipid binding and membrane penetration of polymyxin B derivatives studied in a biomimetic vesicle system. Biochem J 2003;375:405–413.

Kew SJ, Hall EA. pH response of carboxy-terminated colorimetric polydiacetylene vesicles. Anal Chem 2006;78:2231–2238.

Kim HJ, Lee JH, Kim TH, Lyoo WS, Kim DW, Lee C, Lee TS. Synthesis of chromo- and fluorogenic poly(ortho-diaminophenylene) chemosensors for fluoride anion. J Polym Sci A Polym Chem 2007;45:1546–1556.

Kim J-M, Lee J-S, Choi H, Sohn D, Ahn DJ. Rational design and in-situ FTIR analyses of colorimetrically reversible polydiacetylene supramolecules. Macromolecules 2005;38:9366–9376.

Kim J-M, Lee J-S, Lee J-S, Woo S-Y, Ahn DJ. Unique effects of cyclodextrins on the formation and colorimetric transition of polydiacetylene vesicles. Macromol Chem Phys 2005;206:2299–2306.

Kolusheva S, Boyer L, Jelinek R. A colorimetric assay for rapid screening of antimicrobial peptides. Nat Biotechnol 2000;18:225–227.

Kolusheva S, Kafri R, Katz M, Jelinek R. Rapid colorimetric detection of antibody–epitope recognition at a biomimetic membrane interface. J Am Chem Soc 2001;123:417–422.

Kolusheva S, Shahal T, Jelinek R. Cation-selective color sensors composed of ionophore–phospholipid–polydiacetylene mixed vesicles. J Am Chem Soc 2000;122:776–780.

Kolusheva S, Wachtel E, Jelinek R. Biomimetic lipid/polymer colorimetric membranes: molecular and cooperative properties. J Lipid Res 2003;44:65–71.

Kolusheva S, Zadmard R, Schrader T, Jelinek R. Color fingerprinting of proteins by calixarenes embedded in lipid/polydiacetylene vesicles. J Am Chem Soc 2006;128: 13592–13598.

Leclerc M. Optical and electrochemical transducers based on functionalized conjugated polymers. Adv Mater 1999;11:1491–1498.

Lee JK, Na J, Kim TH, Kim Y-S, Park WH, Lee TS. Synthesis of polyhydroxybenzoxazole-based colorimetric chemosensor for anionic species. Mater Sci Eng C 2004;C24:261–264.

Lee NY, Jung YK, Park HG. On-chip colorimetric biosensor based on polydiacetylene (PDA) embedded in photopolymerized poly(ethylene glycol) diacrylate (PEG-DA) hydrogel. Biochem Eng J 2006;29:103–108.

Levesque I, Leclerc M. Ionochromic effects in regioregular ether-substituted polythiophenes. J Chem Soc Chem Commun 1995;2293–2294.

Li C, Numata M, Takeuchi M, Shinkai S. A sensitive colorimetric and fluorescent probe based on a polythiophene derivative for the detection of ATP. Angew Chem Int Ed Engl 2005;44:6371–6374.

Li C, Numata M, Takeuchi M, Shinkai S. Unexpected chiroptical inversion observed for supramolecular complexes formed between an achiral polythiophene and ATP. Chem Asian J 2006;1:95–101.

Lu Y, Yang Y, Sellinger A, Lu M, Huang J, Fan H, Haddad R, Lopez G, Burns AR, Sasaki DY, Shelnutt J, Brinker CJ. Self-assembly of mesoscopically ordered chromatic polydiacetylene/silica nanocomposites. Nature 2001;410:913–917.

Ma G, Cheng Q. Vesicular polydiacetylene sensor for colorimetric signaling of bacterial pore-forming toxin. Langmuir 2005;21:6123–6126.

Ma ZF, Li JR, Jiang L, Cao J, Boullanger P. Influence of the spacer length of glycolipid receptors in polydiacetylene vesicles on the colorimetric detection of *Escherichia coli*. Langmuir 2000;16:7801–7804.

Ma ZF, Li JR, Liu MH, Cao J, Zou ZY, Tu J, Jiang L. Colorimetric detection of *Escherichia coli* by polydiacetylene vesicles functionalized with glycolipid. J Am Chem Soc 1998;120:12678–12679.

Marsella MJ, Swager TM. Designing conducting polymer-based sensors—selective ionochromic response in crown-ether containing polythiophenes. J Am Chem Soc 1993;115: 12214–12215.

Maynor MS, Nelson TL, O'Sullivan C, Lavigne JJ. A food freshness sensor using the multistate response from analyte-induced aggregation of a cross-reactive poly(thiophene). Org Lett 2007;9:3217–3220.

McCullough RD. The chemistry of conducting polythiophenes. Adv Mater 1998;10:93–116.

McCullough RD, Ewbank PC, Loewe RS. Self-assembly and disassembly of regioregular, water soluble polythiophenes: chemoselective ionchromatic sensing in water. J Am Chem Soc 1997;119:633–634.

Mcquade DT, Pullen AE, Swager TM. Conjugated polymer-based chemical sensors. Chem Rev 2000;100:2537–2574.

Nelson TL, O'Sullivan C, Greene NT, Maynor MS, Lavigne JJ. Cross-reactive conjugated polymers: analyte-specific aggregative response for structurally similar diamines. J Am Chem Soc 2006;128:5640–5641.

Nie QL, Zhang Y, Zhang J, Zhang MQ. Immobilization of polydiacetylene onto silica microbeads for colorimetric detection. J Mater Chem 2006;16:546–549.

Okada S, Peng S, Spevak W, Charych D. Color and chromism of polydiacetylene vesicles. Acc Chem Res 1998;31:229–239.

Oren Z, Ramesh J, Avrahami D, Suryaprakash N, Shai Y, Jelinek R. Structures and mode of membrane interaction of a short alpha helical lytic peptide and its diastereomer determined by NMR, FTIR, and fluorescence spectroscopy. Eur J Biochem 2002;269:3869–3880.

Orynbayeva Z, Kolusheva S, Livneh E, Lichtenshtein A, Nathan I, Jelinek R. Visualization of membrane processes in living cells by surface-attached chromatic polymer patches. Angew Chem Int Ed 2005;44:1092–1096.

Pan JJ, Charych D. Molecular recognition and colorimetric detection of cholera toxin by poly (diacetylene) liposomes incorporating GM1 ganglioside. Langmuir 1997;13:1365–1367.

Peek BM, Callahan JH, Namboodiri K, Singh A, Gaber BP. Effect of vesicle size on the polymerization of a diacetylene lipid. Macromolecules 1994;27:292–297.

Peng H, Tang J, Yang L, Pang J, Ashbaugh HS, Brinker CJ, Yang Z, Lu Y. Responsive periodic mesoporous polydiacetylene/silica nanocomposites. J Am Chem Soc 2006;128:5304–5305.

Potisatityuenyong A, Tumcharern G, Dubas ST, Sukwattanasinitt M. Layer-by-layer assembly of intact polydiacetylene vesicles with retained chromic properties. J Colloid Interface Sci 2006;304:45–51.

Rangin M, Basu A. Lipopolysaccharide identification with functionalized polydiacetylene liposome sensors. J Am Chem Soc 2004;126:5038–5039.

Reichert A, Nagy JO, Spevak W, Charych D. Polydiacetylene liposomes functionalized with sialic-acid bind and colorimetrically detect influenza-virus. J Am Chem Soc 1995;117: 829–830.

Rozner S, Kolusheva S, Cohen Z, Dowhan W, Eichler J, Jelinek R. Detection and analysis of membrane interactions by a biomimetic colorimetric lipid/polydiacetylene assay. Anal Biochem 2003;319:96–104.

Sadagopan K, Sawant SN, Kulshreshtha SK, Jarori GK. Physical and chemical characterization of enolase immobilized polydiacetylene Langmuir–Blodgett film. Sensors Actuat B 2006;115:526–533.

Satchell DP, Sheynis T, Shirafuji Y, Kolusheva S, Ouellette AJ, Jelinek R. Interactions of mouse Paneth cell alpha-defensins and alpha-defensin precursors with membranes.

Prosegment inhibition of peptide association with biomimetic membranes. J Biol Chem 2003;278:13838–13846.

Schott M. The colors of polydiacetylenes: a commentary. J Phys Chem B 2006;110: 15864–15868.

Scindia Y, Silbert L, Volinsky R, Kolusheva S, Jelinek R. Colorimetric detection and fingerprinting of bacteria by glass-supported lipid/polydiacetylene films. Langmuir 2007;23:4682–4687.

Sheynis T, Sykora J, Benda A, Kolusheva S, Hof M, Jelinek R. Bilayer localization of membrane-active peptides studied in biomimetic vesicles by visible and fluorescence spectroscopies. Eur J Biochem 2003;270:4478–4487.

Shim HY, Lee SH, Ahn DJ, Ahn KD, Kim JM. Micropatterning of diacetylenic liposomes on glass surfaces. Mater Sci Eng C 2004;24:157–161.

Shtelman E, Tomer A, Kolusheva S, Jelinek R. Imaging membrane processes in erythrocyte ghosts by surface fusion of a chromatic polymer. Anal Biochem 2006;348:151–153.

Silbert L, Ben Shlush I, Israel E, Porgador A, Kolusheva S, Jelinek R. Rapid chromatic detection of bacteria by use of a new biomimetic polymer sensor. Appl Environ Microbiol 2006;72:7339–7344.

Song J, Cheng Q, Kopta S, Stevens RC. Modulating artificial membrane morphology: pH-induced chromatic transition and nanostructural transformation of a bolaamphiphilic conjugated polymer from blue helical ribbons to red nanofibers. J Am Chem Soc 2001;123:3205–3213.

Song J, Cisar JS, Bertozzi CR. Functional self-assembling bolaamphiphilic polydiacetylenes as colorimetric sensor scaffolds. J Am Chem Soc 2004;126:8459–8465.

Su YL. Assembly of polydiacetylene vesicles on solid substrates. J Colloid Interface Sci 2005;292:271–276.

Su YL, Li JR, Jiang L. Chromatic immunoassay based on polydiacetylene vesicles. Colloids Surf B Biointerfaces 2004;38:29–33.

Su YL, Li JR, Jiang L, Cao J. Biosensor signal amplification of vesicles functionalized with glycolipid for colorimetric detection of *Escherichia coli*. J Colloid Interface Sci 2005;284:114–119.

Tang Y, Feng F, He F, Wang S, Li Y, Zhu D. Direct visualization of enzymatic cleavage and oxidative damage by hydroxyl radicals of single-stranded DNA with a cationic polythiophene derivative. J Am Chem Soc 2006;128:14972–14976.

Valenta C, Steininger A, Auner BG. Phloretin and 6-ketocholestanol: membrane interactions studied by a phospholipid/polydiacetylene colorimetric assay and differential scanning calorimetry. Eur J Pharm Biopharm 2004;57:329–336.

Vetrichelvan M, Nagarajan R, Valiyaveettil S. Carbazole-containing conjugated copolymers as colorimetric/fluorimetric sensor for iodide anion. Macromolecules 2006;39:8303–8310.

Volinsky R, Kliger M, Sheynis T, Kolusheva S, Jelinek R. Glass-supported lipid/polydiacetylene films for colour sensing of membrane-active compounds. Biosens Bioelectron 2007;22:3247–3251.

Wang C, Ma Z. Colorimetric detection of oligonucleotides using a polydiacetylene vesicle sensor. Anal Bioanal Chem 2005;382:1708–1710.

Wang C, Ma Z, Su Z. Facile method to detect oligonucleotides with functionalized polydiacetylene vesicles. Sens Actuat B Chem 2006;B113:510–515.

Wegner G. Topochemical reactions of monomers with conjugated triple bonds. VI. Topochemical polymerization of monomers with conjugated triple bonds. Makromol Chem 1972;154:35–48.

Yamanaka SA, Charych DH, Loy DA, Sasaki DY. Solid phase immobilization of optically responsive liposomes in sol–gel materials for chemical and biological sensing. Langmuir 1997;13:5049–5053.

Yang Y, Lu Y, Lu M, Huang J, Haddad R, Xomeritakis G, Liu N, Malanoski AP, Sturmayr D, Fan H, Sasaki DY, Assink RA, Shelnutt JA, van Swol F, Lopez GP, Burns AR, Brinker CJ. Functional nanocomposites prepared by self-assembly and polymerization of diacetylene surfactants and silicic acid. J Am Chem Soc 2003;125:1269–1277.

Yoon J, Chae SK, Kim J-M. Colorimetric sensors for volatile organic compounds (VOCs) based on conjugated polymer-embedded electrospun fibers. J Am Chem Soc 2007;129:3038–3039.

Zhang Y, Fan Y, Sun C, Shen D, Li Y, Li J. Functionalized polydiacetylene–glycolipid vesicles interacted with *Escherichia coli* under the TiO2 colloid. Colloids Surf B Biointerfaces 2005;40:137–142.

Zhou G, Cheng Y, Wang L, Jing X, Wang F. Novel polyphenylenes containing phenol-substituted oxadiazole moieties as fluorescent chemosensors for fluoride ion. Macromolecules 2005;38:2148–2153.

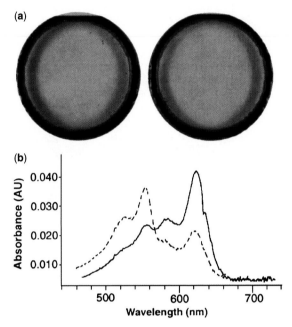

Figure 12.3 Colorimetric detection of influenza by sialoside bilayer assembly (2% sialoside lipid **2** and 98% matrix lipid **1**). (a) The colorimetric response of the film, supported on a glass microscope slide, is readily visible to the naked eye for qualitative evaluation of the presence of virus. The film on the left (blue) has been exposed to a blank solution of PBS. The film on the right (red) has been exposed to 100 hemagglutinin units (HAU) of virus (CR = 77%). (b) The visible absorption spectrum of a bilayer assembly (—) prior to and (—) after viral incubation. Reprinted from Charych et al. (1993). Copyright 1993 American Association for the Advancement of Science.

Figure 12.4 Colorimetric detection of influenza virus using polymerized liposomes containing sialic acid. (a) Photograph of liposomes to which have been added increasing amounts (from left to right) of influenza virus. Liposomes were **1** and 95% **5**. To each well was added the following amounts of influenza virus (left to right): 0, 8, 16, and 32 hemagglutinin units (HAU). Reprinted from Charych et al. (1996). Copyright 1996 Elsevier Science.

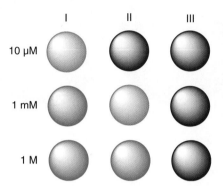

Figure 12.5 Patterns generated upon addition of a membrane surface-binding molecule (I), membrane penetrating compound (II), and a compound that does not interact with membranes (III). Adapted from Katz et al. (2006). Copyright 2006 Springer.

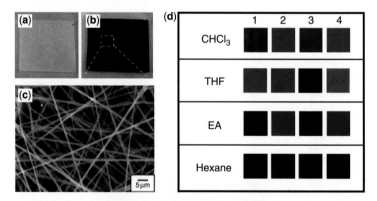

Figure 12.7 Photographs of electrospun fiber mats embedded with **1** (a) before and (b) after 254-nm UV irradiation ($1 \, \text{mW/cm}^2$) for 3 min. (c) Scanning electron microscopy image of the microfibers containing polymerized **1**. (c) Photographs of the polydiacetylene-embedded electrospun fiber mats prepared with various diacetylene monomers after exposure to organic solvent. Reprinted from Yoon et al. (2007). Copyright 2007 American Chemical Society.

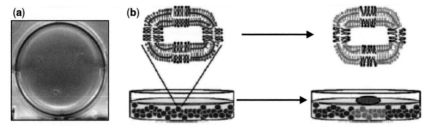

Figure 12.8 (a) Image of diacetylene liposome-embedded agar plate with three *E. coli* BL colonies after 18 h of growth. (b) Chromatic sensor concept. (Left) Agar scaffold containing vesicular nano-particles composed of phospholipids (green) and PDA (blue). (Right) Bacterial proliferation (green oval) on the agar surface causes a blue to red transformation of embedded vesicles due to bacterially secreted compounds that diffuse through the agar. Reprinted from Silbert et al. (2006). Copyright 2006 American Society for Microbiology.

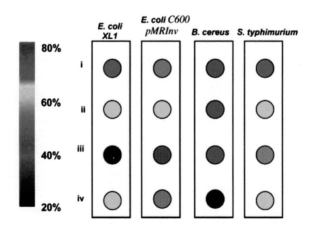

Figure 12.9 Colorimetric bacterial fingerprinting. The color combination for each bacterium reflects the chromatic transitions (RCS) recorded 7 h after the start of growth at a bacterial concentration of 1×10^9/mL. The RCS color key is shown on the left: (i) DOPE/PDA (1:9 mole ratio), (ii) sphingomyelin/cholesterol/PDA (7:3:90), (iii) DMPC/PDA (1:9), and (iv) POPG/PDA (1:9). Reprinted from Scindia et al. (2007). Copyright 2007 American Chemical Society.

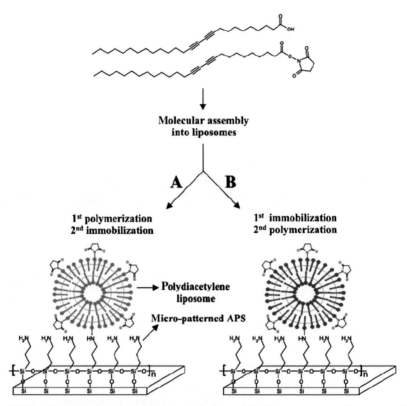

Figure 12.11 Strategies for immobilization of diacetylene liposomes on micropatterned glass substrates: (a) immobilization after polymerization of monomeric liposomes and (b) polymerization after immobilization of monomeric liposomes. Reprinted from Shim et al. (2004). Copyright 2004 Elsevier Science.

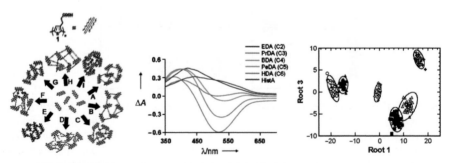

Figure 12.15 (a) Schematic representation of the multidimensional response from polymer **9** (yellow rods) upon analyte-induced aggregation with different amines (A–I). (b) Changes in the absorption spectra of polymer **9** (1 mM) upon addition of different diamines (1 mM each) in aqueous HEPES buffer (40 mM, pH 7.4). (c) Two-dimensional linear discriminant analysis plot of polymer **9** response to various diamines. Reprinted from Maynor et al. (2007). Copyright 2007 American Chemical Society.

CHAPTER 13

GLYCODENDRIMERS AND OTHER MACROMOLECULES BEARING MULTIPLE CARBOHYDRATES

MARY J. CLONINGER

13.1. INTRODUCTION

Multivalent protein–carbohydrate interactions are implicated in a wide variety of intercellular recognition processes including cancer cell aggregation and the metastatic spread of cancer, bacterial adhesion, and the mounting of an immune response (Fig. 13.1; Gorelick et al. 2001; Gabiuset et al. 2004; Rudd et al. 2004). When multivalent protein–carbohydrate interactions occur, sugars are bound into binding clefts on the carbohydrate-binding proteins, which are called lectins. Because the binding sites on the lectins are generally shallow, exposed pockets, the monomeric lectin–carbohydrate interaction is generally weak. In order to achieve physiologically relevant responses, Nature often invokes multivalency in the protein–carbohydrate interaction.

Many lectins have multiple binding sites that are relatively distant from one another; often the binding sites are 3–7 nm apart. Concanavalin A (ConA), for example, is a plant lectin that has been used extensively as a model system for the study of biologically relevant protein–carbohydrate interactions. ConA is a homotetramer at biological pH and has four binding sites approximately 6.5 nm apart (Fig. 13.2a; Derewenda et al. 1989; Naismith et al. 1994). ConA has specificity for the α-pyranose forms of D-mannose and D-glucose; mannose is bound roughly four times better than glucose.

Galectin-3 is a widely studied human lectin that has only one carbohydrate-binding domain, but the collagen-like domain of galectin-3 is believed to cause oligomerization of the lectin (Fig. 13.2b). Like ConA, dimers and higher order oligomers of galectin-3 are likely to have binding sites that are relatively distant from one another (Seetharaman et al. 1998). Like all galectins, galectin-3 shows

Molecular Recognition and Polymers: Control of Polymer Structure and Self-Assembly.
Edited by V. Rotello and S. Thayumanavan
Copyright © 2008 John Wiley & Sons, Inc.

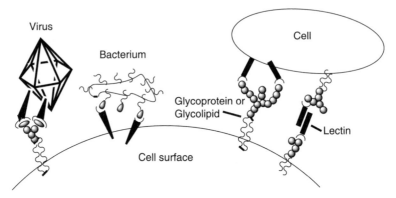

Figure 13.1 A schematic representation of a variety of processes that are mediated by multivalent protein–carbohydrate interactions.

specificity for the pyranose form of β-galactose. The biological ligand for galectin-3 is currently unknown, but *N*-acetyllactosamine has been shown to bind with high affinity. In the mammalian family of galectins, galectins 1, 2, 5, 7, 10, 11, 13, and 14 have one carbohydrate recognition domain but often form noncovalent dimers. Galectins 4, 6, 8, 9, and 12 have two covalently linked carbohydrate-binding domains. Only galectin-3 has the collagen-like domain. Galectins are shown schematically in Figure 13.3 (Leffler et al. 2004).

Efforts to understand and to modulate physiologically relevant protein–carbohydrate interactions have led to the synthesis of many glycosylated frameworks (Zanini and Roy 1998a; Mann and Kiessling 2001; Choi 2004; Wittmann 2004). In order for a synthetic, carbohydrate-functionalized framework to be successfully used for binding to proteins, the carbohydrates on the framework must be appropriately

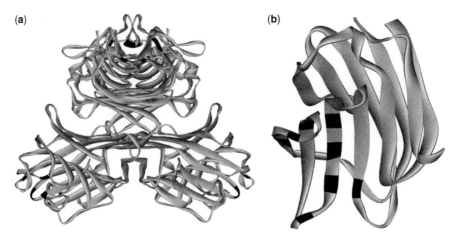

Figure 13.2 The crystal structures of (a) concanavalin A and (b) the carbohydrate recognition domain of galectin-3. Residues in the carbohydrate binding sites are black.

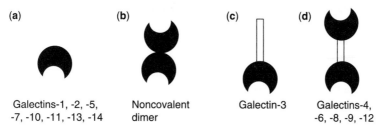

(a) Galectins-1, -2, -5, -7, -10, -11, -13, -14

(b) Noncovalent dimer

(c) Galectin-3

(d) Galectins-4, -6, -8, -9, -12

Figure 13.3 A schematic summary of the galectins. (a) Galectins with one carbohydrate-binding domain and (b) their noncovalent dimers. (c) Galectin-3 has a collagen-like domain and a carbohydrate-binding domain. (d) Galectins with two covalently linked carbohydrate-binding domains.

oriented for interaction with the binding site on the lectin (Gestwicki et al. 2002). The shape of the framework and the length and flexibility of the tether between the framework and the carbohydrate are two critical features of the framework's architecture that must be considered.

Because the carbohydrate-binding sites on lectins with which the synthetic glycosystems are designed to interact may be displayed facing many different directions, many frameworks have been reported. Flexible linear polymers with appended sugars, for example, can readily adjust their shape to interact with a variety of lectins, whereas carbohydrate-functionalized gold nanoparticles should be far more rigid. Other frameworks that have been reported for the study of protein–carbohydrate interactions include star polymers, small glycoclusters, carbohydrate-functionalized viruses, cyclodextrins, pseudopolyrotaxanes, liposomes, micelles, vesicles, proteins, surfaces, and dendrimers (Fig. 13.4).

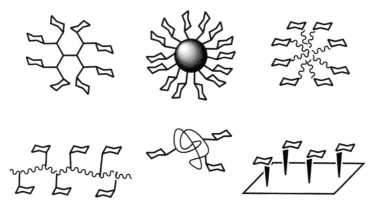

Figure 13.4 A schematic showing some of the frameworks that have been reported for the study of protein–carbohydrate interactions. Carbohydrates are represented as cyclohexane. (Top) Glycodendrimer, carbohydrate-functionalized nanoparticle, and star polymer. (Bottom) Linear glycopolymer, carbohydrate-functionalized protein, and carbohydrate-functionalized surface.

13.2. DENDRIMERS TO GLYCODENDRIMERS

Dendrimers are macromolecules that have a series of branches emanating from a central core. They can be synthesized starting from the core and working out toward the periphery (divergent synthesis). Alternatively, dendrimers can be formed in a top-down approach starting from the outermost residues (convergent synthesis; Fréchet and Tomalia 2001; Newkome et al. 2001). Because dendrimers are built from AB_n-type monomers, each layer or "generation" of branching units doubles or triples ($n = 2$ or 3) the number of peripheral functional groups. The poly(amido amine) (PAMAM) dendrimer is a commercially available framework with highly flexible branches (Fig. 13.5). Tertiary amine units at the branch points result in a doubling of the number of termini for every successive PAMAM generation (Tomalia et al. 1986).

Other dendrimer frameworks have also been widely publicized. Poly(propylene imine) (PPI) dendrimers are like PAMAMs except that no amides are present in their structures. Tertiary amines serve as branch points and primary amines serve as termini for traditional PPI dendrimers (Buhleier et al. 1978; de Brabander-van den Berg and Meijer 1993). The PPI framework is shown in Figure 13.6. Note that PAMAM dendrimers are named using G(0) to describe the species bearing four end groups, but PPI dendrimer nomenclature dictates that a G(1) dendrimer has four end groups. Thus, the G(3) PPI dendrimer shown in Figure 13.6 has the same number of end groups as the G(2) PAMAM shown below.

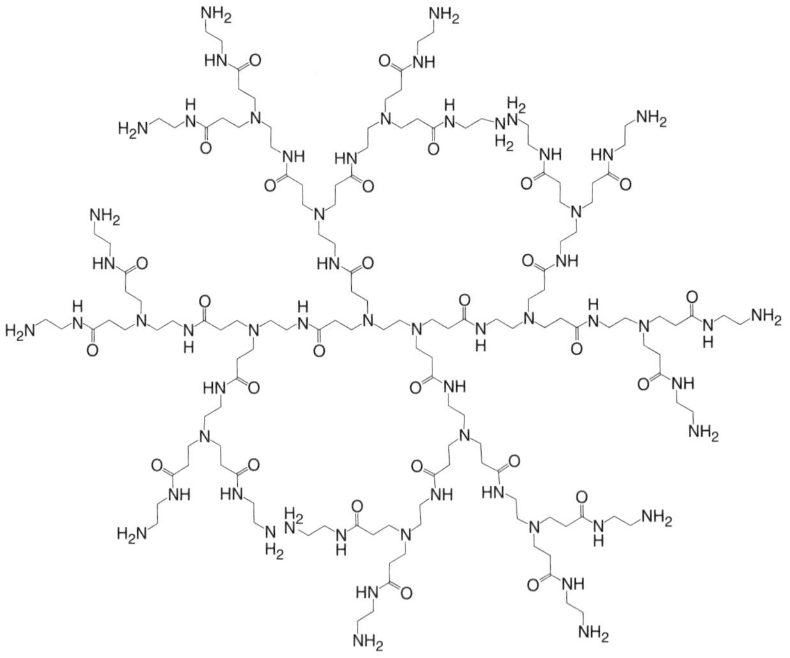

Figure 13.5 The structure of the G(2)-poly(amido amine) dendrimer.

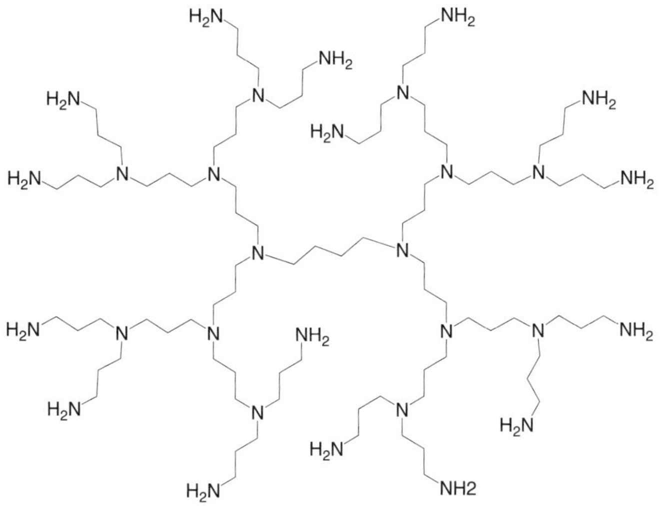

Figure 13.6   The structure of the G(3)-poly(propylene imine) dendrimers.

Fréchet and coworkers have popularized poly(aryl ether) dendrimers and polyester dendrimers (Fig. 13.7), which have been widely used (Hawker and Fréchet 1990; Ihre et al. 1998). Addition of the aromatic rings adds rigidity to the dendrimer framework that is lacking in the PPI and PAMAM dendrimers. The polyester dendrimers may be more like the PPIs and PAMAMs in terms of rigidity, but the ester linkages enable degradation of the polyester dendrimers by esterases in vivo. Thus, the biological stability of the PAMAMs and the polyester frameworks is quite different.

The remaining dendrimer frameworks that are being applied to a wide range of biological problems including protein–carbohydrate interactions include the polylysine dendrimers used by Starpharma Ltd (Melbourne, Australia; Fig. 13.8; Denkewalter et al. 1981) and the phosphorus-containing dendrimers developed by Majoral's group (Fig. 13.9; Launay et al. 1996; Servin et al. 2007). The polylysine systems are elegant in their simplicity, whereas the strength of the Majoral dendrimers is that they achieve an unusually high density of branches. Because they are so highly branched, small generation Majoral dendrimers can be used where larger generation PAMAMs or PPIs would be required to achieve the same degree of functionalization.

Having many dendrimer frameworks available is critical, because dendrimers are currently being used for a variety of biological applications. Different applications will necessitate different levels of structural rigidity and stability, end-group mobility, and end-group density. New dendritic systems, such as the new PrioStar dendrimers (www.dnanotech.com), continue to emerge as the needs of different user communities become clear.

Drug and DNA delivery, photodynamic therapy, boron neutron capture therapy, and magnetic resonance imaging are some areas in which appropriately functionalized dendrimers are being evaluated (Jang and Kataoka 2005; Svenson and Tomalia 2005; Yang and Kao 2006). In one particularly nice example, polylysine

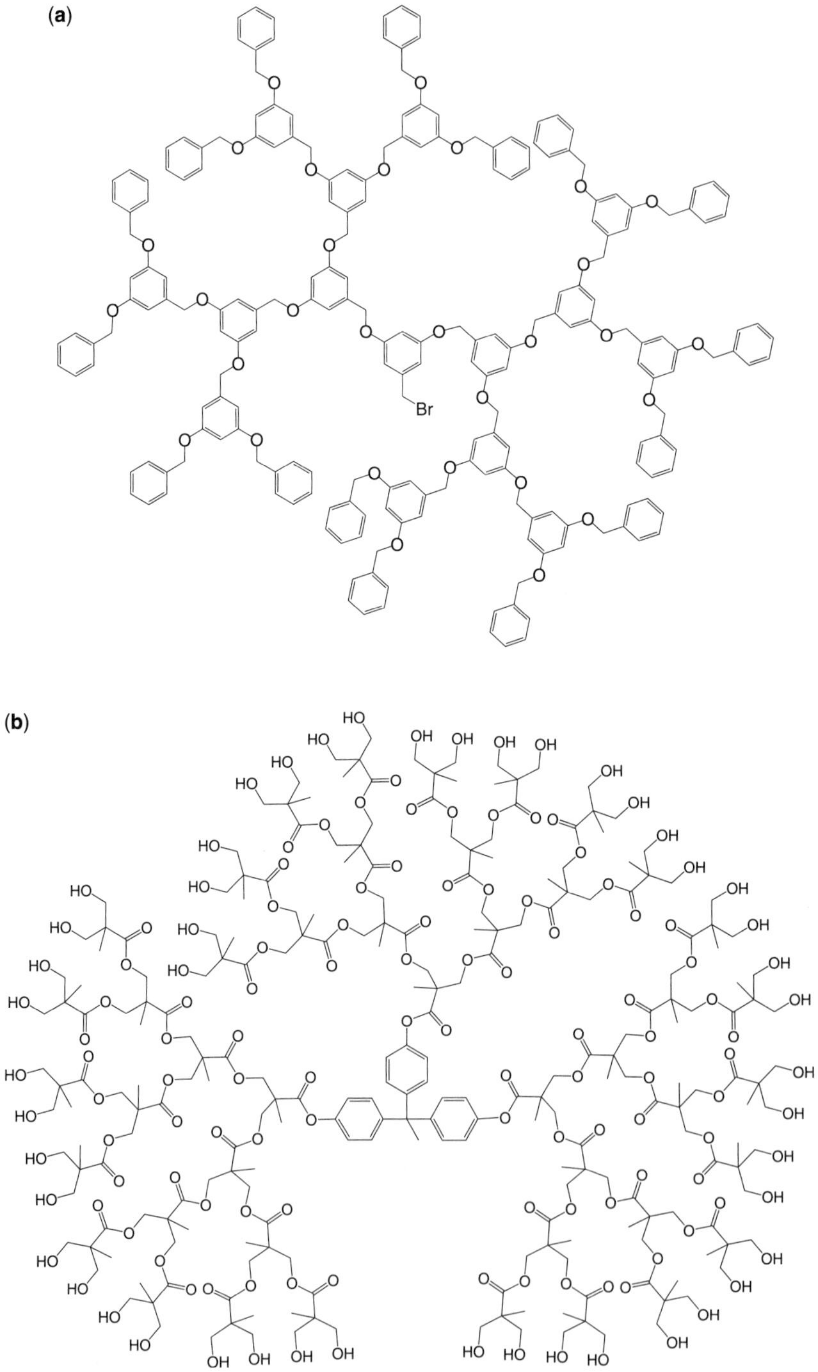

(a)

(b)

Figure 13.7 (a) A poly(aryl ether) dendron and (b) a polyester dendrimer.

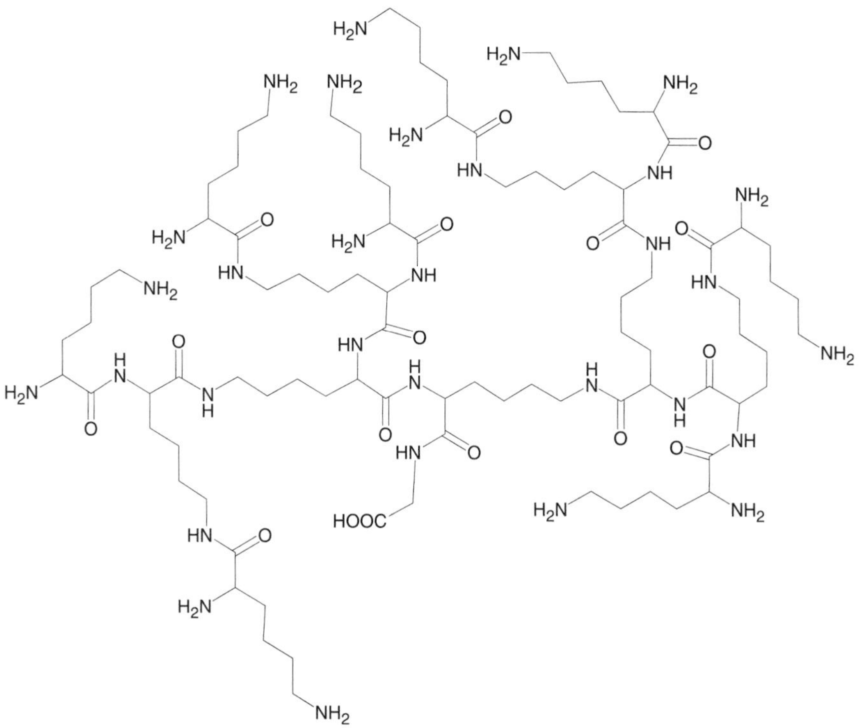

Figure 13.8 A polylysine dendrimer.

dendrimers bearing sulfated naphthyl groups are being developed by Starpharma Ltd to serve as antiviral topicals against herpes simplex virus (Bourne et al. 2000; see also Starpharma Ltd Product focus: VivaGelTM). The biocompatibility and the immunogenicity of the dendrimers are strongly influenced by the nature of the dendrimer end groups, indicating that dendrimers are a viable platform for many biological applications (Duncan 2005).

Because dendrimers have orderly, monodisperse structures (compared to many polymer systems) with reactive termini upon which carbohydrates can readily be appended and because higher generation dendrimers that are large enough to span multiple binding sites on lectins are readily available, glycodendrimers have great utility for the study of multivalent protein–carbohydrate interactions (Roy 2003). In one particularly elegant early example of glycodendrimer research, Zanini and Roy (1996, 1998b) synthesized sialic acid functionalized PAMAM dendrimers and showed that the G(3)-dendrimer was 210-fold (6.7 fold per sialoside) more active than monomeric sialoside in inhibition binding assays with *Limax flavus* and human α1-acid glycoprotein. In studies with ConA, we compared the activity of mannose-functionalized PAMAM dendrimers of different generations and with different degrees of mannose in the hemagglutination inhibition assay. We demonstrated that G(4)- through G(6)-PAMAMs with mannose functionalization of half of their termini (and hydroxyl

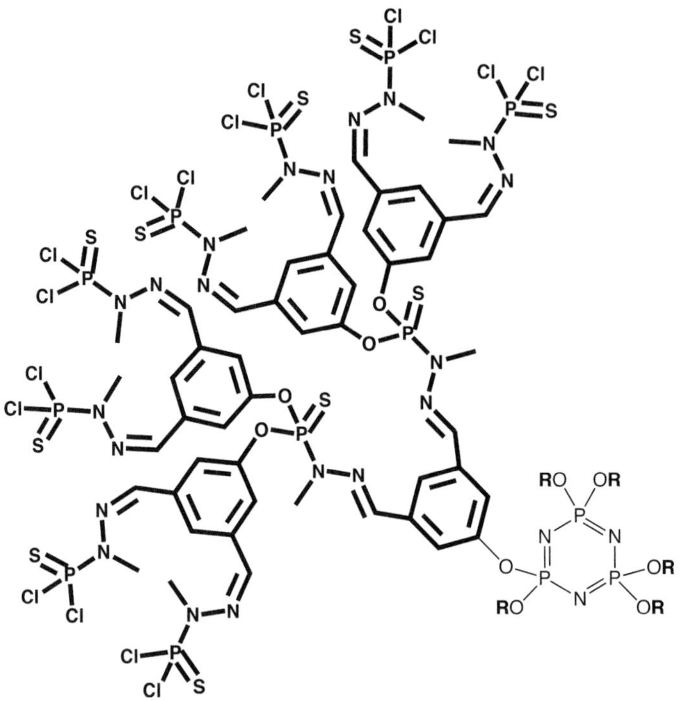

Figure 13.9   Phosphorus-containing dendrimer with 96 terminal chlorides; R is indicated by the bold type in the structure.

groups at the remaining termini) were the most effective and that large generation dendrimers [G(4)–G(6)] had more activity toward ConA than small generation dendrimers [G(1)–G(3); Woller et al. (2003)]. Glycodendrimers have recently been the subject of several excellent reviews, and many more examples of biological systems that have been studied using glycodendrimers are provided in several review articles (Jayaraman et al. 1997; Roy 2003; Srinivas and Narayanaswamy 2005; Tsvetkov and Nifantiev 2005; Matsuoka et al. 2006).

13.3. MULTIVALENCY

As noted, monovalent lectin–carbohydrate interactions are generally too weak to be physiologically relevant. In order to achieve higher affinity interactions, multiple carbohydrates may be clustered around each binding site. The proximity of many carbohydrates to each binding site on the protein would slow the off rate of the interaction of a glycosystem with the lectin. A decrease in the off rate of the interaction will cause an overall increase in the binding affinity of the system. Page and Jencks (1971) refer to chelation effects in a critical discussion in this area, and Lee and Lee (2000) articulate the idea of proximity effects well.

Jencks (1981) also provides a discussion of the energetic terms that likely contribute to a binding interaction involving multiple ligands and multiple receptor binding sites. In addition to the "intrinsic binding energies" of the two binding events, Jencks defines a connection Gibbs energy that represents the change in the probability of binding that results in the connection of two ligands. Other authors have further developed the idea of effective molarity as a critical component of multivalency (Gargano et al. 2001; Mulder et al. 2004). In a seminal review article from the late 1990s, George Whitesides and his group provide an in-depth discussion of how multivalent interactions can be treated mathematically (Mammen et al. 1998). Usually, entropic contributions to multivalent interactions are believed to be of tantamount importance. However, Toone's group report a calcium ion binding study in which enthalpic contributions were more significant (Christensen et al. 2003).

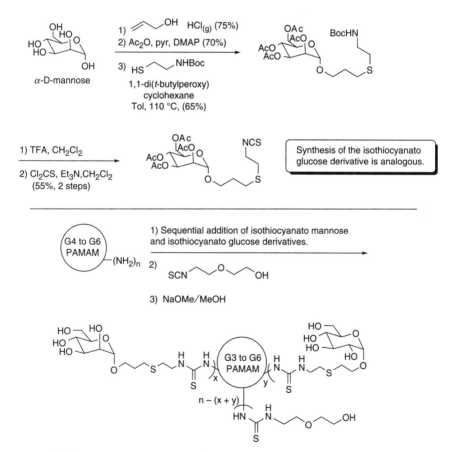

Figure 13.10 Synthesis of mannose-/glucose-functionalized dendrimers. For dendrimer loading, $x + y = 50\%$ loading was used because optimal activity for glycodendrimers with concanavalin A was previously determined to occur at 50% functionalization. Remaining amines from the poly(amido amine) (PAMAM) substrate were capped as alcohols.

The calcium ion system may be most helpful for understanding lectins that have extended binding sites or that have proximal receptor sites and may suggest that enthalpy and entropy each dominate in specific binding situations. A helpful discussion of enthalpy and entropy as they relate to multivalency is found in a recent review article by Handl et al. (2004). A recent review by Whitesides' group provides a discussion of the energetics of multivalency as well as examples of glycosystems that have been used successfully for the study of multivalent interactions of biological relevance (Krishnamurthy et al. 2006).

We functionalized G4, G5, and G6-PAMAM dendrimers with mixtures of mannose and glucose in varying ratios (Fig. 13.10) and evaluated the relative affinities of these compounds for ConA using the hemagglutination assay. We observed a linear trend in the binding constants for all generations that was dependent on the relative degree of functionalization of the dendrimer with mannose and glucose. As the ratio of mannose to glucose increased, the relative activity in the hemagglutination assay (on a per sugar basis) increased linearly (Fig. 13.11). Methyl mannose is bound to ConA with a fourfold higher affinity than methyl glucose; multivalency amplified this trend. For generations 4 and 5, the difference between mannose and glucose-functionalized dendrimers was just under 16, and generation 6 had an only slightly lower value near 11 (Wolfenden and Cloninger 2005).

Sixteen is the difference that is predicted using Eq. (1), which was proposed by Mammen et al. (1998).

$$K_N^{\text{poly}} = (K^{\text{mono}})^{\alpha N} \tag{13.1}$$

where N is the number of receptor–ligand interactions, K^{mono} is the monovalent association constant, K^{poly} is the multivalent binding constant, and α is the cooperativity constant. For the man/glc dendrimers shown in Figure 13.10, no cooperative binding is likely and $\alpha = 1$.

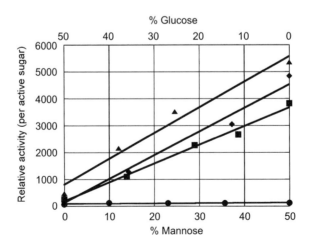

Figure 13.11 Hemagglutination inhibition assay results for the interaction of mannose-/ glucose-functionalized dendrimers with concanvalin A. (●) G(3), (■) G(4), (◆) G(5), and (▲) G(6).

Thus, our work suggests that, although other more complicated methods have also been put forward, the prediction that polyvalent interactions should be equal to monovalent interactions raised to the Nth power, where N is the number of receptor–ligand interactions, is experimentally valid. Our mannose-/glucose-functionalized dendrimer results suggest that the affinity of multivalent associations can be attenuated in predictable, reliable ways based on the monovalent affinities of the ligands (Wolfenden and Cloninger 2005, 2006).

13.4. HETEROMULTIVALENT CARBOHYDRATE SYSTEMS

The majority of the synthetic multivalent frameworks that have been reported have been used to display multiple copies of one carbohydrate ligand. Although the presentation of many copies of a single ligand will in some instances be sufficient, examples are emerging in which the display of multiple carbohydrates will be critical for modulation of cellular processes. An example of work from our laboratory where glycodendrimers functionalized with more than one carbohydrate have significantly different activity in lectin-binding assays is highlighted in two recent studies (Wolfenden and Cloninger 2005, 2006). This section summarizes additional key examples of synthetic carbohydrate-functionalized systems that bear more than one carbohydrate.

In vaccine development for prostate cancer, Danishefsky's research group (Williams et al. 2000; Allen et al. 2001; Ragupathi et al. 2002; Keding and Danishefsky 2004) has recently reported constructs in which up to five different carbohydrate antigens were tethered to a single peptide backbone. Because several different carbohydrate antigens are likely expressed on cancer cell surfaces, they rationalized that inclusion of multiple antigens in one cluster could increase the effectiveness of the vaccine. Globo-H, STn, Tn, TF, and Le^y were incorporated onto a peptide, which was then tethered to a carrier protein (Fig. 13.12). Globo-H is a

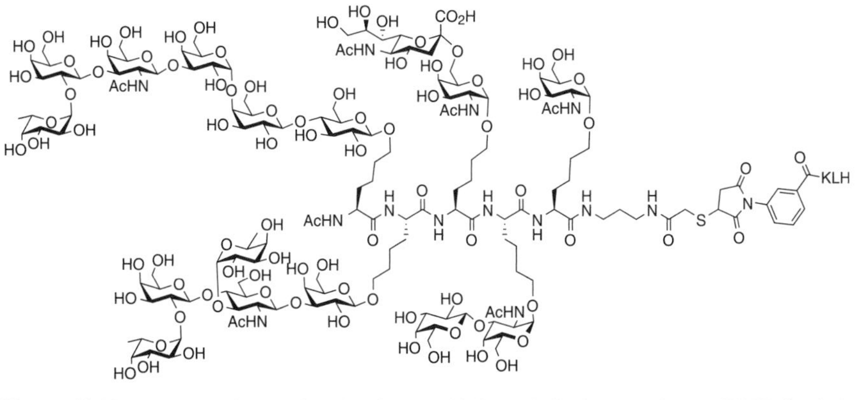

Figure 13.12 A synthetic vaccine bearing multiple carbohydrate antigens; KLH, keyhole limpet hemocyanin.

hexasaccharide that has been utilized in clinical trials for breast, prostate, and ovarian cancer treatments. STn, Tn, TF, and Le^y are blood-group related antigens that have been found in high concentration on tumor cell surfaces. The conjugates, when introduced with an adjuvant, induced an immune response (Williams et al. 2000; Allen et al. 2001; Ragupathi et al. 2002; Keding and Danishefsky 2004).

Danishevsky's work is an excellent example of how the tethering of multiple sugars (rather than just one carbohydrate or one oligosaccharide) in a multivalent fashion produces a biological response that is different from the response that was obtained when only one carbohydrate antigen was used. The same conclusion regarding the importance of multiple carbohydrates was drawn in a recent work describing the binding of galectin-3 by different pectin polysaccharides (Sathisha et al. 2007). Pectin polysaccharides with high arabinose and galactose contents were bound more successfully by galectin-3 than pectin polysaccharides containing other sugars or containing only galactose without arabinose (Sathisha et al. 2007).

The incorporation of more than one functional end group on PAMAM dendrimers for applications in cancer research has been applied very successfully by Baker's group (Majoros et al. 2006). Although carbohydrates were not used, the work from the Baker labs merits mention here because it demonstrates advanced strategies using dendrimers with multiple functionalities. In one recent example, G(5)-PAMAM dendrimers bearing fluoresceins (for fluorescence imaging), folic acids (for cancer cell targeting), paclitaxels (for cytotoxicity), and acetamides and hydroxides (for solubility) were studied (Fig. 13.13). Biological in vitro studies

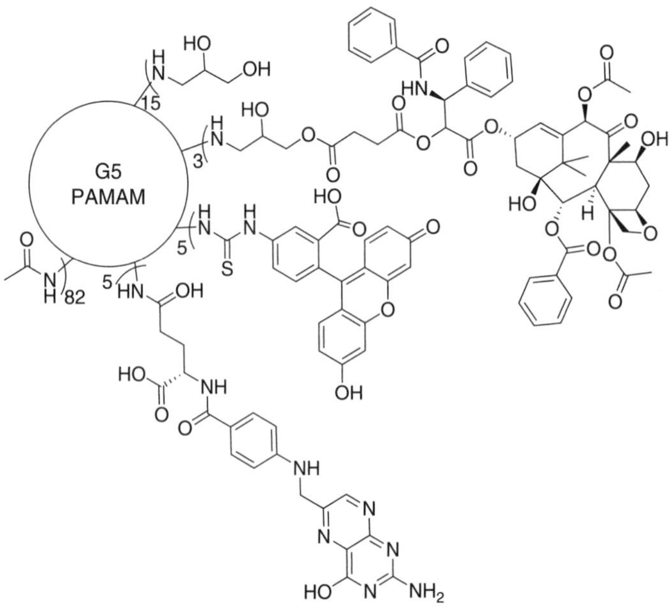

Figure 13.13 Multifunctional anticancer poly(amido amine) (PAMAM) dendrimers. Clockwise from top: diol, taxol, fluorescein isothiocyanate, folic acid, and acetyl groups.

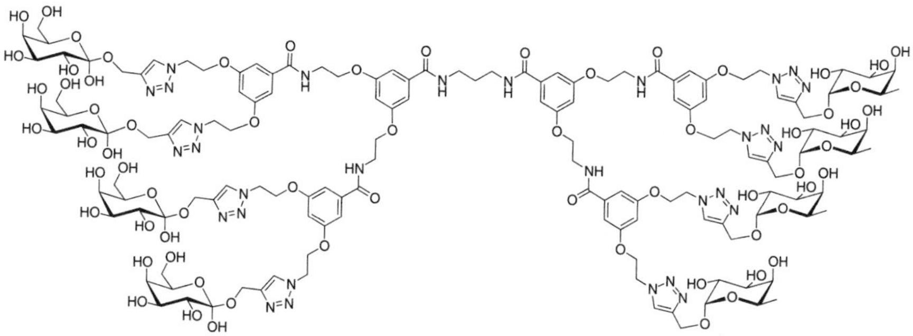

Figure 13.14 Fucose-/galactose-functionalized antibacterial dendrimers.

demonstrated that the multifunctional dendrimers were selectively cytotoxic for KB cells that overexpress folic acid (Majoros et al. 2006). Again, although carbohydrates were not used in this research, the highly multifunctional nature of the dendrimers demonstrates the importance of heterofunctionalized multivalent dendrimers in cancer research.

Dendrimers bearing more than one carbohydrate have been used not only with plant lectins (see previous example from Wolfenden and Cloninger 2005, 2006) but also with bacterial lectins. Deguise et al. (2007) incorporated galactose and fucose residues onto dendrimers (Fig. 13.14). The different sugars were on different halves of the dendrimers, with "click" chemistry as the method of carbohydrate attachment. Dendrimers bearing both carbohydrates were capable of binding to PA-IL and PA-IIL, two lectins from *Pseudomonas aeruginosa*. PA-IL exhibits specificity for D-galactose whereas PA-IIL displays preferential binding to L-fucose, and control experiments indicated that the glycodendrimer/lectin binding occurred through selective lectin–carbohydrate interactions. This report demonstrates that dendrimers bearing more than one carbohydrate can bind to orthogonal targets and represents an important advance for the use of glycodendrimers in multivalent processes (Deguise et al. 2007).

13.5. COMMENTS REGARDING THE SYNTHESIS OF HETEROMULTIVALENT CARBOHYDRATE SYSTEMS

Roy's research group (Deguise et al. 2007) synthesized dendrimers with fucoside on half of the dendrimer and galactoside on the other half using a convergent "outside-in" synthetic procedure. Linear polymers capable of displaying multiple carbohydrates in different segments of the polymer were reported by Pontrello et al. (2005). As demonstrated with Roy's fucose/galactose dendrimers, frameworks that can display blocks of carbohydrates may be preferred for some applications such as those in which binding of more than one lectin is required. In some instances, small molecules bearing two carbohydrates (or two ligands of any kind) may be able to attract larger multivalent systems into appropriate displays to induce desired biological processes. This strategy was recently disclosed by Carlson and colleagues (2007), who used an RGD peptidomimetic to bind to $\alpha_v\beta_3$ integrins, which enabled the α-Gal epitope that was tethered to the peptidomimetic to form a multifunctional display for binding to anti-α-galactosyl antibodies on cells with high integrin levels.

Alternatively, Baker's group (Majoros et al. 2006) synthesized multifunctionalized dendrimers with a random distribution of functional groups simply by controlling the number of equivalents of each terminal group that were added, and linear polymers with a random distribution of mannose and galactose were studied (Cairo et al. 2002; Majoros et al. 2006). Similar to Baker's method, we used controlled equivalents of isothiocyanato sugars to create heterogeneously functionalized glycodendrimers. Routes to heterogeneously functionalized dendrimers where the placement of end groups is rigorously controlled (but where more synthetically intensive routes are required) have been reported by researcher groups including those of Simanek and Thayumanavan (Vutukuri et al. 2003; Hollink and Simanek 2006; Crampton et al. 2007). These systematic efforts will be most important for frameworks with less flexibility (less ability to scramble carefully placed end groups) than the PAMAMs.

Ramström and Lehn (2000) used dynamic combinatorial chemistry (DCC) to form bis-saccharides via a disulfide linkage. Products were templated in the presence of ConA to demonstrate that DCC can be effectively used to tailor generations of synthetic systems bearing multiple carbohydrates (Ramström and Lehn 2000). Although only dimers were formed, their templated DCC could be used for larger frameworks including dendrimers to form idealized heterosaccharide-functionalized systems. Johansson et al. (2007) previously demonstrated combinatorial generation of glycodendrimers. Although fucose was always used as the carbohydrate end group in the current example, a large library of glycodendrimers was synthesized in which the amino acids comprising the dendrimer branches were varied. This methodology could be very well suited to the creation of dendrimers bearing multiple carbohydrates (Johansson et al. 2007).

13.6. ELECTRON PARAMAGNETIC RESONANCE (EPR) CHARACTERIZATION OF HETEROGENEOUSLY FUNCTIONALIZED DENDRIMERS

In order to effectively interpret assays performed with glycodendrimers and other ligand-bearing dendrimers with their receptors, a thorough understanding of the dendrimer itself is required. We performed extensive studies using EPR spectroscopy with spin-labeled dendrimers to determine the relative locations of dendrimer end groups on heterogeneously functionalized dendrimers (Walter et al. 2005). Our work finds its foundation in reports with spin-labeled proteins, where site-directed spin-labeling experiments allowed for the determination of relative distances between two spin-labeled components of proteins or peptides (Miick et al. 1992; Hubbell et al. 1998; Berliner et al. 2001). In addition, an excellent precendent for spin labeling of dendrimers was in place prior to our studies (Ottaviani et al. 1994; Maliakal et al. 2003).

As shown in Figure 13.10, we often functionalize dendrimers using sequential addition of isothiocyanates. The degree of dendrimer functionalization with each component is dictated by the number of equivalents of each isothiocyanate that are added. Characterization is performed using matrix assisted laser desorption ionization (MALDI) time of flight mass spectrometry (MS). Although MALDI provides accurate information regarding the number of end groups that have been added, MS (and NMR) cannot provide information about the relative locations of dendrimer end groups. EPR spectroscopy of spin-labeled dendrimers provides this information.

The synthesis strategy for functionalization of the G(4)-PAMAM dendrimer with 4-isothiocyanato 2,2,6,6-tetramethylpiperidine N-oxide (NCS-TEMPO) is shown in Figure 13.15. We chose to partially functionalize the dendrimer with a spin label while leaving the remaining terminal amino groups unreacted and to add primary, secondary, tertiary, and aromatic and carbohydrate isothiocyanates as the second coupling partner. The degree of dendrimer functionalization with each component was controlled by using different equivalents of isothiocyanates and was verified using MALDI (Samuelson et al. 2004; Walter et al. 2005).

The EPR spectra for the series of dendrimers bearing varying amounts of TEMPO but with the remaining PAMAM primary amines unreacted are provided in Figure 13.16. Dendrimers bearing only a few nitroxides have considerably sharper EPR spectra than dendrimers with higher nitroxide loadings. This is because spin labels that are close to one another in space induce significantly more line broadening in their EPR spectra than the isolated spins. Placing more spins on the dendrimer forces them to reside in proximity, causing an increase in line broadening.

Line-broadening effects for nitroxide-functionalized dendrimers that have unreacted primary amines and line-broadening effects when a second isothiocyanate is reacted with the dendrimer do not vary significantly. One clear indication of

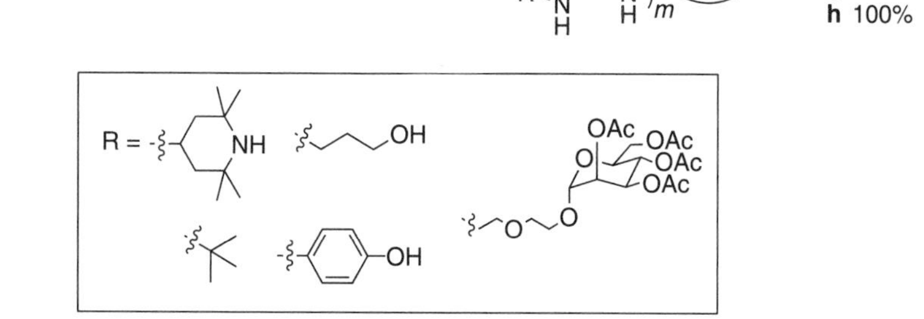

Figure 13.15 Synthesis of 2,2,6,6-tetramethylpiperidine N-oxide (TEMPO) functionalized dendrimers.

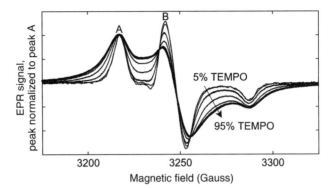

Figure 13.16 Stackplot of electron paramagnetic resonance (EPR) spectra of 2,2,6,6-tetramethylpiperidine *N*-oxide (TEMPO)-functionalized dendrimers with 5, 10, 25, 50, 75, 90, and 95% TEMPO.

line-broadening effects is exhibited by the ratio of peaks in the EPR spectra. The ratio of peak A to peak B in Figure 13.16 is plotted in Figure 13.17 for all of the TEMPO-functionalized dendrimers. In all cases, the A/B ratio of peak heights is independent of the nature of the second isothiocyanate. Thus, for functionalization of dendrimers using a sequential addition strategy, the end-group patterning is not significantly influenced by the nature of the end group.

What, then, is the end-group pattern that emerges from sequential addition of reactive components? We calculated the expected line-broadening effects for randomly functionalized dendrimers and compared them to the experimentally observed line-broadening effects. Figure 13.18 compares the experimental results from the stackplot in Figure 13.16 to the calculated results. Because a wide body of literature now indicates that dendrimer end groups can backfold into the interior of the dendrimer (Ballauff and Likos 2004), we included a volume shell in our calculations.

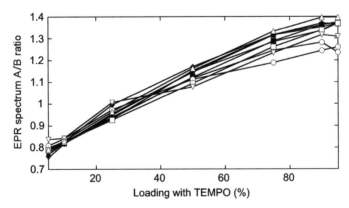

Figure 13.17 Graph of line-broadening effects for 2,2,6,6-tetramethylpiperidine *N*-oxide (TEMPO) and R-4-isothiocyanato (R-NCS) functionalized dendrimers from Figure 13.15. NCS-TEMPO was added first half of the time and R-NCS was added first half of the time.

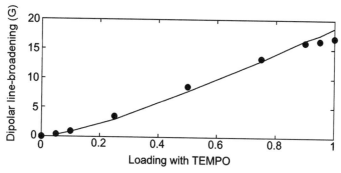

Figure 13.18 Comparison of calculated and experimental results: (●) experimental results from Figure 13.17 and (—) calculated results for a 16 Å volume shell.

Calculated line-broadening effects for end groups occupying at least a 16 Å volume shell show good agreement with the experimental values. Addition of further back-folding volume into the calculation does not change the predicted line-broadening effects significantly (Walter et al. 2005).

Experiments with sequential additions of NCS-TEMPO plus other isothiocyanates indicate that end groups are randomly distributed in relation to one another using this methodology. Even mannose-functionalized dendrimers appear to have a random end-group distribution (see below). As with any random population, it is possible that some carbohydrate-functionalized dendrimers will have higher activity in binding studies than other populations.

To ensure that we are measuring the results of bulk properties rather than amplifying results from a small, idealized subpopulation in our protein-binding studies, we performed affinity chromatography experiments as shown in Figure 13.19. We found that line-broadening effects before and after chromatography with resin-bound ConA were unchanged, and we were unable to isolate any subpopulations of lower and higher affinity dendrimers. From these experiments, we concluded that the randomly carbohydrate-functionalized dendrimers did not have subpopulations where the carbohydrate distribution on the dendrimer was either highly favorable or highly disfavorable for protein binding. Rather, the measured results for protein binding represent the result for the complete population of carbohydrate-functionalized dendrimers (Samuelson et al. 2004).

In our next series of experiments with spin-labeled dendrimers, we attempted to alter the presentation of nitroxides on the dendrimer. Our synthetic strategy is displayed in Figure 13.20. We invoked a boron protecting group strategy to hold three terminal alcohols in proximity. After blocking the remaining alcohols using ethyl chloroformate, we released the boron tether and added a 2,2,5,5-tetramethyl-1-pyrrolidinyloxy (PROXYL) spin label. All compounds were characterized using MALDI, and ^{11}B-NMR was used to characterize the intermediate borate esters. We found that up to three borate esters per dendrimer could be formed. Additional borate triesters could not be formed, presumably because of the highly strained system this would create.

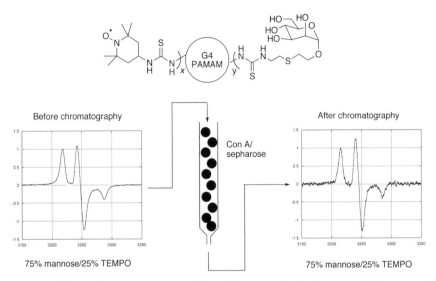

Figure 13.19 Affinity chromatography with 2,2,6,6-tetramethylpiperidine *N*-oxide (TEMPO)-/mannose-functionalized dendrimers. Electron paramagnetic resonance spectra for one TEMPO/mannose experiment are shown.

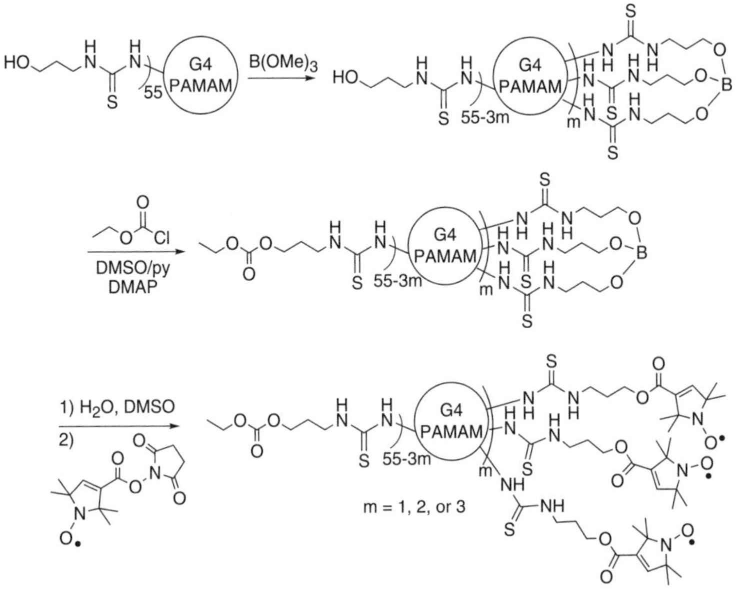

Figure 13.20 Synthesis of 2,2,5,5-tetramethyl-1-pyrrolidinyloxy-functionalized dendrimers.

TABLE 13.1 Comparison of Calculated Line Broadening Effects with Experimentally Determined Line Broadening Effects of PROXYL Functionalized Dendrimers

Spin-label Distribution	A/B Ratio		
	For 3 Spins	For 6 Spins	For 9 Spins
Clustered (calculated)	0.95	1.05[a]	1.1[a]
Random (calculated)	0.72	0.75	0.76
Maximally separated (calculated)	0.72	0.72	0.72
Product in Figure 13.20	0.61 ($m = 3$)	0.63 ($m = 6$)	0.69 ($m = 9$)

The calculated values are taken from Walter et al. (2005).
[a]These computed values are maximal values, because two ($m = 6$) and three ($m = 9$) clusters of three spins may occur, reducing the A/B ratio.

EPR spectra for PROXYL-functionalized dendrimers were obtained, and the line-broadening effects for different PROXYL loadings were evaluated. The line-broadening effects were inconsistent with a clustered distribution pattern. This result indicates that the PAMAM dendrimers are highly flexible frameworks with end groups that can readily reorient in relation to each other. Although the hydroxyl groups upon which PROXYL groups were placed were initially held in proximity by boron, they had redistributed by the time the EPR spectra could be obtained (Han et al. 2007; Table 13.1).

The results that we obtained from EPR characterization of spin-labeled PAMAMs indicate that, to have end groups on a PAMAM dendrimer reside permanently in proximity, covalent linkage between the groups is required. Strategies to synthesize such PAMAMs are currently underway.

The dynamic nature of the PAMAM dendrimers means that appropriate spacings of carbohydrates to achieve optimal binding to multiple lectin binding sites are readily achievable. In addition, because of the highly flexible nature of the functionalized PAMAM, covalent linkage of the carbohydrates is necessary if clustering of carbohydrates is required for optimal biological activity.

13.7. CONCLUSIONS AND OUTLOOK

Multivalent protein–carbohydrate interactions mediate many intercellular recognition processes. In order to study and to modulate multivalent protein–carbohydrate interactions, many carbohydrate-functionalized frameworks and systems have been synthesized. The glycodendrimer is one important glycosystem that has been used to study lectin binding. For our work, glycodendrimers based on the PAMAM framework are ideal because of the inherently flexible nature of PAMAM and because of the ease with which PAMAMs can be functionalized with carbohydrates.

The described examples, where more than one carbohydrate is attached to a synthetic framework, foreshadow the advances that we should expect in research involving multivalent protein–carbohydrate interactions. Functionalization of multivalent frameworks with simple sugars has been instrumental in advancing our understanding of multivalent protein–carbohydrate interactions. Past research with multiple copies of a single sugar have moved toward current research efforts using multivalent displays of multiple carbohydrates for the creation of synthetic glycosystems that can modulate cellular recognition processes. Thus, formation of heterogeneous carbohydrate-functionalized frameworks will be critical to the advancement of applications-driven research in this area.

ACKNOWLEDGMENT

Glycodendrimer research in the Cloninger group is supported primarily by the NIH (NIH RO1 GM62444).

REFERENCES

Allen JR, Harris CR, Danishefsky SJ. Pursuit of optimal carbohydrate-based anticancer vaccines: preparation of a multiantigenic unimolecular glycopeptide containing the Tn, MBr1, and Lewisy antigens. J Am Chem Soc 2001;123:1890–1897.

Ballauff M, Likos CN. Dendrimers in solution: insight from theory and simulation. Angew Chem Int Ed 2004;43:2998–3020.

Berliner LJ, Eaton SS, Eaton GR. Distance measurements in biological systems by EPR; biological magnetic resonance 19. Norwell (MA): Kluwer; 2001.

Bourne N, Stanberry LR, Kern ER, Holan G, Matthews B, Bernstein DI. Dendrimers, a new class of candidate topical microbicides with activity against herpes simplex virus infection. Antimicrob Agents Chemother 2000;44:2471–2474.

Buhleier E, Wehner W, Vögtle F. "Cascade" and "nonskid-chain-like" syntheses of molecular cavity topologies. Synthesis 1978;155–158.

Cairo CW, Gestwicki JE, Kanai M, Kiessling LL. Control of multivalent interactions by binding epitope density. J Am Chem Soc 2002;124:1615–1619.

Carlson CB, Mowery P, Owen RM, Dykhuizen EC, Kiessling LL. Selective tumor cell targeting using low-affinity, multivalent interactions. ACS Chem Biol 2007;2:119–127.

Christensen T, Gooden DM, Kung JE, Toone EJ. Additivity and the physical basis of multivalency effects: a thermodynamic investigation of the calcium EDTA interaction. J Am Chem Soc 2003;125:7357–7366.

Choi S-K. Synthetic multivalent molecules. Hoboken (NJ): Wiley; 2004.

Crampton H, Hollink E, Perez LM, Simanek EE. A divergent route towards single-chemical entity triazine dendrimers with opportunities for structural diversity. New J Chem 2007;31:1283–1290.

de Brabander-van den Berg EMM, Meijer EW. Poly(propylene imine) dendrimers: large scale synthesis via heterogeneously catalyzed hydrogenation. Angew Chem Int Ed Engl 1993;32:1308–1311.

Deguise I, Lagnoux D, Roy R. Synthesis of glycodendrimers containing both fucoside and galactose residues and their binding properties to Pa-IL and PA-IIL lectins from *Pseudomonas aeruginosa*. New J Chem 2007;31:1321–1331.

Denkewalter RG, Kolc J, Lukasavage WJ. 1981. Macromolecular highly branched homogeneous compound based on lysine units. US Patent 4,2889,872. 1981, Sept 15.

Derewenda Z, Yariv J, Helliwell JR, Kalb AJ, Dodson EJ, Papiz MZ, Wan T, Campbell J. The structure of the saccharide-binding site of concanavalin A. EMBO J 1989;8:2189–2193.

Duncan R, Izzo L. Dendrimer biocompatibility and toxicity. Adv Drug Deliv Rev 2005;57:2215–2237.

Fréchet JMJ, Tomalia DA, editors. Dendrimers and other dendritic polymers. West Sussex: Wiley; 2001.

Gabius H-J, Siebert H-C, Andre S, Jimenez-Barbero J, Rudiger H. Chemical biology of the sugar code. ChemBioChem 2004;5:740–764.

Gargano JM, Ngo T, Kim JY, Acheson DWK, Lees WJ. Multivalent inhibition of AB(5) toxins. J Am Chem Soc 2001;123:12909–12910.

Gestwicki JE, Cairo CW, Strong LE, Oetjen KA, Kiessling LL. Influencing receptor–ligand binding mechanisms with multivalent ligand architecture. J Am Chem Soc 2002;124; 14922–14933.

Gorelik E, Galili U, Raz A. On the role of cell surface carbohydrates and their binding proteins (lectins) in tumor metastasis. Cancer Metast Rev 2001;20:245–277.

Han HJ, Sebby KB, Singel DJ, Cloninger MJ. EPR Characterization of heterogeneously-functionalized dendrimers. Macromolecules 2007;40:3030–3033.

Handl HL, Vagner J, Han H, Mash E, Hruby VJ, Gillies RJ. Hitting multiple targets with multimeric ligands. Expert Opin Ther Targets 2004;8:565–586.

Hawker CJ, Fréchet JMJ. Preparation of polymers with controlled molecular architecture. A new convergent approach to dendritic macromolecules. J Am Chem Soc 1990; 112:7638–7647.

Hollink E, Simanek EE. A divergent route to diversity in macromolecules. Org Lett 2006;8:2293–2295.

Hubbell WL, Gross A, Langen R, Lietzow MA. Recent advances in site-directed spin labeling of proteins. Curr Opin Struct Biol 1998;8:649–656.

Ihre H, Hult A, Fréchet JMJ, Gitsov I. Double-stage convergent approach for the synthesis of functionalized dendritic aliphatic polyesters based on 2,2-bis(hydroxymethyl)propionic acid. Macromolecules 1998;31:4061–4068.

Jang W-D, Kataoka K. Bioinspired applications of functional dendrimers. J Drug Deliv Sci Technol 2005;15:19–30.

Jayaraman N, Nepogodiev SA, Stoddart JF. Synthetic carbohydrate-containing dendrimers. Chem Eur J 1997;3:1193–1199.

Jencks WP. On the attribution and additivity of binding energies. Proc Natl Acad Sci USA 1981;78:4046–4050.

Johansson EMV, Kolomiets E, Rosenau F, Jaeger K-E, Darbre T, Reymond J-L. Combinatorial variation of branching length and multivalency in a large (390 625 member) glycodendrimer library: ligands for fucose-specific lectins. New J Chem 2007;31:1291–1299.

Keding SJ, Danishefsky SJ. Prospects for total synthesis: a vision for a totally synthetic vaccine targeting epithelial tumors. Proc Natl Acad Sci 2004;101:11937–11942.

Krishnamurthy VM, Estroff LA, Whitesides GM. Multivalency in ligand design. In: Jahnke W, Erlanson DA, editors. Fragment-based approaches in drug discovery. Weinheim: Wiley; 2006.

Launay N, Slany M, Caminade A-M, Majoral J-P. Phosphorus-containing dendrimers, easy access to new multi-difunctionalized macromolecules. J Org Chem 1996;61:3799–3805.

Lee RT, Lee YC. Affinity enhancement by multivalent lectin–carbohydrate interaction. Glycoconjug J 2000;17:543–551.

Leffler H, Carlsson S, Hedlund M, Qian Y, Poirier F. Introduction to galectins. Glycoconjug J 2004;19:433–440.

Majoros IJ, Myc A, Thomas T, Mehta CB, Baker JR. PAMAM dendrimer-based multifunctional conjugate for cancer therapy: synthesis, characterization, and functionality. Biomacromolecules 2006;7:572–579.

Maliakal AJ, Turro NJ, Bosman AW, Cornel J, Meijer EW. Relaxivity studies on dinitroxide and polynitroxyl functionalized dendrimers: effect of electron exchange and structure on paramagnetic relaxation enhancement. J Phys Chem A 2003;107:8467–8475.

Mammen M, Choi S-K, Whitesides GM. Polyvalent interactions in biological systems: implications for design and use of multivalent ligands and inhibitors. Angew Chem Int Ed Engl 1998;37:2754–2794.

Mann DA, Kiessling LL. The chemistry and biology of multivalent saccharide displays. In: Wang PG, Bertozzi CR, editors. Glycochemistry: principles, synthesis, and applications. New York: Marcel-Dekker; 2001. pp. 221–275.

Matsuoka K, Hatano K, Terunuma D. Glycodendrimers using carbosilanes as core scaffolds. Nanotech Carbohydr Chem 2006;89–102.

Miick SM, Martinez GV, Fiori WR, Todd AP, Millhauser GL. Short alanine-based peptides may form 3(10)-helices and not alpha-helices in aqueous-solution. Nature 1992;359:653–655.

Mulder A, Huskens J, Reinhoudt DN. Multivalency in supramolecular chemistry and nanofabrication. Org Biomol Chem 2004;2:3409–3424.

Naismith JH, Emmerich C, Habash J, Harrop SJ, Helliwell JR, Hunter WN, Raferty J, Kalb-Gilboa AJ, Yariv J. Refined structure of concanavalin-a complexed with α-methyl-D-mannopyranoside at 2.0 angstrom resolution and comparison with the saccharide-free structure. Acta Crystallogr D Biol Crystallogr 1994;50:847–858.

Newkome GR, Moorefield CN, Vögtle F, editors. Dendrimers and dendrons. Weinheim: Wiley–VCH; 2001.

Ottaviani MF, Bossmann S, Turro NJ, Tomalia DA. Characterization of starburst dendrimers by the EPR technique. 1. Copper-complexes in water solution. J Am Chem Soc 1994;116:661–671.

Page MI, Jencks WP. Entropic contributions to rate accelerations in enzymic and intramolecular reactions and the chelate effect. Proc Natl Acad Sci USA 1971;68:1678–1683.

Pontrello JK, Allen MJ, Underbakke ES, Kiessling LL. Solid-phase synthesis of polymers using the ring-opening metathesis polymerization. J Am Chem Soc 2005;127:14536–14537.

Ragupathi G, Coltart DM, Williams LJ, Koide J, Kagan E, Allen J, Harris C, Glunz PW, Livingston PO, Danishefsky SJ. On the power of chemical synthesis: immunological evaluation of models for multiantigenic evaluation of models for multiantigenic carbohydrate-based cancer vaccines. Proc Natl Acad Sci USA 2002;99:13699–13704.

Ramström O, Lehn J-M. In situ generation and screening of a dynamic combinatorial carbohydrate library against concanavalin A. ChemBioChem 2000;1:41–48.

Roy R. A decade of glycodendrimer chemistry. Trends Glycosci Glycotechnol 2003;15: 291–310.

Rudd PM, Wormald MR, Dwek RA. Sugar-mediated ligand–receptor interactions in the immune system. Trends Biotechnol 2004;22:524–530.

Samuelson LE, Sebby KB, Walter ED, Singel DJ, Cloninger MJ. EPR and affinity studies of mannose-TEMPO functionalized PAMAM dendrimers. Org Biomol Chem 2004;2: 3075–3079.

Sathisha UV, Jayaram S, Nayaka MAH, Dharmesh SM. Inhibition of galectin-3 mediated cellular interactions by pectic polysaccharides from dietary sources. Glycoconjug J 2007;24: 497–507.

Seetharaman J, Kanigsberg A, Slaaby R, Leffler H, Barondes SH, Rini JM. X-ray crystal structure of the human galectin-3 carbohydrate recognition domain at 2.1 Å resolution. J Biol Chem 1998;273:13047–13052.

Servin P, Rebout C, Laurent R, Peruzzini M, Caminade A-M, Majoral J-P. Reduced number of steps for the synthesis of dense and highly functionalized dendrimers. Tetrahedron Lett 2007;48:579–583.

Srinivas O, Narayanaswamy J. Synthetic multivalent glycoclusters and their importance to probe intricate carbohydrate–protein interactions. Proc Ind Natl Sci Acad 2005;71A: 187–211.

Svenson S, Tomalia DA. Dendrimers in biomedical applications—reflections on the field. Adv Drug Deliv Rev 2005;57:2106–2129.

Tomalia DA, Baker H, Dewald J, Hall M, Kallos G, Martin S, Roeck J, Ryder J, Smith P. Dendritic macromolecules: synthesis of starburst dendrimers. Macromolecules 1986;19: 2466–2468.

Tsvetkov DE, Nifantiev NE. Dendritic polymers in glycobiology. Russ Chem Bull Int Ed 2005;54:1065–1083.

Vutukuri DR, Sivanandan K, Thayumanavan S. Synthesis of dendrimers with multifunctional periphery using an ABB′ monomer. Chem Commun 2003;796–797.

Walter ED, Sebby KB, Usselman RJ, Singel DJ, Cloninger MJ. Characterization of heterogeneously functionalized dendrimers by mass spectrometry and EPR spectroscopy. J Phys Chem B 2005;109:21532–21538.

Williams LJ, Harris CR, Glunz PW, Danishefsky SJ. In pursuit of an anticancer vaccine: a monomolecular construct containing multiple carbohydrate antigens. Tetrahedron Lett 2000;41:9505–9508.

Wittmann V. Synthetic approaches to study multivalent carbohydrate–lectin interactions. In: Schmuck C, Wennemers H, editors. Highlights in bioorganic chemistry. Weinheim: Wiley–VCH; 2004. pp. 203–213.

Wolfenden ML, Cloninger MJ. Mannose/glucose-functionalized dendrimers to investigate the predictable tunability of multivalent interactions. J Am Chem Soc 2005;127: 12168–12169.

Wolfenden ML, Cloninger MJ. Carbohydrate-functionalized dendrimers to investigate the predictable tunability of multivalent interactions. Bioconjug Chem 2006;17:958–966.

Woller EK, Walter ED, Morgan JR, Singel DJ, Cloninger MJ. Altering the strength of lectin binding interactions and controlling the amount of lectin clustering using mannose/ hydroxyl-functionalized dendrimers. J Am Chem Soc 2003;125:8820–8826.

Yang H, Kao WJ. Dendrimers for pharmaceutical and biomedical applications. J Biomater Sci Polym Ed 2006;17:3–19.

Zanini D, Roy R. Novel dendritic α-sialosides: synthesis of glycodendrimers based on a 3,3′-iminobis(propylamine) core. J Org Chem 1996;61:7348–7354.

Zanini D, Roy R. Architectonic neoglycoconjugates: effects of shapes and valences in multiple carbohydrate–protein interactions. In: Chapleur Y, editor. Carbohydrate mimics. Weinheim: Wiley–VCH; 1998a. pp. 385–415.

Zanini D, Roy R. Practical synthesis of starburst PAMAM α-thiosialodendrimers for probing multivalent carbohydrate–lectin binding properties. J Org Chem 1998b;63:3486–3491.

CHAPTER 14

SUPRAMOLECULAR POLYMERIZATION OF PEPTIDES AND PEPTIDE DERIVATIVES: NANOFIBROUS MATERIALS

HE DONG, VIRANY M. YUWONO, and JEFFREY D. HARTGERINK

14.1. INTRODUCTION

Although there have been much criticism of the final utility of the drive toward nanotechnology (stemming primarily from overblown claims and overly optimistic time lines), there is no doubt that nanotechnology can work. Control over these dimensions will lead to miraculous leaps in medicine and technology: the proof is in every living cell, each a beautifully orchestrated, nanostructured factory. The task is merely to understand the complexity within and apply this knowledge. Unfortunately, although Nature has been kind enough to show us the end result of mastery of this knowledge (life), she has been less forthcoming with respect to the how and why. To tackle this challenge one group of scientists has chosen to attempt to mimic these natural structures. Although most of the biological machines are far beyond what is currently accessible to the synthetic scientist, the synthesis of one subset of these materials, the nanofiber, has seen significant success. Nature utilizes a remarkable array of nanofibrous materials to carry out her intricate designs. Some of their applications include tissue cohesion, cell division, muscle contraction, intracellular molecular trafficking, and cellular locomotion. Most of the fibers responsible for the root of these functions result from precise control of the supramolecular polymerization of comparatively small subunits. Inspiring examples include the intermediate filaments and myosin fibers assembled largely from coiled coils, collagen fibrils assembled from peptide triple helices, and actin filaments and microtubules assembled from globular units. Nanofibers can also result from

Molecular Recognition and Polymers: Control of Polymer Structure and Self-Assembly.
Edited by V. Rotello and S. Thayumanavan
Copyright © 2008 John Wiley & Sons, Inc.

TABLE 14.1 Major Strategies for Peptide Nanofiber Assembly

Motif	Secondary Structure	Sequence Requirements	Method of Assembly	Orientation of Hydrogen Bonding with Respect to Fiber Axis
Coiled-coil	α-Helix	*abcdefg* heptad repeat	Hydrophobic packing (*a* & *d* residues) guided by electrostatics (*e* & *g* residues)	Parallel
Amyloid-like	Antiparallel β-sheet[a]	Alternating hydrophilic/hydrophobic	Hydrophobic nucleation followed by hydrogen bond propagation	Parallel
Collagen-like	Polyproline type II	(X-Y-Gly) repeat	Helix nucleation followed by zippering; fiber formation still a significant challenge	Perpendicular
Peptide-amphiphile/micelle	Parallel β-sheet[b]	Nearly any sequence with terminal alkylation	Hydrophobic aggregation followed by hydrogen bonding	Parallel

[a]Some examples display a parallel hydrogen bonding organization.
[b]Some examples display an anti-parallel hydrogen bonding organization.

loss of control over the assembly process. Most important in this category are the family of amyloid fibers formed from relatively short peptides aggregated into extended β-sheet nanofibers. Regardless of their healthy or diseased source, these fibers variously illustrate the desirable features of controlled assembly, controlled disassembly, nanostructured dimensions, atomically precise localization of chemical functionality, and fiber directionality. All of these features are extremely desirable materials properties. Mimicking these natural examples will help us to understand Nature in more detail and will result in sophisticated new nanostructured materials. This chapter examines some of the exciting results in nanofiber self-assembly (supramolecular polymerization) that have been described in the past decade utilizing peptides and peptide derivatives as their building blocks. The chapter is divided by the underlying peptide architecture used to make the fibers: α-helix, parallel and antiparallel β-sheets, and collagen-like polyproline type II (Table 14.1).

14.2. SELF-ASSEMBLY OF NANOFIBERS BASED ON α-HELICES

α-Helical coiled coils are one of the most common protein structural motifs in Nature. They are utilized to generate higher ordered tertiary and quaternary protein structures or act as recognition elements between two proteins. Because of this ubiquity in Nature, they have been studied intensively and are now one of the few protein motifs for which amino acid sequences can be selected and synthesized that will result in predetermined three-dimensional (3-D) structures.

A single α-helical molecule is stabilized by the intramolecular hydrogen bonding between the (i) and $(i + 4)$ residue along the peptide backbone, allowing for the formation of a right-handed helix with 3.6 amino acids per turn. However, a lone helix is rare. Typically, the helix must be stabilized by interactions with other portions of the protein or, in the case of coiled coils, it may be stabilized by interactions with additional helices.

The resulting structure largely depends on the amphiphilic pattern of the protein primary sequence comprising a variety of hydrophobic and polar residues. If a helix peptide molecule is designed to have all of the polar residues on one face and all of the hydrophobic residues on the other, a specific type of protein tertiary structure can be created with two or more helix chains associated together through hydrophobic interaction to form a coiled coil. The coiled coil was first postulated in 1953 by Crick to explain the X-ray diffraction pattern of α-keratin (Crick 1953). Later it was identified in many natural proteins, including a yeast transcription factor, GCN4 protein, that has been considered, because the X-ray crystal structure has been determined as a parallel coiled coil dimer, as an excellent model system to study protein–protein interactions (O'Shea et al. 1991). The primary sequence of GCN4 (also known as a "leucine zipper") demonstrated a distribution of hydrophobic and polar amino acids along the peptide backbone characterized by a seven-residue repeating unit called a *heptad* repeat, which is denoted *abcdefg*. Many reports (Hodges 1996; Lupas 1997; Oakley and Hollenbeck 2001) have documented the structural features of coiled coils. In order to pattern a peptide into a coiled coil structure, the *a* and *d* positions should be occupied by hydrophobic amino acids to form the interior hydrophobic seam whereas the *e* and *g* positions are populated by charged amino acids to render additional stability through electrostatic interactions. The selection of amino acids at each position controls the oligmerization state and the relative orientation of each helix within a coiled coil, such as dimer versus trimer, tetramer, and higher ordered assembly up to seven helices, or parallel versus antiparallel arrangement. The axial elongation of coiled coil proteins and the side-by-side packing against one another leads to the formation of many types of functional proteins and protein matrices. For example, tropomyosin, in the form of a dimeric coiled coil, regulates the interaction between F-actin filaments and myosin in response to Ca^{2+}. Parallel coiled coil trimers adopted by the HIV virus coat help the entry of HIV into human cells through the fusion of the hydrophobic N-termini with the cell membrane. As a major component of the cytoskeleton, intermediate filaments composed of coiled coil tetramers play an important role in providing mechanical strength to the nucleus and other cellular components. Understanding the rules of self-assembly utilized by Nature has been and will continue to be a major challenge for researchers.

A number of research groups have taken up this challenge and have developed rationally designed peptides adopting coiled coil structures that self-assemble into more complex nanostructures. As a dominating and perhaps the most practical form of nanostructures, nanofiber assembly serves to illustrate the hierarchical molecular self-assembly possible in these systems.

The first report on the formation of α-helix based nanofibers by Kojima and coworkers (1997) described a peptide comprising three heptad repeats that self-assembled into a coiled coil tetramer at near neutral pH in the presence of salts. Analytical ultracentrifugation indicated that not only was atetramer formed but also a species with a high molecular weight, which later was identified by negatively stained transmission electron microscopy (TEM) as nanofibers with 5–10 nm diameters. The same group explored the stability of helices and the resultant nanofibers through sequence reversal design and showed that unexpected enhanced helix stability and fiber elongation was observed with peptides with a primary sequence reversed from the originally designed peptide (Kojima et al. 2005). Hypothetically, the unequal charge distribution after sequence reversal may cause the self-assembly into more stabilized higher ordered aggregation by the formation of salt bridges. Although a more detailed description of the fiber formation and elongation needs to be discussed in terms of the hierarchical design, the work took the initial step toward the development of α-helix based artificial nanofibers.

Following this early study, Woolfson's group designed a series of peptides that have the ability to self-assemble into a dimeric coiled coil utilizing a "sticky-end" assembly mechanism (Pandya et al. 2000). The idea of the rational design of such a structure was motivated by the staggered pattern of subunits leading to the formation of natural fibrous structures, such as intermediate filaments, and the principle utilized by self-assembled DNA systems to create a variety of nanostructures. The molecular design shown in Figure 14.1 involves two four-heptad α-helices (SAF-p1 and SAF-p2) that prefer to form a coiled coil dimer with an overhanging

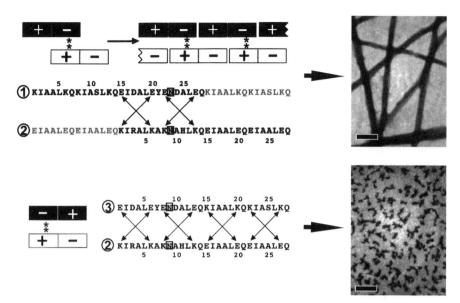

Figure 14.1   (1) SAF-p1, (2) SAF-p2, and (3) SAF-p3. (Top) Sticky-ended coiled coil dimer specified by complementary charge interaction and hydrogen bonding between Asn and Asn, which led to the formation of straight cross-linked nanofibers. (Bottom) Control experiment of the design of a blunt-ended dimer that failed to generate nanofibers. Scale bar = 100 nm for all images. Reprinted from Pandya et al. (2000). Copyright 2000 American Chemical Society.

sticky end at both termini. The structural specificity was gained and controlled by molecularly engineering the dimer interface established by the amino acids at the *a*, *d*, *e*, and *g* positions on both peptides. The selection of isoleucine at the *a* position and leucine at the *d* position favors dimer formation as opposed to other oligomerization states whereas asparagine was included at a different *a* site within each coil to impart the preference of an out-registered structure and parallel orientation. Charge interaction was designed to take place, as indicated in Figure 14.1, between the opposite ends of each building block when they are fixed in a parallel arrangement. The overall structure with the exposure of charged blocks recruited more partner peptides to propagate into elongated nanofibers. Although it is almost impossible to visualize the presence of an out-registered coiled coil dimer that leads to the higher ordered assembly, the significance of the sticky-ended design was illustrated by synthesizing a third peptide (SAF-p3) that was designed to form a stable in-registered structure with SAS-p2 as a control experiment. The result showed that a mixture of SAF-p2 and SAF-p3 forming a blunt-ended dimer failed to generate fibers as opposed to what was observed with the mixed SAF-p1 and SAF-p2 that were programmed to form a sticky-ended dimer. The morphology of these fibers as characterized by TEM was 10-μm length and 50-nm thickness, and X-ray fiber diffraction indicated that the nanofibers are composed of coiled coils.

Concern arises regarding the disagreement between the fiber width determined by TEM measurement and the theoretically calculated size for fibrils made of a single coiled coil dimer. Although important questions can still be raised at this point as to how to control the size of the higher ordered structure, the work presented here definitely proves the feasibility of utilizing de novo designed helical peptides to mimic their natural counterparts. At the same time, it encourages further exploration to seek for more building blocks with diverse molecular and structural information that can be delivered to the resultant self-assemblies. To this end, the same group synthesized several types of nonlinear peptides that are able to coassemble with the originally designed SAF peptides during fibrillogenesis (Ryadnov and Woolfson 2003a, 2003b, 2005). The fiber or fiber network morphology largely depends on the structural feature presented by the coassembled peptides, leading to nanofibers with distinct structural morphology (kinks, twists, branches, segments, polygonal networks).

In addition to morphology control, efforts have also been made to understand the self-assembly in terms of the longitudinal elongation induced by the head–tail packing and lateral aggregation facilitated by the side-by-side packing between fibrils. Rodamine-labeled SAF-p2a (mutant of SAF-p2) was mixed with the preformed fibers composed of SAF-p1 and SAF-p2a. It was found that fluorescent-labeled peptide was only attached on one end of the preformed fibers. This result indicated a unidirectional fiber growth directed by the inherent polarity of the SAF peptides with opposite charge domains dominating N- and C-termini (Smith et al. 2005). The issue of lateral aggregation occurring during fibrillogenesis is a very common phenomenon for natural protein fibers having a repeated pattern with complemented structural features. Considering the highly structural similarity and complementarities within the coiled coil building blocks, it would not be surprising to see the higher ordered packing taking place during self-assembly and fiber formation. Although the complete answer to these questions still awaits more detailed kinetics and molecular structural studies with a variety of spectroscopy methods, two papers published recently by Woolfson's group (Smith et al. 2006; Papapostolou et al. 2007) shed light on the understanding of the molecular and nanostructrual organization within coiled coil fiber bundles and should help the design of coiled coil based nanofibrous materials with engineered nanoscale order and stability. Considering the important role electrostatic interaction plays in the formation and stabilization of self-assembly, redesigned peptides were incorporated with additional ion pairs of Arg-Asp to strength the charge interaction. Collected data from circular dichroism (CD) and TEM all suggests increased stability compared to that of the first generation. It is worth noting that the second generation yielded fibers with higher salt tolerance, a significant improvement if these materials are to be considered in cell culture or other applications requiring physiological salt levels.

The most recent generation of peptides modified from the original SAF-1 and SAF-2 created a striation pattern across the entire length of nanofibers upon self-assembly, revealing a remarkable control on the nanoscale level within each fiber bundle. A molecular model was proposed and supported by the data obtained by TEM and X-ray diffraction analysis, indicating the coiled coil dimer self-assembled into a 3-D hexagonally packed lattice with a size of 1.8 nm and a periodicity of 4.2 nm along the fiber axis. The fact that the specific pattern of striation is reminiscent of what has been observed in natural collagen holds great promise in the development of peptide-based biomimetic materials, and it remains to be seen how cells respond to the new type of peptide-based scaffold.

Attempts to control fiber lateral aggregation have also been made by several other groups through rational design of coiled coil peptides. For instance, Potekhin et al. (2001) reported on the design of an α-helical fibril-forming peptide (aFFP) that was formulated into a coiled coil pentamer (Fig. 14.2) where the interhelical ionic forces between residues at the *f* and *g* positions were highlighted. In contrast, in a coiled coil dimer and trimer, such salt bridges cannot be formed because of the different geometry presented by the self-assembled coiled coils. The structural model proposed in Figure 14.2 accounts for the formation of a staggered coiled coil pentamer. The number of helix strands within a coiled coil is determined by the axial shift made by neighboring helices and the total length of an individual peptide.

The staggered pattern, unlike the design in the SAF peptides modulated by the side chain interaction, was created because of the identical heptad repeats along the entire peptide that tends to form offsets within the multistranded coiled coil assembly. The resultant structure with overhanging ends presents the hydrophobic amino

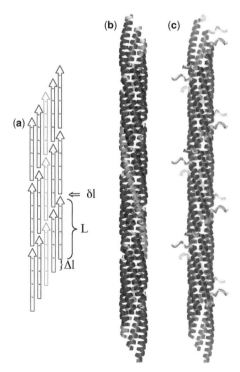

Figure 14.2   (a) Representation of a two-dimensional five-stranded coiled coil assembly with one heptad unit shifting in the axial direction with respect to each other and (b) the corresponding structural model. (c) Illustration of a three-dimensional model with nonheptad units overhanging in the peripheral region showing the structural tolerance of noncoiled coil fragments. Reprinted from Potekhin et al. (2001). Copyright 2001 Cell Press.

acids in the first and last heptad units in a solvent-exposed area, thus providing a driving force in the supramolecular polymerization of coiled coils into elongated fibers. The fibrils that were formed had little tendency to undergo lateral aggregation as evidenced by negatively stained TEM and X-ray analysis. Although this strategy works well to fabricate thinner nanofibers, the self-assembly requires acidic pH and prevents their further use for biomedical applications. The problem was later tackled through mutation experiments in which the glutamic acids at the *g* positions were replaced by neutral hydrophilic glutamine or serine to reduce the charge repulsion at neutral pH (Melnik et al. 2003). Under physiological conditions, the mutant peptide self-assembled into elongated fibrils composed of highly ordered α-helical structures. The same structural model and rules defining the oligomerization state of a staggered coiled coil was utilized by Zimenkov et al. (2004) in the design of a staggered coiled coil dimer that further promotes fibrillization. A six-heptad peptide, YZ1, was designed to form a staggered homodimeric coiled coil (Fig. 14.3). The two individual helices were offset by a three-heptad repeating unit.

Elaborate structural analysis was performed through CD, TEM, X-ray diffraction, and C13 cross-polarization/magic angle spinning NMR, indicating the presence of an α-helical structure that leads to the formation of nanofibers. Recently published work by Zimenkov et al. (2006) followed up this approach and focused on the switchability of nanofiber formation through the change of pH. Three histidine residues were included at the *d* positions to control the peptide secondary conformation in a pH-dependent fashion. The staggered trimeric coiled coil with a two-heptad unit apart from adjacent helices was confined by the specific electrostatic interaction. As shown by TEM, at pH 8.2 both fiber bundles spanning from 40 to 100 nm and thinner fibers at 3.3 nm were observed, whereas at pH 4 no obvious self-assembled

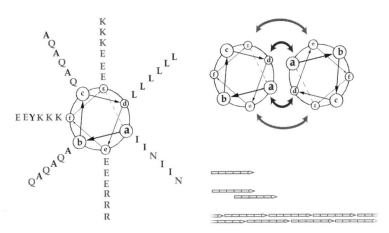

Figure 14.3 Helical wheel diagram of the YZ1 peptide and the schematic representation of the staggered dimer formation with an axial displacement of three heptad repeating units, which promote elongation into coiled coil fibrils. Reprinted from Zimenkov et al. (2004). Copyright 2004 Elsevier Science.

structures were formed when charged histidine prevented the formation of helical structures.

Wagner and coworkers (2005) addressed the issue of fiber maturation in their designed self-assembled nanoropes and nanofilaments. The peptide (CpA) contains two identical GCN4 subunits (each having a two-heptad unit) that are linked by a dipeptide subunit (Ala-Ala). The insertion of Ala-Ala changes the inherent pattern of hydrophobic amino acids on GCN4 that are generally aligned on the same phase in the interior of the coiled coils. The model (Fig. 14.4) instead showed a phase shift of heptad repeats, producing two hydrophobic clusters oriented at 200° with respect to one another. Thus, the self-assembled staggered coiled coil dimer was facilitated and specified by hydrophobic interaction that favors fiber axial growth rather than lateral aggregation.

Because the self-assembly takes place between a pair of two-heptad peptides through hydrophobic interaction, a relatively high concentration of salt needs to be added to enhance the hydrophobic effects. In addition, at the N-terminus, Cys was introduced to promote the association of the dimer and further polymerization.

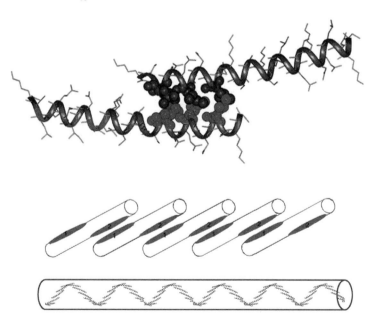

```
       abcdefgabcdefg xx abcdefgabcdefg
CpA :  CKQLEDKIEELLSK AA CKQLEDKIEELLSK
GCN4:  MKQLEDKVEELLSK
```

Figure 14.4 Primary sequence and molecular model of a coiled coil dimer and self-assembled polymer. The final bundle fiber structure is shown at the bottom. Reprinted from Wagner et al. (2005). Copyright 2005 National Academy of Sciences.

The dimension of the resultant structure was visualized by atomic force microscopy (AFM). Indeed, the width of the fibers was reduced compared to that of SAF peptides; however, its length was much more heterogeneous. Most fibers were also shown to be shorter than that of previously described coiled coil nanofibers. Fiber shortening could be related to the 1) weak association between sticky-ended coiled coils and 2) salt effects. It was found that sodium chloride and ammonium sulfate have a distinct effect on the fibril lateral aggregation, leading to short fibers in NaCl and long fibers in ammonium sulfate.

Although the coiled coil stands out as the most common building block in the design of helical nanofibers, other types of helix motifs have also been engineered into self-assembled nanostructures. One example is the helix–turn–helix peptide developed by Meredith's group (Lazar et al. 2005). The primary sequence is derived from apolipoprotein I containing multiple amphiphilic helical fragments connected by turns. The ability of peptides to self-assemble into fibrils was found to depend on the rigidity and structural organization in the turn region. As characterized by CD, FTIR, and X-ray diffraction, the self-assembled fibers were confirmed to be in an α-helical pattern. X-ray diffraction also indicated that the helical axis is perpendicular to the fiber long axis, which is very unusual for helix-based nanofibers and differed from all the mentioned self-assembling helical nanofibers having helices parallel to the fiber axis.

Unlike the conventional heptad motif based coiled coil peptides, Frost et al. (2005) designed a series of amphipathic helical peptides (KIA peptides) that have a consensus sequence of AKAxAAxxKAxAAxxKAGGY where two x positions out of six are occupied by Ala and the rest by Ile. These peptides were created in an attempt to mimic the ridges into grooves packing that has been utilized by many natural proteins (Chothia et al. 1981). A combinatorial library containing 15 short peptides was constructed by varying the location of Ala and Ile. These peptides distinguished themselves from one another in the hydrophobic interface, which in turn affected their folding pattern. Among the 15 peptides, only 2 formed insoluble helical filaments (at a high salt concentration, used to screen the charges between lysine residues). Fiber formation could be reversed by either reducing the salt concentration or through the addition of another peptide (KIA16).

In spite of the great advances in the construction of helix-based nanofibers over the last decade, the application of these materials still remains a large area to explore. A key challenge is to engineer chemical and mechanical stability into the resultant nanofiber scaffold that is important for 3-D cell culture. In this regard, Woolfson's group has taken the first step in targeting the construction of helical nanofibers with tunable thermal stability and nanoscale order (Smith et al. 2006). Potekhin and colleagues (2001) have proven the possibility of attaching a noncoiled coil moiety at the N-terminus of the originally designed aFFP without disrupting fibril formation. This study will encourage development of fibrous scaffolds with incorporation of a variety of biologically active ligands and perhaps other electronically active moieties imparting special electronic and transmitter ability to the "peptide wire." Although in vitro experiments using coiled coil based nanofiber scaffolds are rare, results from Corradin's group are promising in terms of the higher efficiency in promoting cell adhesion by an RGDS containing an α-helical coiled coil compared to fibronectin and vitronection (Villard et al. 2006). In addition to the biomedical applications utilizing fibrous nanostructures, nonfibrous coiled coil assemblies have been reported for templated nanofabrication that has been realized in two coiled coil based systems to pattern gold nanoparticles (Ryadnov et al. 2003; Stevens et al. 2004) and construction of a polynanoreactor derived from a dendrimer-like short leucine-zippers sequences (Ryadnov 2007).

14.3. NANOFIBERS SELF-ASSEMBLED FROM β-SHEETS

β-Sheets also play an important role in the construction of peptide-based functional nanofibrous materials. β-sheets are preferred over α-helices as molecular building blocks in the fabrication of artificial nanostructured materials perhaps because of the growing interest in understanding the self-assembly of two types of natural β-sheet products: silk protein and amyloid-like β-sheets. Furthermore, extended β-sheet conformation is relatively easy to achieve. Indeed, preventing their formation, particularly in high concentration or at high temperature, can be difficult in both synthetic and natural constructs.

Whereas α-helices are held together by *intra*molecular hydrogen bonding, β-sheets are stabilized by *inter*molecular hydrogen bonding perpendicular to the peptide chain. This cross molecular interaction results in a stronger tendency for such peptides to aggregate (or self-assemble) than that of α-helices. In addition to the difference seen in the secondary structure, when they are self-assembled into nanofibers, an additional distinction is made in terms of the orientation of the peptide chains with respect to the fiber long axis. Almost all of the synthetic nanofibers derived from α-helices reported thus far exhibit parallel orientation of individual peptide helices to the fiber axis (an exception is for the fibrous structures formed by apolipoprotein I mimetics) whereas in β-strand nanofibers the peptide chains lie approximately perpendicular to the fiber axis and stack into axially aligned extended β-sheets, a structure well known as a "cross-β" spine.

14.3.1. Helix-Sheet Conversion and Nanofiber Formation

Amyloid-like structures represent a major class of self-assembled β-sheet nanofibers that have common properties, such as elongated unbranched fiber morphology, high resistance toward thermal and chemical unfolding, and tendency to form highly entangled fibrous aggregation. The aggregation formed in the naturally occurring amyloid peptides may be associated with fatal diseases. Although the mechanism of amyloid formation is not fully understood, it has been suggested that a conformational transition from the native helical and globular form to a β-sheet rich structure plays a critical role. Several groups have addressed the issue of the relationship between secondary structural conversion and nanofiber formation with rationally designed model peptides (Takahashi et al. 1998; Ciani et al. 2002; Kammerer et al. 2004; Dong and Hartgerink 2007). This investigation has helped in the understanding of the diseases caused by protein misfolding and development of therapeutic drugs. In addition, new design principles have resulted in linking the primary and secondary structural features and the resultant nanostructures in the fabrication of engineered protein-based nanomaterials. Of particular importance is the work done by Mihara and group (Takahashi et al. 1998, 1999; Takahashi, Ueno et al. 2000; Takahashi, Yamashita et al. 2000) who developed a series of coiled coil peptides modified with hydrophobic domains at the N-terminus that self-assembled into amyloid-like fibers, accompanied by a secondary structural transition from α-helix to β-sheet.

The secondary structural change is thought to be induced by the partially folded helix that exposes the hydrophobic regions of peptides in water, thus leading to the aggregated nanofibers through hydrophobic interaction. Systematic follow-up mutation studies (Matsumura et al. 2004) have demonstrated the ability to control the self-assembly of nanofibers by complementary charge interaction and have resulted in fibers with uniform morphology. Other groups have also shown an analogous helix–sheet transition using different peptide models. Woolfson's group (Ciani et al. 2002) designed a peptide showing "structural duality," namely, the peptide is compatible with both α-helix and β-sheet formations. High β-sheet character amino acids (Thr) were incorporated to enhance the peptide's ability to form β-sheets, whereas the amphiphilic pattern of the hydrophobic and hydrophilic amino acids allowed the peptide to self-assemble into an α-helical coiled coil. The peptide containing Thr readily underwent a thermal-induced formation of β-sheets and nanofibers. A more detailed molecular model for peptides in the fibrillar state was established by Kammerer and coworkers (2004) with a de novo designed trimeric coiled coil. A 17-residue peptide, referred to as ccb, forms a native α-helical structure comprising three stranded helices as determined from X-ray crystal structural analysis. An increase of temperature or ionic strength promoted the transition to β-sheets, which further self-assembled into nanofibrils, showing a laminated, off-register antiparallel cross-β structure. Ala at position 7 hydrogen bonds to Leu at position 14 as evidenced by rotational echo double resonance measurements (Fig. 14.5). Recently, a two-stranded ccb coiled coil dimer was proposed to have the same effect on the structural transition with a temperature change from $4°$ to $70\,°C$ (Kammerer and Steinmetz 2006).

While studying of stability of short coiled coil peptides, our group identified the significance of hydrophobic clusters in promoting the helix–sheet transition and amyloid formation. The amino acids involved in the formation of a tentative hydrophobic cluster (Fig. 14.6 at a, d, and f positions) have been extensively studied for their ability to form either a stable helix or induce the helix to sheet transition in both two-heptad and three-heptad model systems (Dong and Hartgerink 2006, 2007). Five different amino acids (Gln, Ser, Tyr, Leu, and Phe) were incorporated in the f positions. These amino acids were chosen to allow the peptides having different secondary structural propensity and alternating hydrophilic–hydrophobic pattern. Experiments demonstrated that the secondary structural transition is independent of the amino acid secondary structural propensity, but it only relies on the peptides' primary sequence with alternate hydrophobic–hydrophilic residues localized in the central region.

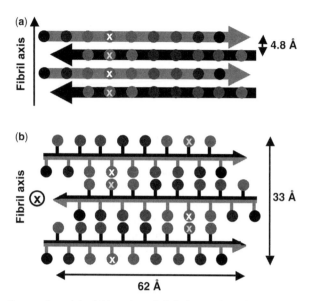

Figure 14.5 Proposed model of (a) antiparallel β-sheet orientation within nanofibers when viewed perpendicular to the β-sheets plane and (b) the cross section. Amino acid side chains are represented as spheres and Ala-7 and Leu-14 are marked by crosses. Reprinted from Kammerer et al. (2004). Copyright 2004 National Academy of Sciences.

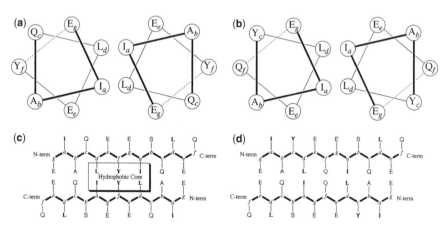

Figure 14.6 Structural arrangement of peptides (a and c) **1** (EIAQLEYEISQLEQ) and (b and d) **3** (EIAYLEQEISQLEQ) in the coiled coil or amyloid-like β-sheet packing motifs. Amino acids are represented with upper case standard single letter code whereas position in the heptad repeat is indicated in lower case italic. Reprinted from Dong and Hartgerink (2007). Copyright 2007 American Chemical Society.

As the last example of a helix–sheet transition, Pagel et al. (2006) demonstrate a coiled coil peptide adopting three types of secondary structures and the transition between random coil, α-helical, and β-sheet fibers can be simply triggered by changing the pH or peptide concentration.

14.3.2. Antiparallel β-Sheet Based Nanofibers

A prominent property of β-sheet forming fibers is their ability to self-organize into a macroscopic, responsive hydrogel. In this regard, Zhang's group took an early start on the development of molecular designed self-assembling peptides that formed entangled and cross-linked fiber networks (Zhang et al. 1993, 1994). The peptide molecule was first discovered from a segment of a left-handed Z-DNA binding protein, Zuotin (Zhang et al. 1992). A series of mutant peptides with similar residue presentation was created demonstrating a unique structural feature with a hydrophobic–hydrophilic alternating pattern and charge complementarities on the hydrophilic face (Fig. 14.7; Zhang et al. 1995; Zhang and Rich 1997; Zhao and Zhang 2006).

EAK16 (Ac-HN-AEAEAKAKAEAEAKAK-CONH$_2$) derived from the segment of the natural protein forms a macroscopic membrane matrix upon the addition of monovalent salts in physiological conditions. Results from the cell culture experiments indicated a nonintegrin mediated cell attachment to the oligopeptide matrices (Zhang et al. 1995). Most recently, a sliding diffusion model was proposed to understand the molecular mechanism of assembly and reassembly of the self-complementary peptide fibers upon mechanical breakage through sonication, suggesting a dynamic reassembly of the nanofiber scaffold (Yokoi et al. 2005). Although the originally designed peptides have no cell adhesion ligands, such as the commonly used integrin-binding peptide RGD, to promote cell attachment and

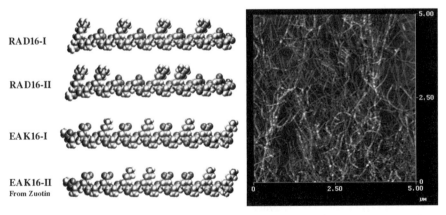

Figure 14.7 (Left) Molecular model of several self-complementary peptides and (right) atomic force microscopy images of nanofiber scaffolds formed by RADA16-I. Reprinted from Zhao and Zhang (2006). Copyright 2006 RSC Publishing.

proliferation, dramatic effects on the induction of neurite outgrowth and synapse formation have been achieved using RAD16-I (Ac-HN-RADARADARADARADA-CONH₂) and RAD16-II (Ac-HN-RARADADARARADADA-CONH₂) that readily self-assembled into hydrogels when exposed to cell culture media or salt solution (Holmes et al. 2000). A mutant peptide KLD-12 (Ac-HN-KLDLKLDLKLDL-CONH₂) was found to promote increased extracellular matrix (ECM) production and chondrocyte proliferation (Kisiday et al. 2002).

Despite the positive effects observed in the cell culture experiments, the molecular mechanism for cells interacting with peptide scaffolds is not clear, although it is possible that the presence of charged residues at every other position plays an important role for the nonspecific interaction between the peptide scaffold and cell surface components. To address the specific cellular response to the 3-D scaffold, a new series of peptides was designed with incorporation of either cell adhesion ligands or short sequences of growth factor peptides at both termini of the pure RADA peptide (Ac(RADA)₄CONH₂) (Horii et al. 2007). AFM showed no obvious difference in fiber morphology for peptide scaffolds with or without functional ligands; however, cell culture studies on osteoblast proliferation and differentiation indicated significantly improved preosteoblast proliferation, alkaline phosphatase activity, and osteocalcin secretion in the designer scaffold with biofunctional ligands. Another impressive result was discovered in an in vivo experiment using a self-assembling peptide (RADA16-I) to achieve hemostasis on wounds made by cuts in the liver and brain. Because hemostasis took less than 15 s, using these materials will have a significant effect on current surgical procedures in dealing with bleeding control (Ellis-Behnke et al. 2006).

Schneider and Pochan (2002) have described a different system of peptides with alternating polar and hydrophobic residues composed of Val as an amino acid with high β-sheet forming propensity and Lys as a modulator of pH-dependent self-assembly. Self-assembly was designed to be triggered by intramolecular folding into a β-hairpin structure based on the stereochemistry of central Pro residues (Fig. 14.8).

Systematic studies were carried out to understand the mechanism of molecular self-assembly into a macroscopic hydrogel structure by a variety of biophysical

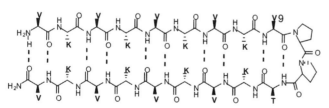
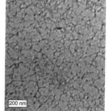

MAX1: VKVKVKVKV^DPPTKVKVKVKV-NH₂

Figure 14.8 (Left) Primary sequence of peptide MAX1 with β-hairpin promoted intramolecular folding, leading to the reversible formation of self-assembled β-sheets. (Right) Cryo-TEM image of self-assembled peptide scaffolds. Scale bar = 200 nm. Reprinted from Schneider et al. (2002). Copyright 2002 American Chemical Society.

and mechanical characterization methods including CD, Fourier transform IR (FTIR), TEM, confocal microscopy, and rheometry (Pochan et al. 2003; Ozbas et al. 2004; Lamm et al. 2005). The results show that the mechanical properties of self-assembled peptides were dramatically increased accompanied by instantaneous gel formation when the pH increased from 5 to 9 when lysine was partially neutralized. The mechanical modulus can be tuned by the added salt concentration. Replacing Val with Thr maintains the peptide's abilities of intramolecular folding and hydrogel formation, but it adjusts the sol–gel transition temperature from 60 to 25 °C, depending on the peptide's overall hydrophobicity. Detailed studies on the kinetics of gel formation were recently reported (Veerman et al. 2006) utilizing microrheological measurements at low peptide concentration to allow temporal control on the formation of a fiber network. Most importantly, as shown by mechanical studies (Haines-Butterick et al. 2007), the response of these materials to mechanical shear and the fast recovery in mechanical rigidity allows them to be used for injectable delivery of cells to the targeted area. This is a very important property and holds promise for the application of these materials as cellular therapeutic agents.

Another example of a pH responsive hydrogel was demonstrated by Aggeli and colleagues (1997) with an 11 amino acid long peptide (DN1). The design principles for pH control of the self-assembly and switching between different macroscopic states (nematic gel, nematic fluid, flocculate, isotropic fluid) was illustrated in a series of peptides containing varying net charges (Aggeli, Bell, Boden et al. 2003; Aggeli, Bell, Carrick et al. 2003). It was discovered that one unit of net charge is needed to stabilize the dispersion of fibrillar structures in nematic gels. Ionic strength has also been taken into consideration in regulating the molecular and macroscopic structures for the same type of peptides. The macroscopic phase transition occurred over a much broader pH range in the presence of NaCl at 130 mM than that of pure peptides (Carrick et al. 2007). These peptides self-assemble as extremely long nanofibers that are physically cross-linked and entangled, as has been seen in many natural amyloids. Because of their great size and entanglement, these fibers form viscous liquids or gels in water and are quite insoluble and difficult to purify.

The infinite nanofibrous structures, on the one hand, help to create a robust scaffold for 3-D cell culture and other applications. On the other hand, the lack of water solubility makes their handling quite difficult. The issue of fiber morphology control has recently been addressed by peptides called multidomain peptides that were designed in our labs (Dong et al. 2007). The peptides consist of three distinct regions that impact self-assembly using the design principle of "molecular frustration." The central fragment comprises a variable number of alternating (Gln-Leu) repeating units to form a facial amphiphile and drive the peptides into elongated β-sheet nanofibers (similar to the described systems) and is flanked by consecutive lysine residues on both termini to work against the peptides' desire to self-assemble into infinitely long fibers, thus aiding in solubilization (Fig. 14.9).

Controlled nanostructural morphology, in this case nanofibers with controlled length, was achieved by varying the ratio of the central repeating number of (Gln-Leu) and the flanking lysine residues. A variety of biophysical characterization methods indicated that the short, dispersed nanofibers formed by $K_2(QL)_6K_2$ reached

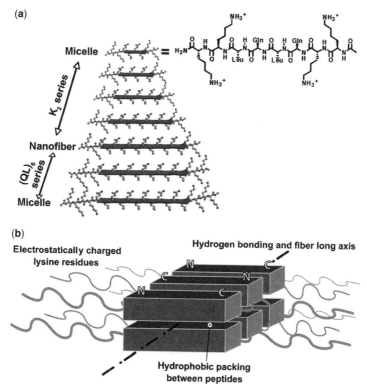

Figure 14.9 (a) Primary sequence of a series of multidomain peptides and (b) proposed molecular model of nanofiber packing indicating hydrophobic packing, axis of hydrogen bonding, and repulsive charge interaction. Reprinted from Dong et al. (2007). Copyright 2007 American Chemical Society.

thermodynamic equilibrium through the balance of forces driving self-assembly (hydrophobic packing and hydrogen bonding) and opposing forces (electrostatic repulsion) favoring solubility. The importance of charge interaction in determining and controlling the fiber length has been corroborated in salt mixing experiments, which showed that the addition of salt dramatically changed the morphology of nanofibers from being short and dispersed to a cross-linked fiber network.

14.3.3. Nanofibers Derived from Short Peptide Sequences

There are numerous reports on the design of building blocks that adopted primarily β-sheets from very short peptides. They can be structurally as simple as di-, tri-, and tetra-amino acid peptides; however, mostly these short peptides were covalently modified by organic functionality to set the conformation in favor of β-sheet packing. For instance, Kelly's group systematically studied the self-assembly of peptidomimetics that comprised two segments of tetrapeptides linked together by a dibenzofuran moiety forming a β-hairpin structure (Lashuel et al. 2000;

Deechongkit et al. 2005). Extensive TEM studies have shown nanofiber formation with distinct morphology (protofibril, twisted fibers, ribbons, lateral associated fibers, etc.) under different sample preparation conditions. The polymorphorism observed in this system has a general significance in understanding the mechanism of protein misfolding that causes many types of neurodegenerative diseases. Banerjee's group established a series of tripeptides differing with each other in the C- or N-terminal protecting group that coordinated peptides to self-assemble into entangled nanofibers composed of antiparallel β-sheets (Das et al. 2004; Maji et al. 2004; Ray, Das et al. 2006; Das et al. 2007). The gelation property was examined in water and a variety of organic solvents by differential scanning calorimetry, TEM, FTIR, single crystal X-ray, and X-ray powder diffraction methods, showing the ability of terminal functionality to alter peptide's gelation properties. The Gazit group reported the formation of peptide nanotubes made of short β-sheet forming dipeptides (Phe-Phe; Reches and Gazit 2003). Other dipeptides (Phe-Trp, Trp-Tyr, Trp-Phe, and Trp-Trp) have also been examined, but none of them showed tubular structures as the primary nanostructures. In contrast, this observation indicates that the specific packing of aromatic residues often found in natural amyloids plays a significant role in the self-assembly of amyloid fibrils. Short peptides comprising β, δ, and ω amino acids have been reported by several groups in the construction of supramolecular self-assembling structures (Banerjee et al. 2005; Dutt et al. 2005; Martinek et al. 2006). Despite the challenge in the synthesis of peptides containing noncanonical amino acids, the distinct secondary structures displayed by these peptides from the commonly seen α amino acid peptides will create a diverse self-assembly pattern and supramolecular nanostructural morphology.

14.3.4. Parallel β-Sheet Nanofibers

Although the majority of β-sheet forming nanofibers adopt an antiparallel orientation because of the preferred orientation of hydrogen bonding between peptide backbone amide linkages, β-sheets can be stabilized in parallel fashion by rational covalent attachment of the hydrophobic domain at the N- or C-terminus. In 2001 a new class of self-assembling molecules known as peptide amphiphiles (PAs) was developed by the Stupp group (Hartgerink et al. 2001), featuring a long hydrophobic aliphatic chain at the N-terminus of a hydrophilic peptide. This simple modification creates an amphiphilic molecule with a propensity to form cylindrical micelles (nanofibers) in aqueous solution. In addition to the amphiphilicity, the stability of the self-assembled structure comes from β-sheets like intermolecular hydrogen bonding in the peptide region, especially those formed by the four amino acid residues closest to the nanofiber core (Paramonov et al. 2006). Spherical, not cylindrical, micelles were observed when the hydrogen bond network was eliminated, for example, through the incorporation of N-methylated residues. Peptides with tails attached to the same terminus form parallel β-sheet.

The nanofibers formed by PAs are microns long and have a uniform diameter of approximately 6–8 nm, depending on the length of the PA molecule. Self-assembly of the nanofibers can be controlled by pH adjustment and/or addition of divalent cations such as Ca^{2+} to neutralize the charges on the amino acid side chains (Hartgerink et al. 2002). For example, the self-assembly of PAs containing Glu can be triggered by lowering the pH or by adding Ca^{2+} to the PA solution at a pH range where the PAs are normally disassembled. Depending on the choice of amino acid residues incorporated in the sequence, we can obtain PA nanofibers in a wide pH range. The presence of multiple cysteines in the sequence allows covalent capture by the formation of intermolecular disulfide bonds after self-assembly, which further stabilizes the nanofibers. PAs form self-supporting gels that can be reversibly disassembled by pH adjustment. However, covalently captured or divalent cation-neutralized nanofibers are not sensitive to pH changes and retain their fibrous morphology.

The formation of PA gels was studied by varying the alkyl tail length, cross-linking region, concentration, and bioactivity (Hartgerink et al. 2002). It was shown that peptides without an alkyl tail or with only a short tail (6 carbons) did not have enough hydrophobic character to drive self-assembly into gels. Peptides connected to C10, C16, and C22 carbon alkyl tails formed gels after pH adjustment. Cross-linking by cysteines is not required for the formation of nanofibers. However, when present, cysteines must be completely in the reduced state prior to self-assembly as random disulfide linkages between PA molecules (or intramolecular cross-links) prevent the structure needed for fiber formation. The PA starting concentration influenced the ability of the fibers to form bundles and 3-D network, but it did not affect the nanostructure of the individual fibers. Variation in cell-binding functionality, including the relatively hydrophobic IKVAV, did not affect the propensity of the PAs to form gels, but TEM revealed changes in the fiber length and stiffness.

Figure 14.10 is a schematic of a PA and the negatively stained nanofibers (Hartgerink et al. 2001). Because the peptide head group is oriented toward the surface of the nanofibers, the chemical functional groups imparted by the amino acid side chains are accessible to the surrounding environment. Furthermore, the peptide region of PAs is easily tailored to accommodate its use in a wide variety of applications. In this example, the PA nanofibers directed the mineralization of hydroxyapatite (HA). Therefore, this particular PA has several features built into the sequence: 1) a hydrophobic tail that drives the assembly into nanofibers, 2) four cysteines for covalent capture, 3) three glycines for flexibility in the molecule, 4) phosphorylated serine for calcium binding, and 5) an Arg-Gly-Asp sequence for cell adhesion. In the HA mineralization studies, the PAs were self-assembled on a holey-carbon TEM grid by exposure to HCl vapor followed by immersion in I_2 for covalent capture. HA mineralization on PA nanofibers was performed by diffusion of $CaCl_2$ and $NaHPO_4$ from opposite sides of the grid. Energy dispersion X-ray fluorescence spectroscopy analysis confirmed the formation of HA on the surface of the nanofibers. TEM imaging and electron diffraction revealed the preferential alignment of the HA crystallographic c-axis with the long axis of the fiber.

Sone and Samuel (2004) continued the studies of mineralization on PA nanofibers by utilizing the same PA described above to nucleate and grow CdS nanocrystals. In this case, the negatively charged phosphate and carboxylate groups bind to Cd^{2+}, and CdS was formed after diffusion of H_2S gas. A low Cd^{2+} to PA ratio led to the formation of CdS nanocrystals that were $3-5\,nm$ in diameter. An intermediate ratio of

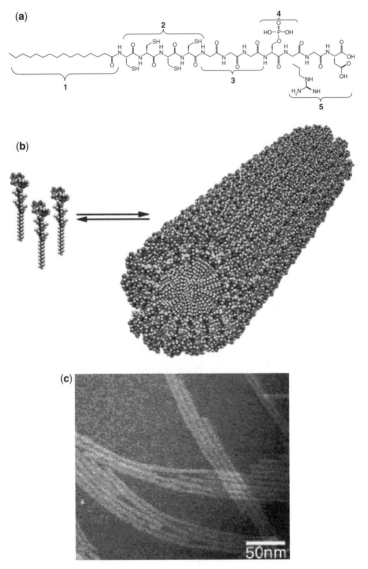

Figure 14.10 Self-assembly of peptide-amphiphiles into nanofibers: (a) a peptide amphiphile molecule with five distinct regions designed for hydroxyapatite mineralization, (b) a schematic of molecular self-assembly, and (c) a negatively stain transmission electron microscopy image of the nanofibers. Reprinted from Hartgerink et al. (2001). Copyright 2001 American Association for the Advancement of Science.

Cd^{2+} to PA produced a continuous polycrystalline CdS layer on the nanofibers. TEM analysis, which showed a lighter core surrounded by a darker layer, indicated that the hydrophilic portion of the PA was embedded in CdS whereas the hydrophobic core remained unmineralized.

The Hartgerink group recently extended the application of PA as a biomimetic template in mineralization by synthesizing a series of PAs that were capable of catalyzing the formation of silica from a tetraethoxysilane precursor (Yuwono and Hartgerink 2007). Typically, the sol–gel reaction is catalyzed by ammonium hydroxide. Lys and His residues were incorporated into the peptide sequence to act as surface anchored catalysts through the presence of amine and imidazole groups, respectively, in their side chains. Because of the difference in their pK_a values, PAs containing Lys self-assemble at pH 10 whereas His is at pH 7. Just as with HA and CdS, silica was successfully formed on the surface of the nanofibers at their corresponding self-assembly pH. Calcination of the nanocomposite material yields hollow silica nanotubes. The thickness of the silica layer can be tuned by modulating the length of the PAs. In contrast, the same experiments with Glu residues did not lead to silica tube formation, because of the lack of catalytic groups on the surface of the fibers.

PA nanofibers are also widely used in biological applications. Self-assembled PA gels have desirable properties as ECM mimics in tissue engineering, which include biocompatibility, mechanical strength tunability, and versatility of the peptide region to incorporate diverse bioactive functionality, which are demonstrated in the next example. A peptide amphiphile designed to function as an ECM mimic incorporated relevant features in its peptide region: 1) a matrix metalloproteinase (MMP)-2 sensitive sequence (GTAGLIGQ) to allow cell-mediated proteolytic degradation, 2) a glutamic acid and C-terminal carboxylic acid for solubility and calcium binding, and 3) a cell adhesive sequence (RGDS; Jun et al. 2005). Proteolytic degradation of the gel is necessary for cell migration and cell remodeling of the network. The density of cell adhesive sites on the nanofiber surface can be controlled by mixing the PA with another PA containing a nonadhesive sequence (RDGS) at various ratios. The viscoelasticity and gelation behavior at various Ca^{2+} to PA ratios were studied by rheometry. Hydrogel networks were obtained when the ratio of Ca^{2+}/PA was 0.5 and the storage modulus continued to increase until the Ca^{2+}/PA was equal to 2. Further addition of calcium caused the nanofibers to form large parallel bundles that phase separated and decreased the mechanical strength of the gel. The PA gels were incubated in type IV collagenase to evaluate their enzymatic degradation behavior. The gels lost 50% of their weight in 1 week and were completely degraded in 1 month. Under TEM, the gels' nanostructure was observed to have changed from long 8-nm diameter fibers before incubation in an enzyme to egg-shaped aggregates and twisted ribbons. The data suggested that cleavage of PA at the MMP-2 sensitive site created defects in the fibers that undermined the structural integrity of the fibers and led to the formation of irregular aggregates. The biocompatibility, bioactivity, and biodegradability of the PA gel were evaluated by encapsulation of MMP-2 producing cells, which were rat maxillary incisor pulp cells. The density of cell binding ligands was modulated by mixing PAs with RGDS and RDGS sequences at various ratios. The cells were able to interact with the PA as indicated by cell spreading and elongation. Cells were also able to proliferate and remodel the scaffold.

Neural progenitor cells were successfully encapsulated in vitro (Silva et al. 2004) in a manner similar to the previous case, but in this case the PA contained the bioactive moiety IKVAV, which is known to direct neurite growth. The scaffold induced rapid cell differentiation into neurons but inhibited differentiation into astrocytes. It is interesting that in order for the cell differentiation to proceed, the cells have to be entrapped in the IKVAV nanofiber matrix. No differentiation was observed when the cells were grown in PA gels without a bioactive sequence, even when the IKVAV PA was introduced at a later time.

PAs with the heparin binding site LRKKLGKA for application in medicinal therapy to control angiogenesis have also been synthesized (Rajangam et al. 2006). This peptide self-assembles into a nanofiber network in the presence of heparin. The heparin is conveniently displayed on the surface of the nanofibers and becomes accessible for angiogenic growth factor binding. The release of rhodamine-labeled fibroblast growth factor 2 (FGF-2) from heparin-PA gels was monitored and compared to PA gels without heparin. This experiment showed that PA gels containing heparin delayed the FGF-2 release, compared to an almost immediate release in PA gels without heparin.

The heparin-PA gel was subjected to a rat cornea angiogenesis assay to evaluate its performance in vivo. The heparin-PA gel with vascular endothelial growth factor (VEGF) and FGF-2 was placed into a surgically created pocket in a rat cornea. Heparin-PA gel with growth factors showed significant neovascularization as quantified by the ratio of the corneal area involved in the response, the average length of the blood vessels, and the maximum length of the blood vessels formed in each eye (both as a ratio of the diameter of the cornea). Several control experiments were also performed using different combinations of materials, including preformed collagen gels with heparin, VEGF, and FGF-2; collagen gels with heparin alone; collagen gels with the growth factors alone; and solutions of PA with growth factors, heparin with growth factors, and growth factors alone. Amazingly, none of the results from the control experiments showed significant neovascularization compared to the heparin-PA gel with growth factors.

Branched PAs attached to biotin were developed to study the packing of the bulkier head group of the modified PAs compared to linear PAs and to increase the accessibility of the bioactive segment to receptors (Guler et al. 2005). Branching in the PA was introduced at a lysine residue by taking advantage of orthogonal protecting group chemistry. The looser packing of biotinylated branched PA increased the biotin availability to avidin, resulting in enhanced recognition binding by avidin protein receptors. Linear PAs having the same epitope formed more compact nanofibers, which caused the approach by avidin to be more sterically hindered.

In another case, PAs were designed to function as magnetic resonance imaging (MRI) contrast agents (Bull et al. 2005) by covalently linking the peptide portion to a derivative of 1,4,7,10-tetraazacyclododecane-1,4,7,10-tetraacetic acid followed by chelation of Gd^{3+} ions by this moiety. The PAs were modified such that self-assembly produce either nanofibers or spherical micelles. The application of this self-assembling system could be extended to noninvasive MRI of PA scaffolds in vivo.

Dip-pen nanolithography (Jiang and Stupp 2005) and soft lithography (Hung and Stupp 2007) have been used to control the placement and orientation of PA nanofibers on two-dimensional substrates. The soft lithographic technique is the more successful one at alignment and patterning of the nanofibers into long-range ordered arrays. In this procedure, PAs were self-assembled by solvent evaporation on the substrate and capped with a poly(dimethylsiloxane) stamp containing patterned grooves while being subjected to continuous sonication for 1 h. After overnight drying, the stamp was removed to reveal fiber bundles oriented in the direction of the grooves, as imaged by AFM.

It is clear from these examples that peptide amphiphiles have great potential. Their tolerance to modifications of their chemical functionality while maintaining fibrous morphology and their capability to form gels are beneficial in many applications. Although most of these applications have focused on utilizing the surface of the nanofibers, several others turned to the hydrophobic core. For instance, PAs have been shown to enhance the solubility of carbon nanotubes (Arnold et al. 2005) in aqueous solution through hydrophobic interactions between their alkyl tail and the surface of carbon nanotubes. In another study, several PAs containing a tryptophan or pyrene chromophore were synthesized to investigate the solvation of different parts of a PA within the nanofiber (Tovar et al. 2005), which is directly related to their accessibility to small molecules. The chromophore was placed at the C- or N-terminus, and in one of the PAs the long alkyl tail was replaced with pyrene. Fluoresence spectroscopy experiments showed that PA nanofibers maintain a high degree of free volume, and individual PA molecules were well solvated when assembled in the nanofiber. Interactions of tryptophan and pyrene probes with fluorescent quenchers in solution show a gradual decrease in response from the exterior to the interior of the nanofiber to the very center of the hydrophobic core.

Nanofibers with a parallel β-sheet character have been found in other systems, most notably in pathogenic Alzheimer's amyloid (Aβ) peptide. Over the years, researchers have studied the self-assembly of Aβ peptide into fibrils. In particular, Aβ peptide segments were synthesized and evaluated with solid-state NMR to study their secondary structure. Depending on their sequence, the peptides form either anti-parallel (Lansbury et al. 1995; Balbach et al. 2000) or parallel β-sheet fibers (Benzinger et al. 1998; Gregory et al. 1998; Antzutkin et al. 2000; Balbach et al. 2002). Peptides with hydrophobic residues concentrated in one of their termini, thereby introducing amphiphilic character to the molecule, tend to form parallel β-sheets (Soreghan et al. 1994). This behavior is reminiscent of peptide-amphiphile systems in which the peptides have hydrophobicity toward one of their termini. However, the hydrophobic character in PAs comes from aliphatic chains or other hydrophobic molecules. In this case, the overall amphiphilicity is inherent in the peptide sequence. Peptides with no pattern of amphiphilicity between their terminals form anti-parallel β-sheets. Without a hydrophobic segment, peptides generally prefer to be in anti-parallel orientation because it produces the most optimum hydrogen bonding interaction between the peptide strands. Gordon et al. (2004) demonstrated that an Aβ peptide derivative that forms anti-parallel β-sheet fibers would self-assemble in parallel orientation after acylation of its N-terminus.

The parallel β-sheet is also the secondary structural motif of the pathogenic form of ataxin-3, a 42-kDa peptide expressed in the brain (Bevivino and Loll 2001). The peptide becomes harmful when the number of glutamine residues near its C-terminus is increased from 12–40 to 55–84 (Cummings and Zoghbi 2000). The excess of glutamine alters the peptide folding behavior into parallel β-sheet fibers. The secondary structure of nonpathogenic and pathogenic forms of ataxin-3 was monitored by CD and other spectroscopic measurements (Bevivino and Loll 2001). The pathogenic form clearly exhibits a β-sheet character, and fibrils were observed under TEM. In contrast, the normal peptide forms an α-helix rather than a β-sheet secondary structure.

There have been several efforts at designing peptides that form parallel β-sheets. The general strategy is to impart an amphiphilic nature into the peptide by using hydrophobic amino acids as seen in the natural systems described earlier. Interactions between certain amino acid pairs in adjacent peptide strands have been found to be favorable for parallel β-sheet formation (Fooks et al. 2006). Electrostatic interaction between positive and negatively charged amino acids is the most favored in a parallel β-sheet, followed by Asn-Asn pairing and interactions between hydrophobic residues. Interaction between Cys pairs is also energetically favored.

Another approach is to include unnatural amino acids, such as γ-aminobutyric acid (Ray, Drew, et al. 2006) or δ-aminovaleric acid residues (Banerjee et al. 2005) at the peptide N-terminus. Characterization by FTIR verified the β-sheet nature of the conformation of the peptides. Depending on the rest of the peptide sequence, the peptides adopt either parallel or anti-parallel β-sheet, which can be determined by single crystal X-ray diffraction studies. TEM and scanning electron microscopy analysis displayed the fibrous morphology of the self-assembled peptides. Parallel β-sheets can also be generated by linking C- or N-termini of two beta strands with turn-inducing moieties such as D-prolyl-1,1-dimethyl-1,2-diaminoethane (Fisk and Gellman 2001).

Potential applications of β-sheet forming nanofibers currently focus on the development of cellular scaffold, templated mineralization for the preparation of composite materials. Excellent examples in the field of biomedical research have been demonstrated in some of the previous discussion. The fabrication of molecular wires is made possible through the "bottom-up" self-assembly approach with β-sheets as templates. The strong affinity of metals toward peptide-based fibrous materials is determined by the wide range of chemical functionalities, especially those capable of coordinating with metals, such as gold, silver, palladium, and others. Lindquist's group has successfully prepared a conducting wire using a fragment of amyloid fibrous protein (Sup35p) as the biotemplate, showing the conductive properties of a solid nanowire, such as low resistance and ohmic behavior (Scheibel et al. 2003). Fu and colleagues (2003) reported another self-assembled amyloid fiber system that directed the formation of single or double helical arrays of metal particles. PAs represent a major class of materials that can be utilized in the construction of inorganic–organic hybrid materials, for example, the nucleation and growth of HA, silica nanotubes, cadmium sulfide, and so forth. As the most recent update in biomineralization of nanofibers, Banejee's group (Bose et al. 2007) described a two-component nongrafting method to fabricate metal particle coated nanofibers formed by a pseudopeptide. These biotemplated nanowires have potential practical applications in the field of electroengineering. Related topics can be found in excellent reviews and articles (Rapaport et al. 2002; Sneer et al. 2004; Mesquida et al. 2005; Whitehouse et al. 2005; Rapaport 2006; Sakurai et al. 2006; Lepere et al. 2007).

14.4. COLLAGEN MIMETICS

Collagen, a class of fibrous protein, is the most abundant protein in the human body, accounting for approximately 25% of the total protein mass. It is the main component of the ECM and serves as a structural protein in connective tissues, such as skin, bone, cartilage, and blood vessels. Twenty-eight types of collagen have been identified in humans to date (Kar et al. 2006). Among these, collagen types I–III are the most abundant.

Collagen fibers have a distinctive triple helical conformation, in which three left-handed polyproline type II helices wrap around one another to form a right-handed super helix. Each peptide strand, also called the α-chain, has a repeating amino acid sequence Gly-X-Y, where X and Y denote positions frequently occupied by the imino acids proline and 4-hydroxyproline. Because of their ring structure, the imino acids stabilize the helical conformation of each α-chain. Glycine, which appears at every third position, enables the formation of the collagen triple helix because its side chain can fit in the interior of the triple helix to allow tight packing of the chains. The collagen molecular assembly is sustained by an inter-molecular hydrogen bonding network between carbonyl oxygen from residue X and amido hydrogen from Gly.

Various approaches have been developed to chemically synthesize triple helical collagen mimics. Because the collagen triple helix is stabilized by the repeating trimeric unit of Gly-Pro-Hyp, this pattern is a good starting point to design self-assembling triple helices. Kar et al. (2006) studied the effects of pH, peptide concentration, temperature, peptide length, and sugar concentration on the self-assembly of (Pro-Hyp-Gly)$_n$, where $n = 7, 8, 10$, and 12. Self-association of (Pro-Hyp-Gly)$_{10}$ was promoted by increasing the temperature and peptide concentration at neutral pH as evidenced by the formation of aggregates in the peptide solution. However, examination of the aggregates under the electron microscope revealed a branched fibrous structure with no long-range order characteristic of its natural counterparts. Because the design of the peptide is just based on the primary amino acid sequence of collagen without any preorganization, the associations between the peptide chains could not be controlled, which led to branching.

Inspired by the self-assembly of double-helix DNA fragments, and similar to the approach used by Woofson for α-helical coiled coils, Raines' group synthesized short Pro-Hyp-Gly based collagen fragments (Kotch and Raines 2006). Each fragment contained three peptide strands that were covalently linked together with disulfide bonds between Cys residues incorporated in the peptide sequence, a strategy also known as covalent knot or cysteine knot. The presence of Cys introduced an offset between the peptide strands and created sticky ends where the fragments connected with one another to form an elongated structure from 30 to >400 nm, as observed under the AFM. Substitution of Hyp by Pro led to more stable and longer fibers as

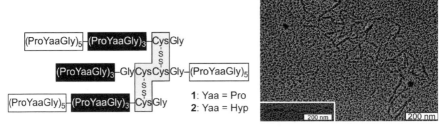

Figure 14.11   (Left) The molecular structure of collagen fragments used as building blocks and (right) the TEM image of self-assembled fragments. Reprinted from Kotch and Raines (2006). Copyright 2006 National Academy of Sciences.

noted in the previous example. This group went a step further and tested the length and thermal stability in aqueous methanol, in which more stable triple helices containing Hyp residues produced longer fibers. The collagen building blocks and the product of their assembly are shown in Figure 14.11.

Recently, Cejas and coworkers (2007) investigated the self-assembly of short peptides with Gly-Pro-Hyp repeating units containing hydrophobic end groups, containing phenyl, pentafluorophenyl, or isopropyl groups. The aromatic end groups were of interest because of potential $\pi-\pi$ stacking noncovalent interactions, which could provide a potential driving force for head–tail self-assembly. Computational molecular modeling of the self-assembly process between two triple helices predicted face–face orientations of the end groups, which result in strong binding between the aromatic groups, especially for pentafluorophenyl/ phenyl pairs, but not for the isopropyl group. The thermal stability of the peptide was studied by CD and NMR, which confirmed its triple helicity at well above room temperature. A newly prepared solution of peptide containing a pentafluorophenyl end group formed a mixture of small (3 nm) and large (190 nm) particles, which aggregated to form fiberlike material that was microns in length and had a 0.26 μm average diameter within 24 h. The peptide was assessed in a human platelet assay and was found to induce platelet aggregation similar to equine type I collagen.

In the previous examples, short collagen-like peptides were designed to have complementary binding sites that led to propagation of self-assembly to form large triple helical fibers or aggregates. The next two examples describe self-assembly of collagen-like peptides with the help of covalent linking of short collagen-like segments.

The work of Kishimoto and cohorts (2005) focused on the synthesis of a high molecular weight collagen-like polypeptide by the direct polycondensation of the collagen characteristic sequence (Pro-Hyp-Gly)$_n$ ($n = 1, 5,$ and 10) using 1-ethyl-3-(3-dimethyl-aminopropyl)-carbodiimide hydrochloride and 1-hydroxybenzotriazole (HOBt). Carbodiimides and HOBt are commonly used coupling reagents in peptide synthesis. The polymerization reaction creates long triple helical chains from short peptides through head–tail propagation. No side reactions were observed by FTIR and ^{1}H-NMR of poly(Pro-Hyp-Gly)$_{10}$. Various concentrations of peptide monomers in phosphate buffer (pH 7.4) and dimethylsulfoxide (DMSO) were tested for the formation of long fibers. Gel permeation chromatography analysis determined that polymerization of (Pro-Hyp-Gly)$_5$ and (Pro-Hyp-Gly)$_{10}$ at a concentration of 2.5 mg/mL in DMSO resulted in the formation of high molecular weight products exceeding 10000 kDa. However, the same peptides yielded products of low molecular weights when the reactions were carried out in phosphate buffer. Of interest, polymerization of Pro-Hyp-Gly at a high concentration (50 mg/mL) in phosphate buffer produced polypeptides with molecular weights over 10000 kDa. CD analysis showed that (Pro-Hyp-Gly)$_5$ and (Pro-Hyp-Gly)$_{10}$ formed more stable triple helices in less polar solvent than in phosphate buffer, whereas Pro-Hyp-Gly did not self-assemble into triple helices in either solvents. This result suggests that preorganization prior to the polymerization lowers the activation energy of the reaction. Polycondensation of Pro-Hyp-Gly only proceeds at high concentration, because the probability of head–tail intermolecular polycondensation significantly increases. Under TEM, poly(Pro-Hyp-Gly)$_{10}$ displayed nanofibers microns in length with a 10-nm diameter, which corresponds to a bundle of about 30 aggregated triple helical chains (Fig. 14.12).

Our group demonstrated an approach involving covalently coupling short peptides to create extended collagen-like peptide polymers by native chemical ligation (Paramonov et al. 2005). Native chemical ligation features the ability to couple two unprotected peptides with a thioester at the C-terminus and a cysteine at the N-terminus. Thioester linkage and subsequent intramolecular reaction forms a native peptide bond between the two peptides. Three peptides were synthesized; each was 30 amino acids long with a Pro-Hyp-Gly repeat and cysteine at the N-terminus followed by the conversion of the C-terminus into an activated thioester.

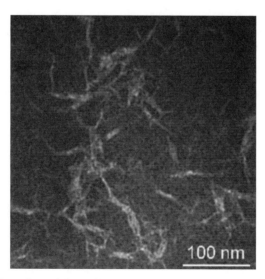

Figure 14.12 Fibers formed by polycondensation of (Pro-Hyp-Gly)$_{10}$. Reprinted from Kishimoto et al. (2005). Copyright 2005 Wiley InterScience.

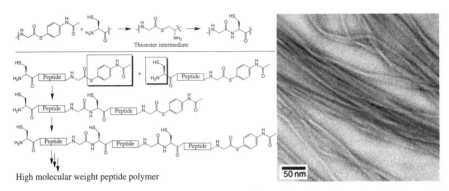

High molecular weight peptide polymer

Figure 14.13 (a) The strategy of creating long collagen-like fibers by native ligation of 30 amino acid peptides and (b) transmission electron microscopy image of the native ligated peptides. Reprinted from Paramonov et al. (2005). Copyright 2005 American Chemical Society.

The peptides were dissolved in phosphate buffered water to allow self-assembly into triple helices prior to native ligation polymerization. The structure of the native ligated peptide polymers was analyzed by TEM, showing fibrous aggregates that were microns in length. We compared the peptide polymerization by the native ligation method to 2-(1H-benzotriazol-1-yl)-1,1,3,3-tetramethyluronium hexafluorophosphate (HBTU)-activated coupling, which is the commonly used method in peptide synthesis. The molecular weights of the polymers were obtained by size exclusion chromatography. The native ligated peptide polymers exhibited molecular weights on the order of hundreds of kilodaltons, which were higher than that produced by HBTU-activated coupling, as observed by TEM (Fig. 14.13).

Most recently, a report by the Conticello and Chaikof groups demonstrated not only fiber assembly but also D-periodic stripping, which has only been previously observed in natural collagen fibrils (Rele et al. 2007). This remarkable result illustrated in Figure 14.14 utilizes designed charge pairing that is formed from the peptide $(PRG)_4(POG)_4(EOG)_4$. Again, this approach has similarities to common sticky-end strategies in DNA and pioneered by Woolfson in α-helical coiled coils.

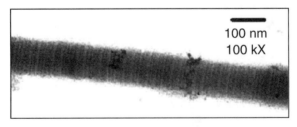

Figure 14.14 A zwitterionic collagen-like peptide is shown to self-assemble into relative thick fibers displaying a native collagen-like D-spacing pattern. Reprinted from Rele et al. (2007). Copyright 2007 American Chemical Society.

14.5. CONCLUSIONS

Our ability to design and synthesize nanofibers from peptidic components has clearly expanded dramatically in the past decade. The challenge is no longer simply the synthesis of a fiber, but what kind of specificity these fibers will have with respect to their dimensions (length is particularly challenging), display of chemical functionality, and control over assembly and disassembly conditions. Applications in the areas of cell culture, drug delivery, regenerative medicine, and tissue engineering appear to be the nearest to fruition, but many other potential applications also await.

REFERENCES

Aggeli A, Bell M, Boden N, Carrick LM, Strong AE. Self-assembling peptide polyelectrolyte beta-sheet complexes form nematic hydrogels. Angew Chem Int Ed 2003;42:5603–5606.

Aggeli A, Bell M, Boden N, Keen JN, Knowles PF, Mcleish TCB, Pitkeathly M, Radford SE. Responsive gels formed by the spontaneous self-assembly of peptides into polymeric β-sheets tapes. Nature 1997;386:259–262.

Aggeli A, Bell M, Carrick LM, Fishwick CWG, Harding R, Mawer PJ, Radford SE, Strong AE, Boden N. pH as a trigger of peptide β-sheet self-assembly and reversible switching between nematic and isotropic phases. J Am Chem Soc 2003;125:9619–9628.

Antzutkin ONB, John J, Leapman RD, Rizzo NW, Reed J, Tycko R. Multiple quantum solid-state NMR indicates a parallel, not antiparallel, organization of β-sheets in Alzheimer's β-amyloid fibrils. Proc Natl Acad Sci USA 2000;97:13045–13050.

Arnold MS, Guler MO, Hersam MC, Stupp SI. Encapsulation of carbon nanotubes by self-assembling peptide amphiphiles. Langmuir 2005;21:4705–4709.

Balbach JJ, Ishii Y, Antzutkin ON, Leapman RD, Rizzo NW, Dyda F, Reed J, Tycko R. Amyloid fibril formation by $A\beta16$–22, a seven-residue fragment of the Alzheimer's β-amyloid peptide, and structural characterization by solid state NMR. Biochemistry 2000;39:13748–13759.

Balbach JJP, Aneta T, Oyler NA, Antzutkin ON, Gordon DJ, Meredith SC, Tycko R. Supramolecular structure in full-length Alzheimer's β-amyloid fibrils: evidence for a parallel β-sheet organization from solid-state nuclear magnetic resonance. Biophys J 2002;83:1205–1216.

Banerjee A, Das AK, Drew MGN, Banerjee A. Supramolecular parallel β-sheet and amyloid-like fibril forming peptides using δ-aminovaleric acid residue. Tetrahedron 2005;61: 5906–5914.

Benzinger TL, Gregory DM, Burkoth TS, Miller-Auer H, Lynn DG, Botto RE, Meredith SC. Propagating structure of Alzheimer's beta-amyloid(10–35) is parallel beta-sheet with residues in exact register. Proc Natl Acad Sci USA 1998;95:13407–13412.

Bevivino AE, Loll PJ. An expanded glutamine repeat destabilizes native ataxin-3 structure and mediates formation of parallel β-fibrils. Proc Natl Acad Sci USA 2001;98: 11955–11960.

Bose PP, Drew MGB, Banerjee A. Nanoparticles on self-assembling pseudopeptide-based nanofiber by using a short peptide as capping agent for metal nanoparticles. Org Lett 2007;9:2489–2492.

Bull SR, Guler MO, Bras RE, Meade TJ, Stupp SI. Self-assembled peptide amphiphile nanofibers conjugated to MRI contrast agents. Nano Lett 2005;5:1–4.

Carrick LM, Aggeli A, Boden N, Fisher J, Ingham E, Waigh TA. Effect of ionic strength on the self-assembly, morphology and gelation of pH responsive β-sheet tape-forming peptides. Tetrahedron 2007;63:7457–7467.

Cejas MA, Kinney WA, Chen C, Leo GC, Tounge BA, Vinter JG, Joshi PP, Maryanoff BE. Collagen-related peptides: self-assembly of short, single strands into a functional biomaterial of micrometer scale. J Am Chem Soc 2007;129:2202–2203.

Chothia C, Levitt M, Richardson D. Helix to helix packing in proteins. J Mol Biol 1981;145:215–250.

Ciani B, Hutchinson EG, Sessions RB, Woolfson DN. A designed system for assessing how sequence affects α to β conformational transitions in proteins. J Biol Chem 2002;277: 10150–10155.

Crick FHC. The packing of α-helixes: simple coiled-coils. Acta Crystallogr 1953;6:689–697.

Cummings CJ, Zoghbi HY. Fourteen and counting: unraveling trinucleotide repeat diseases. Hum Mol Genet 2000;9:909–916.

Das AK, Banerjee A, Drew MGB, Haldar D, Banerjee A. Stepwise self-assembly of a tripeptide from molecular dimers to supramolecular β-sheets in crystals and amyloid-like fibrlis in the solid state. Supramol Chem 2004;16:331–335.

Das AK, Bose PP, Drew MGB, Banerjee A. The role of protecting groups in the formation of organogels through a nano-fibrillar network formed by self-assembling terminally protected tripeptides. Tetrahedron 2007;63:7432–7442.

Deechongkit S, Powers ET, You S-L, Kelly JW. Controlling the morphology of cross β-sheet assemblies by rational design. J Am Chem Soc 2005;127:8562–8570.

Dong H, Hartgerink JD. Short homodimeric and heterodimeric coiled coils. Biomacromolecules 2006;7:691–695.

Dong H, Hartgerink JD. Role of hydrophobic clusters in the stability of α-helical coiled coils and their conversion to amyloid-like beta sheets. Biomacromolecules 2007;8:617–623.

Dong H, Paramonov SE, Aulisa L, Bakota EL, Hartgerink JD. Self-assembly of multi-domain peptides: balancing molecular frustration controls conformation and nanostructure. J Am Chem Soc 2007;129:12468–12472.

Dutt A, Drew MGB, Animesh P. β-Sheet mediated self-assembly of dipeptides of ω-amino acids and remarkable fibrillation in the solid state. Org Biomol Chem 2005;3:2250–2254.

Ellis-Behnke R, Liang Y-X, Tay DKC, Kau PWF, Schneider GE, Zhang S, Wu W, So K-F. Nano hemostat solution: immediate hemostasis at the nanoscale. Nanomedicine 2006;2: 207–215.

Fisk JD, Gellman SH. A parallel β-sheet model system that folds in water. J Am Chem Soc 2001;123:343–344.

Fooks HM, Martin ACR, Woolfson DN, Sessions RB, Hutchinson EG. Amino acid pairing preferences in parallel β-sheets in proteins. J Mol Biol 2006;356:32–44.

Frost DWH, Yip CM, Chakrabartty A. Reversible assembly of helical filaments by de novo designed minimalist peptides. Biopolymers 2005;80:26–33.

Fu X, Wang Y, Huang L, Sha Y, Gui L, Lai L, Tang Y. Assemblies of metal nanoparticles and self-assembled peptide fibrils—formation of double helical and single-chain arrays of metal nanoparticles. Adv Mater 2003;15:902–906.

Gordon DJ, Balbach JJ, Tycko R, Meredith SC. Increasing the amphiphilicity of an amyloido-genic peptide changes the β-sheet structure in the fibrils from antiparallel to parallel. Biophys J 2004;86:428–434.

Gregory DMB, Tammie LS, Burkoth TS, Miller-Auer H, Lynn DG, Meredith SC, Botto RE. Dipolar recoupling NMR of biomolecular self-assemblies: determining inter- and intra-strand distances in fibrilized Alzheimer's β-amyloid peptide. Solid State Nucl Magn Reson 1998;13:149–166.

Guler MO, Soukasene S, Hulvat JF, Stupp SI. Presentation and recognition of biotin on nanofibers formed by branched peptide amphiphiles. Nano Lett 2005;5:249–252.

Haines-Butterick L, Rajagopal K, Branco M, Salick D, Rughani R, Pilarz M, Lamm MS, Pochan DJ, Schneider JP. Controlling hydrogelation kinetics by peptide design for three-dimensional encapsulation and injectable delivery of cells. Proc Natl Acad Sci USA 2007;104:7791–7796.

Hartgerink JD, Beniash E, Stupp SI. Self-assembly and mineralization of peptide-amphiphile nanofibers. Science 2001;294:1684–1688.

Hartgerink JD, Beniash E, Stupp SI. Peptide-amphiphile nanofibers: a versatile scaffold for the preparation of self-assembling materials. Proc Natl Acad Sci USA 2002;99:5133–5138.

Hodges RS. De novo design of alpha-helical proteins: basic research to medical applications. Biochem Cell Biol 1996;74:133–154.

Holmes TC, De Lacalle S, Su X, Liu G, Rich A, Zhang S. Extensive neurite outgrowth and active synapse formation on self-assembling peptide scaffold. Proc Natl Acad Sci USA 2000;97:6728–6733.

Horii A, Wang X, Gelain F, Zhang S. Biological designer self-assembling peptide nanofiber scaffolds significantly enhance osteoblast proliferation, differentiation, and 3-D migration. PLoS One 2007;2:e190.

Hung AM, Stupp SI. Simultaneous self-assembly, orientation, and patterning of peptide-amphiphile nanofibers by soft lithography. Nano Lett 2007;7:1165–1171.

Jiang H, Stupp SI. Dip-pen patterning and surface assembly of peptide amphiphiles. Langmuir 2005;21:5242–5246.

Jun H-W, Yuwono V, Paramonov SE, Hartgerink JD. Enzyme-mediated degradation of peptide-amphiphile nanofiber networks. Adv Mater 2005;17:2612–2617.

Kammerer RA, Kostrewa D, Zurdo J, Detken A, Garcia-Echeverria C, Green JD, Mueller SA, Meier BH, Winkler FK, Dobson CM, Steinmetz MO. Exploring amyloid formation by a de novo design. Proc Natl Acad Sci USA 2004;101:4435–4440.

Kammerer RA, Steinmetz MO. De novo design of a two stranded coiled-coil switch peptide. J Struct Biol 2006;155:146–153.

Kar K, Amin P, Bryan MA, Persikov AV, Mohs A, Wang Y-H, Brodsky B. Self-association of collagen triple helix peptides into higher order structures. J Biol Chem 2006;281: 33283–33290.

Kishimoto T, Morihara Y, Osanai M, Ogata S-I, Kamitakahara M, Ohtsuki C, Tanihara M. Synthesis of poly(Pro-Hyp-Gly)n by direct polycondensation of (Pro-Hyp-Gly)n, where $n = 1$, 5, and 10, and stability of the triple-helical structure. Biopolymers 2005;79: 163–172.

Kisiday J, Jin M, Kurz B, Hung H, Semino C, Zhang S, Grodzinsky AJ. Self-assembling peptide hydrogel fosters chondrocyte extracellular matrix production and cell division implications for cartilage tissue repair. Proc Natl Acad Sci USA 2002;99:9996–10001.

Kojima S, Kuriki Y, Yazaki K, Miura K-I. Stabilization of the fibrous structure of an α-helix-forming peptide by sequence reversal. Biochem Biophys Res Commun 2005;331:577–582.

Kojima S, Kuriki Y, Yoshida T, Yazaki K, Miura K-I. Fibril formation by an amphipathic α-helix-forming polypeptide produced by gene engineering. Proc Jpn Acad Ser B Phys Biol Sci 1997;73B:7–11.

Kotch FW, Raines RT. Self-assembly of synthetic collagen triple helices. Proc Natl Acad Sci USA 2006;103:3028–3033.

Lamm MS, Rajagopal K, Schneider JP, Pochan DJ. Laminated morphology of nontwisting β-sheet fibrils constructed via peptide self-assembly. J Am Chem Soc 2005;127: 16692–16770.

Lansbury PT Jr, Costa PR, Griffiths JM, Simon EJ, Auger M, Halverson KJ, Kocisko DA, Hendsch ZS, Ashburn TT, Spencer RG, Tidor B, Griffin RG. Structural model for the β-amyloid fibril based on interstrand alignment of an antiparallel-sheet comprising a C-terminal peptide. Nat Struct Biol 1995;2:990–998.

Lashuel HA, LaBrenz SR, Woo L, Serpell LC, Kelly JW. Protofilaments, filaments, ribbons and fibrils from peptidomimetic self-assembly implications for amyloid fibril formation and materials science. J Am Chem Soc 2000;122:5262–5277.

Lazar KL, Miller-Auer H, Getz GS, Orgel JPRO, Meredith SC. Helix–turn–helix peptides that form α-helical fibrils: turn sequences drive fibril structure. Biochemistry 2005;44:12681–12689.

Lepere M, Chevallard C, Hernandez J-F, Mitraki A, Guenoun P. Multiscale surface self-assembly of an amyloid-like peptide. Langmuir 2007;23:8150–8155.

Lupas A. Predicting coiled-coil regions in proteins. Curr Opin Struct Biol 1997;7:388–393.

Maji SK, Haldar D, Drew MGB, Banerjee A, Das AK, Banerjee A. Self-assembly of β-turn forming synthetic tripeptides into supramolecular β-sheets and amyloid-like fibrils in the solid state. Tetrahedron 2004;60:3251–3259.

Martinek TA, Hetenyi A, Fulop L, Mandity IM, Toth GK, Dekany I, Fulop F. Secondary structure dependent self-assembly of β-peptide into nanosized fibrils and membranes. Angew Chem Int Ed 2006;45:2396–2400.

Matsumura S, Uemura S, Mihara H. Fabrication of nanofibers with uniform morphology by self-assembly of designed peptides. Chem Eur J 2004;10:2789–2794.

Melnik TN, Villard V, Vasiliev V, Corradin G, Kajava AV, Potekhin SA. Shift of fibril-forming ability of the designed α-helical coiled-coil peptides into the physiological pH region. Protein Eng 2003;16:1125–1130.

Mesquida P, Ammann DL, MacPhee CE, McKendry RA. Microarrays of peptide fibrils created by electrostatically controlled deposition. Adv Mater 2005;17:893–897.

Oakley MG, Hollenbeck JJ. The design of antiparallel coiled coils. Curr Opin Struct Biol 2001;11:450–457.

O'Shea EK, Klemm JD, Kim PS, Alber T. X-ray structure of the GCN4 leucine zipper, a two-stranded, parallel coiled coil. Science 1991;254:539–544.

Ozbas B, Kretsinger J, Rajagopak K, Schneider JP, Pochan DJ. Salt-triggered peptide folding and consequent self-assembly into hydrogels with tunable modulus. Macromolecules 2004;37:7331–7337.

Pagel K, Wagner SC, Samedov K, Von Beripsch H, Boettcher C, Koksch B. Random coils, β-sheet ribbons, and α-helical fibers: one peptide adopting three different secondary structures at will. J Am Chem Soc 2006;128:2196–2197.

Pandya MJ, Spooner GM, Sunde M, Thorpe JR, Rodger A, Woolfson DN. Sticky-end assembly of a designed peptide fiber provides insight into protein fibrillogenesis. Biochemistry 2000;39:8728–8734.

Papapostolou D, Smith AM, Atkins EDT, Oliver SJ, Ryadnov MG, Serpell LC, Woolfson DN. Engineering nanoscale order into a designed protein fiber. Proc Natl Acad Sci USA 2007;104:10853–10858.

Paramonov SE, Gauba V, Hartgerink JD. Synthesis of collagen-like peptide polymers by native chemical ligation. Macromolecules 2005;38:7555–7561.

Paramonov SE, Jun H-W, Hartgerink JD. Self-assembly of peptide-amphiphile nanofibers: the roles of hydrogen bonding and amphiphilic packing. J Am Chem Soc 2006;128: 7291–7298.

Pochan DJ, Schneider JP, Kretsinger J, Ozbas B, Rajagopal K, Haines L. Thermally reversible hydrogels via intramolecular folding and consequent self-assembly of a de novo designed peptide. J Am Chem Soc 2003;125:11802–11803.

Potekhin SA, Melnik TN, Popov V, Lanina NF, Vazina AA, Rigler P, Verdini AS, Corradin G, Kajava AV. De novo design of fibrils made of short α-helical coiled coil peptides. Chem Biol 2001;8:1025–1032.

Rajangam K, Behanna HA, Hui MJ, Han X, Hulvat JF, Lomasney JW, Stupp SI. Heparin binding nanostructures to promote growth of blood vessels. Nano Lett 2006;6:2086–2090.

Rapaport H. Ordered peptide assemblies at interfaces. Supramol Chem 2006;18:445–454.

Rapaport H, Moeller G, Knobler CM, Jensen TR, Kjaer K, Leiserowitz L, Tirrell DA. Assembly of triple-stranded β-sheet peptides at interfaces. J Am Chem Soc 2002;124: 9342–9343.

Ray S, Das AK, Drew MGB, Banerjee A. A short water-soluble self-assembling peptide forms amyloid-like fibrils. Chem Commun 2006;40:4230–4232.

Ray S, Drew MGB, Das AK, Banerjee A. Supramolecular β-sheet and nanofibril formation by self-assembling tripeptides containing an N-terminally located γ-aminobutyric acid residue. Supramol Chem 2006;18:455–464.

Reches M, Gazit E. Casting metal nanowires within discrete self-assembled peptide nanotubes. Science 2003;300:625–627.

Rele S, Song Y, Apkarian RP, Qu Z, Conticello VP, Chaikof EL. D-Periodic collagen-mimetic microfibers. J Am Chem Soc 2007;129:14780–14787.

Ryadnov MG. A self-assembling peptide polynanoreactor. Angew Chem Int Ed 2007;46: 969–972.

Ryadnov MG, Ceyhan B, Niemeyer CM, Woolfson DN. "Belt and braces": a peptide-based linker system of de novo design. J Am Chem Soc 2003;125:9388–9394.

Ryadnov MG, Woolfson DN. Engineering the morphology of a self-assembling protein fiber. Nat Mater 2003a;2:329–332.

Ryadnov MG, Woolfson DN. Introducing branches into a self-assembling peptide fiber. Angew Chem Int Ed 2003b;42:3021–3023.

Ryadnov MG, Woolfson DN. Map peptides: programming the self-assembly of peptide-based mesoscopic matrices. J Am Chem Soc 2005;127:12407–12415.

Sakurai T, Oka S, Kubo A, Nishiyama K, Taniguchi I. Formation of oriented polypeptides on Au(111) surface depends on the secondary structure controlled by peptide length. J Peptide Sci 2006;12:396–402.

Scheibel T, Parthasarathy R, Sawicki G, Lin X-M, Jaeger H, Lindquist SL. Conducting nanowires built by controlled self-assembly of amyloid fibers and selective metal deposition. Proc Natl Acad Sci USA 2003;100:4527–4532.

Schneider JP, Pochan DJ, Ozbas B, Rajagopal K, Pakstis L, Kretsinger J. Responsive hydrogels from the intramolecular folding and self-assembly of a designed peptide. J Am Chem Soc 2002;124:15030–15037.

Silva GA, Czeisler C, Niece KL, Beniash E, Harrington DA, Kessler JA, Stupp SI. Selective differentiation of neural progenitor cells by high-epitope density nanofibers. Science 2004;303:1352–1355.

Smith AM, Acquah SFA, Bone N, Kroto HW, Ryadnov MG, Stevens MSP, Walton DRM, Woolfson DN. Polar assembly in a designed protein fiber. Angew Chem Int Ed 2005;44: 325–328.

Smith AM, Banwell EF, Edwards WR, Pandya MJ, Woolfson DN. Engineering increased stability into self-assembled protein fibers. Adv Funct Mater 2006;16:1022–1030.

Sneer R, Weygand MJ, Kjaer K, Tirrell DA, Rapaport H. Parallel β-sheet assemblies at interface. ChemPhysChem 2004;5:747–750.

Sone EDS, Samuel I. Semiconductor-encapsulated peptide-amphiphile nanofibers. J Am Chem Soc 2004;126:12756–12757.

Soreghan B, Kosmoski J, Glabe C. Surfactant properties of Alzheimer's Aβ peptides and mechanism of amyloid aggregation. J Biol Chem 1994;269:28551–28554.

Stevens MM, Flynn NT, Wang C, Tirrell DA, Langer R. Coiled-coil peptide-based assembly. Adv Mater 2004;16:915–918.

Takahashi Y, Ueno A, Mihara H. Design of a peptide undergoing α–β structrual transition and amyloid fibrillogenesis by the introduction of a hydrophobic defect. Chem Eur J 1998;4: 2475–2484.

Takahashi Y, Ueno A, Mihara H. Optimization of hydrophobic domains in peptides that undergo transformation from α-helix to β-fibril. Bioorg Med Chem 1999;7:177–185.

Takahashi Y, Ueno A, Mihara H. Mutational analysis of designed peptides that undergo structural transition from α helix to β sheet and amyloid fibril formation. Structure 2000;8:915–925.

Takahashi Y, Yamashita T, Ueno A, Mihara H. Construction of peptides that undergo structrual transition from α-helix to β-sheet and amyloid fibril formation by the introduction of N-terminal hydrophobic amino acids. Tetrahedron 2000;56:7011–7018.

Tovar JD, Claussen RC, Stupp SI. Probing the interior of peptide amphiphile supramolecular aggregates. J Am Chem Soc 2005;127:7337–7345.

Veerman C, Rajagopal K, Palla CS, Pochan DJ, Schneider JP, Furst EM. Gelation kinetics of β-hairpin peptide hydrogel networks. Macromolecules 2006;39:6608–6614.

Villard V, Kalyuzhniy O, Riccio O, Potekhin S, Melnik TN, Kajava AV, Ruegg C, Corradin G. Synthetic RGD-containing α-helical coiled coil peptides promote integrin-dependent cell adhesion. J Peptide Sci 2006;12:206–212.

Wagner DE, Phillips CL, Ali WM, Nybakken GE, Crawford ED, Schwab AD, Smith WF, Fairman R. Toward the development of peptide nanofilaments and nanoropes as smart materials. Proc Natl Acad Sci USA 2005;102:12656–12661.

Whitehouse C, Fang J, Aggeli A, Bell M, Brydson R, Fishwick CWG, Henderson JR, Knobler CM, Owens RW, Thomson NH, Smith DA, Boden N. Adsorption and self-assembly of peptides on mica substrates. Angew Chem Int Ed 2005;44:1965–1968.

Yokoi H, Kinoshita T, Zhang S. Dynamic reassembly of peptide RADA16 nanofiber scaffold. Proc Natl Acad Sci USA 2005;102:8414–8419.

Yuwono VM, Hartgerink JD. Peptide amphiphile nanofibers template and catalyze silica nanotube formation. Langmuir 2007;23:5033–5038.

Zhang S, Lockshin C, Herbert A, Winter E, Rich A. Zuotin, a putative Z-DNA binding protein in *Saccharomyces cerevisiae*. EMBO J 1992;11:3787–3796.

Zhang S, Holmes T, Lockshin C, Rich A. Spontaneous assembly of a self-complementary oligopeptide to form a stable macroscopic membrane. Proc Natl Acad Sci USA 1993;90: 3334–3338.

Zhang S, Lockshin C, Cook R, Rich A. Unusually stable β-sheet formation in an ionic self-complementary oligopeptide. Biopolymers 1994;34:663–672.

Zhang S, Holmes TC, DiPersio CM, Hynes RO, Su X, Rich A. Self-complementary oligopeptide matrices support mammalian cell attachment. Biomaterials 1995;16: 1385–1393.

Zhang S, Rich A. Direct conversion of an oligopeptide from a β-sheet to an α-helix: a model for amyloid formation. Proc Natl Acad Sci USA 1997;94:23–28.

Zhao X, Zhang S. Molecular designer self-assembling peptides. Chem Soc Rev 2006;35: 1105–1110.

Zimenkov Y, Conticello VP, Guo L, Thiyagarajan P. Rational design of a nanoscale helical scaffold derived from self-assembly of a dimeric coiled coil motif. Tetrahedron 2004;60: 7237–7246.

Zimenkov Y, Dublin SN, Ni R, Tu RS, Breedvild V, Apkarian RP, Conticello VP. Rational design of a reversible pH-responsive switch for peptide self-assembly. J Am Chem Soc 2006;128:6770–6771.

CHAPTER 15

MOLECULAR IMPRINTING FOR SENSOR APPLICATIONS

XIANGYANG WU and KEN D. SHIMIZU

15.1. INTRODUCTION TO SENSING PLATFORMS

Chemical sensors and biosensors are very powerful tools in modern analytical sciences (Eggins 2002). The new demands of clinical diagnostics, environmental analysis, food analysis, and pharmaceutical monitoring, as well as the detection of illicit drugs, explosives, and chemical warfare agents has driven the development of more selective, more sensitive, and lower cost chemical sensing systems. An essential component of every sensing system is a recognition platform that is able to selectively bind a target analyte in the presence of competing analytes. Both synthetic and biologically derived recognition platforms have been successfully incorporated into molecular sensors. Synthetic recognition systems include synthetic molecular receptors (Schrader and Hamilton 2005) and functionalized polymers (Senaratne et al. 2005). Synthetic recognition platforms generally have the advantages of enhanced stability and lower cost than biological systems. Biological recognition systems include immobilized antibodies, enzymes, DNA, and aptamers. The biological recognition platforms generally display higher levels of affinity and selectivity than their synthetic counterparts and can be more easily targeted toward specific analytes.

This chapter will introduce the field of sensors based on molecular imprinted polymers (MIPs). MIPs are highly cross-linked polymers that are formed with the presence of a template molecule (Haupt and Mosbach 2000; Wulff 2002). The removal of the template molecule from the polymer matrix creates a binding cavity that is complementary in size and shape to the template molecule and is lined with appropriately positioned recognition groups (Scheme 15.1).

Molecular Recognition and Polymers: Control of Polymer Structure and Self-Assembly.
Edited by V. Rotello and S. Thayumanavan
Copyright © 2008 John Wiley & Sons, Inc.

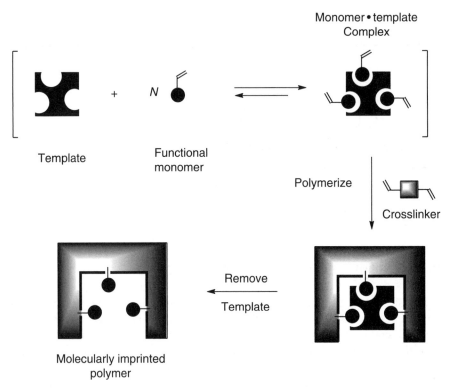

Scheme 15.1   Schematic representation of the molecular imprinting process.

MIPs share the advantages of both biological and synthetic recognition platforms. Like other synthetic polymer systems, MIPs possess excellent mechanical, chemical, and thermal stability (Svenson and Nicholls 2001). However, they also possess the ability to be tailored like biological systems with specificity for a molecule of interest. For example, MIPs have been prepared with recognition properties for nucleotide bases (Shea et al. 1993), pesticides (Yamazaki et al. 2001), amino acids (Lepisto and Sellergren 1989), peptides (Hart and Shea 2002), pharmaceuticals (Vlatakis et al. 1993), steroids (Cheong et al. 1997), sugars (Striegler 2003), and metal ions (Uezu et al. 1997).

There are already several excellent reviews and monographs on MIPs (Wulff 1995; Sellergren 2001b; Yan and Ramstrom 2005; Alexander et al. 2006). Thus, our focus will be on practical aspects of preparing and developing MIPs for sensing applications (Takeuchi and Haginaka 1999; Haupt and Mosbach 2000; Haupt 2001). This chapter will be divided into two parts. The first half (Sections 15.2–15.4) gives a general introduction to MIPs with a focus on preparation strategies, recognition and material properties, and MIP formats. The second half (Section 15.5) presents examples of MIP-based sensors. These include MIP binding assays and mass sensitive, optical, and electrochemical sensors.

15.1.1. Utility of MIPs

Perhaps the most persuasive demonstration of the molecular imprinting effect is enantioselective MIPs by imprinting a chiral template. These enantioselective MIPs have been primarily used as MIP-based chiral stationary phases (CSPs). Accordingly, the majority of the early MIP literature is dedicated to developing CSPs (Ramstrom and Ansell 1998; Sellergren 2001a). MIPs prepared using a single enantiomer of a chiral template typically show a pronounced selectivity for the templated enantiomer. The enantioselectivities of MIP-CSPs often exceed those of commercially available CSPs with the added benefit that they can be tailored for a specific analyte or isomer. The ability of the MIPs to perform as stationary phases in high-performance liquid chromatography (HPLC) columns also demonstrates the excellent mechanical and chemical stability of MIPs (Svenson and Nicholls 2001). For example, MIP-CSPs can maintain their enantioselectivity even after 5 years of use in an HPLC column (Wulff 1995). MIPs have also been successfully applied in an array of other applications that require tailored recognition platforms. These include catalysis (Beach and Shea 1994), solid-phase extraction (Lanza and Sellergren 2001), drug release (Bures et al. 2001), and sensing (Takeuchi and Haginaka 1999; Haupt and Mosbach 2000; Haupt 2001). MIP-based solid-phase extraction columns are currently sold by Aldrich under the SupelMIP brand name. An MIP version of the cholesterol lowering polymeric drug, Cholestagel, has been developed (Huval et al. 2001). Sensing has also been one of the most active areas of research for MIP-based applications and wis the focus of this chapter. There are now many examples of MIP-based sensors, and these studies have demonstrated the viability of MIPs as synthetic alternatives to antibodies and DNA in sensing applications.

There are two major challenges in developing MIP sensors. The first is tailoring and optimizing the selectivity of the polymer matrix for the analyte of interest via the molecular imprinting process. Thus, Section 15.2 will introduce methods for preparation, and Sections 15.3 and 15.4 will address issues relating to the recognition and materials properties of MIP-based sensors. The second challenge is coupling the recognition event with a signal transduction event. Section 15.5 will address issues relating to the integration of MIPs with a signal transduction. This will include examples of different types of MIP-based sensing mechanisms including optical, mass, and electrochemical sensing systems.

15.2. SYNTHESIS OF MIPs

15.2.1. General Types of Imprinted Polymers (Covalent and Noncovalent)

Many of the unique characteristics and advantages of MIPs are a direct consequence of the efficient molecular imprinting process. Therefore, this first section will introduce the general classes and methods for preparing MIPs. The first report about molecular imprinting was published in 1931 (Polyzkov 1931), but the modern study of the molecular imprinting process began with the preparation of the first organic molecularly imprinted polymer in 1972 by Wulff and Sarhan (1972). Their initial imprinted polymer system contained all of the components of modern MIPs, which are depicted in Scheme 15.1. In the imprinting process, the template molecule is coupled to multiple polymerizable functional monomers to form a prepolymerization complex. The prepolymerization complex is copolymerized with a cross-linker that forms a rigid polymer matrix around the monomer–template complex. Upon removal of the templating agent, a molecule-specific recognition site is formed in the polymer matrix that has both shape and functional group complementarity to the template molecule.

The efficiency of the imprinting process is dependent on two major factors. The first is the fidelity with which the monomer–template complex is formed. Thus, conditions or strategies that stabilize the monomer–template complexes yield MIPs with enhanced binding properties. The second factor is the ability of the polymer matrix to retain the shape and the positioning of the complementary recognition groups to the template molecule after its removal from the polymer matrix. Thus, factors that yield a more rigid, nonswellable matrix also yield MIPs with higher selectivities.

15.2.2. Types and Preparation of MIPs

There are two general classes of imprinted polymers: covalent and noncovalent MIPs. These two categories refer to the types of interactions between the functional monomer and the template in the prepolymerization complex. There are also hybrid MIPs that utilize a combination of covalent and noncovalent interactions in the preparation and rebinding events (Klein et al. 1999). Covalent MIPs utilize reversible covalent interactions to bind the template to the functional monomers. In contrast, noncovalent MIPs rely on weaker noncovalent functional monomer–template interactions. Each type has specific advantages and disadvantages with respect to sensing applications that will be addressed in subsequent sections.

Covalent MIPs. The covalent imprinting approach was first developed by Wulff and colleagues in the 1970s (Wulff and Sarhan 1972; Wulff et al. 1973). Covalent MIPs can utilize either reversible covalent or dative metal–ligand bonds. After polymerization, these monomer–template bonds must be chemically broken in order to remove the template from the polymer matrix. The affinity of the covalent MIP for the target molecule can come from either reforming covalent bonds or from forming new noncovalent interactions between the template and the functional groups lining the binding site. In the initial system (Scheme 15.2) reported by Wulff, boronic acid functional monomers were condensed with a carbohydrate template. The boronic ester prepolymerization complex was then copolymerized with divinyl benzene (DVB) under free-radical conditions. Finally, the carbohydrate template was removed by hydrolysis in methanol/water. Reversible covalent bonds that have been used in the covalent molecular imprinting process include carboxylic (Sellergren and Andersson 1990) and boronic esters (Wulff, Grobe-Einsler et al. 1977; Wulff, Vesper et al. 1977), imine (Wulff and Akelah 1978), and ketals (Shea and Sasaki 1991; Table 15.1).

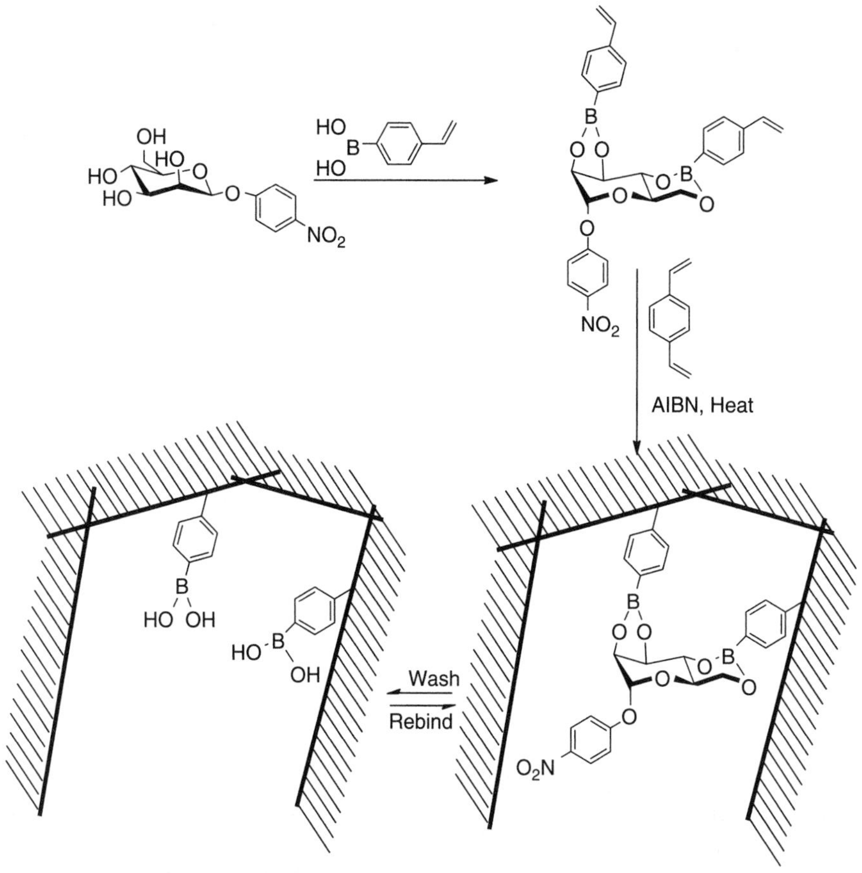

Scheme 15.2 Covalent imprinting of α-phenyl-D-mannopyranoside in a divinyl benzene/4-vinylphenylboronic acid matrix, via the formation of covalent boronic ester linkages between the 4-vinylphenylboronic acid and the carbohydrate. Adapted from Wulff, Vesper, et al. (1977). Copyright 1977 Wiley InterScience.

The covalent nature of the monomer–template complex yields greater control over the imprinting process. The kinetic stability of the prepolymerization complex ensures that the binding sites are more structurally homogeneous and are lined with the same number of recognition groups. The primary disadvantage of the covalent imprinting process is that it requires additional synthetic steps. First, the covalent bonds in the monomer–template complex must be formed. Second, the polymer–template bonds in the polymer must be cleaved after polymerization. Another disadvantage with respect to their use in molecular sensors is their slow binding kinetics that are due to the necessity to reform the covalent bonds in the binding event. Thus, covalent MIPs are often better suited for use in single-use sensor systems such as dosimeters or binding assays.

TABLE 15.1 Examples of Functional Monomers Used in Covalent Molecular Imprinting Process

Monomer	Covalent Bond	Template	References
	Boronic ester	Diols, carbohydrates	Wulff and Sarhan (1972), Wulff, Vesper et al. (1977)
	Carboxylic ester	Alcohols	Sellergren and Andersson (1990)
	Acetal	Ketones, aldehydes	Shea and Sasaki (1991)
	Imine	Amines	Wulff et al. (1978)

Noncovalent MIPs. The more popular noncovalent imprinting process was developed by Arshady and Mosbach (1981) in the 1980s. These systems are characterized by noncovalent interactions between the template and functional monomers such as hydrogen bonds, hydrophobic interactions, $\pi-\pi$ interactions, electrostatic interactions, charge transfer, and van der Waals forces. An example of the noncovalent imprinting process is provided in Scheme 15.3. First, methacrylic acid (MAA) functional monomers were complexed with an L-phenylalanine anilide template via hydrogen bonds and electrostatic interactions. The noncovalent complex was polymerized in the presence of the cross-linker ethylene glycol dimethylcrylate (EGDMA) to form a rigid polymer matrix. Subsequently, the polymer was ground and washed with acetonitrile to remove the template.

The noncovalent imprinting process has quickly become the most popular method for preparing MIPs because it is synthetically more efficient and more versatile and adaptable. First, the monomer–template complex can be formed in situ during the polymerizing process, because the noncovalent interactions do not require any additional reagents or catalysts. Thus, the entire imprinting process can be carried out efficiently in a single reaction vessel. Second, a single functional monomer can form noncovalent interactions with many different template molecules and can effectively imprint a wide range of templating agents. This versatility is exemplified by the most common noncovalent functional monomer, MAA, which has been used to

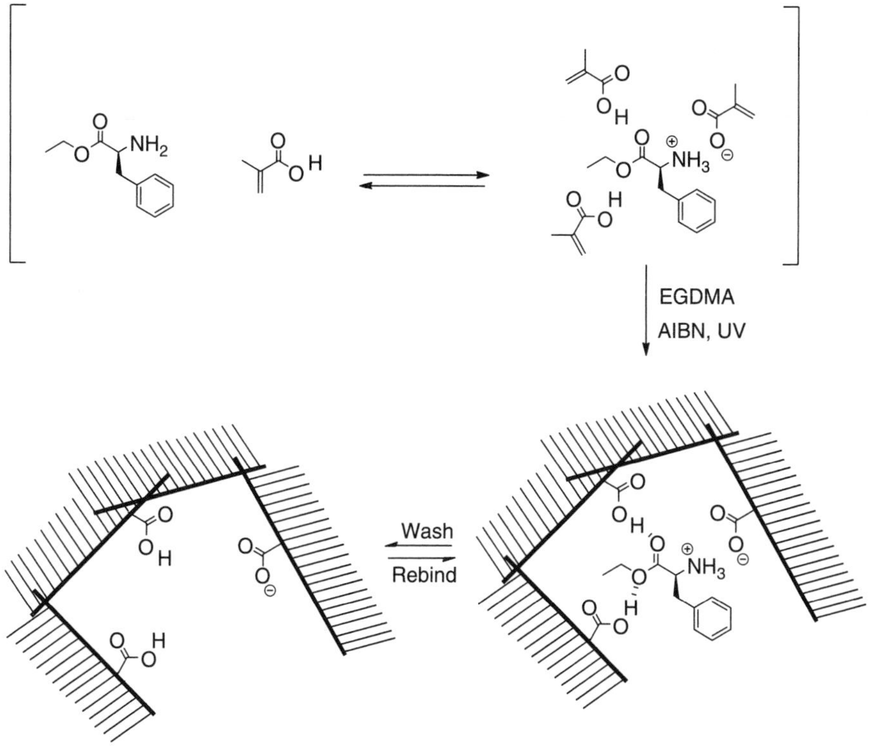

Scheme 15.3 Noncovalent imprinting of L-phenylalanine anilide in a methacrylic acid (MAA)/ethylene glycol dimethylcrylate (EGDMA) polymer matrix. Adapted from Sellergren et al. (1998). Copyright 1988 American Chemical Society.

imprint a diverse array of basic organic template molecules that can form complementary hydrogen bonding and electrostatic interactions (Table 15.2).

Disadvantages of the noncovalent imprinting process arise from the instability of the noncovalent monomer–template complexes. This instability reduces the fidelity of the imprinting process because monomer–template complexes of varying stoichiometries and structures are present in the prepolymerization mixtures. In addition, typical noncovalent imprinting conditions utilize a large excess of functional monomers in order to favor the formation of weak monomer–template complexes. Consequently, the majority of functional monomers are not incorporated into the binding site, which leads to a very high percentage of low-affinity and low-selectivity binding sites.

Hybrid MIPs. More recently, hybrid imprinting strategies have been developed that combine the advantages of both the covalent and noncovalent imprinting strategies (Whitcombe et al. 1995; Lubke et al. 2000). A recent example by Whitcombe's group is presented in Scheme 15.4 (Klein et al. 1999). The monomer–template prepolymerization complex is formed via a combination of covalent amide bonds and

TABLE 15.2 Structural and Functional Building Blocks in Molecular Imprinting Monomers Used in Noncovalent Imprinting

Monomer	Non-covalent Interaction	Template	References
MAA	Hydrogen bond and electrostatic	Basic amines and amides	Kempe and Mosbach (1994), Kempe (1996), Sellergren et al. (1993)
VP	Hydrogen bonding and electrostatic interactions	Carboxylic acids, alcohols, and primary/secondary amides	Kempe et al. (1993)
HEMA	Hydrogen bonding	Amides, esters, alcohols, amines	Dirion et al. (2003), Oral and Peppas (2006)
MA	Hydrogen bonding	Amides, esters, alcohols, amines, carboxylates	Bereczki et al. (2001)
	Hydrogen bonding	Carboxylates	Urraca et al. (2006)

noncovalent hydrogen bonds to 2-vinylpyridine. After polymerization, the template is cleaved from the matrix by hydrolysis of the ester linkages. Although the imprinting process utilized both covalent and noncovalent interactions, this hybrid MIP binds the guest molecule by only noncovalent hydrogen bonding and electrostatic interactions. Thus, this hybrid imprinting strategy combines the greater imprinting fidelity of the covalent imprinting process with the faster binding kinetics of noncovalent MIPs.

15.2.3. Polymerization Conditions

The recognition and material properties of MIPs are strongly dependent on the polymerization conditions. Variation of the polymerization temperature; solvent, template, and monomer concentrations; and cross-linker percentage attenuates the fidelity of the imprinting process by changing the structure and stability of the prepolymerization complex as well as the templated binding sites. The noncovalent imprinting

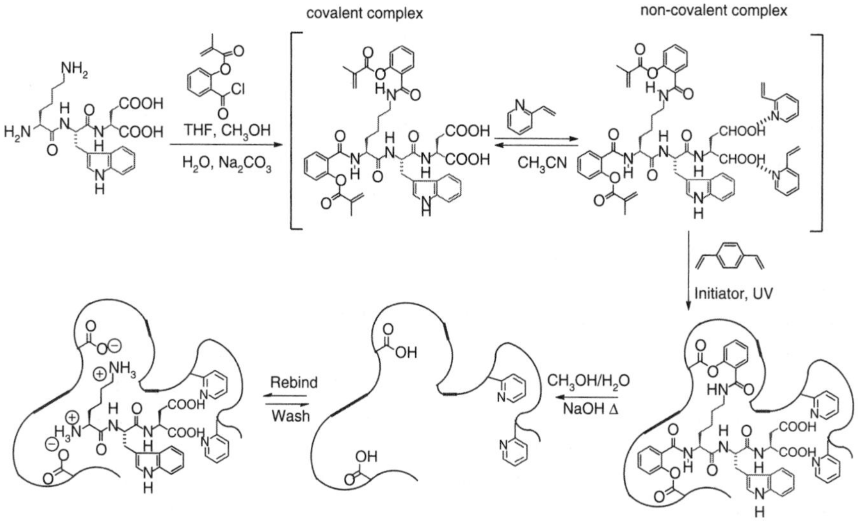

Scheme 15.4 Molecular imprinting of the tripeptide Lys-Trp-Asp using both covalent and non-covalent interactions. Adapted from Klein et al. (1999). Copyright 1999 Wiley InterScience.

process is particularly sensitive to changes in polymerization conditions because of the instability and reversible nature of noncovalent monomer–template complexes. Generally, conditions that favor the formation of the noncovalent monomer–template complex yield MIPs with higher capacities and selectivities. For example, lower temperatures, higher monomer and template concentrations, and less polar solvents enhance the fidelity of the imprinting process. The polymerization conditions can also control material properties such as the porosity, surface area, and rigidity of the polymer matrix. Some of the more important variables and their influences on the imprinting process are discussed here.

Polymerization Methods. The majority of MIPs have been prepared via free-radial polymerization of vinyl monomers. Radical polymerization conditions are favored because they are mild, can utilize a large pool of commercially available monomers, and are compatible with most functional groups. This last attribute enables the use of monomers containing polar, aromatic, acidic, basic, and charged recognition groups. Polymerizations are carried out with 1–5% of a radical initiator such as AIBN, benzophones, or α-dialkoxy-aceto-phenones. Both thermal and UV irradiation polymerization conditions have been used. The thermal irradiation conditions are more general and can be applied to a broader array of monomers, cross-linkers, and templates and yield more uniform polymers. UV irradiation conditions can be carried out at lower temperatures that are more favorable for noncovalent imprinting conditions. MIPs have also been prepared using other polymerization methods. The largest and most successful of these are the imprinted sol gels (Katz and Davis 2000; Chang et al. 2002). MIPs have also been prepared using electropolymerization (Gong et al. 2004), condensation (Asanuma et al. 1997), step growth (Beinhoff et al. 2006), and metathesis polymerizations (Patel et al. 2003).

Cross-linker. The cross-linker plays an important role in determining the recognition and material properties of MIPs. MIPs are commonly prepared with a high percentage of cross-linking agent, typically 50–80%. These high cross-linking percentages are required to maintain the structural integrity of the imprinted binding sites (Biffis et al. 2001). Despite these high cross-linking percentages, the majority of the template (>90%) can usually be removed because of the macroporous nature of the polymer matrix (Siemann et al. 1997). The most common cross-linkers are EGDMA (Vlatakis et al. 1993) and DVB (Fig. 15.1; Wulff and Sarhan 1972; Sellergren and Andersson 1990). Other cross-linkers include trimethylolpropane trimethacrylate (Glad et al. 1995) and *N,N'*-methylenebisacrylamide (MBA; Wang et al. 1997; Suedee et al. 2006). Although the primary role of the cross-linker is to form a rigid matrix, the cross-linker can also contain recognition groups and thus participate directly in recognition events. In principle, the recognition groups in the cross-linker become more rigidly fixed into the polymer matrix and thus can form higher affinity binding sites. For example, cross-linkers with hydrogen bonding urea (Hall et al. 2003), amide (Sibrian-Vazquez and Spivak 2004), *N,N'*-diacyl-2,6-diaminopyridine

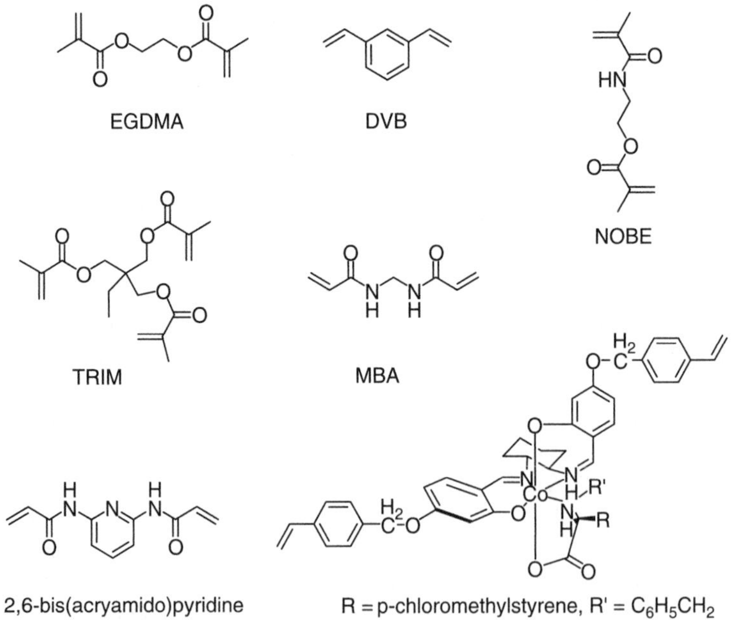

EGDMA   DVB   NOBE

TRIM   MBA

2,6-bis(acryamido)pyridine   R = p-chloromethylstyrene, R' = C₆H₅CH₂

Figure 15.1 Examples of common cross-linkers used in the preparation of molecular imprinted polymers: ethylene glycol dimethylcrylate (EGDMA); divinyl benzene (DVB), trimethylolpropane trimethacrylate (TRIM), *N,N'*-methylenebisacrylamide (MBA), and *N,O*-bismethacryloyl ethanolamine (NOBE).

(Tanabe et al. 1995), and a metal containing complex (Fujii et al. 1985) have been used to imprint dicarboxylates, amine, and imides.

Solvent/Porogen. The important role that a solvent serves in the imprinting process is as a porogen that controls the morphology and porosity of the polymer matrix. MIP polymerizations are usually carried out at very high concentrations in which the solvent occupies 33–50% of the reaction volume. Under these conditions, the growing polymer chains ideally phase separate from the solution phase during polymerization, leading to a macroporous polymer monolith with surface areas of $100-1000 \, \text{m}^2/\text{g}$ (Shea et al. 1990; Santora et al. 2001). The solvent accessibilities of these macroporous monoliths are excellent as evidenced by their use as chromatographic stationary phases (Svec and Frechet 1996; Peters et al. 1999). As noted in the previous section, the solvent can also accentuate or disrupt the noncovalent interactions in the prepolymerization complex. This dual role of the solvent in the imprinting process highlights the complexity of optimizing. The variables are all highly interdependent and attenuate not only the efficiency of the imprinting process but also the polymer morphology and structure. Because of this complexity, multivariate experimental design procedures have recently been applied to optimize the imprinting process (Davies et al. 2004; Rosengren et al. 2005).

15.3. RECOGNITION PROPERTIES OF MIPs

The most attractive characteristic of MIPs is the ability to tailor their binding selectivities in a manner similar to biological recognition systems. The average recognition properties of MIPs, however, usually fall short of those of biological recognition systems. This is understandable considering the relative simplicity of the imprinting process. However, it is not generally appreciated that the recognition properties of MIPs are very complex and highly concentration dependent. Thus, under optimized conditions, MIPs can achieve levels of affinity and selectivity equal or even surpassing those of antibodies (Vlatakis et al. 1993; Andersson et al. 1995). Therefore, an understanding of the recognition properties of MIPs is vital to the development and optimization of MIP-based sensors.

15.3.1. Binding Site Heterogeneity

The source of the complex recognition behavior of MIPs arises from their binding site heterogeneity. Individual MIPs contain an array of different binding sites of varying size, shape, binding affinities, and binding selectivities (Rampey et al. 2004). In contrast, binding sites in synthetic molecular receptors and biological recognition systems are structurally homogeneous because every binding site possesses the same affinity and selectivity. The characteristic binding site heterogeneity in MIPs is depicted in Figure 15.2. The binding sites line the channels of the macroporous MIP matrix and vary widely in their shape and functional group complementarity to the template molecule. The result of the structural heterogeneity is that only a

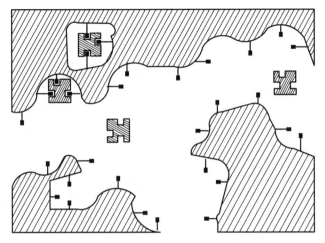

Figure 15.2 Depiction of the binding site heterogeneity of a molecular imprinted polymer.

small fraction of the functional monomers form productive binding sites. Instead the majority of functional monomers are randomly incorporated into the walls of the channel or are inaccessible within the polymer matrix.

The binding affinities within a single MIP are similarly heterogeneous and span the entire range from low to high affinity. The typical binding affinity distribution observed in an MIP is shown in Figure 15.3 (Umpleby et al. 2000, 2001). The distribution is strongly weighted toward the nonspecific low-affinity sites. Thus, the average binding affinities of MIPs are much lower than antibodies and aptamers (Houk et al. 2003). However, the binding site distribution does asymptotically tail out into the high-affinity region (Fig. 15.3). Thus, there is a small but accessible subset of binding sites with high affinity and selectivity. This asymmetric distribution helps to explain the strong concentration dependence of the binding properties of MIPs. At higher analyte concentrations, both the low- and high-affinity sites are

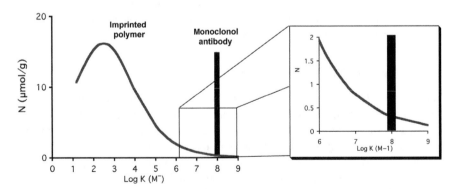

Figure 15.3 Typical heterogeneous distribution of binding site affinities found in molecular imprinted polymers compared with the homogenous distribution of a typical monoclonal antibody. Reprinted from Umpleby et al. (2001). Copyright 2001 American Chemical Society.

able to bind the analyte. Under these conditions, the binding properties are dominated by the more numerous low-affinity low-selectivity binding sites. However, at lower analyte concentrations, only the high-affinity sites are able to bind the analyte. Thus, MIPs show much higher binding affinities and selectivities at lower analyte concentrations, but they also have much lower capacities. The enhanced selectivities of MIPs at low analyte concentrations enable their use in sensing applications. Sensing applications also do not require high capacities and thus the small subset of high-affinity binding sites in MIPs are sufficient for most sensing applications. It is interesting that the asymmetric heterogeneous distribution in MIPs is not well matched to applications that required selectivity over a very broad concentration range such as chromatography, which has been one of the most popular applications for MIPs.

15.3.2. Limitations of the Recognition Properties of MIPs

There are important limitations to consider when using MIPs in sensing applications. These include their recognition properties in water, binding kinetics, and slow leaching of the template from the polymer matrix.

Recognition Properties in Water. Another advantage of MIPs is that they can perform in almost any solvent system including nonpolar organic, polar organic, and even in water. The recognition properties of MIPs in aqueous environments deserve additional comment, because water is the most important solvent system for sensing applications. The majority of MIPs are formed and tested in nonpolar organic solvents. This is because most MIPs rely on hydrogen bonding and electrostatic interactions that are strongest in nonpolar organic solvents. What is not commonly appreciated is that MIPs formed in organic solvents often retain their selectivity in aqueous solvent systems. This has been demonstrated with the most common MAA/EGDMA imprinting matrix with multiple template molecules (Karlsson et al. 2001; Dirion et al. 2003). The selectivities of the MIPs are typically lower when tested in aqueous solvent systems. This is not surprising because it has been shown that MIPs usually show optimal recognition properties when they are tested in the same solvent as the one used in their preparation (Spivak et al. 1997; Kim and Guiochon 2005a). An important consideration when using MIPs in aqueous solvent systems is the extremely high background binding that is due to the hydrophobic nature of the organic polymer matrix. These nonspecific interactions can be minimized by the addition of small amounts of organic modifiers such as ethanol, dimethylsulfoxide, or surfactants to the solvent systems (Karlsson et al. 2001). Another important consideration is the protonation state of ionizable recognition groups. For example, the most common functional monomer, MAA, exists as a carboxylate in water and forms electrostatic interactions with positively charged guests. Thus, the binding interactions of MIPs in water are greatly attenuated by changing the pH using buffers (Sellergren and Shea 1993; Chen et al. 2001).

There has also been considerable research and development of imprinted polymer systems that can be both prepared and used in an aqueous environment. These second-generation MIPs utilize functional monomers that form strong covalent, coordination, and hydrophobic interactions that are effective in water and can bind peptides (Asanuma et al. 2000; Hart and Shea 2001), carbohydrates (Wulff and Schauhoff 1991; Striegler 2003), and proteins (Bossi et al. 2001; Takeuchi et al. 2007). Perhaps the most notable of these is the report of a peptide imprinted polymer system that can effectively distinguish peptides in water that differ by one single amino acid (Nishino et al. 2006).

Binding Kinetics. Another important consideration in adapting MIPs for sensing applications is the optimization of their binding kinetics. MIP binding processes are often quite slow in reaching equilibrium and are on the order of minutes to hours (Norell et al. 1998; Chen et al. 1999). This is further complicated by the binding site heterogeneity of MIPs, which leads to a wide variation in the rebinding kinetics for different binding sites within an MIP (Khasawneh et al. 2001; Garcia-Calzon and Diaz-Garcia 2007). Unfortunately, the more desirable high-affinity sites show the slowest binding kinetics (Sellergren 1989). The binding kinetics are also highly solvent, concentration, and temperature dependent. One strategy that has been used to circumvent this is the implementation of MIPs in sensing systems that do not need to reach equilibrium such as continuous flow systems.

A related consideration is the slow leaching of the template molecule from the imprinted polymer matrix (Ellwanger et al. 2001). The majority of the template used in the imprinting process is readily removed from the polymer matrix during the washing step. However, a small fraction of template remains encapsulated in the polymer matrix and is slowly released into solution. Under normal conditions, the leaching of the template from the polymer does not impact the binding properties. However, at low analyte concentrations that are common to sensing applications, the concentration of template leaching from the matrix becomes competitive with the analyte concentration. This limits the use of MIPs in trace analysis. One solution is to imprint a structural analog to the analyte of interest (Andersson et al. 1997; Kim and Guiochon 2005b). Alternatively, the binding of the analyte can be monitored indirectly using a competitive radioligand or spectroscopically labeled ligand assays (Vlatakis et al. 1993; Haupt et al. 1998).

Low Average Affinities and Capacities. Despite improvements in the imprinting process and better understanding of the optimal binding conditions for MIPs, the low binding affinities and selectivities of MIPs are still a critical problem for sensing applications. The problem of the template slowly leaching from the matrix, in particular, does not allow the analysis using MIPs to be easily carried out at the submicromolar concentrations where MIPs show the highest selectivities. A number of strategies have been applied to improve the binding properties of MIPs. These can be divided into pre- and postpolymerization strategies. The prepolymerization strategies have focused on improving the efficiency of the imprinting process. One of the most popular strategies has been to develop functional monomers that form much stronger noncovalent interactions. Examples of these "stoichiometric imprinting" monomers include the carboxylate and phosphate binding amidine functional monomers (Wulff and Knorr 2001), an ampicillin binding oxyanion receptor-based functional monomer (Lubke et al. 2000), and an amine binding functional monomer that forms strong charge transfer complexes (Lubke et al. 2000). Another prepolymerization strategy is the previously mentioned hybrid imprinting strategy. The combination of covalent and noncovalent monomer–template interactions yields polymers with higher association constants and selectivities.

The postpolymerization strategies to improve the selectivities of MIPs include using optimized rebinding conditions and, more recently, arrays of imprinted polymers (Hirsch et al. 2003; Greene et al. 2004). Sensor arrays have been widely used in sensing applications for compound detection and identification (Albert et al. 2000). MIPs are particularly well suited for use in an array format. First, the molecular imprinting process is synthetically efficient and it is relatively easy to prepare an array of polymers with different selectivities in several days simply by preparing each MIP using a different template molecule. In contrast, the preparation of an array of different synthetic receptors or antibodies may take weeks or months. Second, the array format is able to compensate for the low selectivities and high cross-reactivities of the individual MIPs. Thus, compounds are distinguished based on their unique response patterns to the array, which are compared with previously measured patterns of standards (Hirsch et al. 2003; Greene et al. 2004). The efficacy of MIP arrays has been demonstrated to differentiate nucleosides and pharmaceutically active amines. In the latter example, diastereomeric ephedrine and pseudoephedrine were easily differentiated by the MIP array whereas individual MIPs did not show sufficient selectivity to accurately distinguish these diastereomers. One disadvantage of the array format is that the concentration of the analytes is more difficult to ascertain because the analysis measures a pattern as opposed to the magnitude of an individual signal. Despite these limitations, the development of MIP arrays appears to be a very promising strategy for developing sensors based on MIPs.

15.4. POLYMER FORMATS AND MORPHOLOGIES

In addition to the recognition properties of an MIP, the polymer format and morphology are important in sensing applications because they also affect the binding kinetics and integration with a signal transduction element. The difficulty with MIPs is that they are highly cross-linked materials and thus have very limited processibility. Therefore, the morphology and format must be set during the polymerization process. There are three general polymer formats that have been used to make MIP sensors. These are ground particles, membrane or thin films, and self-assembled monolayers. Methods and examples of each of these formats will be presented in the following section.

15.4.1. Ground MIP Particles

The traditional method for preparing MIPs yields a porous monolith that is ground and sieved into a fine powder. It is important to note that the capacity and selectivity of an MIP monolith are independent of the particle size. This is because the particles are macroporous and thus the majority of the binding sites are on the interior surface of each particle. Smaller particles do, however, display faster binding kinetics. Although the ground monolith particles cannot be easily processed, they can still be utilized in sensing devices. The key is to immobilize the MIP particles on the transduction element's surface without restricting the analyte's access to the particles. An early example that directly used the ground MIP particles was an amperometric sensor for the detection of morphine by Kriz and Mosbach (1995). The MIP particles were suspended in an aqueous agarose solution that was cross-linked to form an agarose layer on the transducer surface. Kroger et al. (1999) also used agarose gel immobilized particles in electrochemical MIP sensors for voltammetric determination of herbicides. MIP particles were utilized in optical sensing systems as well. The MIP particles were held in a porous screen against the quartz window of an optical fiber for the detection of an amino acid derivatized with a fluorescent labeling group, dansyl-L-phenylalanine (Kriz et al. 1995). The MIP-based fiber optic sensor allowed measurement of the enantioselectivity of the MIP for the templated L-enantiomer (Fig. 15.4). MIP particles were also immobilized in plastic membranes such as poly(vinyl chloride) (PVC). A MIP-based bulk acoustic wave (BAW) sensor was developed with this method for detection of aminopyrine in water (Tan, Nie, et al. 2001). The sensing layer was prepared by suspending the MIP particles in a tetrahydrofuran solution containing dissolved PVC, which was spin coated on an Ag surface. This BAW-MIP sensor showed high selectivity and sensitivity to aminopyrine with a detection limit of 2.5×10^{-8} M in water.

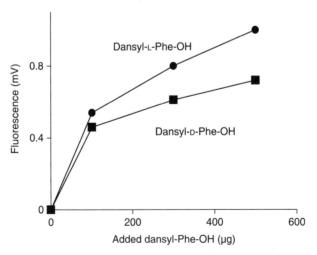

Figure 15.4 Fiber optic measurements of the dansyl-L-Phe-OH molecular imprinted polymers (L-MIP) in the presence of increasing amounts of enantiomers. (●) Dansyl-L-Phe-OH and (■) dansyl-D-Phe-OH. Reprinted from Kriz et al. (1995). Copyright 1995 American Chemical Society.

15.4.2. Thin Films and Membranes

Thin films or membranes are traditionally better suited for sensing applications because of their more uniform morphology and better optical properties (Schneider et al. 2005). However, the preparation of MIP thin films or membranes is challenging. Polymerization of the traditional MIP formulation yields irregular cracked films because of the inability of the highly cross-linked matrix to accommodate the contraction of the polymer matrix induced by polymerization and porogen evaporation. Thus, new polymerization methods had to be developed to prepare MIP thin films. These methods include surface initiated polymerizations, polymer grafting, polymerizations using nonvolatile porogens, and electrochemical polymerizations.

Surface Initiated Polymerizations. Thin MIP coatings have been prepared on a variety of surfaces via the covalent anchoring of initiator or monomer units to the surface (Quaglia et al. 2001; Ruckert et al. 2002; Piacham et al. 2005). The most common are on silica surfaces that can be functionalized with organo orthosilicates. For example, a propranolol imprinted polymer coating was synthesized in a fused silica capillary column by surface initiation (Schweitz 2002). This $0.15-2\ \mu m$ thick MIP coating was demonstrated to be enantioselective for the template (*S*-propranolol). The method of surface initiated atom transfer radical polymerization was utilized for the preparation of an MIP nanotube membrane using a porous anodic alumina oxide (Wang et al. 2006). The β-estradiol imprinted polymer nanotube membrane has high affinity and selectivity for the template molecule.

Polymer Grafting. Surface grafted MIPs are prepared by attaching preformed imprinted polymers to a surface or reinitiating polymerization under imprinting conditions from a polymer surface (Sellergren et al. 2002; Fairhurst et al. 2004; Hattori et al. 2004; Yang et al. 2005; Titirici and Sellergren 2006). For example, the grafting of an epinephrine imprinted polymer film onto the polystyrene surface of a microplate was realized through oxidation of 3-aminophenylboronic acid by ammonium persulfate (Piletsky, Piletska, et al. 2000). This MIP-coated microplate displayed excellent stability and reproducibility. Polypropylene membranes were imprinted with desmetryn by photografting with 2-acrylamido-2-methylpropanesulfonic acid (functional monomer) and MBA (cross-linker) in water with benzophenone as the photoinitiator (Piletsky, Matuschewski, et al. 2000). These MIP membranes possess good selectivity for triazine herbicides in water. The surface grafting approach has several advantages: 1) it has the possibility of modifying an inert surface, 2) the synthesis and immobilization of the MIP membrane can be performed in one step, 3) the grafted polymers usually have fast response times because they are very thin, and 4) the thickness of the polymer membrane can be regulated by the polymerization time or initiator efficiency.

Polymerizations Using Nonvolatile Porogens. Most recently, MIP membranes and thin films have been prepared under conditions that limit the evaporation of the porogen from the polymer matrix during the exothermic polymerization process. One key is to slow the rate of the polymerization process in order to allow the mechanical stresses to dissipate. For example, a MAA/EGDMA imprinted polymer membrane, imprinted with 9-ethyladenine, was prepared between two glass plates for the selective transport of nucleosides (MathewKrotz and Shea 1996). A nonvolatile porogen, dimethylformamide, was utilized during the polymerization, which resulted in a homogeneous optical transparent membrane. These membranes exhibited good stability and mechanical strength, as well as good selectivity. The imprinted membrane facilitated the transport of adenosine in a methanol/chloroform (6:94, v/v) solution with an adenosine/guanosine selectivity factor of 3.4. In a second example, a propranolol imprinted polymer film was prepared through spin coating by Schmidt et al. (2004). The prepolymerization mixture was spin coated on a surface and then polymerized by UV initiation to form a solid polymer film. This process used a combination of a polymeric porogen, poly(vinyl acetate), and a low volatility solvent, diethylene glycol dimethyl ether, to prevent cracking of the film during the spin-coating and polymerization steps. The thickness and porosity of the films could be controlled by varying the concentration of the polymer porogen. This MIP film retained the high capacity and selectivity for propranolol that was demonstrated using the traditional MIP monoliths.

Electrochemical Polymerizations. Electropolymerization is considered to be one of the most direct methods for interfacing the MIP with a transducer surface (Ulyanova et al. 2006). An early electrosynthesized MIP was prepared in 1989 by Lapkowski's group for the electrochemical detection of adenosine triphosphate (Boyle et al. 1989). An MIP-based quartz crystal microbalance (QCM) sensor was also developed by electropolymerization of poly(o-phenylenediamine) onto the conducting surface of a QCM in the presence of a neutral template, glucose (Malitesta et al. 1999). The first MIP-based capacitive sensor was prepared by electrochemical polymerization (Panasyuk et al. 1999). The MIP multilayer was prepared by electropolymerizing phenol on gold electrodes with phenylalanine as the template molecule. This capacitive sensor displayed selectivity for phenylanaline over other compounds such as amino acids and phenol (Fig. 15.5). Although electropolymerization can deposit the recognition element directly on the transducer surface, the requirement of electrochemically active monomers and polymers has limited its utility.

15.4.3. Self-Assembled Monolayers

The self-assembly of an imprinted layer on the surface of a transducer was realized through the adsorption of the template on gold, SiO_2, or InO_2 surfaces followed by treatment with an alkylthiol or organosilane (Hirsch et al. 2003). The first example of this type of sensor was reported in 1987 by Tabushi and coworkers (1987). Octadecylchlorosilane was chemisorbed in the presence of n-hexadecane onto tin dioxide or silicon dioxide for electrochemical detection of phylloquinone, menaquinone, topopherol, cholesterol, and adamantane. Another MIP-based sensor was

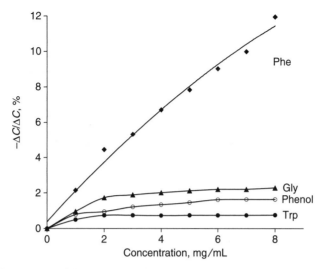

Figure 15.5   Changes of the sensor capacitance on exposure to phenylalanine (Phe), glycine (Gly), phenol, and tryptophan (Trp). Reprinted from Panasyuk et al. (1999). Copyright 1999 American Chemical Society.

developed using a similar approach for detection of phylloquinone (Andersson et al. 1988). The self-assembly technique was applied to prepare an electrochemical sensor with a nano-TiO_2 film self-assembled on a glassy carbon electrode for selective determination of parathion (Li et al. 2006). This MIP-based sensor had a detection limit of $1.0 \times 10^{-8} M^{-1}$ and was successfully applied to parathion concentrations in spiked vegetable samples. This approach has the advantages of easy preparation and fast senor response. This approach does not yield a three-dimensional polymer membrane, which limits their selectivity and stability and has limited the application of this approach in MIP sensors.

15.5. APPLICATION OF MIPs IN SENSING

15.5.1. General Considerations

MIPs are primarily recognition platforms and do not possess any innate signaling properties. Thus, an important challenge in developing MIP-based chemical sensors is interfacing the MIP with a signal transduction device or mechanism (Scheme 15.5). The second half of this chapter will present examples of different signal transduction mechanisms and strategies that have been successfully employed in MIP sensors. These include competitive binding assays (Section 15.5.2), mass-sensitive devices (Section 15.5.3), and optical (Section 15.5.4) and electrochemical sensors (Section 15.5.5).

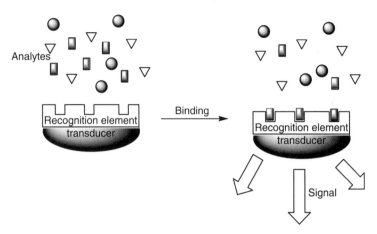

Scheme 15.5   The principle of a chemical sensor.

15.5.2. Competitive Binding Assays

Binding assays are widely used in medical diagnostics and typically utilize an immobilized antibody as the recognition element (Hage 1999; Weller 2000). Binding assays are particularly well matched for use with MIP recognition elements because they do not require integration of a signal transduction platform. For example, most binding assays are monitored via chemical methods such as competitive binding or enzyme-linked assays. In addition, fast response times and multiple-use cycles are not essential requirements for successful binding assays. Thus, many successful MIP binding assays have been reported in the literature (Ansell et al. 1996; Andersson 2000). These examples have demonstrated that MIP-based binding assays can have comparable selectivities and sensitivities to immunoassays with the added benefits of longer shelf-lives, a wider range of testing conditions, and lower costs.

The first report of an MIP binding assay was in 1993 by Vlatakis and colleagues (1993). Noncovalent MIPs were prepared against theophylline and diazepam in chloroform and were used in competitive radioligand binding assays to analyze aqueous serum samples. The assay was able to accurately measure the theophylline in blood in the concentration range of $10–200\ \mu M$, which is satisfactory for therapeutic monitoring of the drug. The MIP assay displayed binding and cross-reactivity profiles similar to that of the commercial enzyme-multiplied immunoassay technique (Fig. 15.6). The excellent selectivity of this MIP assay and ability to function in highly competitive aqueous environments can be attributed to the very low analyte concentrations that were examined. The assay was monitored using radioligands that allowed measurements of submillimolar analyte concentrations where MIPs display the highest selectivity. Another example of the effectiveness of MIP assays in aqueous environments was an automated flow-through assay for the mycotoxin zearalenone (Navarro-Villoslada et al. 2007). The assay used an MIP imprinted with a zearalenone analogue, cyclododecyl 2,4-dihydroxybenzoate, and was monitored by the displacement of a pyrene labled fluorescent tracer from the MIP.

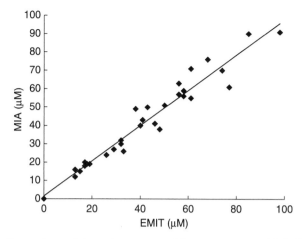

Figure 15.6 Comparison of the values measured by an enzyme-multiplied immunoassay technique (EMIT) assay and a molecularly imprinted sorbent assay (MIA) for determination of theophylline in 32 patient serum samples. The correlation coefficient was 0.98. Reprinted from Vlatakis et al. (2003). Copyright 1993 Macmillan Ltd.

The assay was utilized to measure zearlenone in food samples with a detection limit of 25 μM.

MIP assays can also be utilized in synthetic organic applications. For example, MIP-based assays have been used to measure the chiral purity of samples in organic solvents. An L-phenylalanine anilide (L-PAA) imprinted polymer was utilized as a recognition element to measure the enantiomeric excess (ee) of PAA samples (Chen and Shimizu 2002). The MIP displays greater capacity for L-PAA versus D-PAA samples of similar concentration, and this difference was used to estimate enantiomeric excess. The enantiomeric excess of an unknown solution was determined by comparing the UV absorbance of the PAA remaining in solution after equilibration against a calibration curve. This MIP assay was demonstrated to be rapid and accurate with a standard error of $\pm 5\%$ ee.

15.5.3. Mass-Sensitive Sensors

In mass-sensitive sensors, the binding of an analyte to a surface generates a signal due to the change in mass of the surface. There are two general types of mass sensitive transducers. QCMs measure changes in oscillation frequency, and surface acoustic wave devices (SAWs) measure density related changes (Benes et al. 1995). Mass sensing platforms do not require the addition of responsive monomers or reformulation of the MIP recognition matrix. Thus, device fabrication is relatively straightforward and inexpensive. At the same time, these systems are also highly sensitive. QCMs can provide mass resolutions in the range of 1 ng, and SAWs can accurately measure mass differences of 1 pg and lower (Dickert et al. 1998).

An early example of an MIP-QCM sensor was a glucose monitoring system by Malitesta et al. (1999). A glucose imprinted poly(o-phenylenediamine) polymer was electrosynthesized on the sensor surface. This QCM sensor showed selectivity for glucose over other compounds such as ascorbic acid, paracetamol, cysteine, and fructose at physiologically relevant millimolar concentrations. A unique QCM sensor for detection of yeast was reported by Dickert and coworkers (Dickert et al. 2001; Dickert and Hayden 2002). Yeast cells were imprinted in a sol–gel matrix on the surface of the transducer. The MIP-coated sensor was able to measure yeast cell concentrations in situ and in complex media. A QCM sensor coated with a thin permeable MIP film was developed for the determination of L-menthol in the liquid phase (Percival et al. 2001). The MIP-QCM sensor displayed good selectivity and good sensitivity with a detection limit of 200 ppb (Fig. 15.7). The sensor also displayed excellent enantioselectivity and was able to easily differentiate the L- and D-enantiomers of menthol.

A final example of a mass-sensitive MIP device is a BAW sensor for determination of phenacetin in human serum and urine (Tan, Peng, et al. 2001). A phenacetin imprinted polymer was synthesized and used as the artificial recognition element on a piezoelectric element.

15.5.4. Optical MIP Sensors

Optical sensors have the advantage of an easily measured signal that can be seen by the naked eye in some cases. Optical detection methods include fluorescence, surface plasmon resonance spectroscopy, Raman, IR, and chemiluminescence (Fabbrizzi and Poggi 1995; deSilva et al. 1997). However, the fabrication and development of optical MIP sensors requires that a colored, emissive, or fluorescent monomer

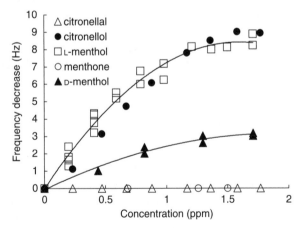

Figure 15.7 Response of the molecular imprinted polymer quartz crystal microbalance (MIP-QCM) sensor to monoterpene analogues: (□) L-menthol, (▲) D-menthol, (●) citronellol, (△) citronellal, and (○) menthone. Reprinted from Percival et al. (2001). Copyright 2001 American Chemical Society.

be integrated into the MIP. These changes in the imprinting process can often diminish the recognition properties of the MIP. Alternatively, the spectroscopically responsive component can be in the solution phase either as the analyte itself or as a competitive analog.

An early example of an MIP-based fluorescent sensor was developed by Piletsky et al. (1999) for the detection of dansyl-L-phenylalanine (Fig. 15.8). An optical fiber was used to monitor the binding of the fluorescent analyte to the MIP surface. The fluorescent sensor displayed stereoselectivity for the templated L-enantiomer in acetonitrile. An exciting example of a sensitive fiber optic MIP sensor was constructed by Jenkins and colleagues for the detection of nerve agents (Jenkins et al. 1999, 2001). A luminescent lanthanide ion, Eu^{3+}, was coordinated into the polymer matrix, which selectively bound the phosphonate hydrolysis products of Sarin and Soman. When the analyte coordinated to the Eu^{3+} in the polymer, the change in the luminescence spectrum could be monitored by the fiber optic devices. This MIP-based optical sensor showed good selectivity and excellent sensitivity with a detection limit of 100 ppt. A fluorescent functional monomer, 2,6-bis(acrylamido)pyridine (Fig. 15.1; Kubo et al. 2003), was synthesized for the preparation of an MIP imprinted with cyclobarbital. This polymer can selectively bind the template, leading to an enhancement in the fluorescence intensity. Another interesting optically responsive monomer, 1-(4-styryl)-3-(3-nitrophenyl)urea, was synthesized for the imprinting of z-(D or L)glutamate (Manesiotis et al. 2004). The z-glutamate imprinted polymer showed strong affinity for carboxylates in aqueous solvents that was accompanied by a color change that could be monitored by the naked eye.

Other detection methods have been used in optical MIP sensing systems. An MIP-based chemiluminescent flow-through sensor was developed for the detection of 1,10-phenanthroline (Lin and Yamada 2001). A metal complex was used to catalyze the decomposition of hydrogen peroxide and form the superoxide radical ion that can

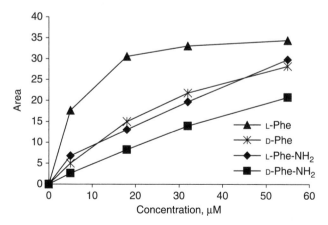

Figure 15.8 Dependence of the peak area of dye displaced by chiral analytes from blank polymer at varying analyte concentrations. Reprinted from Piletsky et al. (1999). Copyright 2004 Springer Science and Business Media.

react with 1,10-phenanthroline to yield a chemiluminescent signal. An ammonia imprinted polymer was coated on the surface of a semiconductor to discriminate ammonia and trimethyamine (Nickel et al. 2001). The bare Cd-Se surface displayed a similar photoluminescent enhancement with the addition of ammonia and mono-, di-, and trimethyamine. However, a selective response for ammonia was observed when the Cd-Se surface was coated with the ammonia imprinted PAA film.

15.5.5. Electrometrical MIP Sensors

Electrometrical transducers that have been utilized for the preparation of MIP-based chemical sensors can be classified as amperometric (monitor the current at a fixed voltage), potentiometric (monitor the voltage at zero current), or conductometric (measure conductivity or impedance changes; Piletsky et al. 2002; Blanco-Lopez et al. 2004). Amperometric sensors measure the current flowing between a counter-electrode and a working electrode that is coated with the recognition element. They have the advantages of simplicity, ease of production, and low cost and are considered the most popular electrochemical sensor format. An early example of an amperometric morphine sensor for a competitive format was constructed in 1995 based on a morphine imprinted MAA/EGDMA copolymer (Kriz and Mosbach 1995). Morphine in the concentration range of $0.1-10$ pg/ml could be detected by the amperometric sensor based on competitive binding (Fig. 15.9). This amperometric sensor demonstrates autoclave compatibility, long-term stability, and resistance to harsh chemical environments, although the response of the sensor was slow, which was due to the time required for equilibration. For nonelectrochemical active species, amperometric sensors are used for detection through a displacement

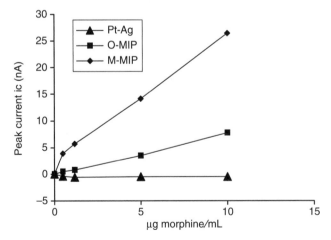

Figure 15.9 Competitive morphine sensor response as a function of the morphine concentration $(0-10 \, \mu g/mL)$ present in the solution. Three sensor types were examined: morphine molecular imprinted polymer (M-MIP), reference (O-MIP), and agarose-covered platinum electrode (Pt-Ag). Reprinted from Kriz and Mosbach (1995). Copyright 1995 Elsevier Science.

or competitive step coupled with an electrochemical reaction (Kroger et al. 1999). Normally, the analyte should be electrochemically active to use this transducer type. For example, molecularly imprinted polypyrrole was electrochemically deposited on the surface of platinum electrodes as the recognition elements to detect glycoprotein gp51 (Ramanaviciene and Ramanavicius 2004). Analytical signals were generated directly by rebinding gp51 to the MIP with good selectivity, although the reusability of the sensor was not satisfactory.

In potentiometric sensors, an electrical potential between the working electrode and a reference electrode is measured at zero current conditions in a solution containing ions that exchange with the surface. The first potentiometric MIP sensor was prepared in 1992 by Vinokurov (1992). The substrate-selective polyaniline electrode was electrosynthesized with polypyrrole, polyaniline, and aniline-*p*-aminophenol copolymers. The development of an MIP-based potentiometric sensor was reported in 1995 by Hutchins and Bachas (1995). This potentiometric sensor has high selectivity for nitrite with a low detection limit of $(2 \pm 1) \times 10^{-5}$ M (Fig. 15.10).

Conductometric sensors are based on measuring the change in conductivity of a selective layer, which interacts with the analyte. The conductivity and capacitance increase with the binding of analytes to the MIP whereas the impedance decreases during the measurement. The conductometric sensor has the lowest sensitivity in electrochemical transducers because a high concentration can often mask a low concentration of analyte ions. An early conductometric sensor was successfully developed in 1996 for measurement of benzyltriphenylphosphonium (Kriz et al. 1996). This MIP-based conductometric sensor gave a significantly higher conductivity reading than the reference sensor when exposed to the analyte in acetonitrile. An atrazine-sensitive conductometric sensor was developed with high selectivity and sensitivity with a detection limit of 5 nM for atrazine (Sergeyeva et al. 1999;

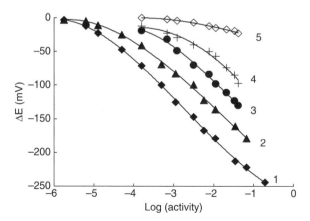

Figure 15.10 Selectivity pattern of $PPy(NO_3^-)$ electrodes. The anions tested include (1) nitrate, (2) bromide, (3) perchlorate, (4) salicylate, and (5) phosphate; ΔE is the difference between the steady-state potential and the starting potential. Reprinted from Hutchins and Bachas (1995). Copyright 1995 American Chemical Society.

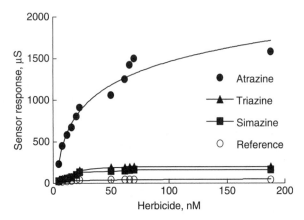

Figure 15.11 Sensor response to herbicide concentration obtained with a polymer imprinted with atrazine. Reprinted from Sergeyeva et al. (1999). Copyright 1999 Elsevier Science.

Fig. 15.11). The sensor has response times of 6–15 min, depending on the membrane thickness, and the polymer membranes maintained their recognition characteristics over 6 months.

15.6. CONCLUSIONS AND OUTLOOK

The use of MIPs as artificial recognition elements for the chemical sensor is practical and promising. MIPs can be easily synthesized for sensing molecules when biological receptors can not be easily obtained. Their low cost, ease of preparation, and robustness, coupled with the wide choice of templates and functional monomers make MIPs appropriate for the recognition element in sensing applications. Great efforts have been made to develop the combination of MIPs and transducing devices over the past two decades.

The development of MIPs with a more homogeneous distribution of binding sites, which also possess high affinity to the target molecule and good selectivity in aqueous environments, will attract broad interest for developing MIP-based chemical sensors. Then, MIP-based chemical sensors will find great opportunities in the areas where other natural or synthetic receptors struggle to find reliable applications.

REFERENCES

Albert KJ, Lewis NS, Schauer CL, Sotzing GA, Stitzel SE, Vaid TP, Walt DR. Cross-reactive chemical sensor arrays. Chem Rev 2000;100:2595–2626.

Alexander C, Andersson HS, Andersson LI, Ansell RJ, Kirsch N, Nicholls IA, O'Mahony J, Whitcombe MJ. Molecular imprinting science and technology: a survey of the literature for the years up to and including 2003.J Mol Recognit 2006;19:106–180.

Andersson LI. Molecular imprinting for drug bioanalysis. a review on the application of imprinted polymers to solid-phase extraction and binding assay. J Chromatogr B Biomed Sci Appl 2000;739:163–173.

Andersson LI, Mandenius CF, Mosbach K. Studies on guest selective molecular recognition on an octadecyl silylated silicon surface using ellipsometry. Tetrahedron Lett 1988;29: 5437–5440.

Andersson LI, Muller R, Vlatakis G, Mosbach K. Mimics of the binding-sites of opioid receptors obtained by molecular imprinting of enkephalin and morphine. Proc Natl Acad Sci USA 1995;92:4788–4792.

Andersson LI, Paprica A, Arvidsson T. A highly selective solid phase extraction sorbent for pre-concentration of sameridine made by molecular imprinting. Chromatographia 1997;46:57–62.

Ansell RJ, Kriz D, Mosbach K. Molecularly imprinted polymers for bioanalysis: chromatography, binding assays and biomimetic sensors. Curr Opin Biotechnol 1996;7:89–94.

Arshady R, Mosbach K. Synthesis of substrate-selective polymers by host–guest polymerization. Makromol Chem 1981;182:687–692.

Asanuma H, Hishiya T, Komiyama M. Tailor-made receptors by molecular imprinting. Adv Mater 2000;12:1019–1030.

Asanuma H, Kakazu M, Shibata M, Hishiya T, Komiyama M. Molecularly imprinted polymer of beta-cyclodextrin for the efficient recognition of cholesterol. Chem Commun 1997;1971–1972.

Beach JV, Shea KJ. Designed catalysts. A synthetic network polymer that catalyzes the dehydrofluorination of 4-fluoro-(p-nitrophenyl)butan-2-one. J Am Chem Soc 1994;116: 379–380.

Beinhoff M, Appapillai AT, Underwood LD, Frommer JE, Carter KR. Patterned polyfluorene surfaces by functionalization of nanoimprinted polymeric features. Langmuir 2006;22:2411–2414.

Benes E, Groschl M, Burger W, Schmid M. Sensors based on piezoelectric resonators. Sens Actuat A 1995;48:1–21.

Bereczki A, Tolokan A, Horvaia G, Horvath V, Lanza F, Hall AJ, Sellergren B. Determination of phenytoin in plasma by molecularly imprinted solid-phase extraction. J Chromatogr A 2001;930:31–38.

Biffis A, Graham NB, Siedlaczek G, Stalberg S, Wulff G. The synthesis, characterization and molecular recognition properties of imprinted microgels. Macromol Chem Phys 2001;202: 163–171.

Blanco-Lopez MC, Lobo-Castanon MJ, Miranda-Ordieres AJ, Tunon-Blanco P. Electrochemical sensors based on molecularly imprinted polymers. Trends Anal Chem 2004;23:36–48.

Bossi A, Piletsky SA, Piletska EV, Righetti PG, Turner AP. Surface-grafted molecularly imprinted polymers for protein recognition. Anal Chem 2001;73:5281–5286.

Boyle A, Genies EM, Lapkowski M. Application of electronic conducting polymers as sensors—polyaniline in the solid-state for detection of solvent vapors and polypyrrole for detection of biological ions in solutions. Synth Met 1989;28:C769–C774.

Bures P, Huang Y, Oral E, Peppas NA. Surface modifications and molecular imprinting of polymers in medical and pharmaceutical applications. J Controlled Release 2001;72: 25–33.

Chang DK, Chul O, Oh S-G, Chang JY. The use of a thermally reversible bond for molecular imprinting of silica spheres. J Am Chem Soc 2002;124:14838–14839.

Chen Y, Shimizu KD. Measurement of enantiomeric excess using molecularly imprinted polymers. Org Lett 2002;4:2937–2940.

Chen YB, Kele M, Quinones I, Sellergren B, Guiochon G. Influence of the pH on the behavior of an imprinted polymeric stationary phase—supporting evidence for a binding site model. J Chromatogr A 2001;927:1–17.

Chen YB, Kele M, Sajonz P, Sellergren B, Guiochon G. Influence of thermal annealing on the thermodynamic and mass transfer kinetic properties of D- and L-phenylalanine anilide on imprinted polymeric stationary phases. Anal Chem 1999;71:928–938.

Cheong SH, McNiven S, Rachkov K, Levi R, Yano K, Karube I. Testosterone receptor binding mimic constructed using molecular imprinting. Macromolecules 1997;30: 1317–1322.

Davies MP, De Biasi V, Perrett D. Approaches to the rational design of molecularly imprinted polymers. Anal Chim Acta 2004;504:7–14.

deSilva AP, Gunaratne HQN, Gunnlaugsson T, Huxley AJM, McCoy CP, Rademacher JT, Rice TE. Signaling recognition events with fluorescent sensors and switches. Chem Rev 1997;97:1515–1566.

Dickert FL, Forth P, Lieberzeit P, Tortschanoff M. Molecular imprinting in chemical sensing—detection of aromatic and halogenated hydrocarbons as well as polar solvent vapors. Fresenius J Anal Chem 1998;360:759–762.

Dickert FL, Hayden G. Bioimprinting of polymers and sol–gel phases. Selective detection of yeasts with imprinted polymers. Anal Chem 2002;74:1302–1306.

Dickert FL, Hayden O, Halikias KP. Synthetic receptors as sensor coatings for molecules and living cells. Analyst 2001;126:766–771.

Dirion B, Cobb Z, Schillinger E, Andersson LI, Sellergren B. Water-compatible molecularly imprinted polymers obtained via high-throughput synthesis and experimental design. J Am Chem Soc 2003;125:15101–15109.

Eggins BR. Chemical sensors and biosensors. Hoboken, NJ: Wiley–VCH; 2002.

Ellwanger A, Berggren C, Bayoudh S, Crecnzi C, Karlsson L, Owens PK, Ensing K, Cormack P, Sherrington D, Sellergren B. Evaluation of methods aimed at complete removal of template from molecularly imprinted polymers. Analyst 2001;126:784–792.

Fabbrizzi L, Poggi A. Sensors and switches from supramolecular chemistry. Chem Soc Rev 1995;24:197–202.

Fairhurst RE, Chassaing C, Venn RF, Mayes AG. A direct comparison of the performance of ground, beaded and silica-grafted MIPs in HPLC and turbulent flow chromatography applications. Biosens Bioelectron 2004;20:1098–1105.

Fujii Y, Matsutani K, Kikuchi K. Formation of a specific coordination cavity for a chiral amino-acid by template synthesis of a polymer Schiff–base cobalt(III) complex. Chem Commun 1985;415–417.

Garcia-Calzon JA, Diaz-Garcia ME. Characterization of binding sites in molecularly imprinted polymers. Sens Actuat B 2007;123:1180–1194.

Glad M, Reinholdsson P, Mosbach K. Molecularly imprinted composite polymers based on trimethylolpropane trimethacrylate (TRIM) particles for efficient enantiomeric separations. React Polym 1995;25:47–54.

Gong JL, Gong FC, Kuang Y, Zeng GM, Shen GL, Yu RQ. Capacitive chemical sensor for fenvalerate assay based on electropolymerized molecularly imprinted polymer as the sensitive layer. Anal Bioanal Chem 2004;379:302–307.

Greene NT, Morgan SL, Shimizu KD. Molecularly imprinted polymer sensor arrays. Chem Commun 2004;1172–1173.

Hage DS. Immunoassays. Anal Chem 1999;71:294R–304R.

Hall AJ, Achilli L, Manesiotis P, Quaglia M, De Lorenzi E, Sellergren B. A substructure approach toward polymeric receptors targeting dihydrofolate reductase inhibitors. 2. Molecularly imprinted polymers against Z-L-glutamic acid showing affinity for larger molecules. J Org Chem 2003;68:9132–9135.

Hart BR, Shea KJ. Synthetic peptide receptors: molecularly imprinted polymers for the recognition of peptides using peptide–metal interactions. J Am Chem Soc 2001;123: 2072–2073.

Hart BR, Shea KJ. Molecular imprinting for the recognition of N-terminal histidine peptides in aqueous solution. Macromolecules 2002;35:6192–6201.

Hattori K, Hiwatari M, Iiyama C, Yoshimi Y, Kohori F, Sakai K, Piletsky SA. Gate effect of theophylline-imprinted polymers grafted to the cellulose by living radical polymerization. J Membr Sci 2004;233:169–173.

Haupt K. Molecularly imprinted polymers in analytical chemistry. Analyst 2001;126: 747–756.

Haupt K, Mayes AG, Mosbach K. Herbicide assay using an imprinted polymer-based system analogous to competitive fluoroimmunoassays. Anal Chem 1998;70:3936–3939.

Haupt K, Mosbach K. Molecularly imprinted polymers and their use in biomimetic sensors. Chem Rev 2000;100:2495–2504.

Hirsch T, Kettenberger H, Wolfbeis OS, Mirsky VM. A simple strategy for preparation of sensor arrays: molecularly structured monolayers as recognition elements. Chem Commun 2003;432–433.

Houk KN, Leach AG, Kim SP, Zhang X. Binding affinities of host–guest, protein–ligand, and protein-transition-state complexes. Angew Chem Int Ed 2003;42:4872–4897.

Hutchins RS, Bachas LG. Nitrate-selective electrode developed by electrochemically mediated imprinting doping of polypyrrole. Anal Chem 1995;67:1654–1660.

Huval CC, Bailey MJ, Braunlin WH, Holmes-Farley SR, Mandeville WH, Petersen JS, Polomoscanik SC, Sacchiro RJ, Chen X, Dhal PK. Novel cholesterol lowering polymeric drugs obtained by molecular imprinting. Macromolecules 2001;34: 1548–1550.

Jenkins AL, Uy OM, Murray GM. Polymer-based lanthanide luminescent sensor for detection of the hydrolysis product of the nerve agent Soman in water. Anal Chem 1999;71:373–378.

Jenkins AL, Yin R, Jensen JL. Molecularly imprinted polymer sensors for pesticide and insecticide detection in water. Analyst 2001;126:798–802.

Karlsson JG, Andersson LI, Nicholls IA. Probing the molecular basis for ligand-selective recognition in molecularly imprinted polymers selective for the local anaesthetic bupivacaine. Anal Chim Acta 2001;435:57–64.

Katz A, Davis ME. Molecular imprinting of bulk, microporous silica. Nature 2000;403: 286–289.

Kempe M. Antibody-mimicking polymers as chiral stationary phases in HPLC. Anal Chem 1996;68:1948–1953.

Kempe M, Fischer L, Mosbach K. Chiral separation using molecularly imprinted heteroaromatic polymers. J Mol Recognit 1993;6:25–29.

Kempe M, Mosbach K. Chiral recognition of N alpha-protected amino acids and derivatives in non-covalently molecularly imprinted polymers. Int J Peptide Protein Res 1994;44: 603–606.

Khasawneh MA, Vallano PT, Remcho VT. Affinity screening by packed capillary high performance liquid chromatography using molecular imprinted sorbents. II. Covalent imprinted polymers. J Chromatogr A 2001;922:87–97.

Kim H, Guiochon G. Thermodynamic studies of the solvent effects in chromatography on molecularly imprinted polymers. 3. Nature of the organic mobile phase. Anal Chem 2005a;77:2496–2504.

Kim H, Guiochon G. Adsorption on molecularly imprinted polymers of structural analogues of a template. Single-component adsorption isotherm data. Anal Chem 2005b;77:6415–6425.

Klein JU, Whitcombe MJ, Mulholland F, Vulfson EN. Template-mediated synthesis of a polymeric receptor specific to amino acid sequences. Angew Chem Int Ed 1999;38:2057–2060.

Kriz D, Kempe M, Mosbach K. Introduction of molecularly imprinted polymers as recognition elements in conductometric chemical sensors. Sens Actuat B 1996;33:178–181.

Kriz D, Mosbach K. Competitive amperometric morphine sensor-based on an agarose immobilized molecularly imprinted polymer. Anal Chim Acta 1995;300:71–75.

Kriz D, Ramström O, Svensson A, Mosbach K. Introducing biomimetic sensors based on molecularly imprinted polymers as recognition elements. Anal Chem 1995;67:2142–2144.

Kroger S, Turner APF, Mosbach K, Haupt K. Imprinted polymer based sensor system for herbicides using differential-pulse voltammetry on screen printed electrodes. Anal Chem 1999;71:3698–3702.

Kubo H, Nariai H, Takeuchi T. Multiple hydrogen bonding-based fluorescent imprinted polymers for cyclobarbital prepared with 2,6-bis(acrylamido)pyridine. Chem Commun 2003;2792–2793.

Lanza F, Sellergren B. The application of molecular imprinting technology to solid phase extraction. Chromatographia 2001;53:599–611.

Lepisto M, Sellergren B. Discrimination between amino-acid amide conformers by imprinted polymers. J Org Chem 1989;54:6010–6012.

Li CY, Wang CF, Wang CH, Hu SS. Development of a parathion sensor based on molecularly imprinted nano-TiO2 self-assembled film electrode. Sens Actuat B 2006;117:166–171.

Lin JM, Yamada M. Chemiluminescent flow-through sensor for 1,10-phenanthroline based on the combination of molecular imprinting and chemiluminescence. Analyst 2001;126:810–815.

Lubke C, Lubke M, Whitcombe MJ, Vulfson EN. Imprinted polymers prepared with stoichiometric template–monomer complexes: efficient binding of ampicillin from aqueous solutions. Macromolecules 2000;1433:5098–5105.

Malitesta C, Losito I, Zambonin PG. Molecularly imprinted electrosynthesized polymers: new materials for biomimetic sensors. Anal Chem 1999;71:1366–1370.

Manesiotis P, Hall AJ, Emgenbroich M, Quaglia M, De Lorenzi E, Sellergren B. An enantio-selective imprinted receptor for Z-glutamate exhibiting a binding induced color change. Chem Commun 2004;2278–2279.

MathewKrotz J, Shea KJ. Imprinted polymer membranes for the selective transport of targeted neutral molecules. J Am Chem Soc 1996;118:8154–8155.

Navarro-Villoslada F, Urraca JL, Moreno-Bondi MC, Orellana G. Zearalenone sensing with molecularly imprinted polymers and tailored fluorescent probes. Sens Actuat B 2007;121:67–73.

Nickel AML, Seker F, Ziemer BP, Ellis AB. Imprinted poly (acrylic acid) films on cadmium selenide. A composite sensor structure that couples selective amine binding with semiconductor substrate photoluminescence. Chem Mater 2001;13:1391–1397.

Nishino H, Huang CS, Shea KJ. Selective protein capture by epitope imprinting. Angew Chem Int Ed 2006;45:2392–2396.

Norell MC, Andersson HS, Nicholls IA. Theophylline molecularly imprinted polymer dissociation kinetics: a novel sustained release drug dosage mechanism. J Mol Recognit 1998;11:98–102.

Oral E, Peppas NA. Hydrophilic molecularly imprinted poly(hydroxyethyl-methacrylate) polymers. J Biomed Mater Res A 2006;78:205–210.

Panasyuk TL, Mirsky VM, Piletsky SA, Wolfbeis OS. Electropolymerized molecularly imprinted polymers as receptor layers in a capacitive chemical sensor. Anal Chem 1999;71:4609–4613.

Patel A, Fouace S, Steinke JH. Enantioselective molecularly imprinted polymers via ring-opening metathesis polymerisation. Chem Commun 2003;88–89.

Percival CJ, Stanley S, Galle TM, Braithwaite A, Newton MI, McHale G, Hayes W. Molecular-imprinted, polymer-coated quartz crystal microbalances for the detection of terpenes. Anal Chem 2001;73:4225–4228.

Peters EC, Svec F, Frechet JMJ. Rigid macroporous polymer monoliths. Adv Mater 1999;11:1169–1181.

Piacham T, Josell A, Arwin H, Prachayasittikul V, Ye L. Molecularly imprinted polymer thin films on quartz crystal microbalance using a surface bound photo-radical initiator. Anal Chim Acta 2005;536:191–196.

Piletsky SA, Matuschewski H, Schedler U, Wilpert A, Piletska EV, Thiele TA, Ulbricht M. Surface functionalization of porous polypropylene membranes with molecularly imprinted polymers by photograft copolymerization in water. Macromolecules 2000;33:3092–3098.

Piletsky SA, Piletska EV, Chen B, Karim K, Weston D, Barrett G, Lowe P, Turner AP. Chemical grafting of molecularly imprinted homopolymers to the surface of microplates. Application of artificial adrenergic receptor in enzyme-linked assay for beta-agonists determination. Anal Chem 2000;72:4381–4385.

Piletsky SA, Piletska EV, Karim K, Freebairn KW, Legge CH, Turner APF. Polymer cookery: influence of polymerization conditions on the performance of molecularly imprinted polymers. Macromolecules 2002;35:7499–7504.

Piletsky SA, Terpetschnig E, Andersson HS, Nicholls IA, Wolfbeis OS. Application of non-specific fluorescent dyes for monitoring enantio-selective ligand binding to molecularly imprinted polymers. Fresenius J Anal Chem 1999;364:512–516.

Polyzkov MV. Adsorption properties and structure of silica gel. Zhur Fiz Khim 1931;2: 799–805.

Quaglia M, De Lorenzi E, Sulitzky C, Massolini G, Sellergren B. Surface initiated molecularly imprinted polymer films: a new approach in chiral capillary electrochromatography. Analyst 2001;126:1495–1498.

Ramanaviciene A, Ramanavicius A. Molecularly imprinted polypyrrole-based synthetic receptor for direct detection of bovine leukemia virus glycoproteins. Biosens Bioelectron 2004;20:1076–1082.

Rampey AM, Umpleby RJ, Rushton GT, Iseman JC, Shah RN, Shimizu KD. Characterization of the imprint effect and the influence of imprinting conditions on affinity, capacity, and heterogeneity in molecularly imprinted polymers using the Freundlich isotherm-affinity distribution analysis. Anal Chem 2004;76:1123–1133.

Ramstrom O, Ansell RJ. Molecular imprinting technology: challenges and prospects for the future. Chirality 1998;10:195–209.

Rosengren AM, Karlsson JG, Andersson PO, Nicholls IA. Chemometric models of template—molecularly imprinted polymer binding. Anal Chem 2005;77:5700–5705.

Ruckert B, Hall AJ, Sellergren B. Molecularly imprinted composite materials via iniferter-modified supports. J Mater Chem 2002;12:2275–2280.

Santora BP, Gagne MR, Moloy KG, Radu NS. Porogen and cross-linking effects on the surface area, pore volume distribution, and morphology of macroporous polymers obtained by bulk polymerization. Macromolecules 2001;34:658–661.

Schmidt RH, Mosbach K, Haupt K. A simple method for spin-coating molecularly imprinted polymer films of controlled thickness and porosity. Adv Mater 2004;16:719.

Schneider F, Piletsky S, Piletska E, Guerreiro A, Ulbricht M. Comparison of thin-layer and bulk MIPs synthesized by photoinitiated in situ crosslinking polymerization from the same reaction mixtures. J Appl Polym Sci 2005;98:362–372.

Schrader T, Hamilton AD. Functional synthetic receptors. Weinheim: Wiley–VCH; 2005.

Schweitz L. Molecularly imprinted polymer coatings for open-tubular capillary electrochromatography prepared by surface initiation. Anal Chem 2002;74:1192–1196.

Sellergren B. Molecular imprinting by noncovalent interactions: tailor-made chiral stationary phases of high selectivity and sample load capacity. Chirality 1989;1:63–68.

Sellergren B. Imprinted chiral stationary phases in high-performance liquid chromatography. J Chromatogr A 2001a;906:227–252.

Sellergren B. Molecularly imprinted polymers: man-made mimics of antibodies and their applications in analytical chemistry. New York: Elsevier; 2001b.

Sellergren B, Andersson L. Molecular recognition in macroporous polymers prepared by a substrate–analog imprinting strategy. J Org Chem 1990;55:3381–3383.

Sellergren B, Lespisto M, Mosbach K. Highly enantioselective and substrate-selective polymers obtained by molecular imprinting utilizing noncovalent interactions. NMR and chromatographic studies on the nature of recognition. J Am Chem Soc 1988;110: 5853–5860.

Sellergren B, Ruckert B, Hall AJ. Layer-by-layer grafting of molecularly imprinted polymers via iniferter modified supports. Adv Mater 2002;14:1204.

Sellergren B, Shea KJ. Chiral ion-exchange chromatography—correlation between solute retention and a theoretical ion-exchange model using imprinted polymers. J Chromatogr A 1993;654:17–28.

Senaratne W, Andruzzi L, Ober CK. Self-assembled monolayers and polymer brushes in biotechnology: current applications and future perspectives. Biomacromolecules 2005;6: 2427–2448.

Sergeyeva TA, Piletsky SA, Brovko AA, Slinchenko EA, Sergeeva LM, El'skaya AV. Selective recognition of atrazine by molecularly imprinted polymer membranes. Development of conductometric sensor for herbicides detection. Anal Chim Acta 1999;392:105–111.

Shea KJ, Sasaki DY. An analysis of small-molecule binding to functionalized synthetic polymers by 13C CP/MAS NMR and FT-IR spectroscopy. J Am Chem Soc 1991;113: 4109–4120.

Shea KJ, Spivak DA, Sellergren B. Polymer complements to nucleotide bases. Selective binding of adenine derivatives to imprinted polymers. J Am Chem Soc 1993;115: 3368–3369.

Shea KJ, Stoddard GJ, Shavelle DM, Wakui F, Choate RM. Synthesis and characterization of highly cross-linked polyacrylamides and polymethacrylamides—a new class of macroporous polyamides. Macromolecules 1990;23:4497–4507.

Sibrian-Vazquez M, Spivak DA. Molecular imprinting made easy. J Am Chem Soc 2004;126:7827–7833.

Siemann M, Andersson LI, Mosbach K. Separation and detection of macrolide antibiotics by HPLC using macrolide-imprinted synthetic polymers as stationary phases. J Antibiot 1997;50:89–91.

Spivak D, Gilmore MA, Shea KJ. Evaluation of binding and origins of specificity of 9-ethyladenine imprinted polymers. J Am Chem Soc 1997;119:4388–4393.

Striegler S. Carbohydrate recognition in cross-linked sugar-templated poly(acrylates). Macromolecules 2003;36:1310–1317.

Suedee R, Seechamnanturakit V, Canyuk B, Ovatlarnporn C, Martin GP. Temperature sensitive dopamine-imprinted (*N,N*-methylene-bis-acrylamide cross-linked) polymer and its potential application to the selective extraction of adrenergic drugs from urine. J Chromatogr A 2006;1114:239–249.

Svec F, Frechet JMJ. Pore-size specific modification as an approach to separation media for single-column, two-dimensional HPLC. Am Lab 1996;28:25.

Svenson J, Nicholls IA. On the thermal and chemical stability of molecularly imprinted polymers. Anal Chim Acta 2001;435:19–24.

Tabushi I, Kurihara K, Naka K, Yamamura K, Hatakeyama H. Supramolecular sensor based on SNO2 electrode modified with octadecylsilyl monolayer having molecular-binding sites. Tetrahedron Lett 1987;28:4299–4302.

Takeuchi T, Goto D, Shinmori H. Protein profiling by protein imprinted polymer array. Analyst 2007;132:101–103.

Takeuchi T, Haginaka J. Separation and sensing based on molecular recognition using molecularly imprinted polymers. J Chromatogr B 1999;728:1–20.

Tan YG, Nie LH, Yao SZ. A piezoelectric biomimetic sensor for aminopyrine with a molecularly imprinted polymer coating. Analyst 2001;126:664–668.

Tan YG, Peng H, Liang CD, Yao SZ. A new assay system for phenacetin using biomimic bulk acoustic wave sensor with a molecularly imprinted polymer coating. Sens Actuat B 2001;73:179–184.

Tanabe K, Takeuchi T, Matsui J, Ikebukuro K, Yano K, Karube I. Recognition of barbiturates in molecularly imprinted copolymers using multiple hydrogen bonding. Chem Commun 1995;2303–2304.

Titirici MM, Sellergren B. Thin molecularly imprinted polymer films via reversible addition–fragmentation chain transfer polymerization. Chem Mater 2006;18:1773–1779.

Uezu K, Nakamura H, Kanno J, Sugo T, Goto M, Nakashio F. Metal ion-imprinted polymer prepared by the combination of surface template polymerization with postirradiation by gamma-rays. Macromolecules 1997;30:3888–3891.

Ulyanova YV, Blackwell AE, Minteer SD. Poly(methylene green) employed as molecularly imprinted polymer matrix for electrochemical sensing. Analyst 2006;131: 257–261.

Umpleby II RJ, Baxter SC, Chen Y, Shah RN, Shimizu KD. Characterization of molecularly imprinted polymers with the Langmuir–Freundlich isotherm. Anal Chem 2001;73: 4584–4591.

Umpleby II RJ, Bode M, Shimizu KD. Measurement of the continuous distribution of binding sites in molecularly imprinted polymers. Analyst 2000;125:1261–1265.

Urraca JL, Hall AJ, Moreno-Bondi MC, Sellergren B. A stoichiometric molecularly imprinted polymer for the class-selective recognition of antibiotics in aqueous media. Angew Chem Int Ed 2006;45:5158–5161.

Vinokurov IA. A new kind of redox sensor based on conducting polymer-films. Sens Actuat B 1992;10:31–35.

Vlatakis G, Andersson LI, Müller R, Mosbach K. Drug assay using antibody mimics made by molecular imprinting. Nature 1993;361:645–647.

Wang HJ, Zhou WH, Yin XF, Zhuang ZX, Yang HH, Wang XR. Template synthesized molecularly imprinted polymer nanotube membranes for chemical separations. J Am Chem Soc 2006;128:15954–15955.

Wang HY, Kobayashi T, Fujii N. Surface molecular imprinting on photosensitive dithiocarbamoyl polyacrylonitrile membranes using photograft polymerization. J Chem Technol Biotechnol 1997;70:355–362.

Weller MG. Immunochromatographic techniques—a critical review. Fresenius J Anal Chem 2000;366:635–645.

Whitcombe MJ, Rodriguez ME, Villar P, Vulfson EN. A new method for the introduction of recognition site functionality into polymers prepared by molecular imprinting: synthesis and characterization of polymeric receptors for cholesterol. J Am Chem Soc 1995;117:7105–7111.

Wulff G. Molecular imprinting in cross-linked materials with the aid of molecular templates— a way towards artificial antibodies. Angew Chem Int Ed Engl 1995;34:1812–1832.

Wulff G. Enzyme-like catalysis by molecularly imprinted polymers. Chem Rev 2002;102: 1–27.

Wulff G, Akelah A. Enzyme-analog built polymers. 6. Synthesis of 5-vinylsalicylaldehyde and a simplified synthesis of some divinyl derivatives. Makromol Chem 1978;179: 2647–2651.

Wulff G, Grobe-Einsler R, Vesper W, Sarhan A. Enzyme-analog built polymers. 5. The specificity distribution of chiral cavities prepared in synthetic polymers. Makromol Chem 1977;178:2817–2825.

Wulff G, Knorr K. Stoichiometric noncovalent interaction in molecular imprinting. Bioseparation 2001;10:257–276.

Wulff G, Sarhan A. Use of polymers with enzyme-analogous structures for the resolution of racemates. Angew Chem Int Ed Engl 1972;11:341.

Wulff G, Sarhan A, Zabrocki. K. Enzyme-analogue built polymers and their use for the resolution of racemates. Tetrahedron Lett 1973;44:4329–4332.

Wulff G, Schauhoff S. Enzyme-analog-built polymers. 27. Racemic-resolution of free sugars with macroporous polymers prepared by molecular imprinting—selectivity dependence on the arrangement of functional-groups versus spatial requirements. J Org Chem 1991;56:395–400.

Wulff G, Vesper W, Grobe-Einsler R, Sarhan A. Enzyme-analog built polymers. 4. The synthesis of polymers containing chiral cavities and their use for the resolution of racemates. Makromol Chem 1977;178:2799–2816.

Yamazaki T, Yilmaz E, Mosbach K, Sode K. Towards the use of molecularly imprinted polymers containing imidazoles and bivalent metal complexes for the detection and degradation of organophosphotriester pesticides. Anal Chim Acta 2001;435:209–214.

Yan M, Ramstrom O. Molecularly imprinted materials: science and technology. New York: Marcel Dekker; 2005.

Yang H, Lazos D, Ulbricht M. Thin, highly crosslinked polymer layer synthesized via photo-initiated graft copolymerization on a self-assembled-monolayer-coated gold surface. J Appl Polym Sci 2005;97:158–164.

INDEX

Molecular Recognition and Polymers: Control of Polymer Structure and Self-Assembly.
Edited by V. Rotello and S. Thayumanavan
Copyright © 2008 John Wiley & Sons, Inc.